STATISTICAL TOLERANCE REGIONS

STATISTICAL TOLERANCE REGIONS
Theory, Applications, and Computation

K. KRISHNAMOORTHY

THOMAS MATHEW

A JOHN WILEY & SONS, INC., PUBLICATION

To view supplemental material for this book, please visit ftp://ftp.wiley.com/public/sci_tech_ med/statistical_tolerance.

Published by John Wiley & Sons, Inc., Hoboken, New Jersey.
Published simultaneously in Canada.

For general information on our other products and services or for technical support, please contact our Customer Care Department within the United States at (800) 762-2974, outside the United States at (317) 572-3993 or fax (317) 572-4002.

Wiley also publishes its books in a variety of electronic formats. Some content that appears in print may not be available in electronic format. For information about Wiley products, visit our web site at www.wiley.com.

Library of Congress Cataloging-in-Publication Data:

Krishnamoorthy, K. (Kalimuthu)
 Statistical tolerance regions : theory, applications, and computation / K. Krishnamoorthy, Thomas Mathew.
  p. cm. — (Wiley series in probability and statistics)
 Includes bibliographical references and index.
 ISBN 978-0-470-38026-0 (cloth)
 1. Statistical tolerance regions. I. Mathew, Thomas, 1955– II. Title.
 QA276.74.K75 2009
 519.5—dc22   2008047069

Printed in the United States of America.

10 9 8 7 6 5 4 3 2 1

To the memory of my parents
KK

*To my father K. T. Mathew, and to the memory of my
mother Aleyamma Mathew*
TM

Contents

Contents

List of Tables

Preface

The theory of statistical tolerance intervals and tolerance regions has undergone vigorous development during the last three decades. In particular, the derivation of satisfactory tolerance intervals in the context of random effects models, and satisfactory simultaneous tolerance intervals for regression models, has been carried out only during the 1980s and 1990s. Furthermore, the construction of satisfactory tolerance regions for multivariate normal populations and multivariate regression models was accomplished only within the last ten years. The bibliography collected by Jilek (1981) lists around 270 articles on the topic, and the one by Jilek and Ackerman (1989) lists an additional 130 articles. The literature on the topic has grown considerably since the publication of the latter bibliography. However, no book-length treatment of the topic has been available since Guttman's (1970) monograph. The present book was conceived based on the perceived need to have a single source that brings together the recent developments as well as the earlier results on the topic of tolerance intervals and tolerance regions.

As opposed to a confidence interval that provides information concerning an unknown population parameter, a tolerance interval provides information on the entire population; to be specific, a tolerance interval is expected to capture a certain proportion or more of the population, with a given confidence level. For example, an upper tolerance limit for a univariate population is such that with a given confidence level, a specified proportion or more of the population will fall below the limit. This proportion is referred to as the *content* of a tolerance interval. A lower tolerance limit, or a tolerance interval having both lower and upper limits, satisfies similar conditions. For multivariate populations, we analogously have tolerance regions. The applications of tolerance intervals and tolerance regions are varied. They include clinical and industrial applications, including quality control, applications to environmental monitoring, to the assessment of agreement between two methods or devices, and applications in industrial hygiene. As suggested by the title, this book discusses the theoretical derivation of tolerance intervals and tolerance regions in a wide variety of scenarios, along with applications and examples, and also illustrates the computational procedures. Analytic formulas for the tolerance limits are available in only simple cases, for example, for the upper or lower tolerance limit for a univariate normal population. Thus it becomes necessary to use numerical methods or approximations in order to derive tolerance intervals for many populations. The book discusses the various approximations available for different tolerance interval problems, and also discusses comparisons among the different approximations, making recommendations regarding the choice of the approximation for practical use. When it comes to random or mixed effects models, the book provides the available procedures for the balanced as well as the unbalanced data situations. Furthermore, for situations where the tolerance intervals have to be numerically obtained, the book includes extensive tables providing the necessary tolerance factors for various combinations of the sample size, content and confidence level.

The book has twelve chapters and gives a rather broad coverage of its topic. Chapter 1 gives the basic concepts and definitions, and also gives some of the technical results used throughout the book. The ideas of generalized p-values and generalized confidence intervals are extensively used in some of the later chapters, and these are also described in Chapter 1. Chapter 2 gives a thorough discussion of the various tolerance intervals that have been constructed in the context of the univariate normal distribution. Chapter 3 is on the univariate linear regression model, where we describe the construction of tolerance intervals and simultaneous tolerance intervals. Chapters 4–6 are on the construction of tolerance intervals in mixed effects and random effects models. The one-way random model is given special emphasis, and is the topic covered in both Chapter 4 (the case of balanced data) and Chapter 5 (the case of unbalanced data). Other mixed and random effects models are taken up in Chapter 6. The computation of tolerance intervals for some continuous distributions other than the normal is the topic of Chapter 7. The lognormal, gamma, exponential and Weibull distributions are considered in this chapter. Non-parametric tolerance intervals form the topic of Chapter 8. Chapter 9 and Chapter 10 deal with multivariate populations; Chapter 9 is on the computation of tolerance regions for a multivariate normal distribution, and Chapter 10 addresses the problem in the context of a multivariate linear regression model. Bayesian approaches are described in Chapter 11. Some special topics not covered in the previous chapters are discussed in Chapter 12. The topics covered in this chapter include the derivation of β-expectation tolerance intervals, sample size determination, tolerance intervals for the ratio of normal random variables, tolerance intervals for binomial and Poisson distributions, and tolerance intervals based on censored data. In Chapter 3 and Chapter 10, the calibration problem is also included, since the computation of a multiple use confidence interval or region in the calibration problem can be accomplished using appropriate tolerance intervals and regions.

In each chapter of the book, the theoretical derivations are described in detail, along with the computational procedures. In fact computational algorithms are given throughout the book. However, we have not emphasized any particular software. The computational algorithms can be easily coded in any machine language (Fortran, C, SAS®, etc.). In each chapter, the results are all illustrated with data analysis based on real examples. Most of the data sets used are included in the relevant chapter. Some data sets are also given in Appendix A at the end of the book. Appendix B gives table values of tolerance factors.

The book is appropriate for a graduate level course on tolerance intervals, the prerequisite being a basic knowledge of ANOVA, mixed models, regression and multivariate analysis. In fact each chapter includes a set of exercises. For a researcher interested in the topic, the book provides the state of the art in the field. For an applied statistician or a consultant who encounter problems that call for the use of tolerance intervals, the book is expected to be a valuable resource. In a Technometrics article, Carroll and Ruppert (1991, p. 199) mention that "It appears to us that tolerance intervals should be more widely understood and used." It is hoped that this book will serve this purpose.

The authors acknowledge the support and the facilities received from the Department of Mathematics, University of Louisiana at Lafayette, and the Department of Mathematics and Statistics, University of Maryland Baltimore County. The authors are grateful to Dr. Ionut Bebu for his assistance with some of the numerical computations on Bayesian tolerance intervals. Krishnamoorthy is thankful to his wife Usha and sons Prathap and Tharany for their enduring love and moral support. Mathew wishes to express his appreciation to his wife Ruby, and daughters Stacy and Betsy for their continued affection and support.

K. KRISHNAMOORTHY
University of Louisiana at Lafayette

THOMAS MATHEW
University of Maryland Baltimore County

Chapter 1

Preliminaries

1.1 Introduction

Statistical intervals computed based on a random sample have wide applicability, for the purpose of quantifying the uncertainty about a scalar quantity associated with a sampled population. The type of interval to be computed obviously depends on the underlying problem and application. A confidence interval based on a random sample is used to provide bounds for an unknown scalar population parameter such as the population mean, standard deviation, percentile, tail probability, etc. A prediction interval based on a random sample is used to provide bounds for one or more future observations from a univariate sampled population. For multivariate populations, we have correspondingly confidence regions and prediction regions. The topic of this book is a third type of interval and region, namely tolerance intervals and tolerance regions. For a univariate population, a tolerance interval is an interval, based on a random sample, that is expected to contain a specified proportion or more of the sampled population. A tolerance region is similarly defined for a multivariate population.

Here is a simple example to illustrate the differences among a confidence interval, prediction interval and tolerance interval. The application deals with the assessment of air lead levels in a laboratory. The data are given in Chapter 2 (see Table 2.1) and represent air lead levels collected by the National Institute of Occupational Safety and Health (NIOSH) at a laboratory for health hazard evaluation purpose. The air lead levels were collected from $n = 15$ different areas within the facility. It was noted that the log-transformed lead levels fitted a normal distribution well (that is, the data are from a lognormal distribution). Let μ and σ^2, respectively, denote the population mean and variance for the log-

1

transformed data. If X denotes the corresponding random variable, we thus have $X \sim N(\mu, \sigma^2)$. We note that $\exp(\mu)$ is the median air lead level. A confidence interval for μ can be constructed the usual way, based on the t-distribution; this in turn will provide a confidence interval for the median air lead level. If $\bar{X}$ and S denote the sample mean and standard deviation of the log-transformed data for a sample of size n, a 95% confidence interval for μ is given by $\bar{X} \pm t_{n-1;.975} \frac{S}{\sqrt{n}}$, where $t_{m;1-\alpha}$ denotes the $1 - \alpha$ quantile of a t-distribution with m degrees of freedom. It may also be of interest to derive a 95% upper confidence bound for the median air lead level. Such a bound for μ is given by $\bar{X} + t_{n-1;.95} \frac{S}{\sqrt{n}}$. Consequently, a 95% upper confidence bound for the median air lead level is given by $\exp(\bar{X} + t_{n-1;.95} \frac{S}{\sqrt{n}})$. Now suppose we want to predict the air lead level at a particular area within the laboratory. A 95% upper prediction limit for the log-transformed lead level is given by $\bar{X} + t_{n-1;.95} S \sqrt{1 + \frac{1}{n}}$. A two-sided prediction interval can be similarly computed. The meaning and interpretation of these intervals are well known. For example, if the confidence interval $\bar{X} \pm t_{n-1;.975} \frac{S}{\sqrt{n}}$ is computed repeatedly from independent samples, 95% of the intervals so computed will include the true value of μ, in the long run. In other words, the interval is meant to provide information concerning the parameter μ only. A prediction interval has a similar interpretation, and is meant to provide information concerning a single lead level only. Now suppose we want to use the sample to conclude whether or not at least 95% of the population lead levels are below a threshold. The confidence interval and prediction interval cannot answer this question, since the confidence interval is only for the median lead level, and the prediction interval is only for a single lead level. What is required is a tolerance interval; more specifically, an upper tolerance limit. The upper tolerance limit is to be computed subject to the condition that at least 95% of the population lead levels is below the limit, with a certain confidence level, say 99%. Once such an upper tolerance limit is computed, we can verify if it is less than the threshold value.

We shall now give the precise definitions of tolerance intervals. This will be followed by a summary of several preliminary concepts and results, to be used for the derivation of tolerance intervals in some of the later chapters.

1.1.1 One-Sided Tolerance Intervals

Let X be a continuous random variable with cumulative distribution function (cdf) $F_X(x) = P(X \leq x)$. For a given p $(0 < p < 1)$, the inverse cdf is defined by

$$F_X^{-1}(p) = \inf\{x : F_X(x) \geq p\}. \tag{1.1.1}$$

The quantity $F_X^{-1}(p)$ is obviously the p quantile or $100p$ percentile of the distribution F_X. We shall also denote the p quantile by q_p. Notice that a proportion p of the population (with distribution function F_X) is less than or equal to q_p. If $F_X(x)$ is a strictly increasing function of x (this is the case for many commonly used distributions), then $F_X^{-1}(p)$ is the value of x for which $F_X(x) = P(X \leq x) = p$.

Let X_1, X_2, ..., X_n be a random sample from $F_X(x)$, and write $\boldsymbol{X} = (X_1, X_2, ..., X_n)$. In order to define a tolerance interval, we need to specify its *content* and *confidence level*. These will be denoted by p and $1 - \alpha$, respectively, and the tolerance interval will be referred to as a p content and $(1 - \alpha)$ coverage (or p content and $(1 - \alpha)$ confidence) tolerance interval or simply a $(p, 1 - \alpha)$ tolerance interval $(0 < p < 1, 0 < \alpha < 1)$. In practical applications, p and $1 - \alpha$ usually take values from the set $\{0.90, 0.95, 0.99\}$. The interval will be constructed using the random sample $\boldsymbol{X}$, and is required to contain a proportion p or more of the sampled population, with confidence level $1 - \alpha$. Formally, a $(p, 1 - \alpha)$ one-sided tolerance interval of the form $(-\infty, U(\boldsymbol{X})]$ is required to satisfy the condition

$$P_{\boldsymbol{X}} \left\{ P_X \left(X \leq U(\boldsymbol{X}) \middle| \boldsymbol{X} \right) \geq p \right\} = 1 - \alpha, \qquad (1.1.2)$$

where X also follows F_X, independently of $\boldsymbol{X}$. That is, $U(\boldsymbol{X})$ is to be determined such that at least a proportion p of the population is less than or equal to $U(\boldsymbol{X})$ with confidence $1 - \alpha$. The interval $(-\infty, U(\boldsymbol{X})]$ is called a one-sided tolerance interval, and $U(\boldsymbol{X})$ is called a one-sided upper tolerance limit. Note that based on the definition of the p quantile q_p, we can write (1.1.2) as

$$P_{\boldsymbol{X}} \left\{ q_p \leq U(\boldsymbol{X}) \right\} = 1 - \alpha. \qquad (1.1.3)$$

It is clear from (1.1.3) that $U(\boldsymbol{X})$ is a $1 - \alpha$ upper confidence limit for the p quantile q_p.

A $(p, 1 - \alpha)$ one-sided lower tolerance limit $L(\boldsymbol{X})$ is defined similarly. Specifically, $L(\boldsymbol{X})$ is determined so that

$$P_{\boldsymbol{X}} \left\{ P_X \left(X \geq L(\boldsymbol{X}) \middle| \boldsymbol{X} \right) \geq p \right\} = 1 - \alpha,$$

or equivalently,

$$P_{\boldsymbol{X}} \left\{ L(\boldsymbol{X}) \leq q_{1-p} \right\} = 1 - \alpha.$$

Thus, $L(\boldsymbol{X})$ is a $1 - \alpha$ lower confidence limit for q_{1-p}.

1.1.2   Tolerance Intervals

There are two types of two-sided tolerance intervals. One is constructed so that it would contain at least a proportion p of the population with confidence $1 - \alpha$, and is simply referred to as the tolerance interval. A second type of tolerance interval is constructed so that it would contain at least a proportion p of the *center* of the population with confidence $1 - \alpha$, and is usually referred to as an equal-tailed tolerance interval.

A $(p, 1-\alpha)$ two-sided tolerance interval $(L(\boldsymbol{X}), U(\boldsymbol{X}))$ satisfies the condition

$$P_{\boldsymbol{X}} \left\{ P_X \left(L(\boldsymbol{X}) \leq X \leq U(\boldsymbol{X}) \middle| \boldsymbol{X} \right) \geq p \right\} = 1 - \alpha, \qquad (1.1.4)$$

or equivalently,

$$P_{\boldsymbol{X}} \left\{ F_X(U(\boldsymbol{X})) - F_X(L(\boldsymbol{X})) \geq p \right\} = 1 - \alpha. \qquad (1.1.5)$$

In other words, the interval $(L(\boldsymbol{X}), U(\boldsymbol{X}))$ is constructed so that it would contain at least a proportion p of the population with confidence $1 - \alpha$. The quantities $L(\boldsymbol{X})$ and $U(\boldsymbol{X})$ are referred to as the tolerance limits. It is important to note that the computation of $L(\boldsymbol{X})$ and $U(\boldsymbol{X})$ *does not* reduce to the computation of confidence limits for certain percentiles.

In order to define an equal-tailed tolerance interval, assume that $p > 0.5$. A $(p, 1 - \alpha)$ equal-tailed tolerance interval $(L(\boldsymbol{X}), U(\boldsymbol{X}))$ is such that, with confidence $1 - \alpha$, no more than a proportion $\frac{1-p}{2}$ of the population is less than $L(\boldsymbol{X})$ and no more than a proportion $\frac{1-p}{2}$ of the population is greater than $U(\boldsymbol{X})$. This requirement can be stated in terms of percentiles. Note that the condition $L(\boldsymbol{X}) \leq q_{\frac{1-p}{2}}$ is equivalent to no more than a proportion $\frac{1-p}{2}$ of the population being less than $L(\boldsymbol{X})$, and the condition $q_{\frac{1+p}{2}} \leq U(\boldsymbol{X})$ is equivalent to no more than a proportion $1 - \frac{1+p}{2} = \frac{1-p}{2}$ of the population being greater than $U(\boldsymbol{X})$. Consequently, for $(L(\boldsymbol{X}), U(\boldsymbol{X}))$ to be a $(p, 1 - \alpha)$ equal-tailed tolerance interval, the condition to be satisfied is

$$P_{\boldsymbol{X}} \left(L(\boldsymbol{X}) \leq q_{\frac{1-p}{2}} \text{ and } q_{\frac{1+p}{2}} \leq U(\boldsymbol{X}) \right) = 1 - \alpha. \qquad (1.1.6)$$

Apart from the one-sided and two-sided $(p, 1 - \alpha)$ tolerance intervals introduced above, intervals have also been constructed so as to contain a proportion β of the population, on the average. Such intervals are referred to as β-expectation tolerance intervals. It has been noted that these are simply $100\beta\%$ prediction intervals for a future observation from the population, constructed using a random sample from the population.

Early works on the tolerance interval problem are due to Wilks (1941, 1942), Wald (1943) and Wald and Wolfowitz (1946). The book by Guttman (1970) gives a concise treatment of tolerance intervals and regions, both $(p, 1 - \alpha)$ tolerance intervals as well as β-expectation tolerance intervals; see also the book by Aitchison and Dunsmore (1975, Chapters 5 and 6). Extensive bibliographies of the literature on tolerance intervals and regions are given in the articles by Jilek (1981) and Jilek and Ackerman (1989). Reviews of the literature on the topic are provided in Patel (1986, 1989), and in the book by Hahn and Meeker (1991). For brief introductions and review, we refer to Guttman (1988) and Vangel (2008a, b). Several articles on tolerance intervals and regions also provide tables of tolerance factors that facilitate easy computation of the required intervals and regions, and the book by Odeh and Owen (1980) gives the required factors in the context of the normal distribution. The PC calculator *StatCalc* (Krishnamoorthy, 2006) can also be conveniently used to compute tolerance factors for univariate and multivariate normal populations.

1.1.3 Survival Probability and Stress-Strength Reliability

Estimation of Survival Probability

In many applications it is desired to estimate the probability that a random variable exceeds a specified value. For example, in lifetime data analysis, it is of interest to assess the probability that the lifetime of an item exceeds a value; this probability is commonly referred to as the survival probability. In industrial hygiene, it is of interest to estimate the probability that the exposure level (level of exposure to a contaminant in a workplace) of a worker exceeds the occupational exposure limit (OEL; usually set by the Occupational Safety and Health Administration). This is referred to as an *exceedance probability*. To assess the lifetime of an item, a lower confidence limit for the survival probability is warranted. Such a lower confidence limit can be easily deduced from a suitable lower tolerance limit, as shown below.

Let X be a continuous random variable with the distribution function $F_X(x)$. For a given t, define the survival probability $S_t = P(X > t) = 1 - F_X(t)$. Let $\boldsymbol{X}$ be a sample from F_X, and $L(\boldsymbol{X}) = L(\boldsymbol{X}; p)$ be a $(p, 1 - \alpha)$ lower tolerance limit for the distribution of X. Being a $(p, 1 - \alpha)$ lower tolerance limit, we have

$$P_{\boldsymbol{X}} \left\{ P_X \left(X \geq L(\boldsymbol{X}; p) | \boldsymbol{X} \right) \geq p \right\} = 1 - \alpha.$$

That is,

$$P_{\boldsymbol{X}} \left\{ S_{L(\boldsymbol{X}; p)} \geq p \right\} = 1 - \alpha.$$

If $L(\boldsymbol{X};p) \geq t$, then we obviously have $S_t \geq S_{L(\boldsymbol{X};p)}$. Furthermore, if $S_{L(\boldsymbol{X};p)} \geq p$, we can conclude that $S_t \geq p$ whenever $L(\boldsymbol{X};p) \geq t$. Consequently the maximum value of p for which $L(\boldsymbol{X};p) \geq t$ gives a $1-\alpha$ lower bound, say p_l, for S_t. That is,

$$p_l = \max\{p : L(\boldsymbol{X};p) \geq t\}. \tag{1.1.7}$$

In general, $L(\boldsymbol{X};p)$ is a decreasing function of p, and so the maximum in (1.1.7) is attained when $L(\boldsymbol{X};p) = t$. That is, p_l is the solution of $L(\boldsymbol{X};p) = t$. A lower tolerance limit can also be used to test one-sided hypotheses concerning S_t. If it is desired to test

$$H_0 : S_t \leq p_0 \quad \text{vs.} \quad H_a : S_t > p_0$$

at a level α, then H_0 will be rejected if a $(p_0, 1-\alpha)$ lower tolerance limit is greater than t.

An upper confidence limit for an exceedance probability is often used to assess the exposure level (exposure to pollution or contaminant) in a workplace. For instance, if t denotes the occupational exposure limit and X denotes the exposure measurement for a worker, then the exceedance probability is defined by $P(X > t)$. If $U(\boldsymbol{X};p)$ is a $(p, 1-\alpha)$ upper tolerance limit, and is less than or equal to t, then we can conclude that $P(X > t)$ is less than $1 - p$. Arguing as in the case of (1.1.7), we conclude that if $p_u = \max\{p : U(\boldsymbol{X};p) \leq t\}$, then $1-p_u$ is a $1-\alpha$ upper confidence limit for the exceedance probability. In general, $U(\boldsymbol{X};p)$ is a nondecreasing function of p, and so p_u is the solution of the equation $U(\boldsymbol{X};p) = t$.

Stress-Strength Reliability

The classical stress-strength reliability problem concerns the proportion of the time the strength X of a component exceeds the stress Y to which it is subjected. If $X \leq Y$, then either the component fails or the system that uses the component may malfunction. If both X and Y are random, then the reliability R of the component can be expressed as $R = P(X > Y)$. A lower limit for R is commonly used to assess the reliability of the component. Writing $R = P(X - Y > 0)$, we see that R can be considered as a survival probability. Therefore, the procedures given for estimating survival probability can be applied to find a lower confidence limit for R. More specifically, if it is desired to test

$$H_0 : R \leq R_0 \quad \text{vs.} \quad H_a : R > R_0$$

at a level α, then the null hypothesis will be rejected if a $(R_0, 1 - \alpha)$ lower tolerance limit for the distribution of $X - Y$ is greater than zero.

1.2   Some Technical Results

In this section, we shall give a number of technical results to be used in later chapters. In particular, the first result has important applications in the derivation of tolerance intervals for a univariate normal distribution and other normal based models.

Result 1.2.1   Let $X \sim N(0, c)$ independently of $Q \sim \frac{\chi_m^2}{m}$, where χ_ν^2 denotes a chi-square random variable with degrees of freedom (df) m. Let $0 < p < 1$, $0 < \gamma < 1$, and let Φ denote the standard normal distribution function.

(i) The factor k_1 that satisfies

$$P_{X,Q}\left(\Phi\left(X + k_1 \sqrt{Q}\right) \geq p\right) = \gamma \qquad (1.2.1)$$

is given by

$$k_1 = \sqrt{c} \times t_{m;\gamma}\left(\frac{z_p}{\sqrt{c}}\right), \qquad (1.2.2)$$

where z_p denotes the p quantile of a standard normal distribution, and $t_{\nu;\eta}(\delta)$ denotes the η quantile of a noncentral t distribution with df ν and noncentrality parameter δ.

(ii) The factor k_2 that satisfies

$$P_{X,Q}\left(\Phi(X + k_2 \sqrt{Q}) - \Phi(X - k_2 \sqrt{Q}) \geq p\right) = \gamma \qquad (1.2.3)$$

is the solution of the integral equation

$$\sqrt{\frac{2}{\pi c}} \int_0^\infty P_Q\left(Q \geq \frac{\chi_{1;p}^2(x^2)}{k_2^2}\right) e^{-\frac{x^2}{2c}}\, dx = \gamma, \qquad (1.2.4)$$

where $\chi_{\nu;\eta}^2(\delta)$ denotes the η quantile of a noncentral chi-square distribution with df ν and noncentrality parameter δ.

(iii)   An approximation to k_2 that satisfies (1.2.4) is given by

$$k_2 \simeq \left(\frac{m\chi_{1;p}^2(c)}{\chi_{m;1-\gamma}^2}\right)^{\frac{1}{2}}, \qquad (1.2.5)$$

where $\chi_{\nu;\eta}^2$ denotes the η quantile of a chi-square distribution with df ν.

Proof.

(i) Note that the inner probability inequality in (1.2.1) holds if and only if $X + k_1\sqrt{Q} \ge z_p$. So we can write (1.2.1) as

$$
\begin{aligned}
P_{X,Q}\left(X + k_1\sqrt{Q} \ge z_p\right) &= P_{X,Q}\left(\frac{X - z_p}{\sqrt{Q}} \ge -k_1\right) \\
&= P_{X,Q}\left(\sqrt{c}\,\frac{X/\sqrt{c} + z_p/\sqrt{c}}{\sqrt{Q}} \le k_1\right) \\
&= \gamma.
\end{aligned}
\tag{1.2.6}
$$

To get the second step of (1.2.6), we used the fact that X and $-X$ are identically distributed. Because $X/\sqrt{c} \sim N(0,1)$ independently of $Q \sim \frac{\chi_m^2}{m}$, we have

$$
\frac{X/\sqrt{c} + z_p/\sqrt{c}}{\sqrt{Q}} \sim t_m\left(\frac{z_p}{\sqrt{c}}\right),
$$

where $t_\nu(\delta)$ denotes a noncentral t random variable with degrees of freedom ν and noncentrality parameter δ. Therefore, k_1 that satisfies (1.2.6) is given by (1.2.2).

(ii) Note that, for a fixed X, $\Phi(X + r) - \Phi(X - r)$ is an increasing function of r. Therefore, for a fixed X, $\Phi(X + k_2\sqrt{Q}) - \Phi(X - k_2\sqrt{Q}) \ge p$ if and only if $k_2\sqrt{Q} > r$ or $Q > \frac{r^2}{k_2^2}$, where r is the solution of the equation

$$
\Phi(X + r) - \Phi(X - r) = p,
\tag{1.2.7}
$$

or equivalently,

$$
P_Z\left((Z - X)^2 \le r^2 | X\right) = p,
\tag{1.2.8}
$$

Z being a standard normal random variable (see Exercise 1.5.3). For a fixed X, $(Z - X)^2 \sim \chi_1^2(X^2)$, where $\chi_m^2(\delta)$ denotes a noncentral chi-square random variable with noncentrality parameter δ. Therefore, conditionally given X^2, r^2 that satisfies (1.2.8) is the p quantile of $\chi_1^2(X^2)$, which we denote by $\chi_{1;p}^2(X^2)$. Using these results, and noticing that r is a function of X^2 and p, we have

$$
\begin{aligned}
P_Q\left(\Phi(X + k_2\sqrt{Q}) - \Phi(X - k_2\sqrt{Q}) > p\,\Big|\,X\right) &= P_Q\left(Q > \frac{r^2}{k_2^2}\,\Big|\,X\right) \\
&= P_Q\left(Q > \frac{\chi_{1;p}^2(X^2)}{k_2^2}\,\Big|\,X\right).
\end{aligned}
\tag{1.2.9}
$$

Taking expectation with respect to the distribution of X, we see that the factor k_2 satisfies

$$E_X\left[P\left(Q > \frac{\chi^2_{1;p}(X^2)}{k_2^2}\right)\right] = \gamma. \qquad (1.2.10)$$

Since $X \sim N(0, c)$, we can write (1.2.10) as

$$\frac{1}{\sqrt{2\pi c}} \int_{-\infty}^{\infty} P\left(Q > \frac{\chi^2_{1;p}(x^2)}{k_2^2}\right) e^{-\frac{x^2}{2c}}\, dx$$

$$= \sqrt{\frac{2}{\pi c}} \int_0^{\infty} P\left(Q > \frac{\chi^2_{1;p}(x^2)}{k^2}\right) e^{-\frac{x^2}{2c}}\, dx$$

$$= \gamma. \qquad (1.2.11)$$

(iii) An approximation for k_2 can be obtained from (1.2.10) as follows. Let $V = X^2$ and $g(V) = P_Q\left(Q > \frac{\chi^2_{1;p}(V)}{k_2^2}\right)$. Using a Taylor series expansion around $V = E(V) = c$, we have

$$g(V) = g(c) + (V - c)\frac{\partial g(V)}{\partial V}\bigg|_{V=c} + \frac{(V - c)^2}{2!}\frac{\partial^2 g(V)}{\partial V^2}\bigg|_{V=c} + \dots \qquad (1.2.12)$$

Noting that $\frac{V}{c} \sim \chi^2_1$, and taking expectation on both sides, we get

$$\begin{aligned} E(g(V)) &= g(c) + c^2\frac{\partial^2 g(V)}{\partial V^2}\bigg|_{V=c} + \dots \\ &= g(c) + O(c^2). \end{aligned} \qquad (1.2.13)$$

Thus, $E(g(V)) \simeq g(c)$, and using this approximation in (1.2.10), we get

$$P_Q\left(Q > \frac{\chi^2_{1;p}(c)}{k_2^2}\right) \simeq \gamma.$$

As $Q \sim \frac{\chi^2_m}{m}$, it follows from the above expression that $\frac{\chi^2_{1;p}(c)}{k_2^2} \simeq \frac{\chi^2_{m;1-\gamma}}{m}$, and solving for k_2, we get part (iii). □

Result 1.2.2 (*Satterthwaite Approximation*) Let $Q_1, ..., Q_k$ be independent random variables with $Q_i \sim \sigma_i^2\frac{\chi^2_{m_i}}{m_i}$, $i = 1, ..., k$. Let $c_1, ..., c_k$ be positive constants. Then

$$V = \frac{\sum_{i=1}^{k} c_i Q_i}{\sum_{i=1}^{k} c_i \sigma_i^2} \sim \frac{\chi^2_f}{f}, \quad \text{approximately,} \qquad (1.2.14)$$

where

$$f = \frac{\left(\sum_{i=1}^{k} c_i \sigma_i^2\right)^2}{\sum_{i=1}^{k} c_i^2 \sigma_i^4 / m_i}. \tag{1.2.15}$$

Proof. The above approximation can be obtained by the *moment matching* method. Specifically, the degrees of freedom f is obtained by matching the mean and variance of V with those of $\frac{\chi_f^2}{f}$. Toward this, we note that $E(\chi_n^2) = n$ and $\mathrm{Var}(\chi_n^2) = 2n$. Using these results, we see that $E(V) = E\left(\frac{\chi_f^2}{f}\right) = 1$, and $\mathrm{Var}(V) = \mathrm{Var}\left(\frac{\chi_f^2}{f}\right)$ when f is as defined in (1.2.15). $\square$

It should be noted that in many applications the σ_i^2's are unknown (see Exercise 1.5.6); in these situations, the estimate

$$\widehat{f} = \frac{\left(\sum_{i=1}^{k} c_i Q_i\right)^2}{\sum_{i=1}^{k} c_i^2 Q_i^2 / m_i},$$

of the df f is commonly used. Note that $\widehat{f}$ is obtained by replacing σ_i^2 by its unbiased estimator Q_i, $i = 1, ..., k$.

The result (1.2.14) (with f replaced by $\widehat{f}$) can be conveniently used to obtain approximate confidence intervals for the linear combination $\sum_{i=1}^{k} c_i \sigma_i^2$, provided $c_i > 0$. For example, an approximate $1 - \alpha$ upper confidence bound for $\sum_{i=1}^{k} c_i \sigma_i^2$ is given by

$$\frac{\widehat{f} \sum_{i=1}^{k} c_i Q_i}{\chi_{\widehat{f};\alpha}^2},$$

where $\chi_{m;\alpha}^2$ denotes the α quantile of a chi-square distribution with m degrees of freedom.

Result 1.2.3 (*Bonferroni Inequality*) Let $A_1, ..., A_k$ be a set of events. Then

$$P(A_1 \cap A_2 \cap ... \cap A_k) \geq 1 - \sum_{i=1}^{k} P(A_i^c), \tag{1.2.16}$$

where A_i^c denotes the complement of the event A_i, $i = 1, ..., k$.

Proof. This inequality essentially follows from the well-known result that, for a collection of events $E_1, ..., E_k$,

$$P(E_1 \cup ... \cup E_k) \leq \sum_{i=1}^{k} P(E_i).$$

So

$$P(A_1 \cap A_2 \cap \ldots \cap A_k) \;=\; 1 - P(A_1^c \cup \ldots \cup A_k^c)$$

$$\geq\; 1 - \sum_{i=1}^{k} P(A_i^c),$$

which completes the proof.   □

The above inequality is useful for constructing conservative simultaneous confidence intervals for a set of parameters $\theta_1, \ldots, \theta_k$, provided a confidence interval for each θ_i is available. For instance, let I_j be a $1 - \frac{\alpha}{k}$ confidence interval for θ_j, $j = 1, \ldots, k$. Then

$$P(I_1 \text{ contains } \theta_1, \ldots, I_k \text{ contains } \theta_k) \;\geq\; 1 - \sum_{j=1}^{k} P(I_j \text{ does not contain } \theta_j)$$

$$=\; 1 - k\left(\frac{\alpha}{k}\right)$$

$$=\; 1 - \alpha.$$

Thus, simultaneously the intervals $I_1, \ldots, I_k$ contain $\theta_1, \ldots, \theta_k$, respectively, with probability at least $1 - \alpha$.

1.3 The Modified Large Sample (MLS) Procedure

Consider the set up of the Satterthwaite approximation (Result 1.2.2), so that we have independent random variables $Q_1, \ldots, Q_k$ with $Q_i \sim \sigma_i^2 \frac{\chi_{m_i}^2}{m_i}$, $i = 1, \ldots, k$. As already noted, the Satterthwaite approximation (1.2.14) can be used, in particular, to obtain approximate confidence intervals for a linear combination $\sum_{i=1}^{k} c_i \sigma_i^2$ of the variance components σ_i^2, $i = 1, 2, \ldots, k$. However, the c_i's are required to be positive in order to be able to use the chi-square approximation (1.2.14). If some of the c_i's are negative, the modified large sample (MLS) procedure can be used to obtain approximate confidence intervals for $\sum_{i=1}^{k} c_i \sigma_i^2$.

We shall first describe the MLS procedure for obtaining an upper confidence bound for $\sum_{i=1}^{k} c_i \sigma_i^2$ when all the c_i's are positive. Since $Q_i \sim \sigma_i^2 \frac{\chi_{m_i}^2}{m_i}$, $\hat{\sigma}_i^2 = Q_i$ is an unbiased estimator for σ_i^2, and a $1 - \alpha$ upper confidence bound for σ_i^2 is $\frac{m_i \hat{\sigma}_i^2}{\chi_{m_i;\alpha}^2}$. Note that this is an exact upper confidence bound, and the bound can be rewritten as

$$\hat{\sigma}_i^2 + \sqrt{\hat{\sigma}_i^4 \left(\frac{m_i}{\chi_{m_i;\alpha}^2} - 1\right)^2}. \tag{1.3.1}$$

Since $E(\widehat{\sigma}_i^2) = \sigma_i^2$ and $V(\widehat{\sigma}_i^2) = 2\frac{\sigma_i^4}{m_i}$, we have

$$E\left(\sum_{i=1}^{k} c_i\widehat{\sigma}_i^2\right) = \sum_{i=1}^{k} c_i\sigma_i^2, \text{ and } V\left(\sum_{i=1}^{k} c_i\widehat{\sigma}_i^2\right) = 2\sum_{i=1}^{k} c_i^2\frac{\sigma_i^4}{m_i}.$$

Thus, an estimator of $V\left(\sum_{i=1}^{k} c_i\widehat{\sigma}_i^2\right)$, is given by

$$\widehat{V}\left(\sum_{i=1}^{k} c_i\widehat{\sigma}_i^2\right) = 2\sum_{i=1}^{k} c_i^2\frac{\widehat{\sigma}_i^4}{m_i}.$$

Assuming an asymptotic normal distribution for $\sum_{i=1}^{k} c_i\widehat{\sigma}_i^2$, an approximate $1-\alpha$ upper confidence bound for $\sum_{i=1}^{k} c_i\sigma_i^2$ is given by $\sum_{i=1}^{k} c_i\widehat{\sigma}_i^2 + z_{1-\alpha}\sqrt{\widehat{V}(\sum_{i=1}^{k} c_i\widehat{\sigma}_i^2)}$, which simplifies to

$$\sum_{i=1}^{k} c_i\widehat{\sigma}_i^2 + \sqrt{2\sum_{i=1}^{k} z_{1-\alpha}^2 c_i^2\frac{\widehat{\sigma}_i^4}{m_i}}. \qquad (1.3.2)$$

In the above, $z_{1-\alpha}$ denotes the $1-\alpha$ quantile of a standard normal distribution. When all the c_i's are positive, the MLS procedure consists of imitating (1.3.1), and replacing $2\frac{z_{1-\alpha}^2}{m_i}$ in (1.3.2) with the quantity $\left(\frac{m_i}{\chi_{m_i;\alpha}^2} - 1\right)^2$. Thus the $1-\alpha$ MLS upper confidence bound for $\sum_{i=1}^{k} c_i\sigma_i^2$, when all the c_i's are positive, is

$$\sum_{i=1}^{k} c_i\widehat{\sigma}_i^2 + \sqrt{\sum_{i=1}^{k} c_i^2\widehat{\sigma}_i^4\left(\frac{m_i}{\chi_{m_i;\alpha}^2} - 1\right)^2}. \qquad (1.3.3)$$

In other words, the confidence bound in (1.3.3) has been obtained by modifying a large sample confidence bound for $\sum_{i=1}^{k} c_i\sigma_i^2$; hence the name *modified large sample* procedure. It can be shown that the coverage probability of the above MLS confidence interval approaches $1-\alpha$ as $m_i \to \infty$ for all i.

If some of the c_i's are negative, the $1-\alpha$ MLS upper confidence bound for $\sum_{i=1}^{k} c_i\sigma_i^2$ is obtained as

$$\sum_{i=1}^{k} c_i\widehat{\sigma}_i^2 + \sqrt{\sum_{i=1}^{k} c_i^2\widehat{\sigma}_i^4\left(\frac{m_i}{u_i} - 1\right)^2}, \qquad (1.3.4)$$

where

$$u_i = \begin{cases} \chi_{m_i;\alpha}^2, & c_i > 0, \\ \chi_{m_i;1-\alpha}^2, & c_i < 0. \end{cases}$$

Furthermore, the $1 - \alpha$ MLS lower confidence bound for $\sum_{i=1}^{k} c_i \sigma_i^2$ is obtained as

$$\sum_{i=1}^{k} c_i \widehat{\sigma}_i^2 + \sqrt{\sum_{i=1}^{k} c_i^2 \widehat{\sigma}_i^4 \left(\frac{m_i}{l_i} - 1\right)^2}, \qquad (1.3.5)$$

where

$$l_i = \left\{ \begin{array}{l} \chi^2_{m_i;1-\alpha}, \ c_i > 0, \\ \chi^2_{m_i;\alpha}, \ c_i < 0. \end{array} \right.$$

The limits of a $1 - \alpha$ two-sided MLS confidence interval for $\sum_{i=1}^{k} c_i \sigma_i^2$ are obtained as the $\left(1 - \frac{\alpha}{2}\right)$ MLS lower confidence bound and the $\left(1 - \frac{\alpha}{2}\right)$ MLS upper confidence bound.

The idea of the MLS confidence interval was proposed by Graybill and Wang (1980). The book by Burdick and Graybill (1992) gives a detailed treatment of the topic along with numerical results regarding the performance of such intervals.

1.4 The Generalized P-value and Generalized Confidence Interval

The generalized p-value approach for hypothesis testing has been introduced by Tsui and Weerahandi (1989) and the generalized confidence interval by Weerahandi (1993). Together, these are referred to as the generalized variable approach or generalized inference procedure. The concepts of generalized p-values and generalized confidence intervals have turned out to be extremely fruitful for obtaining tests and confidence intervals for some complex problems where standard procedures are difficult to apply. The generalized variable approach has been used successfully to develop tests and confidence intervals for "non-standard" parameters, such as lognormal mean (Krishnamoorthy and Mathew, 2003) and normal quantiles (Weerahandi, 1995a). Applications of this approach include inference for variance components (Zhou and Mathew (1994), Khuri, Mathew and Sinha (1998), Mathew and Webb (2005)), ANOVA under unequal variances (Weerahandi, 1995b), growth curve model (Weerahandi and Berger, 1999), common mean problem (Krishnamoorthy and Lu, 2003), tolerance limits for the one-way random effects model (Krishnamoorthy and Mathew, 2004) and multivariate Behrens-Fisher problem (Gamage, Mathew and Weerahandi, 2004). For the computation of tolerance limits in the context of mixed and random effects models, the generalized variables approach will be used rather extensively in this

book. For more applications and a detailed discussion, we refer to the books by
Weerahandi (1995a, 2004).

In the following we shall describe the concepts of generalized confidence
intervals and generalized p-values in a general setup.

1.4.1 Description

Let $\boldsymbol{X}$ be a random sample from a distribution $F_X(x; \theta, \delta)$, where θ is a scalar
parameter of interest, and δ is a nuisance parameter. Let $\boldsymbol{x}$ be an observed
sample; that is, $\boldsymbol{x}$ represents the data.

Generalized Pivotal Quantity (GPQ)

A generalized confidence interval for θ is computed using the percentiles of a
generalized pivotal quantity (GPQ), say $G(\boldsymbol{X}; \boldsymbol{x}, \theta)$, a function of $\boldsymbol{X}$, $\boldsymbol{x}$, and θ
(and possibly the nuisance parameter δ) satisfying the following conditions:

(i) For a given $\boldsymbol{x}$, the distribution of $G(\boldsymbol{X}; \boldsymbol{x}, \theta)$ is free of all unknown parameters.

(ii) The "observed value" of $G(\boldsymbol{X}; \boldsymbol{x}, \theta)$, namely its value at $\boldsymbol{X} = \boldsymbol{x}$, is θ, the
parameter of interest.   (C.1)

The conditions given above are a bit more restrictive than what is actually
required. However, we shall assume the above conditions since they are met in
our applications. Regarding condition (ii) above, note that θ is not observable,
since it is an unknown parameter. The term "observed value" refers to the
simplified form of $G(\boldsymbol{X}; \boldsymbol{x}, \theta)$ when $\boldsymbol{X} = \boldsymbol{x}$. When the conditions (i) and (ii) in
(C.1) hold, appropriate quantiles of $G(\boldsymbol{X}; \boldsymbol{x}, \theta)$ form a $1 - \alpha$ confidence interval
for θ. For example, if G_p is the p quantile of $G(\boldsymbol{X}; \boldsymbol{x}, \theta)$, then $\left(G_{\frac{\alpha}{2}}, G_{1-\frac{\alpha}{2}}\right)$ is
a $1 - \alpha$ confidence interval for θ. Such confidence intervals are referred to as
generalized confidence intervals. As already pointed out, $G(\boldsymbol{X}; \boldsymbol{x}, \theta)$ may also
depend on the nuisance parameters. Even though this is not made explicit in
our notation, this will be clear from the examples given later.

Generalized Test Variable

A generalized test variable , denoted by $T(\boldsymbol{X}; \boldsymbol{x}, \theta)$, is a function of $\boldsymbol{X}$, $\boldsymbol{x}$, and θ
(and possibly the nuisance parameter δ) and satisfies the following conditions:

(i) For a given x, the distribution of $T(X; x, \theta)$ is free of the nuisance parameter δ.

(ii) The "observed value" of $T(X; x, \theta)$, namely its value at $X = x$, is free of any unknown parameters.

(iii) For a fixed x and δ, the distribution of $T(X; x, \theta)$ is stochastically monotone in θ. That is, $P(T(X; x, \theta) \geq a)$ is an increasing function of θ, or is a decreasing function of θ, for every a.   (C.2)

Let $t = T(x; x, \theta)$, the value of $T(X; x, \theta)$ at $X = x$. Suppose we are interested in testing

$$H_0 : \theta \leq \theta_0 \quad \text{vs.} \quad H_a : \theta > \theta_0, \tag{1.4.1}$$

where θ_0 is a specified value. If $T(X; x, \theta)$ is stochastically increasing in θ, the generalized p-value for testing the hypotheses in (1.4.1) is given by

$$\sup_{H_0} P(T(X; x, \theta) \geq t) = P(T(X; x, \theta_0) \geq t),$$

and if $T(X; x, \theta)$ is stochastically decreasing in θ, the generalized p-value for testing the hypotheses in (1.4.1) is given by

$$\sup_{H_0} P(T(X; x, \theta) \leq t) = P(T(X; x, \theta_0) \leq t). \tag{1.4.2}$$

Note that the computation of the generalized p-value is possible because the distribution of $T(X; x, \theta_0)$ is free of the nuisance parameter δ, and $t = T(x; x, \theta_0)$ is free of any unknown parameters. However, the nuisance parameter δ may be involved in the definition of $T(X; x, \theta)$, even though this is not made explicit in the notation.

In many situations, the generalized test variable $T(X; x, \theta)$ is the GPQ minus the parameter of interest; that is $T(X; x, \theta) = G(X; x, \theta) - \theta$. It should be noted that in general, there is no unique way of constructing a GPQ for a given problem. Consideration of sufficiency and invariance may simplify the problem of finding a GPQ.

Numerous applications of generalized confidence intervals and generalized p-values have appeared in the literature. Several such applications are given in the books by Weerahandi (1995a, 2004). It should however be noted that generalized confidence intervals and generalized p-values may not satisfy the usual repeated sampling properties. That is, the actual coverage probability of a 95% generalized confidence interval could be different from 0.95, and the coverage could in fact depend on the nuisance parameters. Similarly, the generalized p-value may not have a uniform distribution under the null hypothesis. Consequently, a test carried out using the generalized p-value at a 5% significance level can have actual type I error probability different from 5%, and the type I error probability may depend on the nuisance parameters. The asymptotic accuracy of a class of generalized confidence interval procedures has recently been established by Hannig, Iyer and Patterson (2005). However, the small sample accuracy of any procedure based on generalized confidence intervals and generalized p-values has to be investigated at least numerically, before they can be recommended for practical use. While applying the generalized confidence interval idea for the derivation of tolerance limits in later chapters, we shall comment on their accuracy based on simulations.

1.4.2   GPQs for a Location-Scale Family

A continuous univariate distribution is said to belong to the location-scale family if its probability density function (pdf) can be expressed in the form

$$f(x;\mu,\sigma) = \frac{1}{\sigma}g\left(\frac{x-\mu}{\sigma}\right), \quad -\infty < x < \infty, \ -\infty < \mu < \infty, \ \sigma > 0, \qquad (1.4.3)$$

where g is a completely specified pdf. Here μ and σ are referred to as the location and scale parameters, respectively. As an example, the family of normal distributions belongs to the location-scale family because the pdf can be expressed as

$$f(x;\mu,\sigma) = \frac{1}{\sigma}\phi\left(\frac{x-\mu}{\sigma}\right) \quad \text{with} \ \ \phi(x) = \frac{1}{\sqrt{2\pi}}e^{-\frac{x^2}{2}}.$$

Let $X_1, ..., X_n$ be a sample from a distribution with the location parameter μ and scale parameter σ. Estimators $\widehat{\mu}(X_1, ..., X_n)$ of μ and $\widehat{\sigma}(X_1, ..., X_n)$ of σ are said to be equivariant if for any constants a and b with $a > 0$,

$$\begin{aligned} \widehat{\mu}(aX_1 + b, ..., aX_n + b) &= a\widehat{\mu}(X_1, ..., X_n) + b \\ \widehat{\sigma}(aX_1 + b, ..., aX_n + b) &= a\widehat{\sigma}(X_1, ..., X_n). \end{aligned} \qquad (1.4.4)$$

For example, the sample mean $\bar{X}$ and the sample variance S^2 are equivariant estimators for a normal mean and variance respectively.

Result 1.4.1 Let $X_1, ..., X_n$ be a sample from a continuous distribution with the pdf of the form in (1.4.3). Let $\widehat{\mu}(X_1, ..., X_n)$ and $\widehat{\sigma}(X_1, ..., X_n)$ be equivariant estimators of μ and σ, respectively. Then

$$\frac{\widehat{\mu}-\mu}{\sigma}, \quad \frac{\widehat{\sigma}}{\sigma} \quad \text{and} \quad \frac{\widehat{\mu}-\mu}{\widehat{\sigma}}$$

are all pivotal quantities. That is, their distributions do not depend on any parameters.

Proof. Let $Z_i = \frac{X_i - \mu}{\sigma}$, $i = 1, ..., n$. As the sample is from a location-scale distribution, the joint distribution of Z_i's are free of unknown parameters. Since $\widehat{\mu}$ and $\widehat{\sigma}$ are equivariant, we have

$$\frac{\widehat{\mu}(X_1, ..., X_n) - \mu}{\sigma} = \widehat{\mu}\left(\frac{X_1 - \mu}{\sigma}, ..., \frac{X_n - \mu}{\sigma}\right)$$
$$= \widehat{\mu}(Z_1, ..., Z_n)$$

and

$$\frac{\widehat{\sigma}(X_1, ..., X_n)}{\sigma} = \widehat{\sigma}\left(\frac{X_1 - \mu}{\sigma}, ..., \frac{X_n - \mu}{\sigma}\right)$$
$$= \widehat{\sigma}(Z_1, ..., Z_n).$$

Thus, $\frac{\widehat{\mu} - \mu}{\sigma}$ and $\frac{\widehat{\sigma}}{\sigma}$ are pivotal quantities. Furthermore, $\frac{\widehat{\mu} - \mu}{\widehat{\sigma}} = \left(\frac{\widehat{\mu} - \mu}{\sigma}\right)\left(\frac{\sigma}{\widehat{\sigma}}\right)$ is also a pivotal quantity. □

GPQs for μ and σ based on the above pivotal quantities can be constructed as follows. Let $\widehat{\mu}_0$ and $\widehat{\sigma}_0$ be observed values of the equivariant estimators $\widehat{\mu}$ and $\widehat{\sigma}$, respectively. Recall from condition (ii) of (C.1) that the value of a GPQ of μ at $(\widehat{\mu}, \widehat{\sigma}) = (\widehat{\mu}_0, \widehat{\sigma}_0)$ should be μ. Keeping this in mind, we construct a GPQ for μ as

$$G_\mu(\widehat{\mu}, \widehat{\sigma}; \widehat{\mu}_0, \widehat{\sigma}_0) = \widehat{\mu}_0 - \left(\frac{\widehat{\mu} - \mu}{\widehat{\sigma}}\right)\widehat{\sigma}_0. \tag{1.4.5}$$

Notice that the value of $G_\mu(\widehat{\mu}, \widehat{\sigma}; \widehat{\mu}_0, \widehat{\sigma}_0)$ at $(\widehat{\mu}, \widehat{\sigma}) = (\widehat{\mu}_0, \widehat{\sigma}_0)$ is μ. For a given $(\widehat{\mu}_0, \widehat{\sigma}_0)$, the distribution of G_μ doest not depend on any unknown parameters because $\frac{\widehat{\mu} - \mu}{\widehat{\sigma}}$ is a pivotal quantity. Thus, G_μ in (1.4.5) satisfies both conditions in (C.1).

A GPQ for σ^2 can be obtained similarly, and is given by

$$G_{\sigma^2}(\widehat{\sigma}^2; \widehat{\sigma}_0^2) = \frac{\sigma^2}{\widehat{\sigma}^2}\widehat{\sigma}_0^2. \tag{1.4.6}$$

Again, it is easy to see that $G(\widehat{\sigma}^2; \widehat{\sigma}_0^2)$ satisfies both conditions of (C.1).

1.4.3 Some Examples

Generalized Pivotal Quantities for Normal Parameters

Let $X_1, ..., X_n$ be a random sample from a $N(\mu, \sigma^2)$ distribution. Define the sample mean $\bar{X}$ and sample variance S^2 by

$$\bar{X} = \frac{1}{n}\sum_{i=1}^{n} X_i \quad \text{and} \quad S^2 = \frac{1}{n-1}\sum_{i=1}^{n}(X_i - \bar{X})^2. \tag{1.4.7}$$

Let $\bar{x}$ and s^2 denote the observed values of $\bar{X}$ and S^2, respectively. Then $\hat{\mu} = \bar{X}$ and $\hat{\sigma}^2 = S^2$, and $\hat{\mu}_0 = \bar{x}$ and $\hat{\sigma}_0^2 = s^2$.

A GPQ for a Normal Mean

It follows from (1.4.5) that

$$
\begin{aligned}
G_\mu = G_\mu(\hat{\mu}, \hat{\sigma}; \hat{\mu}_0, \hat{\sigma}_0) &= \hat{\mu}_0 - \frac{\hat{\mu} - \mu}{\hat{\sigma}} \hat{\sigma}_0 \\
&= \hat{\mu}_0 - \frac{\sqrt{n}(\bar{X} - \mu)/\sigma}{S/\sigma} \frac{s}{\sqrt{n}} \\
&= \hat{\mu}_0 - \frac{Z}{U} \frac{s}{\sqrt{n}} \\
&= \bar{x} + \frac{Z}{U} \frac{s}{\sqrt{n}},
\end{aligned}
\tag{1.4.8}
$$

where $Z = \frac{\sqrt{n}(\bar{X}-\mu)}{\sigma} \sim N(0,1)$ independently of $U^2 = \frac{S^2}{\sigma^2} \sim \frac{\chi_{n-1}^2}{n-1}$. We have also used the property that Z and $-Z$ have the same distribution in order to get the last step of (1.4.8).

Noticing that $\frac{Z}{U} \sim t_{n-1}$, we can also write

$$
G_\mu = G_\mu(\bar{X}, S; \bar{x}, s) = \bar{x} + t_{n-1} \frac{s}{\sqrt{n}}.
\tag{1.4.9}
$$

The generalized confidence interval is then given by

$$
\left(\bar{x} + t_{n-1;\frac{\alpha}{2}} \frac{s}{\sqrt{n}}, \ \bar{x} + t_{n-1;1-\frac{\alpha}{2}} \frac{s}{\sqrt{n}} \right),
$$

which is the usual t-interval. Here $t_{n-1;\gamma}$ denotes the γ quantile of a central t distribution with $n-1$ df.

Now consider the testing problem

$$
H_0 : \mu \le \mu_0 \quad \text{vs.} \quad H_a : \mu > \mu_0.
\tag{1.4.10}
$$

A generalized test variable , say $T_\mu(\bar{X}, S; \bar{x}, s)$, for the above testing problem is given by

$$
T_\mu(\bar{X}, S; \bar{x}, s) = G_\mu(\bar{X}, S; \bar{x}, s) - \mu.
\tag{1.4.11}
$$

It is easy to see that the generalized test variable $T_\mu = T_\mu(\bar{X}, S; \bar{x}, s)$ satisfies the first two conditions in (C.2). Furthermore, for a given $(\bar{x}, s)$, T_μ is stochastically decreasing in μ because, for a fixed $(\bar{x}, s)$, the distribution of $G_\mu(\bar{X}, S; \bar{x}, s)$ does not depend on μ and $T_\mu = G_\mu(\bar{X}, S; \bar{x}, s) - \mu$. So the generalized p-value for testing (1.4.10) is given by (see (1.4.2))

$$
\begin{aligned}
\sup_{H_0} P(T_\mu \leq 0) &= \sup_{H_0} P(G_\mu \leq \mu) \\
&= P(G_\mu \leq \mu_0) \\
&= P\left(\bar{x} + t_{n-1}\frac{s}{\sqrt{n}} \leq \mu_0\right) \\
&= P\left(t_{n-1} \geq \frac{\bar{x} - \mu_0}{s/\sqrt{n}}\right),
\end{aligned}
\tag{1.4.12}
$$

which is the usual p-value based on the t-test for a normal mean. To get the third step of (1.4.12), we have used the result that t_m and $-t_m$ are identically distributed.

A GPQ for the Normal Variance

A GPQ for σ^2 is given by

$$
G_{\sigma^2} = G_{\sigma^2}(\hat{\sigma}^2; \hat{\sigma}_0^2) = \frac{\sigma^2}{S^2}s^2 = \frac{s^2}{U^2},
\tag{1.4.13}
$$

where $U^2 = \frac{S^2}{\sigma^2} \sim \frac{\chi_{n-1}^2}{n-1}$. It can be easily verified that G_{σ^2} satisfies the two conditions in (C.2). Furthermore, the generalized confidence interval

$$
\left(G_{\sigma^2;\frac{\alpha}{2}}, G_{\sigma^2;1-\frac{\alpha}{2}}\right) = \left(\frac{(n-1)s^2}{\chi_{n-1;1-\frac{\alpha}{2}}^2}, \frac{(n-1)s^2}{\chi_{n-1;\frac{\alpha}{2}}^2}\right)
$$

is also the usual $1 - \alpha$ confidence interval for σ^2.

Thus, we see that the generalized variable method is conducive to get exact inferential procedures for normal parameters, and the solutions reduce to the respective standard solutions. An appealing feature of this approach is that the GPQ for any function of (μ, σ^2) can be easily obtained by substitution. Indeed, if it is desired to make inference for a function $f(\mu, \sigma^2)$, then the GPQ is given by $f(G_\mu, G_{\sigma^2})$. A particular case is described below.

A GPQ for a Lognormal Mean

We first note that if Y follows a lognormal distribution with parameters μ and σ^2, then $X = \ln(Y)$ follows a $N(\mu, \sigma^2)$ distribution. Furthermore, note that the lognormal mean is given by $E(Y) = \exp\left(\mu + \frac{\sigma^2}{2}\right)$, and so the GPQs for μ and σ^2 derived above can be readily applied to get a GPQ for $E(Y)$ or for $\eta = \mu + \frac{\sigma^2}{2}$.

Let $\bar{X}$ and S^2 denote the mean and variance of the log-transformed sample from a lognormal distribution. Then a GPQ for η, say G_η, based on the GPQs of μ and σ^2, is given by

$$
\begin{aligned}
G_\eta &= G_\mu + \frac{G_{\sigma^2}}{2} \\
&= \bar{x} + \frac{Z}{U}\frac{s}{\sqrt{n}} + \frac{1}{2}\frac{(n-1)s^2}{U^2},
\end{aligned}
$$

where Z and U^2 are the quantities in (1.4.8). Being a function of the independent random variables $Z \sim N(0,1)$ and $U^2 \sim \frac{\chi^2_{n-1}}{n-1}$, it is straightforward to estimate the percentiles of G_η by Monte Carlo simulation. In other words, the computation of a generalized confidence interval for η is quite simple. We refer to Krishnamoorthy and Mathew (2003) for further details. An exact confidence interval for η is available, based on a certain conditional distribution; see Land (1973). Krishnamoorthy and Mathew (2003) showed that the results based on G_η are practically equivalent to the exact ones by Land (1973).

The reader of this book will notice the rather heavy use of the generalized variable approach in this book, especially in Chapters 4, 5 and 6, dealing with the derivation of tolerance intervals for mixed and random effects models. The reason for this is that this approach has turned out to be particularly fruitful for obtaining satisfactory tolerance intervals in such models. It also appears that solutions to several such tolerance interval problems have been possible due to the availability of the generalized variable approach. To a reader who is new to this topic, a careful reading of Section 1.4 is recommended. Further insight and experience on this topic can be gained by solving the related problems in the exercise set given below.

1.5   Exercises

1.5.1. Let $\boldsymbol{X}$ be a sample from a continuous distribution F_X, and let q_p denote the p quantile of F_X. Define $U(\boldsymbol{X})$ such that

$$
P_{\boldsymbol{X}}\left\{P_X\left(X \leq U(\boldsymbol{X})|\boldsymbol{X}\right) \geq p\right\} = 1 - \alpha.
$$

Show that $U(\boldsymbol{X})$ is a $1-\alpha$ upper confidence limit for q_p.

1.5.2. Let q_p denote the p quantile of a continuous distribution F_X. Let $L_l\left(\boldsymbol{X};p,\frac{\alpha}{2}\right)$ and $L_u\left(\boldsymbol{X};p,1-\frac{\alpha}{2}\right)$ respectively be one-sided lower and upper confidence limits for q_p. Furthermore, let p_l and p_u be such that $L_l\left(\boldsymbol{X};p_l,\frac{\alpha}{2}\right) = t$ and $L_u\left(\boldsymbol{X};p_u,1-\frac{\alpha}{2}\right) = t$, where t is a specified value.

Show that (p_l, p_u) is a $1 - \alpha$ confidence interval for the survival probability $S_t = 1 - F_X(t)$.

1.5.3. Let Φ denote the standard normal cdf. For a given p in $(0, 1)$, show that the solution of $\Phi(a) - \Phi(-a) = p$, is given by $a = \chi_{1:p}^2$. For a constant $c \neq 0$, show that $\Phi(c + a) - \Phi(c - a) = p$ when $a = \chi_{1:p}^2(c^2)$.

1.5.4. Consider a distribution $F_X(x|\theta)$ that depends only on a single parameter θ, and is stochastically increasing in θ. Let $q_p(\theta)$ be the p quantile of the distribution, and let (θ_l, θ_u) be a $1 - 2\alpha$ confidence interval for θ.

(a) Show that $q_p(\theta_u)$ is a $(p, 1 - \alpha)$ upper tolerance limit for the distribution.

(b) Show that $(q_{1-p}(\theta_l), q_p(\theta_u))$ is a $(p, 1 - 2\alpha)$ equal-tailed tolerance interval for the distribution.

1.5.5. Let X_1 and X_2 be independent continuous random variables, and let $R = P(X_1 > X_2)$. Consider the testing problem $H_0 : R \leq R_0$ vs. $H_a : R > R_0$, where R_0 is a specified number in $(0, 1)$. Show that the test that rejects H_0 whenever a $(R_0, 1 - \alpha)$ lower tolerance limit for the distribution of $X_1 - X_2$ is greater than zero, is a level α test.

1.5.6. Let $(\bar{X}_i, S_i^2)$ denote the (mean, variance) based on a sample of size n_i from a $N(\mu_i, \sigma_i^2)$ distribution, $i = 1, 2$.

(a) Show that

$$\frac{S_1^2}{n_1} + \frac{S_2^2}{n_2} \sim \frac{\chi_f^2}{f} \text{ approximately, with } f = \frac{\left(\frac{\sigma_1^2}{n_1} + \frac{\sigma_2^2}{n_2}\right)^2}{\left(\frac{\sigma_1^4}{n_1^2(n_1-1)} + \frac{\sigma_2^4}{n_2^2(n_2-1)}\right)}.$$

(b) Using part (a), show that for testing $H_0 : \mu_1 = \mu_2$, the test statistic

$$\frac{\bar{X}_1 - \bar{X}_2}{\sqrt{\frac{S_1^2}{n_1} + \frac{S_2^2}{n_2}}} \sim t_{\hat{f}} \text{ approximately, with } \hat{f} = \frac{\left(\frac{S_1^2}{n_1} + \frac{S_2^2}{n_2}\right)^2}{\left(\frac{S_1^4}{n_1^2(n_1-1)} + \frac{S_2^4}{n_2^2(n_2-1)}\right)},$$

when H_0 is true. (This test is known as the Welch approximate degrees of freedom test for equality of the means when the variances are not assumed to be equal).

1.5.7. Let $\bar{X}$ and S denote, respectively, the mean and standard deviation based on a random sample of size n from a $N(\mu, \sigma^2)$ distribution. Explain how you will use the generalized confidence interval methodology to compute an upper confidence limit for $\mu + z_p \sigma$, the p percentile of $N(\mu, \sigma^2)$, where z_p is the p percentile of $N(0, 1)$. Show that the resulting confidence interval is based on a noncentral t distribution, and is exact.

1.5.8. Suppose random samples of sizes n_1 and n_2, respectively, are available from the independent normal populations $N(\mu_1, \sigma^2)$ and $N(\mu_2, \sigma^2)$, having a common variance σ^2. Explain how you will compute a confidence interval for $\mu_1 - \mu_2$ by the generalized confidence interval procedure. Show that the resulting confidence interval coincides with the usual student's t confidence interval.

1.5.9. Suppose random samples of sizes n_1 and n_2, respectively, are available from the independent normal populations $N(\mu_1, \sigma_1^2)$ and $N(\mu_2, \sigma_2^2)$.

(a) Explain how you will compute a confidence interval for $\sigma_1^2 - \sigma_2^2$ by the generalized confidence interval procedure and by the MLS procedure.

(b) Explain how you will compute a confidence interval for σ_1^2 / σ_2^2 by the generalized confidence interval procedure. Show that the resulting confidence interval coincides with the usual confidence interval based on an F distribution.

1.5.10. Let Y_1 and Y_2 be independent random variables with $Y_i \sim \text{lognormal}(\mu_i, \sigma_i^2)$, $i = 1, 2$. Based on random samples from the respective distributions, derive GPQs for computing a confidence interval for the ratio of the two lognormal means, and for the difference between the two lognormal means.
[Krishnamoorthy and Mathew, 2003]

1.5.11. An oil refinery located at the northeast of San Francisco obtained 31 observations on carbon monoxide levels from one of their stacks; the measurements were obtained between April 16 and May 16, 1993, and were submitted to the Bay Area Air Quality Management District (BAAQMD) for establishing a baseline. Nine independent measurements of the carbon monoxide concentration from the same stack were made by the BAAQMD personnel over the period from September 11, 1990 – March 30, 1993. Based on the data, it is decided to test if the refinery overestimated the carbon monoxide emissions (to setup a baseline at a higher level). The data are give below.

Carbon Monoxide Measurements by the Refinery (in ppm)

45	30	38	42	63	43	102	86	99	63	58
34	37	55	58	153	75	58	36	59	43	102
52	30	21	40	141	85	161	86	161	86	71

Carbon Monoxide Measurements by the BAAQMD (in ppm)

$$\overline{12.5, \ 20, \ 4, \ 20, \ 25, \ 170, \ 15, \ 20, \ 15}$$

(a) Based on normal probability plots of the log-transformed data, conclude that lognormality can be assumed for the two samples.

(b) Compute a 95% generalized lower confidence limit for the ratio of the two lognormal means: the population mean carbon monoxide level for the measurements made by the refinery and for the measurements made by the BAAQMD. Use 10,000 simulated values of the GPQ to estimate the required percentile.

(c) Based on the lower confidence limit that you have computed, can you conclude that the refinery overestimated the carbon monoxide emissions?   [Krishnamoorthy and Mathew, 2003]

1.5.12. Let $X_1, ..., X_n$ be a sample from a $N(\mu, \sigma^2)$ distribution. Find $1 - \alpha$ Bonferroni simultaneous confidence intervals, simultaneously for μ and σ^2.

1.5.13. The following is a sample from a $N(\mu, \sigma^2)$ distribution.

−1.11	4.64	−1.14	−0.57	2.61	0.12	0.81	4.18	7.26	4.59
1.90	2.79	2.44	3.74	2.70	4.09	1.42	1.80	−0.82	1.84

For these data the mean $\bar{x} = 2.165$ and the standard deviation $s = 2.225$.

(a) Construct a 95% confidence interval for μ.

(b) Construct a 95% confidence interval for σ^2.

(c) Using the result of Exercise 1.5.12, construct a 95% simultaneous confidence intervals for μ and σ^2.

(d) Compare the confidence intervals in parts (a) and (b) with the corresponding Bonferroni intervals. Explain, why the Bonferroni intervals are wider than the corresponding ones in parts (a) and (b).

1.5.14. Let $Z_1, ..., Z_n$ be a sample from an exponential distribution with the pdf $\frac{1}{\theta} e^{-\frac{(x-\mu)}{\theta}}$, $x > \mu$, $\mu > 0$, $\theta > 0$. The maximum likelihood estimators of μ and θ are given by

$$\hat{\mu} = Z_{(1)} \quad \text{and} \quad \hat{\theta} = \frac{1}{n} \sum_{i=1}^{n} (Z_i - Z_{(1)}) = \bar{Z} - Z_{(1)}, \qquad (1.4.14)$$

where $Z_{(1)}$ is the smallest of the Z_i's, and $\bar{Z}$ is the average of the Z_i's.

(a) Show that $\hat{\mu}$ and $\hat{\theta}$ are equivariant estimators.

(b) It is known that $\widehat{\mu}$ and $\widehat{\theta}$ are independent with

$$\frac{(\widehat{\mu} - \mu)}{\theta} \sim \frac{\chi_2^2}{2n} \quad \text{and} \quad \frac{\widehat{\theta}}{\theta} \sim \frac{\chi_{2n-2}^2}{2n}.$$

Using these distributional results find GPQs for μ and θ.

(c) Show that the $1 - \alpha$ generalized confidence interval for μ is given by

$$\left(\widehat{\mu}_0 - \frac{1}{n-1} F_{2,2n-2;1-\frac{\alpha}{2}} \widehat{\theta}_0, \widehat{\mu}_0 - \frac{1}{n-1} F_{2,2n-2;\frac{\alpha}{2}} \widehat{\theta}_0 \right),$$

where $F_{a,b;p}$ denotes the p quantile of an F distribution with df $=$ (a, b).

1.5.15. Let $X_1, ..., X_n$ be a sample from a Laplace distribution with the pdf

$$f(x|a, b) = \frac{1}{2b} \exp\left[-\frac{|x - a|}{b} \right],$$

$$-\infty < x < \infty, \quad -\infty < a < \infty, \quad b > 0,$$

where a is the location parameter and b is the scale parameter.

(a) Show that the sample median is the MLE of a, and the MLE of b is given by $\widehat{b} = \frac{1}{n} \sum_{i=1}^{n} |X_i - \widehat{a}|$.

(b) Show that the MLEs are equivariant.

(c) Construct a GPQ for a, and obtain a $1 - \alpha$ generalized confidence interval for a.

(d) Using Monte Carlo simulation, check if the coverage probabilities of the confidence interval are close to the nominal level $1 - \alpha$.

Chapter 2

Univariate Normal Distribution

2.1 Introduction

The normal distribution is the most commonly used distribution in practical applications. Early work on the construction of tolerance limits due to Wilks (1941, 1942), Wald (1943) and Wald and Wolfowitz (1946) are all for the normal distribution. Exact methods for computing one-sided tolerance limits, two-sided tolerance intervals, and two-sided tolerance intervals controlling both tails, are available for the normal distribution. Factors for constructing tolerance intervals for normal distributions have been tabulated for a wide range of sample sizes, and software packages that compute tolerance factors are also available. Because of its popularity and applicability, normal based tolerance limits are routinely used in acceptance sampling plans and for setting tolerance specifications for engineering products.

Note that normal based methods are applicable to a non-normal distribution if it has a one-one relation with a normal distribution. For example, if X follows a lognormal distribution, then $\ln(X)$ follows a normal distribution. Therefore, the approaches that we shall describe in the following sections can be used to construct tolerance intervals for a lognormal distribution as well. Specifically, if the data are from a lognormal distribution, then normal based methods for constructing tolerance intervals can be used after taking logarithmic transformation of the data.

In the following sections, we describe methods for constructing one-sided tolerance limits, two-sided tolerance intervals, and two-sided tolerance intervals controlling both tails, for a normal distribution. In addition, some approximate

25

methods for setting tolerance limits for the distribution of the difference between two independent normal random variables will be given. Also described are procedures for testing whether a specified proportion of the data are within tolerance specifications, and exact procedures for finding factors for simultaneous tolerance limits.

2.2 One-Sided Tolerance Limits for a Normal Population

Let $X_1, ..., X_n$ be a sample from a $N(\mu, \sigma^2)$ population with unknown mean μ and unknown variance σ^2. The sample mean $\bar{X}$ and sample variance S^2 are defined by

$$\bar{X} = \frac{1}{n}\sum_{i=1}^{n} X_i \ \text{ and } \ S^2 = \frac{1}{n-1}\sum_{i=1}^{n}(X_i - \bar{X})^2.$$

In this section, we shall describe the computation of one-sided tolerance limits based on $\bar{X}$ and S^2 for a normal population.

Let z_p denote the p quantile of a standard normal distribution. Then the p quantile of $N(\mu, \sigma^2)$ is given by

$$q_p = \mu + z_p\sigma.$$

A $1 - \alpha$ upper confidence limit for q_p is a $(p, 1 - \alpha)$ one-sided upper tolerance limit for the normal population (see Section 1.2, Chapter 1). In most practical applications, an upper limit for q_p is desired if $p > .5$ and a lower limit for q_p is desired if $p < .5$.

The Classical Approach

We shall assume that the $(p, 1 - \alpha)$ upper tolerance limit is of the form $\bar{X} + k_1 S$. The factor k_1, referred to as a *tolerance factor*, is to be determined such that at least a proportion p of the population measurements are less than $\bar{X} + k_1 S$ with confidence $1 - \alpha$. That is,

$$P_{\bar{X},S}\{P(X < \bar{X} + k_1 S | \bar{X}, S) > p\} = 1 - \alpha, \qquad (2.2.1)$$

where $X \sim N(\mu, \sigma^2)$. Letting $Z = \frac{X-\mu}{\sigma} \sim N(0, 1)$, $Z_n = \frac{\bar{X}-\mu}{\sigma} \sim N\left(0, \frac{1}{n}\right)$, $U^2 = \frac{S^2}{\sigma^2} \sim \frac{\chi_{n-1}^2}{n-1}$, where χ_m^2 denotes a chi-square random variable with m df, we can write (2.2.1) as

$$P_{Z_n,U}\left\{P(Z < Z_n + k_1 U | Z_n, U) > p\right\} = P_{Z_n,U}\left\{\Phi(Z_n + k_1 U) > p\right\}$$
$$= 1 - \alpha. \qquad (2.2.2)$$

As $Z_n \sim N\left(0, \frac{1}{n}\right)$ independently of $U^2 \sim \frac{\chi^2_{n-1}}{n-1}$, we can apply Result 1.2.1(ii) of Chapter 1 with $c = \frac{1}{n}$, $\gamma = 1 - \alpha$ and $m = n - 1$, to get

$$k_1 = \frac{1}{\sqrt{n}} t_{n-1;1-\alpha}(z_p \sqrt{n}), \qquad (2.2.3)$$

where $t_{m;1-\alpha}(\delta)$ denotes the $1 - \alpha$ quantile of a noncentral t distribution with degrees of freedom m, and the noncentrality parameter δ. Thus, a $(p, 1 - \alpha)$ upper tolerance limit is given by

$$\bar{X} + k_1 S = \bar{X} + t_{n-1;1-\alpha}(z_p \sqrt{n}) \frac{S}{\sqrt{n}}. \qquad (2.2.4)$$

The same factor k_1 can be used to obtain $(p, 1 - \alpha)$ lower tolerance limit, and the limit is given by $\bar{X} - k_1 S$ (see Exercise 2.6.1).

The Generalized Variable Approach

Since the computation of a $(p, 1-\alpha)$ upper tolerance limit for $N(\mu, \sigma^2)$ is equivalent to the computation of a $1 - \alpha$ upper confidence limit for $q_p = \mu + z_p \sigma$, we shall now describe the generalized confidence interval approach for obtaining such a confidence limit. A generalized pivotal quantity (GPQ) for $q_p = \mu + z_p \sigma$ can be constructed as follows (see also Section 1.4 of Chapter 1). Let $\bar{x}$ and s denote the observed values of $\bar{X}$ and S, respectively. That is, $\bar{x}$ and s are the numerical values of $\bar{X}$ and S based on an observed sample. A GPQ for $q_p = \mu + z_p \sigma$ is given by

$$\begin{aligned} G_{q_p} &= G_\mu + z_p \sqrt{G_{\sigma^2}} \\ &= \bar{x} + \frac{Z}{U} \frac{s}{\sqrt{n}} + z_p \frac{s}{U} \\ &= \bar{x} + \frac{Z + z_p \sqrt{n}}{U} \frac{s}{\sqrt{n}} \\ &= \bar{x} + \frac{1}{\sqrt{n}} t_{n-1}(z_p \sqrt{n}) s, \qquad (2.2.5) \end{aligned}$$

where $Z = \frac{\bar{X} - \mu}{\sigma/\sqrt{n}}$ follows a standard normal distribution independently of $U^2 = \frac{S^2}{\sigma^2}$ which is distributed as $\frac{\chi^2_{n-1}}{n-1}$, and G_μ and G_{σ^2} are given in (1.4.8) and (1.4.13), respectively. These distributional results and a definition of the noncentral t variable are used to get the last step of the above equation from its preceding step. It can be easily verified that the GPQ in (2.2.5) satisfies both conditions in (C.1) of Section 1.4. Specifically, using step 1 of (2.2.5), we see that the value of G_{q_p} at $(\bar{X}, S) = (\bar{x}, s)$ is q_p, the parameter of interest. Furthermore, for a fixed value of $(\bar{x}, s)$, the distribution of G_{q_p} does not depend on any unknown parameters. Thus, G_{q_p} is a GPQ for q_p, and its percentiles can be used to construct confidence limits for q_p. In particular, the $1 - \alpha$ quantile of G_{q_p} is given by $\bar{x} + t_{n-1;1-\alpha}(z_p\sqrt{n})s$. Thus, it follows from (2.2.4) that the generalized variable approach produces the same exact tolerance limit.

Tolerance factors for computing one-sided tolerance limits are given in Table B1, Appendix B, for values of $p = 0.5, 0.75, 0.80, 0.90, 0.95, 0.99$ and 0.999, $1 - \alpha = 0.90, 0.95$ and 0.99, and n ranging from 2 to 1,000.

Assessing a Survival or Exceedance Probability

A $1 - \alpha$ lower confidence limit for a survival probability $S_t = P(X > t)$, where X is a normal random variable and t is a given number, can be readily obtained following the result of Section 1.1.3. In particular, a $1 - \alpha$ lower confidence limit for S_t is the solution (with respect to p) of the equation

$$t_{n-1;1-\alpha}(z_p\sqrt{n}) = \frac{\bar{X} - t}{S/\sqrt{n}}. \qquad (2.2.6)$$

For a given sample size, p, $1 - \alpha$, $\bar{X}$, S and t, the value of p that satisfies (2.2.6) can be obtained by first solving for $z_p\sqrt{n}$ and then solving the resulting equation for p. The PC calculator that accompanies the book *StatCalc* by Krishnamoorthy (2006) can be used to solve (2.2.6) for the noncentrality parameter $z_p\sqrt{n}$. The quantity S_t is also referred to as an exceedance probability, since it is simply the probability that X exceeds a specified value t.

For an alternate approach to obtain confidence limits for the exceedance probability, see Exercise 2.6.3.

Example 2.1 (Assessing pollution level)

One-sided upper tolerance limits are commonly used to assess the pollution level in a work place or in a region. In this example, we like to assess the air lead level in a laboratory. The data in Table 2.1 represent air lead levels collected by the National Institute of Occupational Safety and Health (NIOSH) at a laboratory, for health hazard evaluation. The air lead levels were collected from 15 different areas within the facility.

Table 2.1: Air lead levels ($\mu g/m^3$)

200	120	15	7	8	6	48	61
380	80	29	1000	350	1400	110	

A normal distribution fitted the log-transformed lead levels quite well (that is, the sample is from a lognormal distribution; see Section 7.2 for more details). Therefore, we first compute an upper tolerance limit based on the log-transformed data in order to assess the maximum air lead level in the laboratory.

The sample mean and standard deviation of the log-transformed data are computed as $\bar{x} = 4.333$ and $s = 1.739$. A $(0.95, 0.90)$ upper tolerance limit for the air lead level (see (2.2.4)) is $\bar{x} + k_1 s = 4.333 + 2.329(1.739) = 8.383$. The tolerance factor $k_1 = 2.329$ is obtained from Table B1 ($n = 15, 1 - \alpha = 0.90, p = 0.95$) in Appendix B. Thus, $\exp(8.383) = 4372$ is a $(0.95, 0.90)$ upper tolerance limit for the air lead levels. The occupational exposure limit (OEL) for lead exposure set by the Occupational Safety and Health Administration (OSHA) is 50 $\mu g/m^3$. The work place is considered safe if an upper tolerance limit does not exceed the OEL. In this case, the upper limit of 4372 far exceeds the OEL; hence we can not conclude that the workplace is safe.

To assess the probability that the lead level in a randomly chosen location exceeds the OEL, let us compute a 95% lower confidence limit for the exceedance probability $P(X > \text{OEL}) = P(X > 50)$. Using the result of Section 1.1.3 of Chapter 1, we set the $(p, 0.95)$ lower tolerance limit equal to $\ln(50)$, and solve for p. That is, a 95% lower confidence limit for $P(X > 50)$ is the solution of

$$\bar{x} - \frac{1}{\sqrt{n}} t_{n-1;0.95}(z_p \sqrt{n}) s = 4.333 - \frac{1}{\sqrt{15}} t_{14;0.95}(z_p \sqrt{15}) \times 1.739 = \ln(50).$$

To solve the above equation, we first note that $t_{14;0.95}(z_p \sqrt{15}) = 0.9376$. Now, using *StatCalc*, we get $z_p \sqrt{15} = -0.7486$ or $p = 0.423$. Thus, $P(X > 50)$ is at least 0.423 with confidence 0.95.

Tolerance intervals are in fact widely used for the purpose of environmental monitoring and assessment, and for exposure data analysis. For details, examples and further applications, we refer to the books by Gibbons (1994), Gibbons and Coleman (2001) and Millard and Neerchal (2000).

2.3 Two-Sided Tolerance Intervals

Suppose that an item is acceptable for its intended purpose if an associated measurement falls in the interval (L_l, L_u), where L_l and L_u are, respectively, lower and upper specification limits. In general, engineering products are required to meet such specifications (see Example 2.4). If majority of the items (say, a proportion p) in a lot satisfies this requirement, then the lot will be accepted. Acceptability of the lot can be determined using an appropriate two-sided tolerance interval. For example, if a $(0.95, 0.99)$ two-sided tolerance interval is contained in (L_l, L_u) then the lot may be accepted. This is because at least 95% of the items fall within the tolerance interval with a confidence of at least 99%, and the tolerance interval is contained in (L_l, L_u). We now describe two methods of constructing tolerance intervals that would contain at least a proportion p of a normal population with confidence level $1 - \alpha$.

2.3.1 Tolerance Intervals

The two-sided tolerance factor k_2 is determined such that the interval $\bar{X} \pm k_2 S$ would contain at least a proportion p of the normal population with confidence $1 - \alpha$. That is, k_2 is to be determined such that

$$P_{\bar{X},S}\left\{P_X(\bar{X} - k_2 S \leq X \leq \bar{X} + k_2 S | \bar{X}, S) \geq p\right\} = 1 - \alpha, \qquad (2.3.1)$$

where $X \sim N(\mu, \sigma^2)$ independently of $\bar{X}$ and S. The inner probability inequality can be expressed as

$$P_X\left(\frac{\bar{X} - \mu - k_2 S}{\sigma} \leq \frac{X - \mu}{\sigma} \leq \frac{\bar{X} - \mu + k_2 S}{\sigma}\right) \geq p$$

$$\Leftrightarrow \quad \Phi\left(Z_n + k_2 U\right) - \Phi\left(Z_n - k_2 U\right) \geq p, \qquad (2.3.2)$$

where Φ denotes the standard normal cdf, $Z_n = \frac{\bar{X} - \mu}{\sigma} \sim N\left(0, \frac{1}{n}\right)$ independently of $U^2 = \frac{S^2}{\sigma^2} \sim \frac{\chi_m^2}{m}$ with $m = n - 1$. Using (2.3.2), we can write (2.3.1) as

$$P_{Z_n, U}\left(\Phi(Z_n + k_2 U) - \Phi(Z_n - k_2 U) > p\right) = 1 - \alpha. \qquad (2.3.3)$$

Now applying Result 1.2.1(ii) of Section 1.2 with $c = \frac{1}{n}$, we see that k_2 is the solution of the integral equation

$$\sqrt{\frac{2n}{\pi}} \int_0^\infty P\left(\chi_m^2 > \frac{m\chi_{1;p}^2(z^2)}{k_2^2}\right) e^{-\frac{1}{2}nz^2} dz = 1 - \alpha. \qquad (2.3.4)$$

Odeh (1978) has computed the exact tolerance factors k_2 satisfying (2.3.4) for $n = 2(1)98, 100$, $p = 0.75, 0.90, 0.95, 0.975, 0.99, 0.995, 0.999$ and $1 - \alpha = 0.5, 0.75,$ $0.90, 0.95, 0.975, 0.99, 0.995$. Weisberg and Beatty (1960) has also provided factors for computing normal tolerance limits. Using a Fortran program, we computed the tolerance factors k_2 satisfying (2.3.4) for various values of n, for $p = 0.5, 0.75, 0.80, 0.90, 0.95, 0.99, 0.999$ and for $1 - \alpha = 0.90, 0.95, 0.99$. The approximate tolerance factor given in (2.3.5) is used as an initial value for finding the root of the equation (2.3.4) by an iterative method. These exact tolerance factors are given in Table B2, Appendix B.

Remark 2.3.1   There are situations where the degrees of freedom m is not necessarily equal to $n - 1$. For example, in a one-way analysis of variance with l groups and sample sizes $n_1, ..., n_l$, the degrees of freedom associated with the pooled variance is $m = N - l$, where $N = \sum_{i=1}^{l} n_i$. If it is desired to find a $(p, 1 - \alpha)$ tolerance interval for a particular group, say the first group, then the required tolerance factor k_2 satisfies (2.3.4) with $n = n_1$ and $m = N - l$. The PC calculator *StatCalc* (Krishnamoorthy, 2006) computes the two-sided tolerance factor for a given sample size n and the df m associated with the sample variance.

An Approximation

Although exact two-sided tolerance factors can be obtained from various sources or can be computed using a computer, it is still worthwhile to point out a simple and satisfactory approximation. Using Result 1.2.1 (iii) of Chapter 1 (with $c = \frac{1}{n}$ and $\gamma = 1 - \alpha$), we can approximate k_2 as

$$k_2 \simeq \left(\frac{m\chi^2_{1;p}(1/n)}{\chi^2_{m;\alpha}} \right)^{\frac{1}{2}}, \tag{2.3.5}$$

where $\chi^2_{m;\alpha}$ denotes the α quantile of a chi-square distribution with df m, and $\chi^2_{m;\alpha}(\delta)$ denotes the α quantile of a noncentral chi-square distribution with df m and noncentrality parameter δ.

The above approximation is known to be very satisfactory even for sample size as small as 3 provided p and $1 - \alpha$ take values from the set $\{0.9, 0.95, 0.99\}$. However, to demonstrate its accuracy, we computed the exact tolerance factors satisfying (2.3.4) and the approximate ones given in (2.3.5) for $n = 3(1)10$, $p = 0.90, 0.95, 0.99$ and $1 - \alpha = 0.90, 0.95$. These values are given in Table 2.2. We observe from the table values that the approximation is, in general, satisfactory even for very small samples. If $n \geq 10$, the differences between the exact tolerance factors and the corresponding approximate ones exceed 0.01 only in a few cases. Furthermore, as already mentioned, these approximate values can be used as initial values to compute the exact tolerance factors satisfying (2.3.4).

Table 2.2: Approximate two-sided tolerance factor given in (2.3.5) and the exact factor satisfying (2.3.4); a = approximate; b = exact

	$1 - \alpha = 0.90$						$1 - \alpha = 0.95$					
	p						p					
	0.90		0.95		0.99		0.90		0.95		0.99	
n	a	b	a	b	a	b	a	b	a	b	a	b
3	5.85	5.79	6.92	6.82	8.97	8.82	8.38	8.31	9.92	9.79	12.9	12.7
4	4.17	4.16	4.94	4.91	6.44	6.37	5.37	5.37	6.37	6.34	8.30	8.22
5	3.49	3.50	4.15	4.14	5.42	5.39	4.28	4.29	5.08	5.08	6.63	6.60
6	3.13	3.14	3.72	3.72	4.87	4.85	3.71	3.73	4.41	4.42	5.78	5.76
7	2.90	2.91	3.45	3.46	4.52	4.50	3.37	3.39	4.01	4.02	5.25	5.24
8	2.74	2.75	3.26	3.27	4.27	4.27	3.14	3.16	3.73	3.75	4.89	4.89
9	2.63	2.64	3.13	3.13	4.10	4.09	2.97	2.99	3.53	3.55	4.63	4.63
10	2.55	2.55	3.02	3.03	3.96	3.96	2.84	2.86	3.38	3.39	4.43	4.44

Example 2.2 (Filling machine monitoring)

A machine is set to fill a liter of milk in plastic containers. At the end of a shift operation, a sample of 20 containers was selected, and the actual amount of milk in each container was measured using an accurate method. The accurate measurements are given in Table 2.3.

Table 2.3: Actual amount of milk (in liters) in containers

| 0.968 | 0.982 | 1.030 | 1.003 | 1.046 | 1.020 | 0.997 | 1.010 | 1.027 | 1.010 |
| 0.973 | 1.000 | 1.044 | 0.995 | 1.020 | 0.993 | 0.984 | 0.981 | 0.997 | 0.992 |

To assess the accuracy of the filling machine, we like to compute a two-sided tolerance interval using the data in Table 2.3. The sample statistics are $\bar{x} = 1.0036$ and $s = 0.0221085$. Using these statistics, let us compute a $(0.99, 0.95)$ two-sided tolerance interval. For this, we get the tolerance factor k_2 satisfying (2.3.4) from Table B2, Appendix B, as 3.621 (use $n = 20$, $p = .99$, $1 - \alpha = .95$). The tolerance interval is $\bar{x} \pm k_2 s = 1.0036 \pm 3.621(0.0221085) = 1.0036 \pm 0.0801$. Thus, at least 99% of containers are filled with amount of milk in the range

0.9235 to 1.0837 liters with confidence 0.95.

2.3.2 Equal-Tailed Tolerance Intervals for a Normal Distribution

We now describe a method of constructing a tolerance interval (I_l, I_u) that would contain at least $100p\%$ of the "center data" of a normal population with $1 - \alpha$ confidence. That is, the interval (I_l, I_u) is constructed using a sample such that at most a proportion $\frac{1-p}{2}$ of the normal data are less than I_l and at most a proportion $\frac{1-p}{2}$ of the normal data are above I_u, with confidence $1 - \alpha$. As argued in Section 1.1.2 of Chapter 1, this amounts to finding the interval (I_l, I_u) such that it would contain the interval $\left(\mu - z_{\frac{1+p}{2}}\sigma, \mu + z_{\frac{1+p}{2}}\sigma\right)$ with confidence $1 - \alpha$. Based on this form of the "population interval", a natural choice for (I_l, I_u) is $(\bar{X} - k_e S, \bar{X} + k_e S)$, where k_e is to be determined such that

$$P_{\bar{X},S}\left(\bar{X} - k_e S < \mu - z_{\frac{1+p}{2}}\sigma \ \text{ and } \ \mu + z_{\frac{1+p}{2}}\sigma < \bar{X} + k_e S\right) = 1 - \alpha. \quad (2.3.6)$$

After standardizing $\bar{X}$ and rearranging the terms, we see that (2.3.6) is equivalent to

$$P_{Z,S}\left(\frac{Z/\sqrt{n} + z_{\frac{1+p}{2}}}{S/\sigma} < k_e \ \text{ and } \ \frac{Z/\sqrt{n} - z_{\frac{1+p}{2}}}{S/\sigma} \geq -k_e\right) = 1 - \alpha, \quad (2.3.7)$$

where $Z = \frac{\sqrt{n}(\bar{X}-\mu)}{\sigma} \sim N(0,1)$. Let $\delta = \sqrt{n} \times z_{\frac{1+p}{2}}$, and $U^2 = \frac{S^2}{\sigma^2}$. In terms of these quantities, we see that (2.3.7) can be expressed as

$$P_{Z,U}\left(Z < -\delta + k_e\sqrt{n}U \ \text{ and } \ Z > \delta - k_e\sqrt{n}U\right) = 1 - \alpha. \quad (2.3.8)$$

Notice that the inequalities in the above probability statement holds only if $\delta - k_e\sqrt{n}U < -\delta + k_e\sqrt{n}U$ or equivalently $U^2 > \frac{\delta^2}{k_e^2 n}$. Thus, (2.3.8) can be expressed as

$$E_U\left[P_Z\left(\delta - k_e\sqrt{n}U < Z < -\delta + k_e\sqrt{n}U \,\middle|\, U^2 > \frac{\delta^2}{k_e^2 n}\right)\right] = 1 - \alpha, \quad (2.3.9)$$

where E_U denotes the expectation with respect to the distribution of U. Because $U^2 \sim \frac{\chi_{n-1}^2}{n-1}$, it follows from (2.3.9) that k_e is the solution of the integral equation

$$\frac{1}{2^{\frac{n-1}{2}}\Gamma\left(\frac{n-1}{2}\right)}\int_{\frac{(n-1)\delta^2}{k_e^2 n}}^{\infty}\left(2\Phi\left(-\delta + \frac{k_e\sqrt{n}x}{\sqrt{n-1}}\right) - 1\right)e^{-x/2}x^{\frac{n-1}{2}-1}dx = 1 - \alpha,$$

$$(2.3.10)$$

where $\Phi(x)$ denotes the standard normal distribution function. To get (2.3.10) from (2.3.9), we have used the relation $\Phi(x) = 1 - \Phi(-x)$.

Owen (1964) computed k_e satisfying (2.3.10) for $p = 0.8, 0.90, 0.95, 0.98$, $1 - \alpha = 0.90$ and for various values of n. We used a numerical integration procedure and a root finding method to compute the values of k_e satisfying (2.3.10). The one-sided tolerance factor $t_{n-1;1-\alpha}\left(z_{\frac{1+p}{2}}\sqrt{n}\right)$, given in 2.2.3, was used as an initial value to find the root of (2.3.10). The exact values of k_e are presented in Table B3, Appendix B for various values of n, $p = 0.5, 0.75, 0.80, 0.90, 0.95, 0.99, 0.999$, and $1 - \alpha = 0.90, 0.95, 0.99$.

Example 2.2 (continued)

To compute the equal-tailed (0.99, 0.95) tolerance interval using the data given in Example 2.2, we find the tolerance factor k_e satisfying (2.3.10) from Table B3, Appendix B as 3.812. This gives the equal-tailed interval $\bar{x} \pm k_e s = 1.0036 \pm 3.812(0.0221085) = 1.0036 \pm 0.0843$. This means that, with 95% confidence, no more than 0.5% of containers are filled with less than 0.9193 liters of milk, and no more than 0.5% of containers are filled with more than 1.0880 liters of milk.

2.3.3   Simultaneous Hypothesis Testing about Normal Quantiles

Owen (1964) proposed an acceptance sampling plan in which a lot will be accepted if the sample data provide sufficient evidence to indicate that no more than a proportion $\frac{1-p}{2}$ of items' characteristics are less than L_l and no more than a proportion $\frac{1-p}{2}$ of items' characteristics are greater than L_u. Notice that this latter acceptance sampling plan not only demands that at least a proportion p of the items are within specification limits but also puts a limit on the proportion of defective items in a single tail of the distribution. If normality is assumed, then the lot will be accepted if we have sample evidence indicating that

$$L_l < \mu - z_{\frac{1+p}{2}}\sigma \quad \text{and} \quad \mu + z_{\frac{1+p}{2}}\sigma < L_u.$$

In a hypothesis testing set up, we have

$$H_0 : L_l \geq \mu - z_{\frac{1+p}{2}}\sigma \quad \text{or} \quad \mu + z_{\frac{1+p}{2}}\sigma \geq L_u$$

vs. $$H_a : L_l < \mu - z_{\frac{1+p}{2}}\sigma \quad \text{and} \quad \mu + z_{\frac{1+p}{2}}\sigma < L_u.$$

The lot is not acceptable if H_0 is true. That is, the lot is not acceptable if either $\mu - z_{\frac{1+p}{2}}\sigma \leq L_l$ or $L_u \leq \mu + z_{\frac{1+p}{2}}\sigma$. The acceptance sampling plan has to be designed such that the probability of accepting a non-acceptable lot (rejecting H_0 when it is true) should not exceed α, where α usually ranges from 0.01 to 0.1.

We see from the above sampling plan that the null hypothesis will be rejected at the level of significance α if

$$L_l \leq \bar{X} - k_h S \quad \text{and} \quad \bar{X} + k_h S \leq L_u,$$

where the critical value k_h is to be determined so that

$$\sup\nolimits_{H_0} P\left(\bar{X} - k_h S \geq L_l \quad \text{and} \quad \bar{X} + k_h S \leq L_u | H_0\right) = \alpha. \tag{2.3.11}$$

Note that the supremum in the above equation is attained at $L_l = \mu - z_{\frac{1+p}{2}}\sigma$ and $L_u = \mu + z_{\frac{1+p}{2}}\sigma$. Because the equations $L_l = \mu - z_{\frac{1+p}{2}}\sigma$ and $L_u = \mu + z_{\frac{1+p}{2}}\sigma$ uniquely determine μ and σ, (2.3.11) is equivalent to

$$P(\bar{X} - k_h S > \mu - z_{\frac{1+p}{2}}\sigma \quad \text{and} \quad \bar{X} + k_h S < \mu + z_{\frac{1+p}{2}}\sigma) = \alpha.$$

Letting $Z_n = \frac{\bar{X}-\mu}{\sigma} \sim N\left(0, \frac{1}{n}\right)$, the above equation can be expressed as

$$P\left(\frac{Z_n + z_{\frac{1+p}{2}}}{S/\sigma} > k_h \quad \text{and} \quad \frac{-Z_n + z_{\frac{1+p}{2}}}{S/\sigma} > k_h\right) = \alpha. \tag{2.3.12}$$

Proceeding as in the derivation of equal-tailed tolerance interval (see Section 2.3.2), we see that the factor k_h is the solution of the integral equation

$$\frac{1}{2^{\frac{n-1}{2}}\Gamma(\frac{n-1}{2})} \int_0^{\frac{(n-1)\delta^2}{k_h^2} \times n} \left(2\Phi\left(\delta - \frac{k_h\sqrt{nu}}{\sqrt{n-1}}\right) - 1\right) e^{-u/2} u^{\frac{n-1}{2}-1} du = \alpha, \tag{2.3.13}$$

where $\delta = \sqrt{n} \times z_{\frac{1+p}{2}}$.

Owen (1964) computed the values of k_h satisfying (2.3.13) for $p = 0.8$, 0.90, 0.95, 0.98, $\alpha = .10$ and for various values of n. Krishnamoorthy and Mathew (2002a) considered the above hypothesis testing problem in a different application (see Example 2.5), and provided tables for $p = 0.90$, 0.95, 0.99, 0.999, $\alpha = 0.01$, 0.05, 0.10, and for various values of n. We used a Fortran program to compute the values of k_h satisfying (2.3.13). The one-sided tolerance factor $t_{n-1,1-\alpha}\left(z_{\frac{1+p}{2}}\sqrt{n}\right)$ given in (2.2.3) was used as an initial value to find the root of (2.3.13). The exact critical values k_h are presented in Table B4, Appendix B, for $p = 0.70$, 0.80, 0.90, 0.95, 0.98, 0.99, $\alpha = 0.01$, 0.05, 0.10, and for various values of n. Our table values are in complete agreement with those in Owen (1964).

Example 2.4 (Shaft diameters)

The quality assurance department of an engineering product manufacturing company wants to check the proportion of shafts (with basic size 1.5 inch; the housing hole diameter at least 1.5 inch) that are within the tolerance specification $(1.4968, 1.4985)^*$. A sample of $n = 24$ shafts was selected randomly, and the diameters were measured. The data are given in Table 2.4. The data satisfy

Table 2.4: Shaft diameters

1.4970	1.4972	1.4970	1.4973	1.4979	1.4978	1.4974	1.4975
1.4981	1.4980	1.4981	1.4984	1.4972	1.4979	1.4974	1.4968
1.4978	1.4973	1.4973	1.4974	1.4974	1.4987	1.4973	1.4971

the normality assumption (Minitab, default method). The sample mean $\bar{x} = 1.497555$ and the sample standard deviation $s = 0.000476$. We shall apply the testing method in Section 2.3.3 to see if at least 95% of the shafts are within the specifications. That is, we want to test

$$H_0 : 1.4968 \geq \mu - z_{\frac{1+.95}{2}} \sigma \quad \text{or} \quad \mu + z_{\frac{1+.95}{2}} \sigma \geq 1.4985$$

vs. $\quad H_a : 1.4968 < \mu - z_{\frac{1+.95}{2}} \sigma \quad \text{and} \quad \mu + z_{\frac{1+.95}{2}} \sigma < 1.4985.$

If we use the nominal level $\alpha = 0.05$, then the necessary factor k_h from Table B4, Appendix B, is 2.424. Using this factor, we have $\bar{x} - k_h s = 1.4964$ and $\bar{x}+k_h s = 1.4988$. Because these two limits are not within the specification limits, we cannot conclude that 95% of the shafts are within the tolerance specification. On the other hand, if we take $p = 0.80$, then $k_h = 1.543$, $\bar{x} - k_h s = 1.4969$ and $\bar{x} + k_h s = 1.4983$. Since $(1.4969, 1.4983)$ is within the tolerance specification $(1.4968, 1.4985)$, we can conclude that no more than 10% of the shafts are below the lower tolerance specification of 1.4968, and no more than 10% of the shafts are above the upper tolerance specification of 1.4985.

The preceding simultaneous hypothesis testing about quantiles can be used to show whether a new measuring device is equivalent to the standard device approved by a government or regulatory agency. For example, the occupational safety and health administration (OSHA) regulations allow the use of an alternate sampling device for exposure monitoring, provided the device has been demonstrated to be equivalent to the standard device. Typically, the OSHA criterion is that 90% of the readings of the sampling device should be within plus or minus 25% of the readings obtained by the standard device. Assuming bivariate lognormal distribution for the measurements, Krishnamoorthy and

---

*http://www.engineersedge.com/

Mathew (2002a) showed that the simultaneous hypothesis testing method can be used to establish the equivalency of the devices based on the OSHA criterion. Their solution is given below for an example.

Example 2.5 (Equivalency of sampling devices)

This example is concerned with establishing equivalency of a new cotton dust sampler (ND) to a standard sampler called Vertical Elutriator (VE) with respect to the OSHA criterion described above. A sample of 60 readings was taken using both devices side-by-side from four different sites, and the data were analyzed by Rockette and Wadsworth (1985) and Krishnamoorthy and Mathew (2002a). For the sake of illustration, we use a subset of 20 pairs of readings from the full data set in Rockette and Wadsworth (1985), and they are given in Table 2.5. Let X and Y, respectively, be random variables denoting the readings using ND and VE. Assume that (X,Y) follows a bivariate lognormal distribution. The above requirement amounts to testing

$$H_0 : \theta \leq 0.90 \quad \text{vs.} \quad H_a : \theta > 0.90, \tag{2.3.14}$$

where

$$\theta = P\left(0.75 \leq \frac{X}{Y} \leq 1.25\right)$$
$$= P\left(\ln(0.75) \leq \ln(X) - \ln(Y) \leq \ln(1.25)\right).$$

Let (X_i, Y_i), $i = 1, ..., n$, denote a random sample of n measurements on (X, Y), and let $D_i = \ln(X_i) - \ln(Y_i)$. Notice that D_i follows a normal distribution, say $N(\mu_d, \sigma^2)$. Instead of (2.3.14) consider the testing problem

$$H_0 : \ln(.75) \geq \mu_d - z_{\frac{1+p}{2}}\sigma \text{ or } \mu_d + z_{\frac{1+p}{2}}\sigma \geq \ln(1.25)$$
vs. $\quad H_a : \ln(.75) < \mu_d - z_{\frac{1+p}{2}}\sigma \text{ and } \mu_d + z_{\frac{1+p}{2}}\sigma < \ln(1.25), \quad (2.3.15)$

where $p = 0.90$. Notice that H_a in (2.3.14) holds if H_a in (2.3.15) holds. Thus the null hypothesis in (2.3.14) will be rejected if H_0 in (2.3.15) is rejected. Let $\bar{D}$ and S_d denote the mean and the standard deviation among the D_i's. These quantities have the observed values $\bar{D} = 0.0042$ and $S_d = 0.0545$. From Table B4 in Appendix B, we get the critical value $k_h = 2.448$ (corresponding to $n = 20$, $p = 0.90$ and $\alpha = 0.05$). Hence $\bar{D} - k_h S_d = 0.0042 - 2.488(0.0545) = -0.1314$ and $\bar{D} + k_h S_d = 0.1380$. Thus, we have $\ln(0.75) = -0.2877 < -0.1314$ and $0.1380 < 0.2231 = \ln(1.25)$. So we reject the null hypothesis at level 0.05, and conclude that at least 90% of the readings of the ND are within plus or minus 25% of the readings obtained by the VE.

Table 2.5: ND and VE readings in $\mu g/m^3$
$$D = \ln(\text{ND}) - \ln(\text{VE})$$

ND	VE	D	ND	VE	D
72	75	−.0408	230	250	−.0834
79	74	.0654	305	270	.1219
93	89	.0440	287	285	.0070
72	78	.0800	329	320	.0277
84	82	.0241	305	320	−.0480
110	118	−.0702	495	480	.0308
132	140	−.0588	640	620	.0317
130	125	.0392	536	525	.0207
120	125	−.0408	630	610	.0323
134	129	.0380	560	547	.0235

Remark 2.3.2 We would like to point out that the tolerance intervals computed in this section (and throughout this chapter) are based on simple random samples from the normal distribution. We refer to Mee (1989) for the computation of tolerance intervals based on a stratified random sample, under the assumption that the variable of interest is normally distributed within each stratum, where the stratum means could be different, but the stratum variances are equal.

2.4 One-Sided Tolerance Limits for the Distribution of the Difference Between Two Independent Normal Random Variables

Let $X_1 \sim N(\mu_1, \sigma_1^2)$ independently of $X_2 \sim N(\mu_2, \sigma_2^2)$, where (μ_1, μ_2) and (σ_1^2, σ_2^2) are unknown means and unknown variances, respectively. Let z_p denote the p quantile of a standard normal distribution, and let

$$L_p = \mu_1 - \mu_2 - z_p \sqrt{\sigma_1^2 + \sigma_2^2} = \mu_1 - \mu_2 + z_{1-p} \sqrt{\sigma_1^2 + \sigma_2^2}. \tag{2.4.1}$$

Note that L_p is the $1-p$ quantile (that is, the lower p quantile) of the distribution of $X_1 - X_2$. Therefore, a $(p, 1-\alpha)$ lower tolerance limit for the distribution of $X_1 - X_2$ is a $1 - \alpha$ lower confidence limit for L_p. An exact method of constructing confidence limits for L_p is available only when the variance ratio $\frac{\sigma_1^2}{\sigma_2^2}$ is known. If the variances are unknown and arbitrary, some satisfactory approximate methods are available. In the following section, we shall describe an exact method of constructing a lower tolerance limit when the variance ratio $\frac{\sigma_1^2}{\sigma_2^2}$ is known. Other approximate methods for constructing tolerance limits when the variances are unknown and arbitrary will be described subsequently.

Let $\bar{X}_i$ and S_i^2, respectively, denote the mean and variance of a random sample of n_i observations from $N(\mu_i, \sigma_i^2)$, $i = 1, 2$. Under the independence assumption in the preceding paragraph, the summary statistics $\bar{X}_1$, $\bar{X}_2$, S_1^2 and S_2^2 are all independent. In the following, we describe the computation of tolerance limits for $X_1 - X_2$ based on these summary statistics.

2.4.1 Exact One-Sided Tolerance Limits for the Distribution of $X_1 - X_2$ When the Variance Ratio Is Known

The method of constructing an exact one-sided lower tolerance limit, when the variance ratio $\frac{\sigma_1^2}{\sigma_2^2}$ is known, is due to Hall (1984). Let $q_1 = \frac{\sigma_1^2}{\sigma_2^2}$ be known, and define

$$S_d^2 = \frac{(1 + q_1^{-1})((n_1 - 1)S_1^2 + (n_2 - 1)q_1 S_2^2)}{(n_1 + n_2 - 2)}. \qquad (2.4.2)$$

Notice that

$$\frac{(n_1 + n_2 - 2)S_d^2}{\sigma_1^2 + \sigma_2^2} = \frac{(n_1 - 1)S_1^2}{\sigma_1^2} + \frac{(n_2 - 1)S_2^2}{\sigma_2^2},$$

and it follows a chi-square distribution with degrees of freedom $n_1 + n_2 - 2$, independently of $\bar{X}_1 - \bar{X}_2$. Also, we can write

$$\frac{\bar{X}_1 - \bar{X}_2 - L_p}{\sqrt{\sigma_1^2/n_1 + \sigma_2^2/n_2}} = Z + z_p\sqrt{\nu_1}, \quad \text{where } \nu_1 = \frac{n_1(1 + q_1)}{q_1 + n_1/n_2}, \qquad (2.4.3)$$

with L_p as given in (2.4.1) and $Z = \frac{\bar{X}_1 - \bar{X}_2 - (\mu_1 - \mu_2)}{\sqrt{\sigma_1^2/n_1 + \sigma_2^2/n_2}} \sim N(0, 1)$. Using (2.4.2) and (2.4.3), it can be easily verified that the pivotal quantity

$$\frac{\bar{X}_1 - \bar{X}_2 - L_p}{S_d} \sim \frac{Z + z_p\sqrt{\nu_1}}{\sqrt{\nu_1}\sqrt{\chi_{n_1+n_2-2}^2/(n_1 + n_2 - 2)}}. \qquad (2.4.4)$$

Because Z defined above is independent of the chi-square random variable $\frac{(n_1+n_2-2)S_d^2}{\sigma_1^2+\sigma_2^2}$, we have

$$\frac{\bar{X}_1 - \bar{X}_2 - L_p}{S_d} \sim \frac{1}{\sqrt{\nu_1}} t_{n_1+n_2-2}(z_p\sqrt{\nu_1}). \qquad (2.4.5)$$

Thus, using the above exact distributional result, we get a $1 - \alpha$ lower confidence limit for L_p as

$$\bar{X}_1 - \bar{X}_2 - t_{n_1+n_2-2;1-\alpha}(z_p\sqrt{\nu_1})\frac{S_d}{\sqrt{\nu_1}}, \qquad (2.4.6)$$

which is an exact $(p, 1 - \alpha)$ lower tolerance limit for the distribution of $X_1 - X_2$. Here ν_1 is defined in (2.4.3). An exact one-sided upper tolerance limit can be obtained by replacing the negative sign (preceding the noncentral t critical value) in (2.4.6) by the positive sign.

2.4.2   One-Sided Tolerance Limits for the Distribution of $X_1 - X_2$ When the Variance Ratio Is Unknown

If the variance ratio $q_1 = \frac{\sigma_1^2}{\sigma_2^2}$ is unknown, an approximate tolerance limit can be obtained by substituting an estimate for ν_1 in (2.4.3). This approximate limit is, in general, too liberal (actual coverage probability can be much less than the specified confidence level). Therefore, in the following we shall present Hall's (1984) alternative approximate approach.

Hall's Approach

Using the Satterthwaite approximation given in Result 1.2.2, it can be shown that

$$\frac{S_1^2 + S_2^2}{\sigma_1^2 + \sigma_2^2} \sim \frac{\chi_{f_1}^2}{f_1}, \text{ approximately, where } f_1 = \frac{(n_1 - 1)(q_1 + 1)^2}{q_1^2 + (n_1 - 1)/(n_2 - 1)}. \quad (2.4.7)$$

Using this approximation, and proceeding as in Section (2.4.1), it can be shown that

$$\frac{\bar{X}_1 - \bar{X}_2 - L_p}{\sqrt{S_1^2 + S_2^2}} \sim \frac{1}{\sqrt{\nu_1}} \left(\frac{Z + z_p\sqrt{\nu_1}}{\sqrt{\chi_{f_1}^2/f_1}} \right) \text{ approximately,} \quad (2.4.8)$$

where ν_1 is given in (2.4.3). Because $\left(\frac{Z + z_p\sqrt{\nu_1}}{\sqrt{\chi_{f_1}^2/f_1}} \right) \sim t_{f_1}(z_p\sqrt{\nu_1})$, we get a $1 - \alpha$ lower confidence limit for L_p as

$$\bar{X}_1 - \bar{X}_2 - t_{f_1,1-\alpha}(z_p\sqrt{\nu_1})\sqrt{\frac{S_1^2 + S_2^2}{\nu_1}}.$$

Notice that both ν_1 and f_1 involve the unknown variance ratio $q_1 = \frac{\sigma_1^2}{\sigma_2^2}$. To get an approximate limit, Hall suggested using the unbiased estimator $\hat{q}_1 = \frac{S_1^2(n_2-3)}{S_2^2(n_2-1)}$ for q_1. Plugging this estimator of q_1 in (2.4.3) and in (2.4.7), we get estimators

$\widehat{\nu}_1$ of ν_1 and $\widehat{f}_1$ of f_1. Using these estimators, Hall's approximate lower tolerance limit for the distribution of $X_1 - X_2$ can be expressed as

$$\bar{X}_1 - \bar{X}_2 - t_{\widehat{f}_1, 1-\alpha}(z_p\sqrt{\widehat{\nu}_1})\sqrt{\frac{S_1^2 + S_2^2}{\widehat{\nu}_1}}. \tag{2.4.9}$$

A $(p, 1-\alpha)$ upper tolerance limit can be obtained by replacing the minus sign preceding the tolerance factor by the plus sign.

Reiser and Guttman (1986) considered the problem of constructing tolerance limits for the distribution of $X_1 - X_2$ in the context of stress-strength reliability (see Section 2.4.5). The solution given in their paper is essentially Hall's solution given above, except that the variance ratio $q_1 = \sigma_1^2/\sigma_2^2$ is estimated by $\widehat{q} = \frac{S_1^2}{S_2^2}$. Notice that $\widehat{q}$ is a biased estimator of q_1. Using this estimator, we get an approximate $(p, 1-\alpha)$ lower tolerance limit for the distribution of $X_1 - X_2$ as

$$\bar{X}_1 - \bar{X}_2 - t_{\widehat{f}, 1-\alpha}(z_p\sqrt{\widehat{\nu}})\sqrt{\frac{S_1^2 + S_2^2}{\widehat{\nu}}}, \tag{2.4.10}$$

where $\widehat{\nu}$ is ν_1 in (2.4.3) with q_1 replaced by $\widehat{q}$, and $\widehat{f}$ is f_1 in (2.4.7) with q_1 replaced by $\widehat{q}$.

Guo-Krishnamoorthy Approach

Guo and Krishnamoorthy (2004) observed that Hall's (1984) tolerance limits depend on the definition of the variance ratio. Specifically, if the variance ratio is defined as $q_2 = \frac{\sigma_2^2}{\sigma_1^2}$, and it is estimated by $\widehat{q}_2 = \frac{S_2^2(n_1-3)}{S_1^2(n_1-1)}$, then the one-sided lower tolerance limit based on Hall's method is given by

$$\bar{X}_1 - \bar{X}_2 - t_{\widehat{f}_2, 1-\alpha}(z_p\sqrt{\widehat{\nu}_2})\sqrt{\frac{S_1^2 + S_2^2}{\widehat{\nu}_2}}, \tag{2.4.11}$$

where

$$\widehat{\nu}_2 = \frac{n_2(1 + \widehat{q}_2)}{\widehat{q}_2 + n_2/n_1}, \quad \text{and} \quad \widehat{f}_2 = \frac{(n_2 - 1)(\widehat{q}_2 + 1)^2}{\widehat{q}_2^2 + (n_2 - 1)/(n_1 - 1)}.$$

It is clear that the lower tolerance limit in (2.4.11) is different from the one in (2.4.9). Guo and Krishnamoorthy (2004) found via simulation studies that if one of the limits is too liberal for a sample size and parameter configuration, then the other limit is satisfactory at the same configuration. This finding suggests that the tolerance limit given by

$$
\min\left\{\bar{D} - t_{\widehat{f}_1,1-\alpha}(z_p\sqrt{\widehat{\nu}_1})\sqrt{\frac{S_1^2 + S_2^2}{\widehat{\nu}_1}}, \ \bar{D} - t_{\widehat{f}_2,1-\alpha}(z_p\sqrt{\widehat{\nu}_2})\sqrt{\frac{S_1^2 + S_2^2}{\widehat{\nu}_2}}\right\},
$$

$$(2.4.12)$$

where $\bar{D} = \bar{X}_1 - \bar{X}_2$, can be a better lower tolerance limit in terms of coverage probabilities.

The Generalized Variable Approach

Let $(\bar{x}_1, s_1^2, \bar{x}_2, s_2^2)$ be an observed value of $(\bar{X}_1, S_1^2, \bar{X}_2, S_2^2)$. A GPQ for the interval estimation of L_p can be constructed similar to the one-sample case given in Section 2.2, and is given by

$$
\begin{aligned}
G_{L_p} &= \bar{d} - \frac{\bar{D} - (\mu_1 - \mu_2)}{\sqrt{\frac{\sigma_1^2}{n_1} + \frac{\sigma_2^2}{n_2}}}\sqrt{\frac{\sigma_1^2}{n_1}\frac{s_1^2}{S_1^2} + \frac{\sigma_2^2}{n_2}\frac{s_2^2}{S_2^2}} - z_p\sqrt{\frac{\sigma_1^2}{S_1^2}s_1^2 + \frac{\sigma_2^2}{S_2^2}s_2^2} \\
&= \bar{d} - Z\sqrt{\frac{s_1^2}{n_1 U_1^2} + \frac{s_2^2}{n_2 U_2^2}} - z_p\sqrt{\frac{s_1^2}{U_1^2} + \frac{s_2^2}{U_2^2}},
\end{aligned}
$$

$$(2.4.13)$$

where $\bar{d} = \bar{x}_1 - \bar{x}_2$ and Z is a standard normal random variable, $U_i^2 = \frac{S_i^2}{\sigma_i^2} \sim \frac{\chi_{n_i-1}^2}{n_i-1}$, $i = 1, 2$, and all these random variables are mutually independent. We see from the first step of (2.4.13) that the value of G_{L_p} at $(\bar{X}_1, S_1^2, \bar{X}_2, S_2^2) = (\bar{x}_1, s_1^2, \bar{x}_2, s_2^2)$ is L_p in (2.4.1), the parameter of interest. For fixed $(\bar{x}_1, s_1^2, \bar{x}_2, s_2^2)$, it is clear from step 2 of (2.4.13) that the distribution of G_{L_p} is free of any unknown parameters. Thus, G_{L_p} satisfies the required two conditions in (C.1) of Section 1.4 for it to be a valid GPQ, and its percentiles can be used to construct confidence limits for L_p.

Since for fixed $(\bar{x}_1, s_1^2, \bar{x}_2, s_2^2)$, the distribution of G_{L_p} does not depend on any unknown parameters, its percentiles can be estimated using Monte Carlo simulation. This simulation essentially involves generation of random numbers from the standard normal distribution, $\chi_{n_1-1}^2$ distribution and $\chi_{n_2-1}^2$ distribution. Even though Weerahandi and Johnson (1992) provided a numerical method of computing the exact percentiles in a related problem, our own investigation showed that Monte Carlo method is easy to use and produces results (if the number of simulation runs is 100,000 or more) which are practically the same as the ones due to the exact numerical method. This generalized variable approach is often too conservative when the sample sizes are small, even if the exact percentiles are used. Comparison of the different tolerance factors given above is discussed in Section 2.4.4.

2.4.3 Hypothesis Testing About the Quantiles of $X_1 - X_2$

There are some applications where it is of interest to test hypotheses concerning the quantiles of $X_1 - X_2$ (see Section 2.4.5). The confidence limits for $L_p = \mu_1 - \mu_2 - z_p \sqrt{\sigma_1^2 + \sigma_2^2}$ given in the preceding sections can be used to carry out a test at a fixed level α. In practice, one may be interested in computing p-values for testing the hypotheses

$$H_0 : L_p \leq L_{p0} \quad \text{vs.} \quad H_a : L_p > L_{p0}, \tag{2.4.14}$$

where L_{p0} is a specified value of L_p. Assume that the variance ratio $q_1 = \frac{\sigma_1^2}{\sigma_2^2}$ is unknown. Let $(\bar{x}_1, s_1^2, \bar{x}_2, s_2^2)$ be an observed value of $(\bar{X}_1, S_1^2, \bar{X}_2, S_2^2)$. Then, the p-value for testing the above hypotheses based on Hall's (1984) approach is given by

$$p_1 = P \left(t_{\hat{f}_1}(z_p \sqrt{\hat{\nu}_1}) > \frac{\sqrt{\hat{\nu}_1}(\bar{x}_1 - \bar{x}_2 - L_{p0})}{\sqrt{s_1^2 + s_2^2}} \right). \tag{2.4.15}$$

As pointed out in Guo and Krishnamoorthy (2004), the test based on the above p-value has larger (than the nominal level) Type-I error rates when the sample sizes are small and/or they are drastically different. These authors suggested using the following approach motivated by (2.4.12), which controls the Type-I error rates satisfactorily. Define

$$p_2 = P \left(t_{\hat{f}_2}(z_p \sqrt{\hat{\nu}_2}) > \frac{\sqrt{\hat{\nu}_2}(\bar{x}_1 - \bar{x}_2 - L_{p0})}{\sqrt{s_1^2 + s_2^2}} \right), \tag{2.4.16}$$

where $\hat{f}_2$ and $\hat{\nu}_2$ are as defined in (2.4.11). The test that rejects H_0 whenever

$$\max\{p_1, p_2\} < \alpha \tag{2.4.17}$$

has better size properties than the one based on the p-value in (2.4.15).

Tests for an upper quantile $U_p = \mu_1 - \mu_2 + z_p \sqrt{\sigma_1^2 + \sigma_2^2}$, $p > .5$, can be developed similarly. For example, for testing

$$H_0 : U_p \geq U_{p0} \quad \text{vs.} \quad H_a : U_p < U_{p0}, \tag{2.4.18}$$

the p-value based on Hall's (1984) approach is given by

$$p_1^* = P \left(t_{\hat{f}_1}(z_p \sqrt{\hat{\nu}_1}) < \frac{\sqrt{\hat{\nu}_1}(\bar{x}_1 - \bar{x}_2 - U_{p0})}{\sqrt{s_1^2 + s_2^2}} \right). \tag{2.4.19}$$

In this case, Guo and Krishnamoorthy's (2004) approach rejects H_0 in (2.4.18) when

$$\min\{p_1^*, p_2^*\} < \alpha, \tag{2.4.20}$$

where p_2^* has an expression similar to the p-value in (2.4.19) with $(\widehat{f}_1, \widehat{\nu}_1)$ replaced by $(\widehat{f}_2, \widehat{\nu}_2)$.

We mainly addressed left-tail test for an upper quantile U_p, and right-tail test for a lower quantile L_p, because these are the tests most relevant in practical applications. However, tests for other hypotheses (e.g., right-tail test for an upper quantile of $X_1 - X_2$) can be developed easily using the arguments given above.

To describe the generalized variable approach, we first note that the generalized test variable for testing (2.4.14) is given by

$$T_{L_p} = G_{L_p} - L_p,$$

where G_{L_p} is given in (2.4.13). Notice that given the observed data, the generalized test variable is stochastically decreasing in L_p, and so the generalized p-value is given by

$$\sup_{H_0} P\left(T_{L_p} < 0\right) = P\left(G_{L_p} > L_{p0}\right). \tag{2.4.21}$$

The test rejects the null hypothesis in (2.4.14) whenever the above generalized p-value is less than α.

2.4.4 Comparison of the Approximate Methods for Making Inference about Quantiles of $X_1 - X_2$

Guo and Krishnamoorthy (2004) studied the validity of the approximate methods and the generalized variable method considered in the preceding sections using Monte Carlo simulation. Their studies showed that, in general, the methods due to Hall (1984) and Reiser and Guttman (1986) are liberal (the sizes of the tests are larger than the nominal; the coverage probabilities of the confidence limits are smaller than the nominal confidence level) when the sample sizes are small and/or they are drastically different. These approaches maybe used only when the sample sizes are large and close to each other.

Between the approach due to Guo and Krishnamoorthy (2004) and the generalized variable method, the former is preferable to the latter when the sample sizes are close to each other. For moderate samples, generalized variable method produces results that are very conservative. If both sample sizes are at least 5, and their ratio is between 0.7 and 1.4, then the sizes of the tests based on the suggestion by Guo and Krishnamoorthy are very close to the nominal level. The generalized variable approach can be recommended for practical use if the ratio of the sample sizes is less than 0.7 or greater than 1.4 and both samples are at least 15. We would like to caution the readers that when applied to practical data, the different approaches can lead to very different results and conclusions; see Example 2.6 below. Thus it is important to follow the recommendations given above, and use the appropriate approach.

2.4.5 Applications of Tolerance Limits for X_1-X_2 with Examples

In some practical situations we need to estimate the proportion of times one random variable assumes a larger value than another independent random variable. For example, the classical stress-strength reliability problem concerns the proportion of times the strength X_1 of a component exceeds the stress X_2 to which it is subjected. The component works as long as $X_1 > X_2$. The reliability parameter is defined by $R = P(X_1 > X_2)$. To assess the reliability of the component, inference on R is desired. Suppose we are interested in testing

$$H_0 : R \leq p \ \text{ vs } \ H_a : R > p, \tag{2.4.22}$$

where p is a specified probability, usually close to one. Because $R = P(X_1 - X_2 > 0) > p$ holds if and only if the p lower quantile L_p of $X_1 - X_2$ is greater than zero, the above testing problem is equivalent to testing $H_0 : L_p \leq 0$ vs $H_a : L_p > 0$. A test for the latter consists of rejecting H_0 when a $(p, 1 - \alpha)$ lower tolerance limit for the distribution of $X_1 - X_2$ is greater than zero.

Tolerance limits for the distribution of $X_1 - X_2$ can also be used to establish the equivalence of two treatments or to establish superiority of one treatment over another. For example, let X_1 and X_2, respectively, denote the response times of treatments 1 and 2. If $P(X_1 > X_2) > p$, where $p \in (.5, 1)$, then treatment 2 may be considered superior to treatment 1.

If it is assumed that X_1 and X_2 are independent normal random variables, then the results given in Sections 2.4.2 and 2.4.3 can be readily applied to the aforementioned practical problems. Furthermore, as shown in Section 1.1.3 of Chapter 1, a $1 - \alpha$ lower confidence limit for the reliability parameter can be deduced from a lower tolerance limit for the distribution of X_1-X_2. For example, based on Hall's (1984) lower tolerance limit given in (2.4.9), a $1 - \alpha$ lower limit for R is given by

$$p^* = \max\left\{p : \bar{x}_1 - \bar{x}_2 - t_{\widehat{f}_1;1-\alpha}(z_p\sqrt{\widehat{\nu}_1})\sqrt{\frac{s_1^2 + s_2^2}{\widehat{\nu}_1}} \geq 0\right\}. \qquad (2.4.23)$$

For a fixed degrees of freedom and $1-\alpha$, the noncentral t percentile is increasing with respect to the noncentrality parameter (equivalently, with respect to p); thus the lower tolerance limit is decreasing with increasing p, and p^* is the unique solution (with respect to p) of the equation

$$\bar{x}_1 - \bar{x}_2 - t_{\widehat{f}_1;1-\alpha}(z_p\sqrt{\widehat{\nu}_1})\sqrt{\frac{s_1^2 + s_2^2}{\widehat{\nu}_1}} = 0 \Leftrightarrow t_{\widehat{f}_1;1-\alpha}(z_p\sqrt{\widehat{\nu}_1}) = \frac{\bar{x}_1 - \bar{x}_2}{\sqrt{\frac{s_1^2+s_2^2}{\widehat{\nu}_1}}}. \quad (2.4.24)$$

For given sample sizes and for a given value of $(\bar{x}_1, s_1^2, \bar{x}_2, s_2^2)$, (2.4.24) can be solved numerically for p (see Example 2.6). The root p^* is the $1-\alpha$ lower confidence limit for R based on Hall's (1984) lower tolerance limit in (2.4.9).

Example 2.6 (Simulated data)

We use the following simulated data (Table 2.6) from two normal distributions to show that the results based on the different methods described in the preceding sections could lead to very different conclusions. Suppose we are

Table 2.6: Simulated data from normal distributions

X_1:	10.166	5.889	8.258	7.303	8.757		
X_2:	−0.204	2.578	1.182	1.892	0.786	−0.517	1.156
	0.980	0.323	0.437	0.397	0.050	0.812	0.720

interested in hypothesis testing and interval estimation of the 0.01 quantile of $X_1 - X_2$, which is $\mu_1 - \mu_2 - z_{.99}\sqrt{\sigma_1^2 + \sigma_2^2}$, where $z_{.99} = 2.3264$. The summary statistics are $\bar{x}_1 = 8.075$, $s_1^2 = 2.561$, $\bar{x}_2 = 0.7564$ and $s_2^2 = 0.6512$. Other quantities are

$$\widehat{q}_1 = 3.328,\ \widehat{f}_1 = 6.582,\ \widehat{\nu}_1 = 6.502,$$
$$\widehat{q}_2 = 0.1271,\ \widehat{f}_2 = 5.056,\ \widehat{\nu}_2 = 7.418,$$
$$\widehat{q} = 3.933,\ \widehat{f} = 6.170,\ \text{and}\ \widehat{\nu} = 6.271$$

Given below are the $(0.99, 0.90)$ lower tolerance limits for the distribution of $X_1 - X_2$, based on the different approaches.

Hall's Tolerance Limit in (2.4.9):

$$7.318 - t_{6.582;.90}(5.932) \times 0.7028 = 7.318 - 9.878 \times 0.7028 = 0.376.$$

Hall's Tolerance Limit in (2.4.11):

$$7.318 - t_{5.056;.90}(6.336) \times 0.6580 = 7.318 - 11.437 \times 0.6580 = -0.208.$$

Reiser and Guttman's Limit in (2.4.10):

$$7.318 - t_{6.170;.90}(5.826) \times 0.7157 = 7.318 - 9.899 \times 0.7157 = 0.233.$$

Notice that the tolerance limits are quite different. Specifically, if Hall's approach with variance ratio defined as $q_1 = \frac{\sigma_1^2}{\sigma_2^2}$ is used, then the $(0.99, 0.90)$ lower tolerance limit is 0.376. Since the limit is positive, we conclude that the reliability parameter $R = P(X_1 > X_2)$ is at least 0.99 at the level 0.10. On the other hand, if the variance ratio is defined as $q_2 = \frac{\sigma_2^2}{\sigma_1^2}$, then the lower tolerance limits is -0.208, and based on this value we can not conclude that $R > .99$.

If Guo and Krishnamoorthy's (2004) approach is used, then the $(0.99, 0.90)$ lower tolerance limit is given by $\min\{0.376, -0.208\} = -0.208$.

Suppose we want to test

$$H_0 : \mu_1 - \mu_2 - z_{.99}\sqrt{\sigma_1^2 + \sigma_2^2} \le 0 \quad \text{vs.} \quad H_a : \mu_1 - \mu_2 - z_{.99}\sqrt{\sigma_1^2 + \sigma_2^2} > 0.$$

If the variance ratio is defined as $q_1 = \frac{\sigma_1^2}{\sigma_2^2}$, then the p-value based on Hall's method is given by

$$p_1 = P\left(t_{\widehat{f}_1}(z_p\sqrt{\widehat{\nu}_1}) > \frac{\sqrt{\widehat{\nu}_1}(\bar{x}_1 - \bar{x}_2)}{\sqrt{s_1^2 + s_2^2}}\right) = P(t_{6.582}(5.932) > 10.412) = 0.0788.$$

If the variance ratio is defined as $q_2 = \frac{\sigma_2^2}{\sigma_1^2}$, then the p-value is given by

$$p_2 = P\left(t_{\widehat{f}_2}(z_p\sqrt{\widehat{\nu}_2}) > \frac{\sqrt{\widehat{\nu}_2}(\bar{x}_1 - \bar{x}_2)}{\sqrt{s_1^2 + s_2^2}}\right) = P(t_{5.056}(6.336) > 11.121) = 0.111.$$

The p-value of the Reiser and Guttman's approach is given by

$$P\left(t_{\widehat{f}}(z_p\sqrt{\widehat{\nu}}) > \frac{\sqrt{\widehat{\nu}}(\bar{x}_1 - \bar{x}_2)}{\sqrt{s_1^2 + s_2^2}}\right) = P(t_{6.170}(5.826) > 10.225) = 0.0869.$$

Notice again that the conclusions based on Hall's (1984) tests depend on the definition of the variance ratio.

Example 2.7 (Breakdown voltage - power supply)

This example, taken from Hall (1984), is concerned with the proportion of times the breakdown voltage X_1 of a capacitor exceeds the voltage output X_2 of a transverter (power supply). A sample of $n_1 = 50$ capacitors yielded $\bar{x}_1 = 6.75$ kV and $s_1^2 = 0.123$. The voltage output from $n_2 = 20$ transverters produced $\bar{x}_2 = 4.00$ kV and $s_2^2 = 0.53$. We have $\bar{x}_1 - \bar{x}_2 = 2.75$. Using the methods in Section 2.4.2, we shall compute a $(0.95, 0.95)$ lower tolerance limit for the distribution of $X_1 - X_2$. We computed the required quantities as shown below:

$$\hat{q}_1 = 0.2077, \ \hat{\nu}_1 = 22.3007, \ \hat{f}_1 = 27.2543,$$
$$\hat{q}_2 = 4.1331, \ \hat{\nu}_2 = 22.6471, \ \hat{f}_2 = 28.6559 \ \text{ and } \ z_{.95} = 1.6449.$$

Hall's Tolerance Limit Based on the Variance Ratio $\frac{\sigma_1^2}{\sigma_2^2}$

$$
\begin{aligned}
\bar{x}_1 - \bar{x}_2 - t_{\hat{f}_1;1-\alpha}(z_p\sqrt{\hat{\nu}_1})\sqrt{\frac{s_1^2 + s_2^2}{\hat{\nu}_1}} &= 2.75 - t_{27.2543;.95}(7.7678)\sqrt{\frac{0.123 + 0.53}{22.3007}} \\
&= 2.75 - 10.7169\sqrt{0.0293} \\
&= 0.9156.
\end{aligned}
$$

Hall's Tolerance Limit Based on the Variance Ratio $\frac{\sigma_2^2}{\sigma_1^2}$

$$
\begin{aligned}
\bar{x}_1 - \bar{x}_2 - t_{\hat{f}_2;1-\alpha}(z_p\sqrt{\hat{\nu}_2})\sqrt{\frac{s_1^2 + s_2^2}{\hat{\nu}_2}} &= 2.75 - t_{28.6559;.95}(7.8279)\sqrt{\frac{0.123 + 0.53}{22.6471}} \\
&= 2.75 - 10.7293\sqrt{.0288} \\
&= 0.9292.
\end{aligned}
$$

The 95% lower limit for the reliability parameter $R = P(X_1 > X_2)$ based on the tolerance limit 0.9156 is given by the solution (with respect to p) of the equation

$$
t_{\hat{f}_1;1-\alpha}(z_p\sqrt{\hat{\nu}_1}) = \frac{\bar{x}_1 - \bar{x}_2}{\sqrt{\frac{s_1^2 + s_2^2}{\hat{\nu}_1}}} \Leftrightarrow t_{27.2543;.95}(z_p\sqrt{22.3007}) = \frac{2.75}{\sqrt{0.0293}}.
$$

Solving this equation, we get $z_p = 2.5560$ which yields $p = 0.9947$. The 95% lower limit for R based on the tolerance limit 0.9292 is given by the equation

$$
t_{28.6559;.95}(z_p\sqrt{22.6471}) = \frac{2.75}{\sqrt{0.0288}}.
$$

The above equation yields $z_p = 2.5745$ and $p = 0.9950$.

If we follow Guo and Krishnamoorthy's (2004) suggestion, then we use $\min\{0.9156, 0.9292\} = 0.9156$ as (0.95, 0.95) lower tolerance limit for the distribution of $X_1 - X_2$, and $\min\{0.9947, 0.9950\} = 0.9947$ as 95% lower limit for the reliability parameter R. Finally, the generalized variable approach based on the GPQ in (2.4.13) with 100,000 runs produced the (0.95, 0.95) lower tolerance limit as 0.8870. We see that the tolerance limit based on the generalized variable approach is the smallest among all the limits. This is because this approach, as pointed out in Section 2.4.4, is conservative.

Example 2.8 (Mechanical component data)

The summary statistics for this example are taken from Reiser and Guttman (1986). The data are pertaining to a mechanical component that yielded $\bar{x}_1 = 170,000$, $s_1 = 5,000$, $\bar{x}_2 = 144,500$, $s_2 = 8,900$, for $n_1 = n_2 = 32$. Using the formulas of the preceding sections, we computed

$$\hat{q}_1 = 0.29525, \; \hat{f}_1 = 47.8377, \; \hat{q}_2 = 2.9640, \; \hat{f}_2 = 49.7800, \; \text{and} \; \hat{\nu}_1 = \hat{\nu}_2 = 32.$$

Based on these quantities, we like to test whether the lower 95th percentile of $X_1 - X_2$ is greater than zero. That is,

$$H_0 : \mu_1 - \mu_2 - z_{.95}\sqrt{\sigma_1^2 + \sigma_2^2} \leq 0 \text{ vs. } H_a : \mu_1 - \mu_2 - z_{.95}\sqrt{\sigma_1^2 + \sigma_2^2} > 0,$$

or equivalently $H_0 : R \leq 0.95$ vs $H_a : R > 0.95$.

Hall's (1984) Approach

If the variance ratio defined as $q_1 = \frac{\sigma_1^2}{\sigma_2^2}$, then the p-value is

$$
\begin{aligned}
p_1 &= P\left(t_{\hat{f}_1;1-\alpha}(z_p\sqrt{\hat{\nu}_1}) > \frac{\sqrt{\hat{\nu}_1}(\bar{x}_1 - \bar{x}_2)}{\sqrt{s_1^2 + s_2^2}}\right) \\
&= P\left(t_{47.8377}(9.3050) > 14.1306\right) \\
&= 0.0027.
\end{aligned}
$$

If the variance ratio defined as $q_2 = \frac{\sigma_2^2}{\sigma_1^2}$, then the p-value is

$$
\begin{aligned}
p_2 &= P\left(t_{\hat{f}_2,1-\alpha}(z_p\sqrt{\hat{\nu}_2}) > \frac{\sqrt{\hat{\nu}_2}(\bar{x}_1 - \bar{x}_2)}{\sqrt{s_1^2 + s_2^2}}\right) \\
&= P\left(t_{49.7800}(9.3050) > 14.1306\right) \\
&= 0.0025.
\end{aligned}
$$

The generalized p-value, using (2.4.21) and 100,000 simulation runs, is 0.0042. The p-values of the approximate tests are very close to each other while the generalized p-value is relatively higher than the others. This is once again consistent with the conclusions in Section 2.4.4.

2.5 Simultaneous Tolerance Limits for Normal Populations

We shall now describe methods of constructing $(p, 1 - \alpha)$ simultaneous tolerance intervals for several normal populations. The procedure given in this section is due to Mee (1990a). Let $X_{i1}, ..., X_{in_i}$ be a sample from a $N(\mu_i, \sigma^2)$ population, $i = 1, ..., l$. Let $\bar{X}_i$ and S_i^2, respectively, denote the mean and the variance of the ith sample, $i = 1, ..., l$. The pooled variance estimator of σ^2, say S_c^2, is given by

$$S_c^2 = \frac{\sum_{i=1}^{l}(n_i - 1)S_i^2}{N - l}, \quad \text{where } N = \sum_{i=1}^{l} n_i.$$

2.5.1 Simultaneous One–Sided Tolerance Limits

One-sided $(p, 1 - \alpha)$ tolerance factors k_i are to be determined such that

$$P_{\bar{X}_1, ..., \bar{X}_l, S_c}(\bar{X}_i + k_i S_p \geq \mu_i + z_p \sigma, i = 1, ..., l) = 1 - \alpha. \quad (2.5.1)$$

Letting $Z_i = \frac{\sqrt{n_i}(\bar{X}_i - \mu_i)}{\sigma_i}$, $i = 1, ..., l$, and $U^2 = \frac{S_c^2}{\sigma^2}$, we can express (2.5.1) as

$$P_{Z_1, ..., Z_l, U}(Z_i - \sqrt{n_i} z_p \geq -\sqrt{n_i} k_i U, i = 1, ..., l) = 1 - \alpha.$$

Noticing that Z_i is distributed as $-Z_i$, we can rewrite the above equation as

$$E_W\left[\prod_{i=1}^{l} \Phi\left(\sqrt{n_i}\left(\frac{k_i W}{\sqrt{N - l}} - z_p\right)\right)\right] = 1 - \alpha, \quad (2.5.2)$$

where $W^2 = (N - l)U^2 \sim \chi_{N-l}^2$ distribution. When the sample sizes are equal, it is reasonable to require that $k_1 = ... = k_l$, and this common tolerance factor k_1^* is the solution of the integral equation

$$\frac{1}{2^{\frac{N-l}{2}}\Gamma\left(\frac{N-l}{2}\right)}\int_0^\infty \left[\Phi\left(\sqrt{n}\left(\frac{k_1^*\sqrt{x}}{\sqrt{N-l}} - z_p\right)\right)\right]^l e^{-x/2}x^{\frac{N-l}{2}-1}dx = 1 - \alpha, \quad (2.5.3)$$

where $n = n_1 = ... = n_l$. Mee (1990a) has tabulated the values of k_1^* satisfying (2.5.3) for several values of n ranging from 2 to 1000, $l = 2, 3, 4, 5, 6, 8, 10$, $p = 0.90, 0.95, 0.99$ and $1 - \alpha = 0.90, 0.95, 0.99$. We computed the factors k_1^* for $l = 2(1)10$, $n = 2(1)30, 40, 50, 60, 70, 80, 90, 100, 300, 500, 1000$, $p = 0.90, 0.95, 0.99$ and $1 - \alpha = 0.90, 0.95, 0.99$. These are given in Table B5, Appendix B.

Remark 2.5.1 Mee (1990a) also provided a way of finding approximate tolerance factors $k_1, ..., k_l$ (from the listed table values of k_1^* for the equal sample size case) when the sample sizes are unequal. For example, when $l = 4$, $n_1 = 10$, $n_2 = n_3 = 15$ and $n_4 = 50$, $p = 0.90$ and $1 - \alpha = 0.95$, the approximate tolerance factors k_1, k_2, k_3 and k_4 can be obtained from the table values of k_1^* listed under $p = 0.90, 1 - \alpha = 0.95$ and $l = 4$. Specifically, k_1 is the value of k_1^* when $n = 10$, $k_2 = k_3$ is the value of k^* when $n = 15$, and k_4 is the value of k^* when $n = 50$. From Table B5, Appendix B, we obtained the approximate values as $k_1 = 2.181$, $k_2 = k_3 = 1.989$ and $k_4 = 1.646$. Mee's numerical studies showed that these approximate tolerance factors are satisfactory as long as the degrees of freedom $N - l$ is large.

2.5.2 Simultaneous Tolerance Intervals

To construct $(p, 1 - \alpha)$ simultaneous tolerance intervals for l normal populations with common variance σ^2, we need to determine the tolerance factors $k_1', ..., k_l'$ so that

$$P_{\bar{X}_1, ..., \bar{X}_l, S_c} \left\{ P(\bar{X}_i - k_i' S_c \leq X_i \leq \bar{X}_i + k_i' S_c | \bar{X}_i, S_c) \geq p, i = 1, ..., l \right\} = 1 - \alpha,$$

where $X_1, ..., X_l$ are independent random variables with $X_i \sim N(\mu_i, \sigma^2)$, $i = 1, ..., l$. Furthermore, $X_1, ..., X_l$, $\bar{X}_1, ..., \bar{X}_l$ and S are mutually independent. Let $Y_i = \frac{\bar{X}_i - \mu_i}{\sigma}$, $i = 1, ..., l$ and $U^2 = \frac{S_c^2}{\sigma^2}$. In terms of these variables, we can write the above equation as

$$P_{Y_1, ..., Y_l, U^2} \left\{ P(Y_i - k_i' U \leq Z_i \leq Y_i + k_i' U | Y_i, U^2) \geq p, i = 1, ..., l \right\} = 1 - \alpha, \tag{2.5.4}$$

where $Z_1, ..., Z_l$ are standard normal random variables, $Y_i \sim N\left(0, \frac{1}{n_i}\right)$, $i = 1, ..., l$, and $X_1, ..., X_l$ and $Y_1, ..., Y_l$ are all independent. Thus, we need to determine the factors $k_1', ..., k_l'$ such that

$$E_{Y_1, ..., Y_l} \left\{ P_{U^2} [\Phi(Y_i + k_i' U) - \Phi(Y_i - k_i' U)] \geq p, \ i = 1, ..., l | Y \right\} = 1 - \alpha. \tag{2.5.5}$$

Notice that the left-hand side of (2.5.5) involves an l dimensional integral, and so the above equation is difficult to solve. If it is assumed that $k_1' = ... = k_l' = k'$, then (2.5.5) simplifies to

$$E_Y\left\{P_{U^2}[\Phi(Y+k'U)-\Phi(Y-k'U)]\geq p|Y\right\}=1-\alpha,$$

where $Y=\max\{|Y_1|,...,|Y_l|\}$. Furthermore, if it is assumed that $n_1=...=n_l=n$, then following the lines of the proof of Result 1.2.1(ii) in Chapter 1, we can write the above equation as

$$E_Y\left[P\left(U^2\geq\frac{r^2}{k'^2}\middle|Y\right)\right]=1-\alpha,\qquad\qquad(2.5.6)$$

where r is the solution of the equation $\Phi(Y+r)-\Phi(Y-r)=p$ and $U^2\sim\frac{\chi^2_{N-l}}{N-l}$. If Y is fixed then $r^2=\chi^2_{1;p}(Y^2)$. Notice that $Y\sim\max\{|Z_1|/\sqrt{n},...,|Z_l|/\sqrt{n}\}$, where Z_i's are independent standard normal random variables, and the probability density function of Y is given by

$$g(y)=2l\sqrt{n}[2\Phi(y\sqrt{n})-1]^{l-1}\phi(y\sqrt{n}).$$

Using $g(y)$, we can write (2.5.6) as

$$\int_0^\infty P\left(U^2\geq\frac{\chi^2_{1,p}(y^2)}{k'^2}\right)g(y)dy=1-\alpha,\qquad\qquad(2.5.7)$$

or equivalently

$$2l\int_0^\infty P\left(\chi^2_{N-l}\geq\frac{(N-l)\chi^2_{1,p}(z^2/n)}{k'^2}\right)[2\Phi(z)-1]^{l-1}\phi(z)dz=1-\alpha.\quad(2.5.8)$$

Equation (2.5.8) is obtained from (2.5.7) using the transformation $y\sqrt{n}=z$.

Mee (1990a) has tabulated the values of k' satisfying (2.5.8) for several values of n ranging from 2 to 1000, $l=2,3,4,5,6,8,10$, $p=.90,.95,.99$ and $1-\alpha=0.90,0.95,0.99$. We computed the factors k' for $l=2(1)10$, $n=2(1)30,40,50,60,70,80,90,100,300,500,1000$, $p=0.90,0.95,0.99$ and $1-\alpha=0.90,0.95,0.99$, and these are given in Table B6, Appendix B.

Bonferroni Intervals

As described in Result 1.2.3 of Chapter 1, this approach essentially uses the one-sample tolerance factors but with a higher confidence level. The simultaneous tolerance intervals are constructed such that the ith interval would contain at least a proportion p of the ith normal population with confidence $1 - \frac{\alpha}{l}$. This leads to the tolerance factor for simultaneous one-sided limits as

$$k_{1i}^b = \frac{1}{\sqrt{n_i}} t_{N-l;1-\frac{\alpha}{l}}(z_p\sqrt{n_i}). \qquad (2.5.9)$$

Using the Bonferrnoi inequality, we then have

$$P(\bar{X}_i + k_{1i}^b S_p \geq \mu_i + z_p\sigma, \ i = 1, ..., l) \geq 1 - \alpha.$$

Similarly, the $\left(p, 1 - \frac{\alpha}{l}\right)$ factor for the one-sample tolerance interval can be used to construct conservative simultaneous two-sided tolerance intervals. Let k_{2i}^b denote the factor for a $\left(p, 1 - \frac{\alpha}{l}\right)$ tolerance interval when the sample size is n_i. Then, simultaneously the interval $\bar{X}_i \pm k_{2i}^b S$ contains at least a proportion p of the ith population, $i = 1, ..., l$, with confidence at least $1 - \alpha$.

Example 2.9 (Tensile strengths of bars from castings)

The data in Table 2.7 represent tensile strengths (psi) of samples of bars from three different castings (Hahn and Ragunathan, 1988). The pooled standard

Table 2.7: Tensile strength of bars from three different castings

	Castings		
	1	2	3
	88.0	85.9	94.2
	88.0	88.6	91.5
	94.8	90.0	92.0
	90.0	87.1	96.5
	93.0	85.6	95.6
	89.0	86.0	93.8
	86.0	91.0	92.5
	92.9	89.6	93.2
	89.0	93.0	96.2
	93.0	87.5	92.5
$\bar{x}_i$	90.37	88.43	93.80
s_i	2.869	2.456	1.786

deviation $s_c = 2.412$. Mee (1990a) used the above data to find $(0.95, 0.95)$ one-sided simultaneous lower tolerance limits for the tensile strengths of bars from the three different castings. Noting that $l = 3$, $n_1 = ... = n_3 = 10$, $p = 0.95$ and $1 - \alpha = 0.95$, we find $k_1^* = 2.635$ from Table B5, Appendix B. Thus, the lower tolerance limits are $\bar{x}_i - 2.635(2.412)$, $i = 1, 2, 3$. This leads to lower confidence limits $(84.01, 82.07, 87.44)$, and we conclude that at least 95% of the bars from castings 1, 2 and 3 have tensile strength exceeding 84.01, 82.07 and 87.44 psi, respectively, with confidence level 95%. The factor for constructing (0.95, 0.95) two-sided simultaneous tolerance intervals is $k' = 2.825$ (from Table B6, Appendix B). The tolerance intervals are $\bar{x}_i \pm 2.825(2.412)$, $i = 1, 2, 3$, or 90.37 ± 6.81, 88.43 ± 6.81 and 93.80 ± 6.81.

If we use Bonferroni's simultaneous one-sided intervals, then the limits are given by $\bar{x}_i - t_{N-3;1-\frac{.05}{3}}(z_p\sqrt{n})s_p$, $i = 1, 2, 3$. Noting that $t_{27;0.983}(5.202) = 2.671$, we have $\bar{x}_i - 2.671(2.412) = \bar{x}_i - 6.442$, $i = 1, 2, 3$. Thus, (0.95, 0.95) simultaneous Bonferroni lower tolerance limits are 83.93, 81.99 and 87.3. As expected, these limits are slightly less than the exact ones given in the preceding paragraph, but they are in good agreement. To construct simultaneous two-sided tolerance intervals based on the Bonferroni approach, we found the (0.95, 0.983) factor as 2.929. This factor can be obtained using *StatCalc* by Krishnamoorthy (2006) (normal one-sample factor for tolerance interval) with $n = 10$ and df $= 27$. Using this factor, we computed conservative tolerance intervals as 90.37 ± 7.06, 88.43 ± 7.06 and 93.80 ± 7.06. Again, as expected, these intervals are slightly wider than the corresponding exact ones given in the preceding paragraph. One should certainly expect the Bonferroni procedure to produce even more conservative tolerance intervals as the number of populations become large.

2.6 Exercises

2.6.1. Let $t_{m;\alpha}(\delta)$ denote the α quantile of $t_m(\delta)$, a noncentral t distribution with df $= m$ and the noncentrality parameter δ. Using the representation that

$$t_m(\delta) = \frac{Z + \delta}{\sqrt{\frac{\chi_m^2}{m}}},$$

where $Z \sim N(0, 1)$ independently of χ_m^2, show that $t_{m;1-\alpha}(\delta) = -t_{m;\alpha}(-\delta)$.

2.6.2. Using the representation of the noncentral t random variable in the above exercise, show that $\bar{X} - k_1 S$ is a $(p, 1 - \alpha)$ lower tolerance limit for a $N(\mu, \sigma^2)$ distribution, where $\bar{X}$ and S^2 are respectively the mean and variance based on a sample of size n from the $N(\mu, \sigma^2)$ distribution and the factor k_1 is given in (2.2.3).

2.6.3. Let $\bar{x}$ and s be observed values of the mean and standard deviation of a random sample of size n from a normal distribution. Let $Z \sim N(0, 1)$ independently of $V \sim \sqrt{\frac{\chi_{n-1}^2}{n-1}}$. A $1 - \alpha$ lower confidence limit for $P(X > t)$

is given by the solution (with respect to p) of $\bar{x} - \frac{1}{\sqrt{n}}t_{n-1;1-\alpha}(z_p\sqrt{n})s = t$.

Let $Q = \frac{(\bar{x}-t)V}{s} - \frac{Z}{\sqrt{n}}$.

(a) Using the representation of the noncentral t random variable in Exercise 2.6.1, show that $\Phi(Q_\alpha)$, where Q_α is the α quantile of Q, and Φ is the standard normal cdf, is a $1 - \alpha$ lower confidence limit for $P(X > t)$.

(b) Using the above method, find a 95% lower confidence limit for the exceedance probability in Example 2.1. Notice that the percentile Q_α can be estimated by Monte Carlo simulation.

2.6.4. Show that a $(p, 1 - \alpha)$ upper tolerance limit for $N(\mu, \sigma^2)$ is always larger than the usual $1 - \alpha$ upper limit for μ when $p > 0.5$.

2.6.5. Explain why a two-sided $(p, 1 - \alpha)$ tolerance interval for $N(\mu, \sigma^2)$ may not in general contain the interval $\left(\mu - z_{\frac{1+p}{2}}\sigma, \ \mu + z_{\frac{1+p}{2}}\sigma\right)$.

2.6.6. Suppose the interval $\bar{X} \pm 2S$ is to be used as a two-sided tolerance interval for $N(\mu, \sigma^2)$ for a given sample size n and for a specified value of p.

(a) Show that the corresponding confidence level is given by
$$P_{Z_n,U}\left(\Phi(Z_n + 2U) - \Phi(Z_n - 2U) > p\right),$$
where Z_n and U are as defined below equation (2.3.2).

(b) Explain how you will numerically determine the confidence level using the representation given above.

(c) Explain how you can determine the confidence level approximately, using an approximation similar to (1.2.5).

(d) For $n = 20$ and $p = 0.90$, determine the approximate confidence level using the procedure in (c). Repeat for $n = 30$ and $n = 50$ and $n = 100$. Explain the pattern you notice.

2.6.7. Suppose some machine parts should have a diameter of 1 cm with specification limits of 1 ± 0.03cm. To assess the actual percentage of parts that meet the specification, a sample of 25 parts are measured with the following diameters:

0.9810	1.0102	0.9881	0.9697	0.9970	0.9836	0.9745	0.9793
1.0213	0.9876	0.9935	1.0168	1.0114	0.9914	0.9963	0.9862
1.0103	1.0204	0.9845	0.9840	1.0086	1.0025	0.9776	0.9940
0.9842							

The mean $\bar{x} = 0.99415$ and the standard deviation $s = 0.01479$.

(a) Verify that the data fit a normal distribution.

(b) Show that the exact $(0.90, 0.95)$ two-sided tolerance interval is $(0.96139, 1.02691)$.

(c) Estimate the percentage of parts that fall within the specifications with confidence 0.95.

(d) What percentage of parts that have diameters more than 1.03 cm? Estimate with 95% confidence.

(e) What percentage of parts have diameters less than 0.97 cm? Estimate with 95% confidence.

2.6.8. Consider the data in Exercise 2.6.7.

(a) Show that the $(0.90, 0.95)$ equal-tailed tolerance interval is $(0.95827, 1.03003)$.

(b) Why is the tolerance interval in part (a) wider than the tolerance interval in part (a) of Exercise 2.6.7? Explain.

(c) If we have to use an equal-tailed tolerance interval, then what percentage of machine parts meet the specifications? [That is, find the maximum value of p so that $\bar{x} \pm k_e s$ is close to 1.0 ± 0.03, where k_e is defined in $(2.3.10)$].

2.6.9. Let $X_1, ..., X_n$ be a sample from a $N(\mu, \sigma^2)$ distribution. Furthermore, let $\tilde{X}$ and $\tilde{S}^2$ denote, respectively, the mean and variance of a future random sample of size m from $N(\mu, \sigma^2)$.

(a) Find a $(p, 1 - \alpha)$ two-sided tolerance interval for the distribution of $\tilde{X}$, and interpret its meanings.

(b) Let (L_1, U_1) be a $(p, 1 - \alpha)$ tolerance interval for the $N(\mu, \sigma^2)$ distribution, and let (L_2, U_2) denote the tolerance interval in part (a), both based on $X_1, ..., X_n$. Show that the expected width of (L_2, U_2) approaches zero as $m \to \infty$, while that of (L_1, U_1) approaches a fixed quantity as $n \to \infty$.

(c) Find a $(p, 1 - \alpha)$ lower tolerance limit for the distribution of $\tilde{S}^2$, and interpret its meanings.

2.6.10. Compute the expected value of the $(p, 1 - \alpha)$ upper tolerance limit for $N(\mu, \sigma^2)$, given in $(2.2.4)$.

(a) Suggest an iterative procedure for determining the sample size n so that the expected value is a specified quantity, for given values of p, $1 - \alpha$, μ and σ^2.

(b) Suppose the observed values $\bar{x}$ and s corresponding to the log-transformed data in Example 2.1 can be assumed to be the true values of μ and σ. Determine the sample size n so that a $(0.95, 0.90)$ upper tolerance limit for the log-transformed data has an expected value equal to 6.5.

Chapter 3

Univariate Linear Regression Model

3.1 Notations and Preliminaries

Regression models are used to model the relationship between a response variable and one or more covariates. Consider a group of n individuals or items, and let Y_i denote the response variable for the ith individual or item, and $\mathbf{x}_i$ denote the corresponding $m \times 1$ vector of covariates. The univariate linear regression model assumes that the mean of Y_i is $\mathbf{x}_i'\boldsymbol{\beta}$, where $\boldsymbol{\beta}$ is an $m \times 1$ vector of unknown parameters. In this chapter, we shall consider the univariate linear regression model where we also assume that the Y_i's are normally distributed with a common variance σ^2. If $\boldsymbol{Y} = (Y_1, Y_2, ..., Y_n)'$ denotes the $n \times 1$ vector of observations and $\mathbf{X}$ denotes the $n \times m$ matrix whose ith row is $\mathbf{x}_i'$, the univariate linear regression model, under the normality assumption, can be represented using the usual notation as

$$\boldsymbol{Y} = \mathbf{X}\boldsymbol{\beta} + \boldsymbol{e}, \ \boldsymbol{e} \sim N(\mathbf{0}, \sigma^2 I_n), \tag{3.1.1}$$

where $\boldsymbol{e}$ is an error vector, and $\sigma^2 > 0$ is also an unknown parameter. We assume that the covariates are non-random. If the model contains an intercept term, then the first component of $\boldsymbol{\beta}$ can be taken as the intercept and, in this case, the first column of $\mathbf{X}$ will be a vector of ones. Throughout, we shall assume that $\text{rank}(\mathbf{X}) = m$.

Now let $Y(\mathbf{x})$ denote a future observation corresponding to a covariate vector $\mathbf{x}$. Assume that

$$Y(\mathbf{x}) = \mathbf{x}'\boldsymbol{\beta} + e, \ e \sim N(0, \sigma^2), \tag{3.1.2}$$

where $Y(\mathbf{x})$ is also assumed to be independent of $\boldsymbol{Y}$ in (3.1.1). For a fixed

$m \times 1$ vector $\mathbf{x}$, a $(p, 1 - \alpha)$ tolerance interval for $Y(\mathbf{x})$ is an interval that will contain at least a proportion p of the $Y(\mathbf{x})$–distribution with confidence $1 - \alpha$. The tolerance interval is constructed using the vector of observations $\mathbf{Y}$ in (3.1.1). Note that here $\mathbf{x}$ is a given vector, i.e., $\mathbf{x}$ is a fixed value of the covariates in the regression model. In some applications, the construction of a *simultaneous tolerance interval* is required. A simultaneous tolerance interval is motivated by the fact that future observations may correspond to different values of $\mathbf{x}$. A simultaneous tolerance interval satisfies the following condition. Suppose the same vector of observations $\mathbf{Y}$, following the model (3.1.1), is used a large number of times in order to construct tolerance intervals for a sequence of future observations $Y(\mathbf{x})$, corresponding to possibly different values of $\mathbf{x}$. With a confidence level of $1 - \alpha$, at least a proportion p of the $Y(\mathbf{x})$–distribution is to be contained in the corresponding tolerance interval, simultaneously for every $\mathbf{x}$. In other words, for a simultaneous tolerance interval, the minimum content (with respect to $\mathbf{x}$) is at least p, with confidence level $1 - \alpha$.

Let $\widehat{\boldsymbol{\beta}}$ denote the least squares estimator of $\boldsymbol{\beta}$ and S^2 denote the residual mean square under the model (3.1.1). Then

$$\widehat{\boldsymbol{\beta}} = (\mathbf{X}'\mathbf{X})^{-1}\mathbf{X}'\mathbf{Y} \quad \text{and} \quad S^2 = \frac{(\mathbf{Y} - \mathbf{X}\widehat{\boldsymbol{\beta}})'(\mathbf{Y} - \mathbf{X}\widehat{\boldsymbol{\beta}})}{n - m}, \qquad (3.1.3)$$

where we recall that m is the dimension of $\boldsymbol{\beta}$ and is also the rank of the $n \times m$ matrix $\mathbf{X}$. A two-sided tolerance interval for $Y(\mathbf{x})$ will be taken to be of the form $\mathbf{x}'\widehat{\boldsymbol{\beta}} \pm k(\mathbf{x})S$, where $k(\mathbf{x})$ is the tolerance factor to be determined subject to the content and confidence level requirements. Let $C(\mathbf{x}; \widehat{\boldsymbol{\beta}}, S)$ denote the content of this tolerance interval, given $\widehat{\boldsymbol{\beta}}$ and S. Then

$$C(\mathbf{x}; \widehat{\boldsymbol{\beta}}, S) = P_{Y(\mathbf{x})}\left(\mathbf{x}'\widehat{\boldsymbol{\beta}} - k(\mathbf{x})S \leq Y(\mathbf{x}) \leq \mathbf{x}'\widehat{\boldsymbol{\beta}} + k(\mathbf{x})S \Big| \widehat{\boldsymbol{\beta}}, S \right), \qquad (3.1.4)$$

and the tolerance factor $k(\mathbf{x})$ satisfies the condition

$$P_{\widehat{\boldsymbol{\beta}}, S}\left(C(\mathbf{x}; \widehat{\boldsymbol{\beta}}, S) \geq p \right) = 1 - \alpha. \qquad (3.1.5)$$

On the other hand, if $\mathbf{x}'\widehat{\boldsymbol{\beta}} \pm k(\mathbf{x})S$ is a simultaneous tolerance interval, then the condition to be satisfied is

$$P_{\widehat{\boldsymbol{\beta}}, S}\left(\min_{\mathbf{x}} C(\mathbf{x}; \widehat{\boldsymbol{\beta}}, S) \geq p \right) = 1 - \alpha. \qquad (3.1.6)$$

A function $k(\mathbf{x})$ that satisfies (3.1.6) is called a simultaneous tolerance factor. We note that in practice, the vector $\mathbf{x}$ will be a bounded quantity with known bounds. In (3.1.6), the minimum of $C(\mathbf{x}; \widehat{\boldsymbol{\beta}}, S)$ with respect to $\mathbf{x}$ has to be computed subject to the known bounds for $\mathbf{x}$. One-sided tolerance intervals and one-sided simultaneous tolerance intervals can be similarly defined. For example, an upper tolerance limit for $Y(\mathbf{x})$ can be taken as $\mathbf{x}'\widehat{\boldsymbol{\beta}} + k(\mathbf{x})S$, and a lower tolerance limit for $Y(\mathbf{x})$ can be taken as $\mathbf{x}'\widehat{\boldsymbol{\beta}} - k(\mathbf{x})S$, where the tolerance factor $k(\mathbf{x})$ is once again determined subject to a specified content and confidence level requirement. The corresponding one-sided tolerance intervals are $(-\infty, \mathbf{x}'\widehat{\boldsymbol{\beta}} + k(\mathbf{x})S]$ and $[\mathbf{x}'\widehat{\boldsymbol{\beta}} - k(\mathbf{x})S, \infty)$, respectively.

Another related problem that is of practical interest is *statistical calibration*. The inclusion of this topic in this chapter is motivated by the fact that the derivation of a *multiple use confidence interval* in the calibration problem can be accomplished using a simultaneous tolerance interval. To explain the ideas, consider the special case of (3.1.1) and (3.1.2), where we have the simple linear regression model. Thus, let $Y_1, ..., Y_n$ be independently distributed with

$$Y_i = \beta_0 + \beta_1 x_i + e_i, \ e_i \sim N(0, \sigma^2), \ i = 1, ..., n, \qquad (3.1.7)$$

where the x_i's are known values of a covariate. Also, let

$$Y(x) = \beta_0 + \beta_1 x + e, \ e \sim N(0, \sigma^2) \qquad (3.1.8)$$

denote a future observation corresponding to the value x of the covariate. Now suppose x is unknown. The calibration problem consists of statistical inference concerning x. Here we shall discuss the problem of constructing confidence intervals for x. More specifically, we shall discuss the construction of multiple use confidence intervals. The problem arises when the Y_i's in (3.1.7), referred to as the *calibration data*, are used repeatedly to construct a sequence of confidence intervals for a sequence of unknown and possibly different x–values after observing the corresponding $Y(x)$, following the model (3.1.8). Multiple use confidence intervals are derived subject to the following coverage and confidence level requirements: given that the confidence intervals are constructed using the same calibration data, the proportion of confidence intervals that include the corresponding true x–values is to be at least p. The probability that the calibration data will provide such a set of confidence intervals is to be at least $1 - \alpha$. (The confidence regions that we derive for the calibration problem need not always be intervals; however, we shall simply continue to refer to them as confidence intervals). It turns out that a simultaneous tolerance interval can be used for computing a multiple use confidence interval in the calibration problem. Essentially, a multiple use confidence interval for x is obtained by inverting the simultaneous tolerance interval. For reviews and discussions on the calibration problem, see the paper by Osborne (1991) and the book by Brown (1993).

3.2 One-Sided Tolerance Intervals and Simultaneous Tolerance Intervals

Let $k(\mathbf{x})$ denote the tolerance factor to be computed to obtain one-sided tolerance intervals of the form $(-\infty, \mathbf{x}'\widehat{\boldsymbol{\beta}} + k(\mathbf{x})S]$ and $[\mathbf{x}'\widehat{\boldsymbol{\beta}} - k(\mathbf{x})S, \infty)$. Thus $\mathbf{x}'\widehat{\boldsymbol{\beta}} + k(\mathbf{x})S$ is the upper tolerance limit and $\mathbf{x}'\widehat{\boldsymbol{\beta}} - k(\mathbf{x})S$ is the lower tolerance limit. We shall first consider the derivation of $k(\mathbf{x})$ for obtaining a one-sided tolerance interval for the distribution of $Y(\mathbf{x})$ for a fixed $\mathbf{x}$. This will be followed with the derivation of the factor necessary to obtain simultaneous one-sided tolerance intervals.

3.2.1 One-Sided Tolerance Intervals

The derivation that follows is similar to the derivation in Section 2.2 of Chapter 2 (the "Classical Approach"). The content of the tolerance interval $(-\infty, \mathbf{x}'\widehat{\boldsymbol{\beta}} + k(\mathbf{x})S]$, given $\widehat{\boldsymbol{\beta}}$ and S, is

$$C_1(\mathbf{x}; \widehat{\boldsymbol{\beta}}, S) = P_{Y(\mathbf{x})}\left(Y(\mathbf{x}) \le \mathbf{x}'\widehat{\boldsymbol{\beta}} + k(\mathbf{x})S \mid \widehat{\boldsymbol{\beta}}, S\right),$$

and the factor $k(\mathbf{x})$ should be determined so that

$$P_{\widehat{\boldsymbol{\beta}}, S}\left(C_1(\mathbf{x}; \widehat{\boldsymbol{\beta}}, S) \ge p\right) = 1 - \alpha. \tag{3.2.1}$$

Notice that

$$Z = \frac{Y(\mathbf{x}) - \mathbf{x}'\boldsymbol{\beta}}{\sigma} \sim N(0,1), \quad \mathbf{Z_X} = \frac{\widehat{\boldsymbol{\beta}} - \boldsymbol{\beta}}{\sigma} \sim N(\mathbf{0}, (\mathbf{X}'\mathbf{X})^{-1}),$$

$$\text{and} \quad U^2 = \frac{S^2}{\sigma^2} \sim \frac{\chi^2_{n-m}}{n-m}, \tag{3.2.2}$$

where χ^2_r denotes a central chi-square random variable with r degrees of freedom, and all the random variables are independent. In terms of these random variables, we can write the content as

$$C_1(\mathbf{x}; \widehat{\boldsymbol{\beta}}, S) = P_Z\left(Z \le \mathbf{x}'\mathbf{Z_X} + k(\mathbf{x})U \mid \mathbf{Z_X}, U\right). \tag{3.2.3}$$

Note that $\mathbf{x}'\mathbf{Z_X} \sim N(0, \mathbf{x}'(\mathbf{X}'\mathbf{X})^{-1}\mathbf{x})$. Let

$$d^2 = \mathbf{x}'(\mathbf{X}'\mathbf{X})^{-1}\mathbf{x} \quad \text{and} \quad V = \frac{\mathbf{x}'\mathbf{Z_X}}{d}, \tag{3.2.4}$$

so that $V \sim N(0, 1)$. Using these variables, the content of the tolerance interval can be expressed as

$$C_1(\mathbf{x}; \widehat{\boldsymbol{\beta}}, S) = \Phi\left(dV + k_1(d)U\right) = C_1(d; V, U), \qquad (3.2.5)$$

where Φ denotes the standard normal cdf, and we have used the notation $k_1(d)$ instead of $k(\mathbf{x})$, since $k(\mathbf{x})$ depends on $\mathbf{x}$ only through d defined in (3.2.4). The notation $C_1(d; V, U)$ in the place of $C_1(\mathbf{x}; \widehat{\boldsymbol{\beta}}, S)$ should not cause any confusion. The fact that $k(\mathbf{x})$ depends on $\mathbf{x}$ only through the scalar d is an important and useful simplification. Thus, using (3.2.5) and (3.2.1), we see that the factor $k_1(d)$ satisfies

$$P_{V,U}\left(\Phi\left(dV + k_1(d)U\right) \geq p\right) = 1 - \alpha.$$

Setting $X = dV \sim N(0, d^2)$ and $Q = U^2 \sim \frac{\chi^2_{n-m}}{n-m}$, and then applying Result 1.2.1(i) in Chapter 1, we get

$$k_1(d) = d \times t_{n-m;1-\alpha}(z_p/d), \qquad (3.2.6)$$

where $t_{r;\gamma}(\eta)$ denotes the γ quantile of a noncentral t distribution with r degrees of freedom and noncentrality parameter η. It is easily verified that for both of the one-sided tolerance intervals $(-\infty, \mathbf{x}'\widehat{\boldsymbol{\beta}} + k(\mathbf{x})S]$ and $[\mathbf{x}'\widehat{\boldsymbol{\beta}} - k(\mathbf{x})S, \infty)$, the tolerance factor $k(\mathbf{x})$ is given by $k_1(d)$ given in (3.2.6). Note also that for the simple linear regression model d^2 can be simplified:

$$d^2 = \left(\frac{1}{n} + c^2\right), \quad \text{with} \quad c^2 = \frac{(x - \bar{x})^2}{\sum_{i=1}^{n}(x_i - \bar{x})^2}. \qquad (3.2.7)$$

Example 3.1 (Viscosity data)

This example is taken from Montgomery (2009, p. 394), and deals with a study of the relationship between the viscosity of a polymer and two process variables: reaction temperature and catalyst feed rate. For 16 samples, Table 3.1 gives the data on $Y =$ viscosity (centistokes at 100^{0}C), $x_1 =$ temperature in Celsius, and $x_2 =$ catalyst feed rate (lb/hour).

The regression equation relating the viscosity Y to the temperature x_1, and feed rate x_2 is given by

$$Y = 1566.078 + 7.621x_1 + 8.585x_2, \qquad (3.2.8)$$

with a squared correlation of 0.927. Furthermore, the residual mean square S^2 has value 267.604. The model diagnostics reported in Montgomery (2009, Chapter 10) show that it is reasonable to assume normality. However, the variability

Table 3.1: The viscosity data

Observation	Temperature (x_1)	Catalyst Feed Rate (x_2)	Viscosity (Y)
1	80	8	2256
2	93	9	2340
3	100	10	2426
4	82	12	2293
5	90	11	2330
6	99	8	2368
7	81	8	2250
8	96	10	2409
9	94	12	2364
10	93	11	2379
11	97	13	2440
12	95	11	2364
13	100	8	2404
14	85	12	2317
15	86	9	2309
16	87	12	2328

in the viscosity appears to increase with temperature. We shall ignore this for the purpose of our analysis, and assume that the variance is a constant. Here a tolerance interval for the viscosity is an interval that will contain at least a specified proportion of the viscosity values with a certain confidence level, for a fixed pair of values of the temperature and catalyst feed rate. On the other hand, a simultaneous tolerance interval for the viscosity is an interval that will contain at least a specified proportion of the viscosity values with a certain confidence level, regardless of the (temperature, catalyst feed rate) values, where these values are assumed to be within reasonable bounds resulting from the physical constraints on these variables.

For this example, let us compute the $(0.90, 0.95)$ one-sided upper tolerance limit for the distribution of $Y(\mathbf{x})$ at $\mathbf{x}' = (1, 88, 9)$. We first note that $n = 16$ and $m = 3$, and the fitted model is given in $(3.2.8)$. Other necessary quantities to compute the tolerance limits are

$$(\mathbf{X}'\mathbf{X})^{-1} = \begin{pmatrix} 14.176 & -0.1297 & -0.2235 \\ - & 0.00143 & -0.00005 \\ - & - & 0.02222 \end{pmatrix}, \tag{3.2.9}$$

$$d = \left(\mathbf{x}'(\mathbf{X}'\mathbf{X})^{-1}\mathbf{x}\right)^{\frac{1}{2}} = 0.33289 \quad \text{and} \quad S = 16.3585. \tag{3.2.10}$$

Furthermore, $\mathbf{x}'\widehat{\boldsymbol{\beta}} = 2314.02$, $z_{0.9} = 1.2816$ and $t_{n-m;1-\alpha}(z_p/d) = t_{13;.95}(3.850) = 6.6023$. Thus, the (0.90, 0.95) tolerance factor $k_1(d) = d \times t_{13;.95}(3.850) = 2.1977$ and the corresponding upper tolerance limit is given by

$$\mathbf{x}'\widehat{\boldsymbol{\beta}} + k_1(d)S = 2314.02 + 2.1977 \times 16.3585 = 2349.97.$$

That is, with confidence 95%, at least 90% of the $Y(\mathbf{x})$−measurements are at most 2349.97 when $\mathbf{x}' = (1, 88, 9)$.

Example 3.2 (Blood alcohol concentration)

This example is based on a study that was conducted at Acadiana Criminalistic Laboratory, New Iberia, Louisiana, to compare the breath estimates of blood alcohol concentration (obtained using a breath analyzer) with those determined by a laboratory test. This example is described in Krishnamoorthy, Kulkarni and Mathew (2001). A sample of 15 subjects was used. In Table 3.2, we present the breath estimates Y obtained using Breathalyzer Model 5000 and the results of the laboratory test, denoted by x. These numbers are percentages of alcohol concentration in blood. It turns out that a simple linear regression

Table 3.2: Blood alcohol concentrations data

Subject	Blood alcohol concentration (x)	Breath estimate (Y)
1	.160	.145
2	.170	.156
3	.180	.181
4	.100	.108
5	.170	.180
6	.100	.112
7	.060	.081
8	.100	.104
9	.170	.176
10	.056	.048
11	.111	.092
12	.162	.144
13	.143	.121
14	.079	.065
15	.006	.000

model fits well with a squared correlation of 0.9293. The fitted model is

$$Y = 0.00135 + 0.958x. \qquad (3.2.11)$$

The normal probability plot of the residuals is reasonably linear and hence the distribution of Y can be assumed to be normal.

In many states in the USA the legal maximum limit of blood alcohol concentration while driving is 0.1%. Instead of directly measuring the alcohol concentration in the blood, it is certainly much easier to obtain the breath estimates. Thus the problem is to estimate the blood alcohol concentration x after obtaining the corresponding breath estimate Y, based on the model given above. In other words, we have the calibration problem. A lower or upper confidence limit for the unknown blood alcohol concentration could be of interest. For example, an upper confidence limit can be used to ascertain whether the blood alcohol concentration is below the limit of 0.1%.

Here we shall compute a $(0.90, 0.95)$ lower tolerance limit for the distribution of $Y(\mathbf{x})$ when $\mathbf{x}' = (1, 0.10)$. The fitted model is $Y = 0.00135 + 0.958x$, and the necessary quantities to compute the tolerance limit are

$$(\mathbf{X}'\mathbf{X})^{-1} = \begin{pmatrix} 0.4264 & -3.0538 \\ - & 25.9240 \end{pmatrix} \quad \text{and} \quad d = \left(\mathbf{x}'(\mathbf{X}'\mathbf{X})^{-1}\mathbf{x}\right)^{\frac{1}{2}} = 0.273643.$$

(3.2.12)

Also, $\mathbf{x}'\widehat{\boldsymbol{\beta}} = 0.00135 + 0.958(0.10) = 0.0971$ and $S = 0.01366$. The critical value $t_{n-m;1-\alpha}(z_p/d) = t_{13;.95}(4.6833) = 7.7365$, and the $(0.90, 0.95)$ tolerance factor is $d \times t_{n-m;1-\alpha}(z_p/d) = 2.1172$, where $z_p = z_{0.9} = 1.2816$. The $(0.90, 0.95)$ one-sided lower tolerance limit is given by

$$\mathbf{x}'\widehat{\boldsymbol{\beta}} - k_1(d)S = 0.0971 - 2.1172 \times 0.01366 = 0.068.$$

That is, with confidence 95%, at least 90% of the breath alcohol estimates exceed 0.068 when the blood alcohol level is 0.10.

3.2.2   One-Sided Simultaneous Tolerance Intervals

We shall now describe the derivation of one-sided simultaneous tolerance intervals of the form $(-\infty, \mathbf{x}'\widehat{\boldsymbol{\beta}} + k(\mathbf{x})S]$ and $[\mathbf{x}'\widehat{\boldsymbol{\beta}} - k(\mathbf{x})S, \infty)$, where $k(\mathbf{x})$ is the simultaneous tolerance factor to be determined. The development that follows is taken from the work of Odeh and Mee (1990). Note from (3.1.4) that $C(\mathbf{x}; \widehat{\boldsymbol{\beta}}, S)$ is a function of $\mathbf{x}'\mathbf{Z}_{\mathbf{X}}$ and U, where $\mathbf{Z}_{\mathbf{X}}$ and U are defined in (3.2.2). Now suppose the first m_1 components are common to all the rows of $\mathbf{X}$. Let $\mathbf{x}'_1$ be a $1 \times m_1$ vector denoting this common part. This is the case, for example, when we have models with an intercept term so that the first component of $\mathbf{x}'$ is one. Hence the vectors $\mathbf{x}'$ under consideration are such that $\mathbf{x}' = (\mathbf{x}'_1, \mathbf{x}'_2)$, where $\mathbf{x}'_1$ is the fixed common part, and $\mathbf{x}'_2$ is a $1 \times (m - m_1)$ vector that could be different among the different vectors $\mathbf{x}'$. Now partition the vector $\mathbf{Z}_{\mathbf{X}}$ in (3.2.2) as $(\mathbf{Z}'_1, \mathbf{Z}'_2)'$ similar to the partitioning of $\mathbf{x}'$, and partition $\mathbf{X}$ as $(\mathbf{X}_1, \mathbf{X}_2)$, where

$\mathbf{X}_1$ is an $n \times m_1$ matrix. Furthermore, let

$$d_1^2 = \mathbf{x}_1'(\mathbf{X}_1'\mathbf{X}_1)^{-1}\mathbf{x}_1, \; d_2^2 = d^2 - d_1^2, \; \text{and} \; W_2^2 \sim \chi^2_{m-m_1}, \tag{3.2.13}$$

where the chi-square random variable W_2^2 will be defined later. The derivation that follows is for the one-sided interval $(-\infty, \mathbf{x}'\widehat{\boldsymbol{\beta}}+k(\mathbf{x})S]$; those for the interval $[\mathbf{x}'\widehat{\boldsymbol{\beta}} - k(\mathbf{x})S, \infty)$ are similar.

Using the expression for $C_1(\mathbf{x}; \widehat{\boldsymbol{\beta}}, S)$ in (3.2.3), along with the following inequality (to be proved later in this section)

$$\mathbf{x}_1'\mathbf{Z}_1 + \mathbf{x}_2'\mathbf{Z}_2 \geq d_1 Z_1 - d_2 W_2, \tag{3.2.14}$$

where d_1 and d_2 are as defined in (3.2.13), W_2^2 is a chi-square random variable to be defined later, and $Z_1 \sim N(0,1)$, we conclude that

$$\begin{aligned} C_1(\mathbf{x}; \widehat{\boldsymbol{\beta}}, S) &\geq P_Z\left(Z \leq d_1 Z_1 - d_2 W_2 + k_{1s}(d)U \mid Z_1, W_2, U\right) \\ &= \Phi\left(d_1 Z_1 - d_2 W_2 + k_{1s}(d)U\right), \end{aligned} \tag{3.2.15}$$

where we have assumed that $k(\mathbf{x})$ is a function of d, denoted by $k_{1s}(d)$ in (3.2.15). Odeh and Mee (1990) have computed $k_{1s}(d)$ satisfying the condition

$$P_{Z_1, W_2, U}\left[\min_d \Phi\left(d_1 Z_1 - d_2 W_2 + k_{1s}(d)U\right) \geq p\right] = 1 - \alpha. \tag{3.2.16}$$

Equivalently,

$$P_{Z_1, W_2, U}\left[\min_d \left(d_1 Z_1 - d_2 W_2 + k_{1s}(d)U\right) \geq z_p\right] = 1 - \alpha.$$

Since $Z_1 \sim N(0,1)$, this simplifies to

$$E_{W_2, U}\left[\min_d \Phi\left(\frac{-z_p - d_2 W_2 + k_{1s}(d)U}{d_1}\right)\right] = 1 - \alpha, \tag{3.2.17}$$

where $U^2 \sim \frac{\chi^2_{n-m}}{n-m}$, independently of $W_2^2 \sim \chi^2_{m-m_1}$. Since the functional form of $k_{1s}(d)$ that satisfies (3.2.17) is not unique, Odeh and Mee (1990) assumed the following functional form

$$k_{1s}(d) = \lambda[z_p + (m - m_1 + 3)^{1/2}d], \tag{3.2.18}$$

λ being a scalar to be determined. Thus the problem is to compute λ so that $k_{1s}(d)$ in (3.2.18) satisfies (3.2.17), where the minimum in (3.2.17) is computed with respect to suitable bounds on d. Assuming

$$n^{-\frac{1}{2}} \leq d \leq \sqrt{\frac{1 + \tau^2}{n}},$$

Odeh and Mee (1990) have tabulated the numerical values of λ for $m - m_1 = 1$, 2, 3, $\tau = 2, 3, 4$, $p = 0.75, 0.90, 0.95, 0.99$ and $1 - \alpha = 0.90, 0.95, 0.99$. The values of λ are reproduced in Table B7, Appendix B, for the case $m - m_1 = 1$. The table values indicate that λ is rather insensitive to the value of τ.

Proof of (3.2.14). Note that $\mathbf{X}'\mathbf{X}$ has the partitioned form

$$\mathbf{X}'\mathbf{X} = \begin{pmatrix} \mathbf{X}_1'\mathbf{X}_1 & \mathbf{X}_1'\mathbf{X}_2 \\ \mathbf{X}_2'\mathbf{X}_1 & \mathbf{X}_2'\mathbf{X}_2 \end{pmatrix} = \begin{pmatrix} \mathbf{B}_{11} & \mathbf{B}_{12} \\ \mathbf{B}_{12}' & \mathbf{B}_{22} \end{pmatrix} = \mathbf{B}, \text{ say.} \qquad (3.2.19)$$

Using the expression for the inverse of a partitioned matrix (see Rao (1973, p. 33)), we can write

$$\mathbf{B}^{-1} = \begin{pmatrix} \mathbf{B}_{11}^{-1} & 0 \\ 0 & 0 \end{pmatrix} + \begin{pmatrix} \mathbf{B}_{11}^{-1}\mathbf{B}_{12} \\ -\mathbf{I} \end{pmatrix} \mathbf{B}_{2.1}^{-1} \left(\mathbf{B}_{12}'\mathbf{B}_{11}^{-1}, \; -\mathbf{I} \right), \qquad (3.2.20)$$

where $\mathbf{B}_{2.1} = \mathbf{B}_{22} - \mathbf{B}_{12}'\mathbf{B}_{11}^{-1}\mathbf{B}_{12}$. Since $\mathbf{Z}_{\mathbf{X}} = (\mathbf{Z}_1', \mathbf{Z}_2')' \sim N(0, \mathbf{B}^{-1})$, it is readily verified that

$$\mathbf{Z}_1 + \mathbf{B}_{11}^{-1}\mathbf{B}_{12}\mathbf{Z}_2 \sim N\left(0, \mathbf{B}_{11}^{-1}\right), \qquad (3.2.21)$$

and is independent of $\mathbf{Z}_2$. Hence

$$\mathbf{x}_1' \left(\mathbf{Z}_1 + \mathbf{B}_{11}^{-1}\mathbf{B}_{12}\mathbf{Z}_2 \right) = d_1 Z_1, \qquad (3.2.22)$$

where $Z_1 \sim N(0,1)$. We next note that

$$\begin{aligned} \mathbf{x}_1'\mathbf{Z}_1 + \mathbf{x}_2'\mathbf{Z}_2 &= \mathbf{x}_1'\left(\mathbf{Z}_1 + \mathbf{B}_{11}^{-1}\mathbf{B}_{12}\mathbf{Z}_2\right) + \left(\mathbf{x}_2'\mathbf{Z}_2 - \mathbf{x}_1'\mathbf{B}_{11}^{-1}\mathbf{B}_{12}\mathbf{Z}_2\right) \\ &= d_1 Z_1 + (\mathbf{x}_2' - \mathbf{x}_1'\mathbf{B}_{11}^{-1}\mathbf{B}_{12})\mathbf{Z}_2 \\ &= d_1 Z_1 + (\mathbf{x}_2 - \mathbf{B}_{12}'\mathbf{B}_{11}^{-1}\mathbf{x}_1)'\mathbf{B}_{2.1}^{-1/2}\mathbf{B}_{2.1}^{1/2}\mathbf{Z}_2 \\ &\geq d_1 Z_1 - \left[(\mathbf{x}_2 - \mathbf{B}_{12}'\mathbf{B}_{11}^{-1}\mathbf{x}_1)'\mathbf{B}_{2.1}^{-1}(\mathbf{x}_2 - \mathbf{B}_{12}'\mathbf{B}_{11}^{-1}\mathbf{x}_1)\right]^{1/2} \\ &\quad \times \left(\mathbf{Z}_2'\mathbf{B}_{2.1}\mathbf{Z}_2\right)^{1/2}. \end{aligned}$$

The last inequality was obtained using the Cauchy-Schwartz inequality. Now the inequality (3.2.14) follows from the relation that

$$\begin{aligned} (\mathbf{x}_2 - \mathbf{B}_{12}'\mathbf{B}_{11}^{-1}\mathbf{x}_1)'&\mathbf{B}_{2.1}^{-1}(\mathbf{x}_2 - \mathbf{B}_{12}'\mathbf{B}_{11}^{-1}\mathbf{x}_1) \\ &= (\mathbf{x}_1', \mathbf{x}_2') \begin{pmatrix} \mathbf{B}_{11}^{-1}\mathbf{B}_{12} \\ -\mathbf{I} \end{pmatrix} \mathbf{B}_{2.1}^{-1} \left(\mathbf{B}_{12}'\mathbf{B}_{11}^{-1}, \; -\mathbf{I} \right) \begin{pmatrix} \mathbf{x}_1 \\ \mathbf{x}_2 \end{pmatrix} \\ &= \mathbf{x}'\mathbf{B}^{-1}\mathbf{x} - \mathbf{x}_1'\mathbf{B}_{11}^{-1}\mathbf{x}_1 \\ &= d^2 - d_1^2 \\ &= d_2^2. \end{aligned} \qquad (3.2.23)$$

Since $\mathbf{Z}_2 \sim N\left(\mathbf{0}, \mathbf{B}_{2.1}^{-1}\right)$, it follows that $W_2^2 = \mathbf{Z}_2' \mathbf{B}_{2.1} \mathbf{Z}_2 \sim \chi_{m-m_1}^2$.

Example 3.2 (continued)

We shall now compute a one-sided simultaneous lower tolerance limit for $Y(\mathbf{x})$ at the value $\mathbf{x}' = (1, .10)$. We have already computed $d = \left(\mathbf{x}'(\mathbf{X}'\mathbf{X})^{-1}\mathbf{x}\right)^{\frac{1}{2}} = 0.273643$, $\mathbf{x}'\widehat{\boldsymbol{\beta}} = 0.0971$ and $S = 0.01366$. Furthermore, note that $m - m_1 = 1$ for the simple linear regression model. Thus, using $n = 15$, $p = 0.90$, $1 - \alpha = 0.95$ and $\tau = 2$, we get the value of λ from Table B7 in Appendix B as 1.2618, and so

$$k_{1s}(d) = \lambda[z_p + (m - m_1 + 3)^{1/2}d] = 1.2618[1.2816 + 0.5473] = 2.3077.$$

Thus, a (0.90, 0.95) one-sided simultaneous lower tolerance limit is given by

$$\mathbf{x}'\widehat{\boldsymbol{\beta}} - k_{1s}(d)S = 0.0971 - 2.3077 \times 0.01366 = 0.066.$$

3.3 Two-Sided Tolerance Intervals and Simultaneous Tolerance Intervals

Recall that in the set up of (3.1.1) and (3.1.2), a two-sided tolerance interval is assumed to be of the form $\mathbf{x}'\widehat{\boldsymbol{\beta}} \pm k(\mathbf{x})S$, where $\widehat{\boldsymbol{\beta}}$ and S^2 are defined in (3.1.3). We begin with the derivation of such an interval for the distribution of $Y(\mathbf{x})$ for a fixed $\mathbf{x}$.

3.3.1 Two-Sided Tolerance Intervals

For a fixed $\mathbf{x}$, the tolerance factor $k(\mathbf{x})$ is to be determined subject to the condition (3.1.5), i.e., $k(\mathbf{x})$ satisfies

$$P_{\widehat{\boldsymbol{\beta}}, S}\left(C_2(\mathbf{x}; \widehat{\boldsymbol{\beta}}, S) \geq p\right) = 1 - \alpha, \qquad (3.3.1)$$

where $C_2(\mathbf{x}; \widehat{\boldsymbol{\beta}}, S)$ is the content of the tolerance interval given $\widehat{\boldsymbol{\beta}}$ and S, as defined in the right hand side expression (3.1.4). Let Z, $\mathbf{Z_X}$ and U be as defined in (3.2.2). In terms of these variables, we can write

$$
\begin{aligned}
C_2(\mathbf{x}; \widehat{\boldsymbol{\beta}}, S) &= P_{Y(\mathbf{x})}\left(\mathbf{x}'\widehat{\boldsymbol{\beta}} - k(\mathbf{x})S \leq Y(\mathbf{x}) \leq \mathbf{x}'\widehat{\boldsymbol{\beta}} + k(\mathbf{x})S \mid \widehat{\boldsymbol{\beta}}, S\right) \\
&= P_Z\left(\mathbf{x}'\mathbf{Z}_{\mathbf{X}} - k(\mathbf{x})\frac{S}{\sigma} \leq Z \leq \mathbf{x}'\mathbf{Z}_{\mathbf{X}} + k(\mathbf{x})\frac{S}{\sigma} \;\middle|\; \mathbf{Z}_{\mathbf{X}}, S\right) \\
&= P_Z\left(\mathbf{x}'\mathbf{Z}_{\mathbf{X}} - k(\mathbf{x})U \leq Z \leq \mathbf{x}'\mathbf{Z}_{\mathbf{X}} + k(\mathbf{x})U \mid \mathbf{Z}_{\mathbf{X}}, U\right).
\end{aligned}
$$
$$(3.3.2)$$

Recall that $d^2 = \mathbf{x}'(\mathbf{X}'\mathbf{X})^{-1}\mathbf{x}$, $\mathbf{x}'\mathbf{Z}_{\mathbf{X}} \sim N(0, d^2)$ and $V = \frac{\mathbf{x}'\mathbf{Z}_{\mathbf{X}}}{d} \sim N(0, 1)$. Using these variables, we can simplify $C_2(\mathbf{x}; \widehat{\boldsymbol{\beta}}, S)$ in (3.3.2) as

$$
\begin{aligned}
C_2(\mathbf{x}; \widehat{\boldsymbol{\beta}}, S) &= P_Z\left(dV - k_2(d)U \leq Z \leq dV + k_2(d)U \mid V, U\right) \\
&= \Phi\left(dV + k_2(d)U\right) - \Phi\left(dV - k_2(d)U\right),
\end{aligned}
$$
$$(3.3.3)$$

where Φ denotes the standard normal cdf, and we have used the notation $k_2(d)$ for $k(\mathbf{x})$. From (3.3.1) and (3.3.3), we see that $k_2(d)$ is to be chosen so as to satisfy

$$
P_{V,U}\left(\Phi\left(dV + k_2(d)U\right) - \Phi\left(dV - k_2(d)U\right) \geq p\right) = 1 - \alpha.
$$
$$(3.3.4)$$

Setting $X = dV \sim N(0, d^2)$ and $Q = U \sim \frac{\chi^2_{n-m}}{n-m}$, and noting that X and Q are independent, we can apply Result 1.2.1 (ii) of Chapter 1. Using Result 1.2.1, we see that $k_2(d)$ is the solution of the integral equation

$$
\sqrt{\frac{2}{\pi d^2}} \int_0^\infty P\left(\chi^2_{n-m} > \frac{(n-m)\chi^2_{1;p}(x^2)}{(k_2(d))^2}\right) e^{-\frac{x^2}{2d^2}}\, dx = 1 - \alpha.
$$
$$(3.3.5)$$

A program due to Eberhardt, Mee and Reeve (1989) can be used for computing the tolerance factor $k_2(d)$. Nevertheless, it is certainly desirable to have analytic approximations. We shall now describe three approximations for $k_2(d)$.

Wallis' Approximation

The Wallis (1951) approximation is an extension of the approximation due to Wald and Wolfowitz (1946) for the univariate normal distribution and can be obtained using Result 1.2.1(iii) in Chapter 1. Using Result 1.2.1(iii), we get

$$
k_{2W}(d) = \left(\frac{(n-m)\chi^2_{1;p}(d^2)}{\chi^2_{n-m;\alpha}}\right)^{\frac{1}{2}},
$$
$$(3.3.6)$$

where we have used the notation $k_{2W}(d)$ to emphasize that the above approximation is due to Wallis (1951). In (3.3.6), $\chi^2_{\nu;\gamma}(\delta)$ denotes the γ quantile of a noncentral chi-square distribution with df ν and noncentrality parameter δ, and $\chi^2_{\nu;\gamma}$ denotes the γ quantile of a central chi-square distribution with df ν. Wallis (1951) further used an approximation for $\sqrt{\chi^2_{1;p}(d^2)}$ due to Bowker (1946), and this approximation is given by

$$\sqrt{\chi^2_{1;p}(d^2)} \approx z_{\frac{1+p}{2}}\left[1 + \frac{d^2}{2} - \frac{d^4(2z^2_{\frac{1+p}{2}} - 3)}{24}\right],$$

where z_γ denotes the γ quantile of the standard normal distribution. Using the above approximation in (3.3.6), we finally get the approximate tolerance factor

$$\frac{\sqrt{n-m}}{\sqrt{\chi^2_{n-m;\alpha}}}z_{\frac{1+p}{2}}\left[1 + \frac{d^2}{2} - \frac{d^4(2z^2_{\frac{1+p}{2}} - 3)}{24}\right]. \qquad (3.3.7)$$

It should be noted that (3.3.7) is obtained using two approximations, and as a result it is less accurate than the one in (3.3.6). Note however that there is no longer any need to use the approximation for $\sqrt{\chi^2_{1;p}(d^2)}$, since noncentral chi-square percentiles are now readily available in standard software packages.

Lee and Mathew's Approximation

Lee (1999) and Lee and Mathew (2004) proposed the following approximation which is a special case of two approximations given for the multivariate linear regression model (see Section 10.2, Chapter 10). The approximation is given by

$$k_{2LM}(d) = \frac{ef}{1+\delta}\chi^2_{1;p}(\delta)F_{e,n-m-1,\gamma}, \qquad (3.3.8)$$

where

$$e = \frac{(1+d^2)^2}{d^4}, \quad f = \frac{d^2}{1+d}, \quad \delta = d^2\left[\frac{3d^2 + \sqrt{9d^4 + 6d^2 + 3}}{2d^2 + 1}\right]$$

and $F_{m_1,m_2;\gamma}$ is the γ quantile of an F distribution with df (m_1, m_2).

Approximation Based on One-Sided Tolerance Factors

A one-sided tolerance factor with an adjustment to the confidence level can be used as an approximation to the factor for constructing two-sided tolerance limits. In particular, we consider the one-sided tolerance factor (3.2.6) with confidence level $(1 - \frac{\alpha}{2})$ as an approximation to $k_2(d)$. The resulting tolerance factor will be denoted by $k_{2o}(d)$. That is,

$$k_{2o}(d) = d \times t_{n-m;1-\frac{\alpha}{2}}(z_p/d). \tag{3.3.9}$$

Comparison of the Approximations

Since approximations were used in order to arrive at $k_{2W}(d)$ in (3.3.6), $k_{2LM}(d)$ in (3.3.8) and $k_{2o}(d)$ in (3.3.9), a natural question concerns the accuracy of these approximate tolerance factors. Notice that these approximate tolerance factors and the exact ones based on (3.3.5) depend only on p, $1 - \alpha$, $n - m$ and d^2. In practical applications, d^2 is bounded and typically it will not exceed 1. Krishnamoorthy and Mondal (2008) compared these approximate tolerance factors with the exact one based on (3.3.5) for $n - m = 10, 20$ and 40, and $d^2 = 0.1, 0.3, 0.5, 0.8, 0.9$. To demonstrate the accuracies of these approximations, we compare them with the exact one based on (3.3.5) when $n - m = 10$ and d^2 ranging from 0.1 to 4. The computed tolerance factors are given in Table 3.3 for all possible pairs of $(p, 1 - \alpha)$ from $\{0.90, 0.95, 0.99\}$.

The conclusions from the numerical results is that the Wallis (1951) approximation $k_{2W}(d)$ in (3.3.6) can be substantially below the exact tolerance factor, especially when d^2 is large (i.e., c^2 in (3.2.7) is large). On the other hand, $k_{2LM}(d)$ is somewhat larger than the exact ones most of the time. It is interesting to note that the approximation $k_{2o}(d)$ based on the one-sided factors coincide with the exact ones for large values of d^2. This approximation is not only simple but also very satisfactory for all combinations of content and confidence level considered, as long as $d^2 \geq 0.3$. Thus if an approximation needs to be used, we recommend Wallis' or Lee-Mathew's approach for small values of d^2, and the approximate factor $k_{2o}(d)$ for $d^2 \geq 0.3$.

Example 3.1 (continued)

We shall now compute the (0.90, 0.95) tolerance interval for the distribution of $Y(\mathbf{x})$ when $\mathbf{x}' = (1, 88, 9)$. From (3.2.10), we get $d^2 = 0.11801$, $S = 16.3585$ and $\mathbf{x}'\hat{\boldsymbol{\beta}} = 2314.02$. To compute the Lee and Mathew tolerance factor in (3.3.8), we computed $e = 100.49$, $f = 0.011054$, and $\delta = 0.2064$, and the factor

$$
\begin{aligned}
k_{2LM}(d) &= \left(\frac{ef}{1+\delta} \chi^2_{1;p}(\delta) F_{e,n-m-1;\gamma} \right)^{\frac{1}{2}} \\
&= \left(\frac{(100.49)(0.011054)}{1.2064} (3.2642)(2.2611) \right)^{\frac{1}{2}} \\
&= 2.607.
\end{aligned}
\tag{3.3.10}
$$

The approximation based on the one-sided tolerance factor is given by

$$
d \times t_{n-m;1-\frac{\alpha}{2}} \left(\frac{z_p}{d} \right) = 0.33289 \times t_{13;.975}(3.850) = 2.426.
$$

The Wallis approximation is given by

$$
k_{2W}(d) = \left(\frac{(13)\chi^2_{1;.90}(.11081)}{\chi^2_{13;.05}} \right)^{\frac{1}{2}} = \left(\frac{(13)(3.0061)}{5.8919} \right)^{\frac{1}{2}} = 2.575.
$$

The exact factor using (3.3.5) is evaluated as 2.603. Using these factors, we get the following tolerance intervals. The exact one is 2314.02 ± 42.58; the one based on the approximate one-sided factor is 2314.02 ± 39.69; Lee-Mathew approximation is 2314.02 ± 42.65 and the Wallis approximation is 2314.02 ± 42.12. Recall that these tolerance intervals are obtained using $\mathbf{x}'\hat{\boldsymbol{\beta}} \pm \text{factor} \times S$.

Example 3.2 (continued)

For this example, let us compute a $(0.90, 0.95)$ tolerance interval for the breath alcohol level $Y(\mathbf{x})$ when the blood alcohol level is 0.10. That is, when $\mathbf{x}' = (1, .10)$. From (3.2.12), we get $d^2 = 0.07489$, $S = 0.01366$ and $\mathbf{x}'\hat{\boldsymbol{\beta}} = 0.09715$. To compute the Lee and Mathew tolerance factor in (3.3.8), we computed $e = 206.05$, $f = 0.005216$, and $\delta = 0.1365$, and the factor

$$
\begin{aligned}
k_{2LM}(d) &= \left(\frac{ef}{1+\delta} \chi^2_{1;p}(\delta) F_{e,n-m-1;\gamma} \right)^{\frac{1}{2}} \\
&= \left(\frac{(206.05)(0.005216)}{1.1365} (3.0756)(2.2334) \right)^{\frac{1}{2}} \\
&= 2.549.
\end{aligned}
\tag{3.3.11}
$$

The approximation based on the one-sided tolerance factor is given by

$$
d \times t_{n-m;1-\frac{\alpha}{2}}(z_p/d) = 0.27364 \times t_{13;.975}(4.6835) = 2.328.
$$

The Wallis approximation is given by

$$k_{2W}(d) = \left(\frac{(13)\chi^2_{1;.90}(0.07489)}{\chi^2_{13;.05}} \right)^{\frac{1}{2}} = \left(\frac{(13)(2.9086)}{5.8919} \right)^{\frac{1}{2}} = 2.533.$$

The exact factor using (3.3.5) is evaluated as 2.548. Using these factors, we get the following tolerance intervals. The exact one is 0.0971 ± 0.035; the one based on the approximate one-sided factor is $0.0971 \pm .032$; Lee-Mathew approximation is 0.0971 ± 0.035 and the Wallis approximation is 0.0971 ± 0.035. Notice that all the tolerance intervals, except the one based on the approximate one-sided factor, are nearly the same. As pointed out earlier, the approximate one-sided factor works satisfactorily only when $d^2 \geq 0.3$.

3.3.2 Two-Sided Simultaneous Tolerance Intervals

We once again assume that a two-sided simultaneous tolerance interval is of the form $\mathbf{x}'\widehat{\boldsymbol{\beta}} \pm k(\mathbf{x})S$, where $k(\mathbf{x})$, the simultaneous tolerance factor, satisfies the condition (3.1.6). Several articles have appeared, addressing the computation of a two-sided simultaneous tolerance factor $k(\mathbf{x})$. These include Lieberman and Miller (1963), Lieberman, Miller and Hamilton (1967), Wilson (1967), Miller (1981, Chapter 4), Limam and Thomas (1988a) and Mee, Eberhardt and Reeve (1991). Among the solutions available for the computation of $k(\mathbf{x})$, the procedure due to Mee, Eberhardt and Reeve (1991) appears to provide the narrowest simultaneous tolerance interval; a conclusion based on the numerical results reported by these authors. In this section, we shall describe the Mee, Eberhardt and Reeve (1991) procedure. Note from (3.3.3) that $C(\mathbf{x}; \widehat{\boldsymbol{\beta}}, S)$ is a function of $\mathbf{x}' \mathbf{Z_X}$ and U, where $\mathbf{Z_X}$ and U are defined in (3.3.2). As in Section 3.2.2, we once again assume that the first m_1 components are common to all the rows of $\mathbf{X}$, and for all the vectors $\mathbf{x}'$ under consideration, and this common part will be denoted by the $1 \times m_1$ vector $\mathbf{x}'_1$. Thus we can write $\mathbf{x}' = (\mathbf{x}'_1, \mathbf{x}'_2)$. Note that $C(\mathbf{x}; \widehat{\boldsymbol{\beta}}, S)$ can also be expressed as

$$
\begin{aligned}
C(\mathbf{x}; \widehat{\boldsymbol{\beta}}, S) &= P_Z \left(|\mathbf{x}' \mathbf{Z_X}| - k(\mathbf{x})U \leq Z \leq |\mathbf{x}' \mathbf{Z_X}| + k(\mathbf{x})U \mid \mathbf{Z_X}, U \right) \\
&= P_Z \left(|\mathbf{x}'_1 \mathbf{Z}_1 + \mathbf{x}'_2 \mathbf{Z}_2| - k(\mathbf{x})U \leq Z \right. \\
&\qquad\qquad \left. \leq |\mathbf{x}'_1 \mathbf{Z}_1 + \mathbf{x}'_2 \mathbf{Z}_2| + k(\mathbf{x})U \mid \mathbf{Z_X}, U \right),
\end{aligned}
$$

$$(3.3.12)$$

where Z, $\mathbf{Z_X}$ and U are as defined in (3.2.2), and $\mathbf{Z_X} = (\mathbf{Z}'_1, \mathbf{Z}'_2)'$ is a partitioning similar to the partitioning of $\mathbf{x}$. Let $\mathbf{X} = (\mathbf{X}_1, \mathbf{X}_2)$, where $\mathbf{X}_1$ is $n \times m_1$. We shall now use the following result concerning the distribution of the maximum

Table 3.3: The exact values of $k_2(d)$ and the values of the approximate factors $k_{2W}(d)$, $k_{2LM}(d)$ and $k_{2o}(d)$

$$n - m = 10$$

d^2	Methods	$p = .90$			$p = .95$			$p = .99$		
		$1 - \alpha$			$1 - \alpha$			$1 - \alpha$		
		.90	.95	.99	.90	.95	.99	.90	.95	.99
.1	1	2.49	2.77	3.45	2.95	3.29	4.09	3.86	4.30	5.35
	2	2.29	2.56	3.21	2.83	3.15	3.92	3.87	4.28	5.30
	3	2.49	2.77	3.45	2.96	3.29	4.10	3.87	4.31	5.36
	4	2.47	2.75	3.41	2.94	3.27	4.06	3.86	4.29	5.33
.3	1	2.73	3.08	3.90	3.21	3.60	4.55	4.14	4.62	5.80
	2	2.64	2.99	3.82	3.15	3.54	4.48	4.13	4.61	5.77
	3	2.77	3.11	3.93	3.25	3.65	4.60	4.15	4.66	5.88
	4	2.69	2.99	3.71	3.18	3.54	4.39	4.14	4.60	5.70
.5	1	2.96	3.36	4.31	3.44	3.88	4.95	4.36	4.89	6.18
	2	2.90	3.31	4.28	3.40	3.85	4.92	4.35	4.88	6.16
	3	3.02	3.43	4.38	3.49	3.96	5.07	4.37	4.96	6.35
	4	2.88	3.20	3.97	3.39	3.76	4.67	4.35	4.84	6.00
.8	1	3.25	3.72	4.85	3.73	4.24	5.47	4.64	5.24	6.67
	2	3.22	3.70	4.84	3.70	4.22	5.46	4.64	5.23	6.66
	3	3.33	3.82	4.97	3.78	4.35	5.65	4.64	5.33	6.93
	4	3.13	3.48	4.31	3.64	4.05	5.03	4.62	5.13	6.37
1	1	3.42	3.94	5.17	3.90	4.46	5.78	4.81	5.44	6.96
	2	3.40	3.93	5.16	3.88	4.44	5.77	4.80	5.44	6.96
	3	3.50	4.05	5.31	3.95	4.57	5.99	4.79	5.54	7.27
	4	3.28	3.64	4.52	3.79	4.22	5.23	4.77	5.30	6.58
2	1	3.81	4.42	5.85	4.60	5.33	7.03	5.49	6.29	8.16
	2	3.80	4.41	5.85	4.60	5.32	7.03	5.49	6.28	8.16
	3	3.87	4.53	6.04	4.62	5.44	7.34	5.41	6.38	8.60
	4	3.59	3.99	4.96	4.39	4.87	6.05	5.36	5.96	7.40
4	1	5.18	6.11	8.27	5.63	6.60	8.84	6.50	7.53	9.94
	2	5.18	6.11	8.27	5.63	6.60	8.84	6.50	7.53	9.94
	3	5.18	6.21	8.55	5.59	6.69	9.22	6.35	7.61	10.49
	4	4.70	5.23	6.49	5.23	5.81	7.21	6.20	6.89	8.55

1. Exact tolerance factor based on (3.3.5); 2. the approximate tolerance factor $k_{2o}(d)$ in (3.3.9); 3. Lee and Mathew's approximation $k_{2LM}(d)$ in (3.3.8); 4. Wallis Approximation $k_{2W}(d)$ in (3.3.6)

(with respect to $\mathbf{x}_2$) of $|\mathbf{x}_1' \mathbf{Z}_1 + \mathbf{x}_2' \mathbf{Z}_2|$, subject to the conditions that $\mathbf{x}_1$ is a specified vector and $\mathbf{x}'(\mathbf{X}'\mathbf{X})^{-1}\mathbf{x} = d^2$ is fixed:

$$\max|\mathbf{x}_1' \mathbf{Z}_1 + \mathbf{x}_2' \mathbf{Z}_2| \sim d_1 W_1 + d_2 W_2,$$
$$\text{where} \quad d_1^2 = \mathbf{x}_1'(X_1'X_1)^{-1}\mathbf{x}_1, \ d_2^2 = d^2 - d_1^2, \ W_1^2 \sim \chi_1^2, \ W_2^2 \sim \chi_{m-m_1}^2,$$
$$(3.3.13)$$

and W_1 and W_2 are independently distributed. A proof of (3.3.13) is given later in this section, where the definitions of W_1^2 and W_2^2 are also given. We shall assume that the simultaneous tolerance factor $k(\mathbf{x})$ is a function of d, to be denoted by $k_{2s}(d)$. Then it follows from (3.3.12) and (3.3.13) that

$$\begin{aligned}
C(\mathbf{x}; \widehat{\boldsymbol{\beta}}, S) &\geq P_Z \left(d_1 W_1 + d_2 W_2 - k_{2s}(d)U \leq Z \right. \\
&\qquad \left. \leq d_1 W_1 + d_2 W_2 + k_{2s}(d)U \mid W_1, W_2, U \right) \\
&= C_0(d; W_1, W_2, U), \text{ say,} \qquad\qquad (3.3.14)
\end{aligned}$$

where W_1^2 and W_2^2 have the chi-square distributions given in (3.3.13), U^2 has the chi-square distribution given in (3.2.2), and W_1, W_2 and U are independent random variables. Thus $k_{2s}(d)$ has to be computed subject to the condition

$$P_{W_1^2, W_2^2, U^2}[\min_d C_0(d; W_1, W_2, U) \geq p] = 1 - \alpha. \qquad (3.3.15)$$

If we do not have the information that the first few components are common to all rows of $\mathbf{X}$ (or if we do not use this information), then instead of (3.3.14), we have

$$\begin{aligned}
C(\mathbf{x}; \widehat{\boldsymbol{\beta}}, S) &\geq P_Z \left(dW - k_{2s}(d)U \leq Z \right. \\
&\qquad \left. \leq dW + k_{2s}(d)U \mid W, U \right) \\
&= C_1(d; W, U), \text{ say,} \qquad\qquad (3.3.16)
\end{aligned}$$

where $d^2 = \mathbf{x}'(\mathbf{X}'\mathbf{X})^{-1}\mathbf{x}$ and $W^2 \sim \chi_m^2$. Mee, Eberhardt and Reeve (1991) have computed $k_{2s}(d)$ satisfying the condition

$$P_{W^2, U^2}[\min_d C_1(d; W, U) \geq p] = 1 - \alpha, \qquad (3.3.17)$$

where, similar to (3.2.18), they also assumed that $k_{2s}(d)$ has the functional form

$$k_{2s}(d) = \lambda \left[z_{\frac{1+p}{2}} + (m+2)^{1/2}d \right], \qquad (3.3.18)$$

λ being a scalar to be determined. As in the case of a one-sided simultaneous tolerance interval, the minimum in (3.3.17) is computed with respect to d satisfying reasonable bounds. Mee, Eberhardt and Reeve (1991) have numerically computed λ so that $k_{2s}(d)$ in (3.3.18) satisfies (3.3.17). The numerical results reported in their paper are for the simple linear regression model so that the expressions in (3.2.7) are valid, and they also assumed $n^{-\frac{1}{2}} \leq d \leq \sqrt{(1+\tau^2)/n}$, for the values $\tau = 2$, 3, 4. In Table B8, Appendix B, we have reported values of the factor λ; these values are reproduced from Mee, Eberhardt and Reeve (1991). From the table values we note that λ is rather insensitive to the value of τ. Note that the values in Table B8 correspond to $m = 2$ in (3.3.18), since the values are for a simple linear regression model.

Quoting Oden (1973), Mee, Eberhardt and Reeve (1991, p. 219) have commented that in the case of simple linear regression, even though the first row of $\mathbf{X}$ is a vector of ones, ignoring this aspect makes very little difference in the value of $k_{2s}(d)$. In other words, whether we consider (3.3.15) or (3.3.17) for the determination of $k_{2s}(d)$, the solution appears to be nearly the same. In view of this, and in view of the somewhat involved computation, the table values we have reported (Table B8, Appendix B) correspond to λ in (3.3.18) satisfying the condition (3.3.17), as tabulated in Mee, Eberhardt and Reeve (1991). In other words, we did not carry out the computation and prepare table values corresponding to the condition (3.3.15).

Proof of (3.3.13). It follows from (3.2.22) that

$$W_1^2 = \left[\mathbf{x}_1' \left(\mathbf{Z}_1 + (\mathbf{X}_1'\mathbf{X}_1)^{-1}\mathbf{X}_1'\mathbf{X}_2\mathbf{Z}_2\right)\right]^2 \sim d_1^2\chi_1^2.$$

In other words,

$$\left|\mathbf{x}_1' \left(\mathbf{Z}_1 + (\mathbf{X}_1'\mathbf{X}_1)^{-1}\mathbf{X}_1'\mathbf{X}_2\mathbf{Z}_2\right)\right| = d_1 W_1,$$

where $d_1^2 = \mathbf{x}_1'(\mathbf{X}_1'\mathbf{X}_1)^{-1}\mathbf{x}_1$ as defined in (3.3.13). Let $\mathbf{B}_{ij}$'s be as defined in (3.2.19) for the partitioned form of the matrix $\mathbf{X}'\mathbf{X}$. Now note that

$$
\begin{aligned}
|\mathbf{x}_1'\mathbf{Z}_1 + \mathbf{x}_2'\mathbf{Z}_2| &= |\mathbf{x}_1' \left(\mathbf{Z}_1 + (\mathbf{B}_{11}^{-1}\mathbf{B}_{12}\mathbf{Z}_2) + (\mathbf{x}_2'\mathbf{Z}_2 - \mathbf{x}_1'\mathbf{B}_{11}^{-1}\mathbf{B}_{12}\mathbf{Z}_2)| \\
&\leq |\mathbf{x}_1' \left(\mathbf{Z}_1 + \mathbf{B}_{11}^{-1}\mathbf{B}_{12}\mathbf{Z}_2\right)| + |\mathbf{x}_2'\mathbf{Z}_2 - \mathbf{x}_1'\mathbf{B}_{11}^{-1}\mathbf{B}_{12}\mathbf{Z}_2| \\
&= d_1 W_1 + |(\mathbf{x}_2 - \mathbf{B}_{12}'\mathbf{B}_{11}^{-1}\mathbf{x}_1)'\mathbf{Z}_2| \\
&= d_1 W_1 + |(\mathbf{x}_2 - \mathbf{B}_{12}'\mathbf{B}_{11}^{-1}\mathbf{x}_1)'\mathbf{B}_{2.1}^{-1/2}\mathbf{B}_{2.1}^{1/2}\mathbf{Z}_2| \\
&\leq d_1 W_1 + \left[(\mathbf{x}_2 - \mathbf{B}_{12}'\mathbf{B}_{11}^{-1}\mathbf{x}_1)'\mathbf{B}_{2.1}^{-1}(\mathbf{x}_2 - \mathbf{B}_{12}'\mathbf{B}_{11}^{-1}\mathbf{x}_1)\right]^{1/2} \\
&\quad\times (\mathbf{Z}_2'\mathbf{B}_{2.1}\mathbf{Z}_2)^{1/2}. \qquad\qquad (3.3.19)
\end{aligned}
$$

Thus, (3.3.13) will be established once we show that (i) $W_2^2 = \mathbf{Z}_2'\mathbf{B}_{2.1}\mathbf{Z}_2 \sim \chi_{m-m_1}^2$ and (ii) $\left[(\mathbf{x}_2 - \mathbf{B}_{12}'\mathbf{B}_{11}^{-1}\mathbf{x}_1)'\mathbf{B}_{2.1}^{-1}(\mathbf{x}_2 - \mathbf{B}_{12}'\mathbf{B}_{11}^{-1}\mathbf{x}_1)\right] = d^2 - d_1^2$. Since $\mathbf{Z}_2 \sim N\left(\mathbf{0}, \mathbf{B}_{2.1}^{-1}\right)$, (i) is obvious, and (ii) is already established in (3.2.23).

Example 3.2 (continued)

We shall now compute a $(0.90, 0.95)$ two-sided simultaneous tolerance interval for $Y(\mathbf{x})$ when $\mathbf{x}' = (1, 0.10)$. Recall that $d = \left(\mathbf{x}'(\mathbf{X}'\mathbf{X})^{-1}\mathbf{x}\right)^{\frac{1}{2}} = 0.273643$, $\mathbf{x}'\widehat{\boldsymbol{\beta}} = 0.0971$ and $S = 0.01366$. Furthermore, $n^{-\frac{1}{2}} = 0.25820$ and $\sqrt{(1 + \tau^2)/n} = 0.57735$ when $\tau = 2$. Thus, referring to Table B8, Appendix B for $(n, p, 1 - \alpha, \tau) = (15, 0.90, 0.95, 2)$, we get $\lambda = 1.2405$. So the simultaneous tolerance factor

$$
\begin{aligned}
k_{2s}(d) &= \lambda[z_{\frac{1+p}{2}} + (m + 2)^{1/2}d] \\
&= 1.2405[1.6449 + 2 \times 0.273643] \\
&= 2.7194.
\end{aligned}
$$

Thus, the desired $(0.90, 0.95)$ simultaneous tolerance interval at $\mathbf{x}' = (1, 0.10)$ is $0.0971 \pm 2.6285 \times 0.01366 = 0.0971 \pm 0.0359$.

3.4   The Calibration Problem

This section is on the computation of multiple use confidence intervals for the calibration problem. It turns out that two-sided simultaneous tolerance intervals can be inverted to obtain a conservative solution to this problem. The results in this section are for the simple linear regression models (3.1.7) and (3.1.8). We shall first describe the condition to be satisfied by multiple use confidence intervals in the calibration problem.

Suppose x in the model (3.1.8) is unknown, and a confidence interval is to be computed for x after observing $Y(x)$. Let $\widehat{\beta}_0$ and $\widehat{\beta}_1$ denote the least squares estimators of β_0 and β_1, based on the Y_i's in (3.1.7) and let S^2 denote the residual mean square. Then

$$
\widehat{\beta}_1 = \frac{\sum_{i=1}^{n}(Y_i - \bar{Y})(x_i - \bar{x})}{\sum_{i=1}^{n}(x_i - \bar{x})^2}, \ \widehat{\beta}_0 = \bar{Y} - \widehat{\beta}_1\bar{x} \text{ and } S^2 = \frac{1}{n-2}\sum_{i=1}^{n}(Y_i - \widehat{\beta}_0 - \widehat{\beta}_1 x_i)^2,
$$

where $\bar{Y}$ and $\bar{x}$ denote the averages of the Y_i's and the x_i's ($i = 1, 2, ..., n$). Essentially, it is required to construct confidence intervals for the unknown x-values corresponding to an unlimited sequence of observations $Y(x)$, following the model (3.1.8). The sequence of confidence intervals will be constructed using the same calibration data (or equivalently, using the same estimates of β_0, β_1 and σ^2, namely, $\widehat{\beta}_0$, $\widehat{\beta}_1$ and S^2, respectively). We shall assume that the interval is of the form

$$\{x : \widehat{\beta}_0 + \widehat{\beta}_1 x - k(x)S \le Y(x) \le \widehat{\beta}_0 + \widehat{\beta}_1 x + k(x)S\}, \qquad (3.4.1)$$

where $k(x)$ is a factor to be determined subject to the requirements of a multiple use confidence interval. Let $C(x; \widehat{\beta}_0, \widehat{\beta}_1, S)$ denote the coverage of the confidence interval (3.4.1), given that the same calibration data will be used, i.e., conditionally given $\widehat{\beta}_0$, $\widehat{\beta}_1$ and S. Then

$$C(x; \widehat{\beta}_0, \widehat{\beta}_1, S) = P_{Y(x)}\left[\widehat{\beta}_0 + \widehat{\beta}_1 x - k(x)S \le Y(x) \le \widehat{\beta}_0 + \widehat{\beta}_1 x + k(x)S | \widehat{\beta}_0, \widehat{\beta}_1, S\right].$$

We want the sequence of confidence intervals (3.4.1), corresponding to an unlimited sequence of observations $Y(x)$, to have the following property: the proportion of intervals that will contain the corresponding true x-value is to be at least p, with confidence level $1 - \alpha$. In other words, the factor $k(x)$ is to be chosen so as to satisfy the condition

$$P_{\widehat{\beta}_0, \widehat{\beta}_1, S}\left(\frac{1}{N}\sum_{i=1}^{N} C(x_i; \widehat{\beta}_0, \widehat{\beta}_1, S) \ge p\right) = 1 - \alpha, \qquad (3.4.2)$$

for every N and for every sequence $\{x_i\}$.

The condition (3.4.2) is a rather difficult condition to work with. A sufficient condition for (3.4.2) to hold is

$$P_{\widehat{\beta}_0, \widehat{\beta}_1, S}\left(\min_x C(x; \widehat{\beta}_0, \widehat{\beta}_1, S) \ge p\right) = 1 - \alpha,$$

which is the condition satisfied by a simultaneous tolerance interval. Thus, one approach for deriving a multiple use confidence interval is the following. Obtain the function $k(x)$ so that $\left[\widehat{\beta}_0 + \widehat{\beta}_1 x - k(x)S, \widehat{\beta}_0 + \widehat{\beta}_1 x + k(x)S\right]$ is a simultaneous tolerance interval for $Y(x)$. After observing $Y(x)$, the interval (3.4.1) can be obtained, which will provide a multiple use confidence interval for x. This approach has been exploited by several authors for computing multiple use confidence intervals in the calibration problem; in particular, see Lieberman, Miller and Hamilton (1967), Scheffe (1973), Oden (1973), Mee, Eberhardt and Reeve (1991) and Mee and Eberhardt (1996).

Mee and Eberhardt (1996) conjectured that if $k(x)$ is the tolerance factor (and not a simultaneous tolerance factor), then the confidence interval (3.4.1) computed using such a $k(x)$ will satisfy the condition (3.4.2) required of multiple use confidence intervals. The limited numerical results in Mee and Eberhardt (1996) support this. More extensive numerical computations carried by Lee

(1999) also support the above assertion, even though no theoretical proof is available for this. The above conjecture is clearly of interest since multiple use confidence intervals constructed based on a tolerance factor will be less conservative compared to those constructed using a simultaneous tolerance factor.

If one-sided multiple use confidence intervals are required for x, one can think of using

$$\{x : Y(x) \leq \widehat{\beta}_0 + \widehat{\beta}_1 x + k(x)S\},$$

or

$$\{x : Y(x) \geq \widehat{\beta}_0 + \widehat{\beta}_1 x - k(x)S\}.$$

However, it appears that the sign of β_1 needs to be known before we can compute such intervals. For example, suppose it is known that $\beta_1 > 0$. Then we expect $\widehat{\beta}_1 > 0$, and a multiple use lower confidence limit for x can be obtained as

$$\{x : Y(x) \leq \widehat{\beta}_0 + \widehat{\beta}_1 x + k(x)S\}.$$

This will be illustrated in the following example. It should be noted that while addressing hypothesis testing for calibration problems in a multiple use scenario, Krishnamoorthy, Kulkarni and Mathew (2002) developed procedures that require a knowledge of the sign of β_1. In practical applications, it is certainly realistic to assume that the sign of β_1 is known.

Example 3.2 (continued)

For this example, we shall find (0.90, 0.95) one-sided multiple use lower confidence limit for the unknown x-values. We computed $\widehat{\beta}_0 = 0.00135$, $\widehat{\beta}_1 = 0.958$, $\bar{x} = 0.1178$, $s_{xx} = \sum_{i=1}^{n}(x_i - \bar{x})^2 = 0.03859$ and $S = 0.01366$. Furthermore, when $(n, m - m_1, p, 1 - \alpha) = (15, 1, 0.90, 0.95)$ and $\tau = 2$, the value of λ in (3.2.18) can be found in Table B7, Appendix B, and it is 1.2618. The one-sided upper simultaneous tolerance limit from Section 3.2 is given by

$$
\begin{aligned}
Y(x) \ &\leq \ \widehat{\beta}_0 + x\widehat{\beta}_1 + \lambda[z_p + (m - m_1 + 3)^{\frac{1}{2}} d] \\
&= \ \widehat{\beta}_0 + x\widehat{\beta}_1 + \lambda \left[z_p + 2 \left(\frac{1}{n} + \frac{(x - \bar{x})^2}{s_{xx}} \right)^{\frac{1}{2}} \right]. \quad (3.4.3)
\end{aligned}
$$

Let us consider the observed value $Y(x) = 0.12$. Using this $Y(x)$-value in (3.4.3), and solving the inequality for x, we get the (0.90, 0.95) one-sided multiple use lower limit for x as 0.091; when $Y(x) = 0.10$, this limit is 0.067, and when $Y(x) = 0.08$ it is 0.043.

To compute two sided (0.90, 0.95) multiple use confidence intervals for the x-values, we need to find the set of values of x that satisfies

$$\widehat{\beta}_0 + x\widehat{\beta}_1 - \lambda\,[z_p + 2d] \le Y(x) \le \widehat{\beta}_0 + x\widehat{\beta}_1 + \lambda\,[z_p + 2d] \qquad (3.4.4)$$

When $(n, p, 1 - \alpha) = (15, 1, 0.90, 0.95)$ and $\tau = 2$, the value of λ in (3.3.18) can be found in Table B8, Appendix B, and it is 1.2405. For a given $Y(x)$, the set of values of x, for which the inequality in (3.4.4) holds is a multiple use confidence set for the unknown x-value. Solving the inequality numerically, we obtained the (0.90, 0.95) multiple use confidence interval as (0.084, 0.166) when $Y(x) = 0.12$, (0.061, 0.143) when $Y(x) = 0.10$ and (0.036, 0.121) when $Y(x) = 0.08$.

3.5   Exercises

3.5.1. Consider the simple linear regression model

$$Y_i = \beta_0 + \beta_1 x_i + e_i, \ \ e_i \sim N(0, \sigma^2), \ \ i = 1, ..., n.$$

Let $Y(x) = \beta_0 + \beta_1 x + e$, where $e \sim N(0, \sigma^2)$ independently of the e_i's. Construct a $1 - \alpha$ equal-tailed tolerance interval for the distribution of $Y(x)$ following the approach in Section 2.3.2 of Chapter 2.

3.5.2. Suppose $Y_1, ..., Y_n$ are independent following the simple linear regression model

$$Y_i = \beta_0 + \beta_1 x_i + e_i, \ \ e_i \sim N(0, \sigma^2), \ \ i = 1, ..., n,$$

where the x_i's are known values of a covariate. Also, let

$$Y(x_{0j}) = \beta_0 + \beta_1 x_{0j} + e_{0j}, \ \ e_{0j} \sim N(0, \sigma^2),$$

$j = 1, 2, ..., n_0$, denote n_0 future observations corresponding to the known values x_{0j}, $j = 1, 2, ..., n_0$, of the covariate. Suitably modify the procedures in Section 2.5 of Chapter 2 for computing simultaneous one-sided and two-sided tolerance intervals for $Y(x_{0j})$, $j = 1, 2, ..., n_0$.

3.5.3. For a simple linear regression model, it is decided to compare the responses $Y(x_{01})$ and $Y(x_{02})$ corresponding to two values x_{01} and x_{02}.

   (a) Explain how you will compute one-sided and two-sided tolerance intervals for $Y(x_{01}) - Y(x_{02})$.

   (b) Explain how you will compute an upper confidence limit for $P[Y(x_{01}) > Y(x_{02})]$

3.5.4. Consider the regression models

$$Y_{1i} = \beta_{10} + \beta_{11}x_{1i} + e_{1i}, \; e_{1i} \sim N(0, \sigma_1^2), \; i = 1, ..., n_1$$
$$Y_{2i} = \beta_{20} + \beta_{21}x_{2i} + e_{2i}, \; e_{2i} \sim N(0, \sigma_2^2), \; i = 1, ..., n_2,$$

where the e_{1i}'s and e_{2i}'s are all independent, and the x_{1i}'s and x_{2i}'s denote values of a covariate. Let Y_1 and Y_2 denote future observations based on the first regression model and the second regression model, corresponding to the same value x of the covariate. Explain how you will compute a $(p, 1 - \alpha)$ upper tolerance limit for $Y_1 - Y_2$ using the generalized confidence interval methodology.

3.5.5. Derive the approximation (3.3.8) [*Hint*: Follow the proof of Lemma 10.1 in Chapter 10].

3.5.6. (Fieller's theorem) Consider the models (3.1.7) and (3.1.8), and suppose a single observation $Y(x)$ is available corresponding to an unknown x. Let $\hat{\beta}_0$, $\hat{\beta}_1$ and S^2 be as defined in Section 3.5. Show that

$$\frac{Y(x) - \hat{\beta}_0 - \hat{\beta}_1 x}{S \times \sqrt{1 + \frac{1}{n} + \frac{(x - \bar{x})^2}{\sum_{i=1}^{n}(x_i - \bar{x})^2}}} \sim t_{n-2}.$$

(a) Use the above result to obtain a $1 - \alpha$ confidence region for x.

(b) Show that the region so obtained is a finite interval if the hypothesis $H_0 : \beta_1 = 0$ is rejected by the t test, at significance level α.

3.5.7. (Fieller's theorem for polynomial regression) Let $Y_1, ..., Y_n$ be independently distributed with

$$Y_i = \beta_0 + \sum_{j=1}^{q} \beta_j x_i^j + e_i, \; e_i \sim N(0, \sigma^2), \; i = 1, ..., n, \qquad (3.4.5)$$

where the x_i's are known values of a covariate. Also, let

$$Y(x) = \beta_0 + \sum_{j=1}^{q} \beta_j x^j + e, \; e \sim N(0, \sigma^2) \qquad (3.4.6)$$

denote a future observation corresponding to the value x of the covariate. Explain how you will extend the result in the previous problem to construct a confidence interval for x, based on a pivot statistic that has a t distribution.

3.5.8. In the set up of the polynomial regression model given in the previous problem, explain how you will construct a tolerance interval for $Y(x)$. Also explain how you will construct a multiple use confidence interval for x, based on the tolerance interval, as explained in Section 3.4.

3.5.9. Suppose $Y_1, ..., Y_n$ are independent following the linear regression model

$$Y_i = \beta_0 + \beta_1 x_i + e_i, \ e_i \sim N(0, \sigma^2 x_i), \ i = 1, ..., n,$$

so that the variance is a multiple of the corresponding x_i. Also, let

$$Y(x_0) = \beta_0 + \beta_1 x_0 + e_0, \ e_0 \sim N(0, \sigma^2 x_0).$$

(a) Explain how you will compute one-sided and two-sided tolerance intervals for $Y(x_0)$.

(b) Suppose x_0 is unknown. Explain how you will compute a multiple use confidence interval for x_0.

3.5.10. In the following data, Y represents the electric conductivity measurements corresponding to x, which represents the amount of sodium chloride solution in dionized water. Here the values of x are controlled, and the corresponding electric conductivity measurements were obtained using the Fisher conductivity meter. The purpose is to estimate the sodium chloride solution based on the corresponding electric conductivity measurement. The data are taken from Johnson and Krishnamoorthy (1996) who showed that a linear regression model gives a good fit to the data.

Y	1.6	1.8	2.0	2.2	2.4	2.6	2.8	3.0	3.2	3.4	3.6	3.8
x	0	.5	1.0	1.5	2.0	2.5	3.0	3.5	4.0	4.5	5.0	5.5
Y	3.9	4.1	4.3	4.5	4.6	4.8	5.0	5.1	5.3	5.6	6.0	6.3
x	6.0	6.5	7.0	7.5	8.0	8.5	9.0	9.5	10.0	11.0	12.0	13.0
Y	6.6	6.9	7.2	7.5	7.7	8.2	9.1					
x	14.0	15.0	16.0	17.0	18.0	20.0	24.0					

(a) Verify that the fitted linear regression model is $Y = 1.8904 + 0.3264x$.

(b) Let $Y(x)$ denote an electric conductivity measurement corresponding to an amount x of the sodium chloride solution. Find a $(0.90, 0.95)$ lower tolerance limit for the distribution of $Y(x)$ when $x = 5.3$.

(c) Construct a $(0.95, 0.95)$ tolerance interval for the distribution of $Y(x)$ when $x = 6.5$, and interpret its meaning.

(d) Construct a $(0.95, .095)$ simultaneous tolerance interval for $Y(x)$ when $x = 6.5$, and interpret its meaning.

(e) Find multiple use confidence intervals for x when $Y(x) = 1.8$ and $Y(x) = 3.7$. Use $p = 0.95$ and $1 - \alpha = 0.95$.

Chapter 4

The One-Way Random Model with Balanced Data

4.1   Notations and Preliminaries

Consider an experiment involving a single factor having a levels (assumed to be randomly chosen from a population of levels), with n observations corresponding to each level; i.e., we have balanced data. Let Y_{ij} denote the jth observation corresponding to the ith level, assumed to follow the one-way random model:

$$Y_{ij} = \mu + \tau_i + e_{ij}, \ j = 1, 2, ..., n, \ i = 1, 2, ..., a, \quad (4.1.1)$$

where μ is an unknown general mean, τ_i's represent random effects, and e_{ij}'s represent error terms. It is assumed that τ_i's and e_{ij}'s are all independent having the distributions $\tau_i \sim N(0, \sigma_\tau^2)$ and $e_{ij} \sim N(0, \sigma_e^2)$. Thus, $Y_{ij} \sim N(\mu, \sigma_\tau^2 + \sigma_e^2)$, and σ_τ^2 and σ_e^2 represent the two variance components in the model. Furthermore, Y_{ij} and $Y_{ij'}$ are correlated for $j \neq j'$; in fact the covariance between Y_{ij} and $Y_{ij'}$ is simply the variance component σ_τ^2. Note that the "true value" associated with the ith level is simply $\mu + \tau_i$, having the distribution $\mu + \tau_i \sim N(\mu, \sigma_\tau^2)$. Clearly, the distribution $N(\mu, \sigma_\tau^2 + \sigma_e^2)$ corresponds to the observable random variables, namely the Y_{ij}'s. On the other hand, the distribution $N(\mu, \sigma_\tau^2)$ corresponding to the unobservable true values, namely the $(\mu + \tau_i)$'s. The problems addressed in this chapter are the following:

(i) The computation of lower or upper tolerance limits and two-sided tolerance intervals for the distribution $N(\mu, \sigma_\tau^2 + \sigma_e^2)$, corresponding to the observed values.

(ii) The computation of lower or upper tolerance limits and two-sided tolerance intervals for the distribution $N(\mu, \sigma_\tau^2)$, corresponding to the unobservable true values.

A lower or upper tolerance limit for $N(\mu, \sigma_\tau^2 + \sigma_e^2)$, or for $N(\mu, \sigma_\tau^2)$, is obviously a confidence limit for the appropriate percentile of the corresponding distribution. Thus if p denotes the content and $1 - \alpha$ denotes the confidence level of the tolerance interval, then a $(p, 1 - \alpha)$ lower tolerance limit for $N(\mu, \sigma_\tau^2 + \sigma_e^2)$ is simply a $1 - \alpha$ lower confidence limit for $\mu - z_p\sqrt{\sigma_\tau^2 + \sigma_e^2}$, where z_p denotes the p quantile of the standard normal distribution. Similarly, a $(p, 1 - \alpha)$ lower tolerance limit for $N(\mu, \sigma_\tau^2)$ is simply a $1 - \alpha$ lower confidence limit for $\mu - z_p\sigma_\tau$. The corresponding $(p, 1 - \alpha)$ upper tolerance limits are $1 - \alpha$ upper confidence limits for $\mu + z_p\sqrt{\sigma_\tau^2 + \sigma_e^2}$ and $\mu + z_p\sigma_\tau$, respectively.

Define

$$\bar{Y}_{i.} = \frac{1}{n}\sum_{j=1}^{n} Y_{ij}, \quad \bar{Y}_{..} = \frac{1}{an}\sum_{i=1}^{a}\sum_{j=1}^{n} Y_{ij}, \quad SS_\tau = n\sum_{i=1}^{a}(\bar{Y}_{i.} - \bar{Y}_{..})^2$$

$$\text{and } SS_e = \sum_{i=1}^{a}\sum_{j=1}^{n}(Y_{ij} - \bar{Y}_{i.})^2. \tag{4.1.2}$$

Table 4.1 gives the ANOVA and the expected mean squares under the model (4.1.1).

Table 4.1: ANOVA table for the model (4.1.1)

Source	Sum of squares (SS)	df	Mean square (MS)	Expected mean square
Factor	SS_τ	$a - 1$	$MS_\tau = \frac{SS_\tau}{a-1}$	$n\sigma_\tau^2 + \sigma_e^2$
Error	SS_e	$a(n-1)$	$MS_e = \frac{SS_e}{a(n-1)}$	σ_e^2

We note that $\bar{Y}_{..}$, SS_τ and SS_e are independently distributed with

$$Z = \sqrt{an}\frac{(\bar{Y}_{..} - \mu)}{\sqrt{n\sigma_\tau^2 + \sigma_e^2}} \sim N(0,1),$$

$$U_\tau^2 = \frac{SS_\tau}{n\sigma_\tau^2 + \sigma_e^2} \sim \chi_{a-1}^2,$$

$$\text{and } U_e^2 = \frac{SS_e}{\sigma_e^2} \sim \chi_{a(n-1)}^2, \tag{4.1.3}$$

where χ_r^2 denotes the chi-square distribution with r degrees of freedom. The ANOVA estimators of σ_τ^2 and σ_e^2, say $\hat{\sigma}_\tau^2$ and $\hat{\sigma}_e^2$, respectively, are given by

$$\hat{\sigma}_\tau^2 = \frac{1}{n}\left(MS_\tau - MS_e\right) \quad \text{and} \quad \hat{\sigma}_e^2 = MS_e, \qquad (4.1.4)$$

where MS_τ and MS_e are the mean squares in Table 4.1. The tolerance limits that we shall construct will be functions of $\bar{Y}_{..}$, SS_τ and SS_e, or equivalently, functions of $\bar{Y}_{..}$, $\hat{\sigma}_\tau^2$ and $\hat{\sigma}_e^2$.

4.2   Two Examples

Example 4.1 (Tensile strength measurements)

This example is taken from Vangel (1992), and is on the tensile strength measurements made on different batches of composite materials, used in the manufacture of aircraft components. Lower tolerance limits for the strength distribution are clearly of practical interest. Since the tensile strength can vary from batch to batch, the batch variability needs to be taken into account while computing the tolerance limits. This calls for the use of a model that involves random effects.

The data given below are measurements made on five batches of composite materials, where each batch consists of five specimens. Thus we have $a = 5$ and $n = 5$. The coded tensile strength data are given in Vangel (1992, Table 4), and are reproduced in Table 4.2. Table 4.3 gives the corresponding ANOVA table.

Table 4.2: The tensile strength data

Batch	Observations				
1	379	357	390	376	376
2	363	367	382	381	359
3	401	402	407	402	396
4	402	387	392	395	394
5	415	405	396	390	395

Reprinted with permission from *Technometrics*. Copyright [1992] by the American Statistical Association.

Referring to the model (4.1.1), the Y_{ij} represents the tensile strength measurement of the jth specimen from the ith batch, $i, j = 1, ..., 5$, and the τ_i's represent the random batch effects. It is required to compute a lower tolerance limit for the tensile strength; i.e., for the distribution $N(\mu, \sigma_\tau^2 + \sigma_e^2)$. One can also

Table 4.3: ANOVA table for the tensile strength data in Table 4.2

Source	DF	SS	MS	F-ratio	P-value
Factor	4	4163.4	1040.8	13.19	0.000
Error	20	1578.4	78.9		
Total	24	5741.8			

compute a lower tolerance limit for the true value $\mu + \tau_i$ of the tensile strength; i.e., for the distribution $N(\mu, \sigma_\tau^2)$.

Example 4.2 (Breaking strengths of cement briquettes)

This example is taken from Bowker and Lieberman (1972), and is on the study of breaking strengths (pounds tension) of cement briquettes. Five measurements each were obtained on nine batches of briquettes. Thus $a = 9$, $n = 5$. The data are given in Bowker and Lieberman (1972, p. 439), and Table 4.4 gives the corresponding ANOVA table.

For this example, Y_{ij} represents the breaking strength measurement on the jth briquette from the ith batch, $i = 1, ..., 9$, $j = 1, ..., 5$, and the τ_i's represent the random batch effects. It is required to compute an upper tolerance limit or a two-sided tolerance interval for the distribution of the breaking strength, i.e., for the distribution $N(\mu, \sigma_\tau^2 + \sigma_e^2)$, or for the distribution of the true breaking strength $\mu + \tau_i$, i.e., for the distribution $N(\mu, \sigma_\tau^2)$.

Later in the chapter, we shall return to both of these examples.

Table 4.4: ANOVA table for the breaking strength data (Bowker and Lieberman, 1972, p. 439)

Source	DF	SS	MS	F-ratio	P-value
Factor	8	5037	630	1.20	0.328
Error	36	18918	526		
Total	44	23955			

4.3 One-Sided Tolerance Limits for $N(\mu, \sigma_\tau^2 + \sigma_e^2)$

A motivation for deriving one-sided tolerance limits under the model (4.1.1) is evident in Fertig and Mann (1974), who discuss the point estimation of the percentiles of the Y_{ij}. The application considered by Fertig and Mann (1974) deals with the strength of a material when sampling from various batches of the material, and the one-way random model is appropriate. The first attempt to formally derive a lower tolerance limit for the distribution $N(\mu, \sigma_\tau^2 + \sigma_e^2)$ appears to be due to Lemon (1977). The tolerance limit due to Lemon (1977) turns out to be quite conservative, as noted by Mee and Owen (1983), who also derived a less conservative tolerance limit. Further work in the same direction is due to Vangel (1992), and Krishnamoorthy and Mathew (2004). More recently, Chen and Harris (2006) proposed a numerical approach to compute tolerance limits. A few of the available solutions are reviewed below.

4.3.1 The Mee-Owen Approach

Suppose we are interested in a $(p, 1 - \alpha)$ lower tolerance limit for $N(\mu, \sigma_\tau^2 + \sigma_e^2)$, i.e., a $1 - \alpha$ lower confidence limit for $\mu - z_p\sqrt{\sigma_\tau^2 + \sigma_e^2}$, where z_p denotes the p quantile of the standard normal distribution. Assume that such a tolerance limit is of the form $\bar{Y}_{..} - k\sqrt{\hat{\sigma}_\tau^2 + \hat{\sigma}_e^2}$, where $\hat{\sigma}_\tau^2$ and $\hat{\sigma}_e^2$ are the ANOVA estimators given in (4.1.4) and k is the tolerance factor to be determined. Thus, k should satisfy the condition

$$P_{\bar{Y}_{..}, \hat{\sigma}_\tau^2, \hat{\sigma}_e^2}\left(\bar{Y}_{..} - k\sqrt{\hat{\sigma}_\tau^2 + \hat{\sigma}_e^2} \le \mu - z_p\sqrt{\sigma_\tau^2 + \sigma_e^2}\right) = 1 - \alpha. \tag{4.3.1}$$

We now use the Satterthwaite approximation to get an approximate chi-square distribution associated with $\hat{\sigma}_\tau^2 + \hat{\sigma}_e^2$. Write

$$R = \frac{\sigma_\tau^2}{\sigma_e^2} \quad \text{and} \quad R_0 = \frac{\sigma_\tau^2 + \sigma_e^2}{n\sigma_\tau^2 + \sigma_e^2} = \frac{R + 1}{nR + 1}. \tag{4.3.2}$$

Note that

$$\hat{\sigma}_\tau^2 + \hat{\sigma}_e^2 = \frac{SS_\tau}{n(a - 1)} + \frac{SS_e}{an}$$

and $\frac{SS_\tau}{n\sigma_\tau^2 + \sigma_e^2}$ and $\frac{SS_e}{\sigma_e^2}$ have the independent chi-square distributions given in (4.1.3). Using the Satterthwaite approximation in Result 1.2.2 of Chapter 1, we conclude that

$$\frac{\hat{\sigma}_\tau^2 + \hat{\sigma}_e^2}{\sigma_\tau^2 + \sigma_e^2} \sim \frac{\chi_f^2}{f},$$

where

$$f = \frac{(R + 1)^2}{\left\{\left(R + \frac{1}{n}\right)^2 / (a - 1)\right\} + \left\{\left(1 - \frac{1}{n}\right) / (an)\right\}}. \tag{4.3.3}$$

Thus $\hat{\sigma}_\tau^2 + \hat{\sigma}_e^2$ is distributed as $(\sigma_\tau^2 + \sigma_e^2)\chi_f^2/f$, approximately, and using the definition of Z in (4.1.3), equation (4.3.1) can be approximately expressed as

$$P_{Z,\chi_f^2}\left(\frac{Z + z_p\sqrt{anR_0}}{\chi_f^2/f} \le k\sqrt{anR_0}\right) = 1 - \alpha, \qquad (4.3.4)$$

R_0 being the quantity in (4.3.2). Since $\frac{Z+z_p\sqrt{anR_0}}{\chi_f^2/f}$ follows a noncentral t-distribution (approximately) with f degrees of freedom, and noncentrality parameter $z_p\sqrt{anR_0}$, k satisfying (4.3.4) is given by

$$k = \frac{1}{\sqrt{anR_0}} \times t_{f;1-\alpha}\left(z_p\sqrt{anR_0}\right), \qquad (4.3.5)$$

where $t_{r;1-\alpha}(\delta)$ denotes the $1 - \alpha$ quantile of a noncentral t-distribution with r degrees of freedom and noncentrality parameter δ. The derivation of (4.3.5) is of course similar to that of (2.2.3) in Chapter 2. Note that R_0 in (4.3.2) is a function of R, which is usually unknown; thus the above tolerance factor needs modification before it can be used in practice. Mee and Owen (1983) recommend to replace R with a $1 - \gamma$ upper confidence limit, where γ is to be determined so that coverage probability is at least $1 - \alpha$. Since

$$\frac{a(n-1)}{a-1} \times \frac{SS_\tau/(n\sigma_\tau^2 + \sigma_e^2)}{SS_e/\sigma_e^2} = \frac{1}{nR+1}\frac{MS_\tau}{MS_e}$$

follows a central F-distribution with degrees of freedom $(a - 1, a(n - 1))$, a $1 - \gamma$ upper confidence limit for $nR+1$ is given by $(F_{a-1,a(n-1);\gamma})^{-1}\frac{MS_\tau}{MS_e}$, where $F_{a-1,a(n-1);\gamma}$ denotes the γ quantile of a central F-distribution with $(a-1, a(n-1))$ degrees of freedom, and MS_τ and MS_e denote the mean squares in Table 4.1. Noting that $R \ge 0$, a $1 - \gamma$ upper confidence limit for R is thus given by

$$R^* = \max\left\{0, \frac{1}{n}\left(\frac{1}{F_{a-1,a(n-1);\gamma}}\frac{MS_\tau}{MS_e} - 1\right)\right\}. \qquad (4.3.6)$$

Once a value of γ is decided and R^* is computed using (4.3.6), the Mee-Owen (1983) tolerance factor, say k_{MO}, can be determined from (4.3.5) as

$$k_{MO} = \frac{1}{\sqrt{anR_0^*}} \times t_{f^*,1-\alpha}\left(z_p\sqrt{anR_0^*}\right), \quad \text{where} \quad R_0^* = \frac{R^* + 1}{nR^* + 1}, \qquad (4.3.7)$$

and f^* is an estimate of f that can be obtained by replacing the R in (4.3.3) by R^*. The value of γ depends on p and $1 - \alpha$, and Mee and Owen (1983) have provided a table of values of $1 - \gamma$, numerically determined, and they are reproduced in Table 4.5.

Table 4.5: Values of $1 - \gamma$ for computing R^* in (4.3.6) for $p = 0.90$, 0.95 and 0.99, and $1 - \alpha = 0.90$, 0.95 and 0.99

	$1 - \alpha$		
p	.90	.95	.99
.90	.78	.85	.94
.95	.79	.86	.95
.99	.81	.875	.96

Once the tolerance factor k_{MO} is computed as in (4.3.7), the Mee-Owen (1983) lower tolerance limit for $N(\mu, \sigma_\tau^2 + \sigma_e^2)$ is given by

$$\bar{Y}_{..} - k_{MO}\sqrt{\hat{\sigma}_\tau^2 + \hat{\sigma}_e^2}. \tag{4.3.8}$$

The corresponding upper tolerance limit is given by

$$\bar{Y}_{..} + k_{MO}\sqrt{\hat{\sigma}_\tau^2 + \hat{\sigma}_e^2}. \tag{4.3.9}$$

The values of $1 - \gamma$ in Table 4.5 have been determined by Mee and Owen (1983) so that the tolerance limits are least conservative for a wide range of values of a, n and R.

4.3.2 Vangel's Approach

In order to obtain less conservative tolerance limits, Vangel (1992) uses a different representation for (4.3.2), avoiding the Satterthwaite approximation. Note that

$$
\begin{aligned}
\frac{\hat{\sigma}_\tau^2 + \hat{\sigma}_e^2}{\sigma_\tau^2 + \sigma_e^2} &= \frac{1}{\sigma_\tau^2 + \sigma_e^2}\left[\frac{SS_\tau}{n(a-1)} + \frac{SS_e}{an}\right] \\
&= \frac{1}{\sigma_\tau^2 + \sigma_e^2}\left[\frac{1}{n(a-1)}\frac{U_\tau^2}{}(n\sigma_\tau^2 + \sigma_e^2) + \frac{U_e^2}{an}\sigma_e^2\right] \\
&= \frac{n\sigma_\tau^2 + \sigma_e^2}{\sigma_\tau^2 + \sigma_e^2}\left[\frac{1}{n(a-1)}U_\tau^2 + \frac{U_e^2}{an}\frac{1}{(nR+1)}\right] \\
&= \frac{nR+1}{R+1}\left[\frac{1}{n(a-1)}U_\tau^2 + \frac{U_e^2}{an}\frac{1}{(nR+1)}\right],
\end{aligned}
$$

where U_τ^2 and U_e^2 are the chi-square random variables given in (4.1.3), and R is the variance ratio given in (4.3.2). Note that

$$U_\tau^2 + U_e^2 \sim \chi_{an-1}^2, \text{ and } B = \frac{U_\tau^2}{U_\tau^2 + U_e^2} \sim \text{Beta}\left(\frac{a-1}{2}, \frac{a(n-1)}{2}\right), \tag{4.3.10}$$

and $U_\tau^2 + U_e^2$ and B are independently distributed. The left hand side of (4.3.1) can be simplified as

$$P_{\bar{Y}_{..},\,\hat{\sigma}_\tau^2,\,\hat{\sigma}_e^2}\left[\bar{Y}_{..} - k\sqrt{\hat{\sigma}_\tau^2 + \hat{\sigma}_e^2} \le \mu - z_p\sqrt{\sigma_\tau^2 + \sigma_e^2}\right]$$

$$= P_{Z,\,U_\tau^2,\,U_e^2}\left[\frac{Z + z_p\sqrt{anR_0}}{\sqrt{\frac{aU_\tau^2}{a-1} + \frac{U_e^2}{nR+1}}} \le k\right]$$

$$= P_{Z,\,U_\tau^2,\,U_e^2}\left[\frac{(Z + z_p\sqrt{anR_0})/\sqrt{\frac{U_\tau^2+U_e^2}{an-1}}}{\sqrt{an-1}\sqrt{\frac{a}{a-1}\frac{U_\tau^2}{U_\tau^2+U_e^2} + \frac{U_e^2}{U_\tau^2+U_e^2}\frac{1}{nR+1}}} \le k\right]$$

$$= P_{Z,\,U_\tau^2,\,U_e^2}\left[\frac{(Z + z_p\sqrt{anR_0})/\sqrt{\frac{U_\tau^2+U_e^2}{an-1}}}{\sqrt{an-1}\sqrt{\frac{a}{a-1}B + (1-B)\frac{1}{nR+1}}} \le k\right]$$

$$= P_{Z,\,U_\tau^2,\,U_e^2}\left[\frac{Z + z_p\sqrt{anR_0}}{\sqrt{\frac{U_\tau^2+U_e^2}{an-1}}} \le k\left(\sqrt{an-1}\sqrt{\frac{a}{a-1}B + (1-B)\frac{1}{nR+1}}\right)\right]$$

$$= E_B\left\{\mathcal{F}_{an-1}\left[k\sqrt{an-1}\left(\frac{a}{a-1}B + (1-B)\frac{1}{nR+1}\right)^{\frac{1}{2}}; z_p\sqrt{anR_0}\right]\right\},$$

$$(4.3.11)$$

where R_0 is defined in (4.3.2), B denotes the beta random variable in (4.3.10), $\mathcal{F}_r(x;\delta)$ denotes the cdf of a noncentral t random variable with r degrees of freedom and noncentrality parameter δ, and we have also used the fact that

$$\frac{Z + z_p\sqrt{anR_0}}{\sqrt{\frac{U_\tau^2+U_e^2}{an-1}}} \sim t_{an-1}\left(z_p\sqrt{anR_0}\right).$$

Vangel's (1992) procedure consists of obtaining k as a function of MS_τ/MS_e so that the last expression in (4.3.11) is equal to $1-\alpha$, approximately. Towards this, Vangel (1992) first provides an approximate solution for the tolerance factor k, and then provides a further improvement. The approximate solution is obtained by considering an asymptotic approximation to the tolerance factor as a function of $\frac{MS_\tau}{MS_e}$, and then approximating it further using the expression k_{V1} given below:

$$k_{V1} = \begin{cases} \dfrac{k_{an} - \frac{k_a}{\sqrt{n}} + (k_a - k_{an})W}{1 - \frac{1}{\sqrt{n}}} & \text{for } \frac{MS_\tau}{MS_e} > 1 \\[2ex] k_{an} & \text{for } \frac{MS_\tau}{MS_e} \le 1, \end{cases} \qquad (4.3.12)$$

where

$$W = \left[1 + \frac{n-1}{MS_\tau/MS_e}\right]^{-1/2},$$

and

$$k_r = \frac{1}{\sqrt{r}} t_{r-1;1-\alpha}(z_p \sqrt{r})$$

is the tolerance factor computed for a simple random sample of size r from a normal population (see (2.2.3) of Chapter 2). In (4.3.12) it is worth noting that when $MS_\tau/MS_e \leq 1$, this can be taken as evidence to support the conclusion $\sigma_\tau^2 = 0$. When this is the case, the Y_{ij}'s form a simple random sample from $N(\mu, \sigma_e^2)$. Consequently, the tolerance factor is taken to be k_{an}, the tolerance factor for $N(\mu, \sigma_e^2)$ based on a random sample of size an.

A more accurate tolerance factor has been provided by Vangel (1992) by setting up k as a solution to an integral equation, and then solving it using an iterative scheme based on an approach due to Trickett and Welch (1954), with k_{V1} serving as a starting value. Vangel (1992) has noted that the solution so obtained can be well approximated by a cubic polynomial in the quantity W defined below equation (4.3.12). This gives us the tolerance factor k_{V2} given by

$$k_{V2} = a_1 + a_2 W + a_3 W^2 + a_4 W^3, \qquad (4.3.13)$$

where the coefficients a_1, a_2, a_3 and a_4 (obtained by least squares fit) are tabulated in Vangel's paper for $(p, 1 - \alpha) = (0.90,\ 0.95)$ and $(0.99,\ 0.95)$. We have reproduced the coefficients in Table B9 in Appendix B. The lower tolerance limit due to Vangel is thus given by

$$\bar{Y}_{..} - k\sqrt{\hat{\sigma}_\tau^2 + \hat{\sigma}_e^2}, \qquad (4.3.14)$$

and the corresponding upper tolerance limit is given by

$$\bar{Y}_{..} + k\sqrt{\hat{\sigma}_\tau^2 + \hat{\sigma}_e^2}, \qquad (4.3.15)$$

where the factor k is determined as k_{V1} or k_{V2}, given in (4.3.12) or (4.3.13), respectively. The numerical results in Vangel (1992) show that the use of k_{V2} results in a less conservative tolerance limit compared to the Mee-Owen (1983) tolerance factor.

4.3.3 The Krishnamoorthy-Mathew Approach

The generalized confidence interval idea is used in Krishnamoorthy and Mathew (2004) in order to derive a $1 - \alpha$ lower confidence limit for $\mu - z_p\sqrt{\sigma_\tau^2 + \sigma_e^2}$, which in turn gives a $(p, 1 - \alpha)$ lower tolerance limit for $N(\mu, \sigma_\tau^2 + \sigma_e^2)$. We shall now develop this procedure following the ideas in Section 1.4. Recall that in order to develop a generalized confidence interval, we have to define a generalized pivotal quantity (GPQ) that is a function of the underlying random variables and the corresponding observed values. The random variables of interest to us are $\bar{Y}_{..}$, SS_τ and SS_e; the associated distributions are specified through the quantities Z, U_τ^2 and U_e^2, defined in (4.1.3). Let $\bar{y}_{..}$, ss_τ and ss_e denote the observed values of $\bar{Y}_{..}$, SS_τ and SS_e, respectively. We shall now define a GPQ for $\mu - z_p\sqrt{\sigma_\tau^2 + \sigma_e^2}$. Let

$$
\begin{aligned}
G_1 &= \bar{y}_{..} - \frac{\sqrt{an}(\bar{Y}_{..} - \mu)}{\sqrt{SS_\tau}}\sqrt{\frac{ss_\tau}{an}} \\
&\quad - z_p\left[\left(\frac{\sigma_e^2 + n\sigma_\tau^2}{nSS_\tau}ss_\tau - \frac{\sigma_e^2}{nSS_e}ss_e\right) + \frac{\sigma_e^2}{SS_e}ss_e\right]^{1/2} \\
&= \bar{y}_{..} - \frac{Z}{U_\tau}\sqrt{\frac{ss_\tau}{an}} - \frac{z_p}{\sqrt{n}}\left[\frac{ss_\tau}{U_\tau^2} + (n-1)\frac{ss_e}{U_e^2}\right]^{1/2} \\
&= \bar{x}_{..} + H,
\end{aligned}
$$

$$
\text{where} \quad H = -\frac{Z}{U_\tau}\sqrt{\frac{ss_\tau}{an}} - \frac{z_p}{\sqrt{n}}\left[\frac{ss_\tau}{U_\tau^2} + (n-1)\frac{ss_e}{U_e^2}\right]^{1/2}. \qquad (4.3.16)
$$

From the second expression in (4.3.16), it is clear that the distribution of G_1 is free of any unknown parameters, for fixed values of $\bar{x}_{..}$, ss_τ and ss_e. We also see from the first expression of (4.3.16) that the value of G_1 when $(\bar{Y}_{..}, SS_\tau, SS_e) = (\bar{y}_{..}, ss_\tau, ss_e)$ is $\mu - z_p\sqrt{\sigma_\tau^2 + \sigma_e^2}$, the parameter of interest. In other words, G_1 satisfies the two conditions in (C.1) of Section 1.4. Thus, if $G_{1\alpha}$ denotes the α quantile of G_1, then $G_{1\alpha}$ is a $1 - \alpha$ generalized lower confidence limit for $\mu - z_p\sqrt{\sigma_\tau^2 + \sigma_e^2}$, and hence is also a $(p, 1 - \alpha)$ lower tolerance limit for $N(\mu, \sigma_\tau^2 + \sigma_e^2)$.

The percentile $G_{1\alpha}$ can be estimated by simulation using the following algorithm:

Algorithm 4.1

1. Once the data are obtained, compute the observed values $\bar{y}_{..}$, ss_τ, and ss_e.

2. Let K denote the number of simulation runs. For $i = 1, 2, ..., K$, perform the following steps.

3. Generate independent random variables $Z_i \sim N(0, 1)$, $U_{\tau,i}^2 \sim \chi_{a-1}^2$ and $U_{e,i}^2 \sim \chi_{n(a-1)}^2$.

4. Compute

$$G_{1,i} = \bar{y}_{..} - \frac{Z_i}{U_{\tau,i}} \sqrt{\frac{ss_\tau}{an}} - \frac{z_p}{\sqrt{n}} \left[\frac{ss_\tau}{U_{\tau,i}^2} + (n-1)\frac{ss_e}{U_{e,i}^2} \right]^{1/2}.$$

The α quantile of the $G_{1,i}$ values ($i = 1, 2, ..., K$) gives an estimate of the α quantile of G_1, which in turn gives an estimate of the required $(p, 1-\alpha)$ lower tolerance limit for $N(\mu, \sigma_\tau^2 + \sigma_e^2)$.

It is also possible to compute $T_{1\alpha}$ by numerical integration. Towards this, note from (4.3.16) that

$$G_{1\alpha} = \bar{x}_{..} + \text{the } \alpha \text{ quantile of } H.$$

The quantity H in (4.3.16) can be expressed as

$$H = \left(\frac{ss_\tau}{an(an-1)B} \right)^{\frac{1}{2}} \frac{Z - z_p\sqrt{a}\left(1 + \frac{(n-1)B \times ss_e}{(1-B) \times ss_\tau}\right)^{\frac{1}{2}}}{\sqrt{(U_\tau^2 + U_e^2)/(an-1)}},$$

where B is the beta random variable defined in (4.3.10). Note that B is independent of $U_\tau^2 + U_e^2 \sim \chi_{an-1}^2$. Thus, conditionally given B, H has the representation

$$H = \left(\frac{ss_\tau}{an(an-1)B} \right)^{\frac{1}{2}} t_{an-1}(\delta(B)) \text{ with } \delta(B) = -z_p\sqrt{a}\left(1 + \frac{(n-1)B \times ss_e}{(1-B) \times ss_\tau}\right)^{\frac{1}{2}},$$

and $t_{an-1}(\delta(B))$ denotes the noncentral t random variable with $an-1$ df and noncentrality parameter $\delta(B)$. Let c be the solution of the equation

$$\frac{\Gamma(\frac{an-1}{2})}{\Gamma(\frac{a-1}{2})\Gamma(\frac{a(n-1)}{2})} \int_0^1 P\left(t_{an-1}(\delta(b)) \leq c\sqrt{b}\right) b^{\frac{b-1}{2}-1}(1-b)^{\frac{a(n-1)}{2}-1} db = \alpha.$$

Then $G_{1\alpha} = \bar{x}_{..} + c\sqrt{\frac{ss_\tau}{an(an-1)}}$.

It is also possible to develop an approximation for $G_{1\alpha}$, and this should be of considerable interest from the computational point of view. Towards developing an approximation for $G_{1\alpha}$, note that H in (4.3.16) can also be expressed as

$$H = \sqrt{\frac{ss_T}{a(a-1)n}} \left[\frac{-Z - z_p \left(a + (a-1)\frac{ss_e}{ss_T} \frac{U_T^2/(a-1)}{U_e^2/(a(n-1))} \right)^{1/2}}{\sqrt{U_T^2/(a-1)}} \right]$$

$$= \sqrt{\frac{ss_T}{a(a-1)n}} \left[\frac{-Z - z_p \left(a + (a-1)\frac{ss_e}{ss_T} F \right)^{1/2}}{\sqrt{U_T^2/(a-1)}} \right]$$

$$= \sqrt{\frac{ss_T}{a(a-1)n}} \left[\frac{-Z - z_p \sqrt{a} \left(1 + (n-1)\frac{F}{f_0} \right)^{1/2}}{\sqrt{U_T^2/(a-1)}} \right], \qquad (4.3.17)$$

where $F = \frac{U_T^2/(a-1)}{U_e^2/(a(n-1))}$ follows an F–distribution with $(a-1, a(n-1))$ degrees of freedom, and $f_0 = \frac{ss_T/(a-1)}{ss_e/(a(n-1))}$ is the observed mean square ratio. Based on numerical results, we approximate H in (4.3.17) by replacing the random variable F with $F_{(a-1),a(n-1);\alpha}$, the α quantile of F. Using such an approximation, and after doing straightforward algebra, we get

$$H \overset{d}{\sim} \sqrt{\frac{ss_T}{a(a-1)n}} \left[\frac{-Z - \delta_1}{\sqrt{U_T^2/(a-1)}} \right], \qquad (4.3.18)$$

where $\overset{d}{\sim}$ denotes "approximately distributed as," and

$$\delta_1 = z_p \left(a + (a-1)\frac{ss_e}{ss_T} F_{(a-1),a(n-1);\alpha} \right)^{1/2}. \qquad (4.3.19)$$

This approximation to H has a noncentral t–distribution with $a-1$ df, and noncentrality parameter $-\delta_1$. Hence from (4.3.16) and (4.3.18) it follows that the α quantile of G_1 is approximately equal to

$$\bar{y}_{..} + \sqrt{\frac{ss_T}{a(a-1)n}} t_{a-1;\alpha}(-\delta_1).$$

Since $t_{a-1;\alpha}(-\delta_1) = -t_{a-1;1-\alpha}(\delta_1)$ an approximate $(p, 1-\alpha)$ lower tolerance limit is

$$\bar{y}_{..} - t_{a-1;1-\alpha}(\delta_1)\sqrt{\frac{ss_T}{a(a-1)n}}, \qquad (4.3.20)$$

where δ_1 is given in (4.3.19). Note that the approximation is exact in the limiting case of large σ_T^2; this can be seen by letting the observed mean square ratio ss_T/ss_e become infinite in (4.3.19).

A $(p, 1-\alpha)$ upper tolerance limit for $N(\mu, \sigma_T^2 + \sigma_e^2)$ can be similarly obtained, and has the approximate expression $\bar{y}_{..} + t_{a-1;1-\alpha}(\delta_1)\sqrt{\frac{ss_T}{a(a-1)n}}$.

4.3.4 Comparison of Tolerance Limits

The authors who have investigated the derivation of one-sided tolerance limits reported above, have also numerically investigated the performance of the proposed limits. In fact most of the relevant articles include extensive numerical results on the coverage of the respective tolerance intervals; in some cases they also include the expected values of the tolerance limits. We note that for a lower tolerance limit, the larger the expected value, the better. For an upper tolerance limit, the converse is true. Numerical results on the performance of the different tolerance limits are very often reported as a function of a, n and the value of the intra-class correlation $\sigma_\tau^2/(\sigma_\tau^2 + \sigma_e^2) = \rho$ (say). It is easily verified that the performance of the tolerance limits does not depend on the value of μ. Numerical results indicate that the different tolerance limits exhibit rather similar performance, except the approximate limit (4.3.20). The approximate limit (4.3.20) does not perform satisfactorily for smaller values of ρ, especially when a and/or n is somewhat large. The corresponding tolerance interval tends to be liberal in this case; see the numerical results in Krishnamoorthy and Mathew (2004). The findings in Krishnamoorthy and Mathew (2004) indicate that the approximate limit (4.3.20) is quite satisfactory for $\rho \geq 0.5$.

The examples given below also support the above conclusions. Vangel (1992, pp. 181−182) notes that "For the most part, the differences in the tolerance-limit factors are not large".

4.3.5 Examples

Example 4.1 (continued)

As noted earlier, there are $a = 5$ batches, and each batch consists of $n = 5$ specimens. From the ANOVA Table 4.3, we have $ss_\tau = 4163.4$ and $ss_e = 1578.4$. We also have the observed values

$$\bar{y}_{..} = 388.36, \ \ ms_\tau = 1040.8, \ \ ms_e = 78.9, \ \ \hat{\sigma}_\tau^2 = 192.36, \ \ \text{and} \ \ \hat{\sigma}_e^2 = 78.90,$$

where ms_τ and ms_e are the observed values of MS_τ and MS_e, respectively.

The Mee-Owen approach: From Table 4.5, we get the value of γ as 0.85 when $(p, 1 - \alpha) = (0.90, 0.95)$. The required F critical value to compute R^* in (4.3.6) is $F_{4,20;1-.85} = 0.33665$. Using this critical value, and using the observed values ms_τ and ms_e, we evaluated $R^* = 7.6367$. This gives the value R_0^* in (4.3.7) as 0.22042. The value of f^* in (4.3.7) is 4.8482. Noting that $z_{.90} = 1.2816$, we found the noncentrality parameter in (4.3.7) as 3.0083, and the noncentral t critical value $t_{4.8482;.95}(3.0083) = 7.2118$. Using these values, we computed the tolerance factor k_{MO} in (4.3.7) as $7.2118/\sqrt{5 \times 5 \times 0.22042} = 3.0722$. Thus, the $(0.90, 0.95)$ lower tolerance limit is $\bar{y}_{..} - k_{MO}\sqrt{\hat{\sigma}_\tau^2 + \hat{\sigma}_e^2} = 388.36 - 3.0722 \times 16.4706 = 337.76$.

Vangel's Approach: We first note that $\frac{MS_\tau}{MS_e} = 13.19 > 1$ and $W = 0.87595$. So, we compute the tolerance factor as (see 4.3.12)

$$k_{V1} = \frac{k_{an} - \frac{k_a}{\sqrt{n}} + (k_a - k_{an})W}{1 - \frac{1}{\sqrt{n}}}$$

$$= \frac{1.8381 - \frac{3.4066}{\sqrt{5}} + (3.4066 - 1.8381) \times 0.87595}{1 - \frac{1}{\sqrt{5}}}$$

$$= 3.0546.$$

Thus, the $(0.90, 0.95)$ lower tolerance limit is $\bar{y}_{..} - k_{V1}\sqrt{\hat{\sigma}_\tau^2 + \hat{\sigma}_e^2} = 388.36 - 3.0546\sqrt{192.38 + 78.9} = 338.05$.

To compute the factor k_{V2} in (4.3.13), the coefficients a_1, a_2, a_3 and a_4 can be found from Table B9 in Appendix B, and

$$k_{V2} = 1.598 - 0.638W + 3.984W^2 - 1.537W^3 = 3.0630.$$

This factor leads to the lower tolerance limit $388.36 - 3.0630\sqrt{192.38 + 78.9} = 337.91$.

The Krishnamoorthy-Mathew Approach: To compute the $(0.90, 0.95)$ lower tolerance limit using the generalized variable approach, we need to evaluate the 5th percentile of H in (4.3.17); we evaluated it using Algorithm 4.1 with $K = 10,000$, and 5th percentile of H is estimated as -50.56. Thus, the lower tolerance limit is $388.36 - 50.56 = 337.80$.

To compute the approximate $(0.90, 0.95)$ lower tolerance limit using (4.3.20), we evaluated $\delta_1 = 2.9396$, $t_{4;.95}(2.9396) = 7.7772$ and $\sqrt{\frac{ss_\tau}{a(a-1)n}} = 6.4524$. Thus, the required tolerance limit is $388.36 - 7.7772 \times 6.4524 = 338.18$.

We note that all approaches produced lower tolerance limits that are practically the same.

Example 4.2 (continued)

For this example, note that we have $a = 9$ batches, and each batch consists of $n = 5$ specimens. From ANOVA Table 4.4, we have $ss_\tau = 5037$ and $ss_e = 18918$. Other summary statistics are

$\bar{y}_{..} = 543.2$, $ms_\tau = 630$, $ms_e = 526$ and $\sqrt{\hat{\sigma}_\tau^2 + \hat{\sigma}_e^2} = 23.3838$.

For this example, let us compute (0.95, 0.95) lower tolerance limits using the various approaches.

The Mee-Owen approach: The required γ-value to construct (0.90, 0.95) one-sided tolerance limits is obtained from Table 4.5 as 0.85. The F critical value in (4.3.6) is $F_{8,36;1-.85} = 0.49726$. Using this critical value, and the observed values ms_τ and ms_e, we evaluated $R^* = 0.28173$. This gives the value of 0.53214 for R_0^* in (4.3.7). The value of f^* in (4.3.7) is 35.11. Noting that $z_{.90} = 1.2816$, we found the noncentrality parameter in (4.3.7) as 6.2713, and the noncentral t critical value $t_{35.11;.95}(6.2713) = 8.6779$. Using these values, we computed the tolerance factor k_{MO} in (4.3.7) as $8.6779/\sqrt{9 \times 5 \times 0.53214} = 1.7734$. Thus, the (0.90, 0.95) lower tolerance limit is $\bar{y}_{..} - k_{MO}\sqrt{\hat{\sigma}_\tau^2 + \hat{\sigma}_e^2} = 543.2 - 1.7734 \times 23.3838 = 501.73$.

Vangel's Approach: We first note that $\frac{MS_\tau}{MS_e} = 1.2 > 1$ and $W = 0.4804$. So, we compute the tolerance factor as (see (4.3.12))

$$k_{V1} = \frac{k_{an} - \frac{k_a}{\sqrt{n}} + (k_a - k_{an})W}{1 - \frac{1}{\sqrt{n}}}$$

$$= \frac{1.6689 - \frac{2.4538}{\sqrt{5}} + (2.4538 - 1.6689) \times 0.4804}{1 - \frac{1}{\sqrt{5}}}$$

$$= 1.7160.$$

Thus, the (0.90, 0.95) lower tolerance limit is $\bar{y}_{..} - k_{V1}\sqrt{\hat{\sigma}_\tau^2 + \hat{\sigma}_e^2} = 543.2 - 1.7160 \times 23.3838 = 503.07$.

To compute the factor k_{V2} in (4.3.13), the coefficients a_1, a_2, a_3 and a_4 can be found from Table B9 in Appendix B, and we get

$$k_{V2} = 1.569 - 0.490W + 1.970W^2 - 0.596W^3 = 1.7222.$$

This factor leads to the lower tolerance as $543.2 - 1.7222\sqrt{546.8} = 502.93$.

The Krishnamoorthy-Mathew Approach: To compute the (0.90, 0.95) lower tolerance limit using the generalized variable approach, we need to evaluate the 5th percentile of H in (4.3.17); we evaluated it using Algorithm 4.1 with $K = 10,000$, and the value is -41.72. Thus, the lower tolerance limit is $543.2 - 41.72 = 501.48$.

To compute the approximate (0.90, 0.95) lower tolerance limit using (4.3.20), we evaluated $\delta_1 = 5.5619$, $t_{8;0.95}(5.5619) = 10.1076$ and $\sqrt{\frac{ss_\tau}{a(a-1)n}} = 3.7405$. Thus, the required tolerance limit is $543.2 - 10.1076 \times 3.7405 = 505.39$.

Notice that all the tolerance limits are approximately the same except the one based on the Krishnamoorthy-Mathew approximate approach, which is larger than the other limits. This is anticipated because this approximate approach works satisfactorily only when the intra-class correlation $\rho = \sigma_\tau^2/(\sigma_\tau^2 + \sigma_e^2)$ is somewhat large; otherwise, it provides tolerance limits that are liberal. For this example, the estimated intra-class correlation $\hat{\rho} = \hat{\sigma}_\tau^2/(\hat{\sigma}_\tau^2 + \hat{\sigma}_e^2) = 0.0383$, which is rather small, indicating that ρ is likely to be small. Thus this method produced a tolerance limit that turned out to be larger than the tolerance limits based on the other approaches.

4.3.6   One-Sided Confidence Limits for Exceedance Probabilities

In the context of the one-way random model, an exceedance probability is simply the probability that a random variable Y that follows the one-way random model will exceed a specified limit, say L. Thus if $Y \sim N(\mu, \sigma_\tau^2 + \sigma_e^2)$, then the exceedance probability η is given by $\eta = P(Y > L)$. The parameter η is obviously a function of the parameters μ, σ_τ^2 and σ_e^2 associated with the one-way random model. Here we shall address the problem of computing an upper confidence limit for η. We shall do this using an upper tolerance limit for $N(\mu, \sigma_\tau^2 + \sigma_e^2)$. We are actually using the approach outlined in Section 1.1.3 of Chapter 1; we have already applied this in Chapter 2 in the context of the usual univariate normal distribution $N(\mu, \sigma^2)$.

Inference concerning an exceedance probability is important in industrial hygiene applications. In the area of industrial hygiene, applications of the one-way random effects model for assessing personal exposure levels for a group of workers have been well demonstrated in the literature on occupational exposure assessment; see Krishnamoorthy and Mathew (2002b) and Krishnamoorthy, Mathew and Ramachandran (2007) for references. If multiple measurements are made on each worker, then one should use the random effects model in order to account for the between- and within-worker variability. Inference concerning an exceedance probability is quite important in such applications, since personal exposure levels can be assessed based on the probability that an individual exposure measurement exceeds an occupational exposure limit. In other words, what is required is interval estimation and hypothesis tests concerning an exceedance probability.

Now consider the one-way random model (4.1.1) where we have balanced data Y_{ij}; $j = 1, 2, ..., n$, $i = 1, 2, ..., a$. For $Y \sim N(\mu, \sigma_\tau^2 + \sigma_e^2)$, the exceedance probability simplifies to

$$\eta = P(Y > L) = 1 - \Phi\left(\frac{L - \mu}{\sqrt{\sigma_\tau^2 + \sigma_e^2}}\right),$$

where Φ denotes the standard normal cdf. Here we shall address both the interval estimation and hypothesis testing for η. Suppose we want to test

$$H_0 : \eta \geq A \quad \text{vs} \quad H_1 : \eta < A,$$

for a specified A. Testing this hypothesis is easily seen to be equivalent to testing the following hypotheses concerning the $1 - A$ quantile of $N(\mu, \sigma_\tau^2 + \sigma_e^2)$:

$$H_0 : \mu + z_{1-A}\sqrt{\sigma_\tau^2 + \sigma_e^2} \geq L \quad \text{vs} \quad H_1 : \mu + z_{1-A}\sqrt{\sigma_\tau^2 + \sigma_e^2} < L. \qquad (4.3.21)$$

As noted in Section 1.1.3, the above hypothesis can be tested by comparing an upper tolerance limit for $N(\mu, \sigma_\tau^2 + \sigma_e^2)$ with L, where the content of the tolerance interval is to be $p = 1 - A$, and the confidence level is to be $1 - \alpha$. We reject H_0 if such an upper tolerance limit is less than L. Thus the upper tolerance limits derived earlier in this chapter can be used for carrying out a test for the above hypotheses.

A $1-\alpha$ upper confidence bound for η can be obtained by identifying the set of values of A for which the null hypothesis in (4.3.21) will be accepted. Specifically, the maximum value of A for which the null hypothesis in (4.3.21) is accepted is an upper confidence bound for η. The computation of the upper confidence limit becomes particularly easy for situations where the approximation for the upper tolerance limit, specified in Section 4.3.3 (see equations (4.3.17)–(4.3.19), and the material following these equations), is satisfactory. For a $1 - A$ content and $1 - \alpha$ confidence level upper tolerance limit for $N(\mu, \sigma_\tau^2 + \sigma_e^2)$, the approximation is given by

$$\bar{y}_{..} + t_{a-1;1-\alpha}(\delta_1)\sqrt{\frac{ss_\tau}{a(a-1)n}}, \quad \text{where } \delta_1 = z_{1-A}\left(a + (a-1)\frac{ss_e}{ss_\tau}F_{(a-1),a(n-1);\alpha}\right)^{1/2}. \qquad (4.3.22)$$

Notice that δ_1 given above is a decreasing function of A. Furthermore, $t_{a-1;1-\alpha}(\delta_1)$ is an increasing function of δ_1. As a result, the approximate upper tolerance limit in (4.3.22) is a decreasing function of A. In general, from the definition of the upper tolerance limit, it should be clear that such a limit should be an increasing function of the content (and a lower tolerance limit should be a decreasing function of the content). Thus the required upper confidence bound for η is obtained as the solution (to A) of the equation

$$\bar{y}_{..} + t_{a-1;1-\alpha}(\delta_1)\sqrt{\frac{ss_\tau}{a(a-1)n}} = L, \qquad (4.3.23)$$

where δ_1 is a function of A; see (4.3.22). We note that the above equation can be solved for δ_1, and then for A using the PC calculator *StatCalc* by Krishnamoorthy (2006).

A $100(1-\alpha)\%$ lower confidence bound for η can be similarly obtained as the value of A for which a lower tolerance limit for $N(\mu, \sigma_\tau^2 + \sigma_e^2)$, having content A and confidence level $1 - \alpha$, is equal to L.

If the approximation to the upper tolerance limit is not used, then trial and error is necessary in order to numerically obtain the value of A for which the $(1 - A, 1 - \alpha)$ upper tolerance limit is equal to L. The value of A obtained based on the approximate upper tolerance limit can be used as a starting value for this purpose. The value can be adjusted and the $(1 - A, 1 - \alpha)$ upper tolerance limit can be recomputed until the tolerance limit is approximately equal to the value of L.

Example 4.3 (Styrene exposures at a boat manufacturing plant)

We shall now illustrate the computation of a confidence limit for an exceedance probability using data on styrene exposures (mg/m^3) on laminators at a boat manufacturing plant. The data given below are reproduced from Table C.1 in Lyles, Kupper and Rappaport (1997a); the data reported are the natural logarithm of the exposure measurements. It is of interest to estimate the proportion of workers for whom the exposure exceeds an occupational exposure limit (OEL). For Styrene exposure, the OEL reported in Lyles, Kupper and Rappaport (1997a) is 213 mg/m^3, i.e., 5.3613 on the log-scale. Thus our problem of interest is computation of a lower confidence limit for the exceedance probability $P(Y > 5.3613)$, where Y is the log-transformed exposure measurement that follows a one-way random model. Note that the data reported in Table 4.7 is for $a = 13$ workers, with $n = 3$ observations per worker. The random effect in the model now represents an effect due to the worker. Computations based on the data gave the observed values $\bar{y}_{..} = 4.810$, $ss_\tau = 11.426$ and $ss_e = 14.711$. Choosing $\alpha = 0.05$, we found $F_{12,26;.05} = 0.4015$, and computed δ_1 in (4.3.22) as $\delta_1 = 4.401 \times z_{1-A}$. Using these quantities in (4.3.23), we see that

$$t_{12;.95}(4.401 z_{1-A}) = \frac{L - \bar{y}_{..}}{\sqrt{\frac{ss_\tau}{a(a-1)n}}} = \frac{5.3613 - 4.810}{0.1563} = 3.5272.$$

Solving the above equation for the noncentrality parameter (*StatCalc*, Krishnamoorthy, 2006), we have $4.401 z_{1-A} = 1.450$ or $z_{1-A} = 0.3295$. Thus, $1 - A = \Phi(0.3295)$ or $A = 1 - \Phi(0.3295) = 1 - 0.6291 = 0.3709$, where Φ is the standard normal cdf. Thus, the probability that an exposure measurement exceeds the OEL is at least 0.3709 with confidence 0.95.

Table 4.6: Styrene exposures on laminators at a boat manufacturing plant

Worker	Observations			Worker	Observations		
1	3.071	3.871	2.965	8	4.477	4.807	5.345
2	4.319	4.396	5.045	9	5.060	5.271	5.454
3	5.221	4.876	5.058	10	5.188	4.499	5.340
4	4.572	5.116	5.578	11	5.970	5.660	5.175
5	5.351	3.925	4.217	12	5.619	1.843	5.545
6	5.889	4.893	4.775	13	4.200	5.294	4.945
7	5.192	4.457	5.097				

Reprinted with permission from *Journal of Agricultural, Biological and Environmental Statistics.* Copyright [1997] by the American Statistical Association.

Table 4.7: ANOVA table for the styrene exposure data in Table 4.6

Source	DF	SS	MS	F-ratio	P-value
Factor	12	11.426	0.952	1.68	0.129
Error	26	14.711	0.566		
Total	38	26.136			

4.3.7 One-Sided Tolerance Limits When the Variance Ratio Is Known

In most practical applications, the variance ratio R is unknown; thus a solution for the case of a known R is not of much interest. Nevertheless, it is possible to give a simple exact solution when R is known, and we shall now give such a solution. The situation of a known R is actually addressed by Mee and Owen (1983), Vangel (1992) and Bhaumik and Kulkarni (1996). We simply note that when R is known, the problem can be reduced to that of computing a tolerance factor for a univariate normal distribution, and the solution can then be obtained as discussed in Chapter 2. The solution so obtained is also the solution given in Bhaumik and Kulkarni (1996).

Note that we want to find tolerance intervals for $N(\mu, \sigma_\tau^2 + \sigma_e^2) = N(\mu, \sigma_e^2(R+1))$, and from (4.1.4), we see that $\bar{Y}_{..} \sim N\left(\mu, \frac{\sigma_e^2}{an}(nR+1)\right) = N\left(\mu, \sigma_e^2(R+1)\left(\frac{nR+1}{an(R+1)}\right)\right)$. Furthermore, recall that $\frac{SS_\tau}{n\sigma_\tau^2 + \sigma_e^2} = \frac{SS_\tau}{(nR+1)\sigma_e^2} \sim \chi_{a-1}^2$, and so if we define

$$SS = \frac{SS_\tau}{nR+1} + SS_e,$$

it is clear that $\frac{SS}{\sigma_e^2} \sim \chi_{an-1}^2$. Thus, $\frac{SS(R+1)}{an-1}$ is an unbiased estimator of $\sigma_e^2(R+1)$.

Based on the independent quantities $\bar{Y}_{..}$ and SS, and their distributions, we can now derive a lower or upper tolerance limit for $N(\mu, \sigma_e^2(R+1))$, following the procedures in Chapter 2. Specifically, identifying $\left(\bar{Y}_{..}, \sqrt{\frac{SS(R+1)}{an-1}}, \sigma_e^2(R+1)\right)$ as $(\bar{X}, S, \sigma^2)$ in (2.2.5), we get a $(p, 1-\alpha)$ lower tolerance limit for $N(\mu, \sigma_e^2(R+1))$ as

$$\bar{Y}_{..} - t_{an-1;1-\alpha}\left(z_p\sqrt{\frac{an(R+1)}{nR+1}}\right)\sqrt{\frac{nR+1}{an}}\sqrt{\frac{SS}{an-1}}.$$

A $(p, 1-\alpha)$ upper tolerance limit for $N(\mu, \sigma_e^2(R+1))$ can be obtained by replacing the $-t_{an-1;1-\alpha}$ in the above expression by $+t_{an-1;1-\alpha}$.

4.4 One-Sided Tolerance Limits for $N(\mu, \sigma_\tau^2)$

Recall that a lower tolerance limit for $N(\mu, \sigma_\tau^2)$ is a lower tolerance limit for the distribution of the true values, namely $\mu + \tau_i$ in (4.1.1). Such a problem was considered by Jaech (1984), Mee (1984a) and Wang and Iyer (1994). More recently, Krishnamoorthy and Mathew (2004) have provided a solution based on generalized confidence intervals, and this solution is presented below.

In order to obtain a $(p, 1-\alpha)$ lower tolerance limit for $N(\mu, \sigma_\tau^2)$, we need to obtain a $1-\alpha$ lower confidence limit for $\mu - z_p\sigma_\tau$. Similar to G_1 given in (4.3.16), it is easily verified that G_2 given below is a GPQ for $\mu - z_p\sigma_\tau$:

$$\begin{aligned}
G_2 &= \bar{y}_{..} - \frac{\sqrt{an}(\bar{Y}_{..} - \mu)}{\sqrt{SS_\tau}}\sqrt{\frac{SS_\tau}{an}} - z_p\left[\frac{\sigma_e^2 + n\sigma_\tau^2}{nSS_\tau}SS_\tau - \frac{\sigma_e^2}{nSS_e}SS_e\right]_+^{1/2} \\
&= \bar{y}_{..} - \frac{Z}{\sqrt{U_\tau^2}}\sqrt{\frac{SS_\tau}{an}} + \frac{z_p}{\sqrt{n}}\left[\frac{SS_\tau}{U_\tau^2} - \frac{SS_e}{U_e^2}\right]_+^{1/2}, \qquad (4.4.1)
\end{aligned}$$

where the random variables Z, U_τ^2 and U_e^2 are as defined in (4.1.3), and for any scalar c, $c_+ = \max(c, 0)$. Then $G_{2\alpha}$, the α quantile of G_2, is our $(p, 1-\alpha)$ lower tolerance limit for $N(\mu, \sigma_\tau^2)$. An algorithm similar to Algorithm 4.1 can be used to estimate $G_{2\alpha}$.

As in the previous section, it is possible to develop an approximation for $G_{2\alpha}$. Towards this, note that

$$\frac{Z}{\sqrt{U_\tau^2}}\sqrt{\frac{SS_\tau}{an}} - \frac{z_p}{\sqrt{n}}\left[\frac{SS_\tau}{U_\tau^2} - \frac{SS_e}{U_e^2}\right]_+^{1/2} = \sqrt{\frac{SS_\tau}{k(k-1)n}}\frac{\left[Z - z_p\left\{a - \frac{a-1}{n-1}\frac{ss_e}{ss_\tau}\frac{U_\tau^2/(a-1)}{U_e^2/a(n-1)}\right\}_+^{1/2}\right]}{\sqrt{U_\tau^2/(a-1)}}$$

Arguing as before, the approximate tolerance limit is given by

$$\bar{y}_{..} - t_{a-1;1-\alpha}(\delta_2)\sqrt{\frac{ss_\tau}{a(a-1)n}}, \qquad (4.4.2)$$

where

$$\delta_2 = z_p \left\{ a - \frac{a-1}{n-1}\frac{ss_e}{ss_\tau}F_{a-1,a(n-1);\alpha} \right\}_+^{1/2}, \qquad (4.4.3)$$

and $\{x\}_+ = \max\{x, 0\}$.

Example 4.2 (continued)

Let us compute a $(0.90, 0.95)$ lower tolerance limit for the distribution of true unobservable values $\mu + \tau_i$ using the data in Example 4.2. Recall that $a = 9$, $n = 5$ and from the ANOVA Table 4.4, we have $ss_\tau = 5037$ and $ss_e = 18918$. Other table values are $z_{.90} = 1.2816$ and $F_{8,36;.05} = 0.327324$. Using these values in (4.4.3), we get

$$\delta_2 = 1.2816 \left(9 - \frac{8}{4} \times \frac{18918}{5037} \times 0.327324 \right)^{\frac{1}{2}} = 3.2778.$$

The noncentral critical value in (4.4.2) is $t_{8,.95}(3.2778) = 6.47964$, and $\sqrt{\frac{ss_\tau}{a(a-1)n}} = 3.7405$. Using these quantities in (4.4.2), we computed the $(0.90, 0.95)$ lower tolerance limit as

$$543.2 - 6.47964 \times 3.7405 = 518.96.$$

4.5 Two-Sided Tolerance Intervals for $N(\mu, \sigma_\tau^2 + \sigma_e^2)$

The $(p, 1 - \alpha)$ two-sided tolerance interval is assumed to be of the form $\bar{Y}_{..} \pm k\sqrt{\hat{\sigma}_\tau^2 + \hat{\sigma}_e^2}$, where k is the tolerance factor to be determined. If we consider a random variable Y that follows $N(\mu, \sigma_\tau^2 + \sigma_e^2)$, the condition to be satisfied by k is

$$P_{\bar{Y}_{..}, \hat{\sigma}_\tau^2, \hat{\sigma}_e^2} \left[P_Y \left\{ \bar{Y}_{..} - k\sqrt{\hat{\sigma}_\tau^2 + \hat{\sigma}_e^2} \leq Y \leq \bar{Y}_{..} + k\sqrt{\hat{\sigma}_\tau^2 + \hat{\sigma}_e^2} \,\Big|\, \bar{Y}_{..}, \hat{\sigma}_\tau^2, \hat{\sigma}_e^2 \right\} \geq p \right] = 1 - \alpha,$$

or equivalently,

$$P_{Z, \hat{\sigma}_\tau^2, \hat{\sigma}_e^2} \left[\Phi\left(\frac{Z}{\sqrt{an \times R_0}} + k\sqrt{\frac{\hat{\sigma}_\tau^2 + \hat{\sigma}_e^2}{\sigma_\tau^2 + \sigma_e^2}} \right) \right.$$
$$\left. - \Phi\left(\frac{Z}{\sqrt{an \times R_0}} - k\sqrt{\frac{\hat{\sigma}_\tau^2 + \hat{\sigma}_e^2}{\sigma_\tau^2 + \sigma_e^2}} \right) \geq p \right] = 1 - \alpha, \qquad (4.5.1)$$

where R_0 is defined in (4.3.2), and Φ denotes the standard normal cdf.

The Mee-Owen (1983) approach has been extended by Mee (1984b) to arrive at a two-sided tolerance interval for $N(\mu, \sigma_\tau^2 + \sigma_e^2)$. Later, Beckman and Tietjen (1989) derived a two-sided tolerance interval after replacing the unknown parameter $\frac{1}{anR_0}$ by an upper bound. Note that $\frac{1}{anR_0} = \frac{n\sigma_\tau^2 + \sigma_e^2}{an(\sigma_\tau^2 + \sigma_e^2)}$, and $\frac{1}{a}$ is an upper bound for this quantity. As noted in Beckman and Tietjen (1989), this results in a conservative tolerance interval, unless $\frac{1}{anR_0}$ is close to the upper bound, namely, $\frac{1}{a}$. More recently, Liao and Iyer (2004), and Liao, Lin and Iyer (2005) have derived approximate two-sided tolerance intervals using the generalized confidence interval idea. The approximation used in Liao, Lin and Iyer (2005) appears to be an improvement over that used in Liao and Iyer (2004). It should be noted that the two-sided tolerance interval problem does not reduce to a confidence interval problem concerning percentiles; thus, the Liao, Lin and Iyer (2005) approach is not a straightforward application of the generalized confidence interval idea, even though they succeed in eventually using generalized confidence intervals for a suitable linear combination of the variance components σ_τ^2 and σ_e^2. Here we shall discuss the solutions due to Mee (1984b) and Liao, Lin and Iyer (2005).

4.5.1   Mee's Approach

Mee (1984b) once again uses the Satterthwaite approximation

$$\frac{\widehat{\sigma}_\tau^2 + \widehat{\sigma}_e^2}{\sigma_\tau^2 + \sigma_e^2} \sim \frac{\chi_f^2}{f},$$

where f is given in (4.3.3). Thus, as an approximation, (4.5.1) can be written as

$$P_{Z,\,\chi_f^2}\left[\Phi\left(\frac{Z}{\sqrt{anR_0}} - k\sqrt{\frac{\chi_f^2}{f}}\right) - \Phi\left(\frac{Z}{\sqrt{anR_0}} + k\sqrt{\frac{\chi_f^2}{f}}\right) \geq p\right] = 1 - \alpha. \quad (4.5.2)$$

If R, and hence R_0, are known, then (4.5.2) is the same as equation (1.2.3) in Chapter 1 for a normal distribution. Specifically, an approximate tolerance factor is given by (1.2.5) where c in (1.2.5) is replaced with $\frac{1}{anR_0}$, m is replaced with f, and γ in (1.2.5) is replaced with $1 - \alpha$.

However, since R is unknown, Mee (1984b) recommends using a $1 - \gamma$ level upper confidence bound in the place of R, just as in the case of the lower or upper tolerance limit discussed in Section 4.3. Based on numerical results, the recommended values of $1 - \gamma$ are 0.85, 0.905 and 0.975, corresponding to $1 - \alpha$

$= 0.90$, 0.95 and 0.99, respectively. A $1 - \gamma$ level upper confidence bound for R is the quantity R^* given in (4.3.6), and let R_0^* be as defined in (4.3.7). Then k is computed from

$$P_{Z, \chi_f^2}\left[\Phi\left(\frac{Z}{\sqrt{anR_0^*}} - k\sqrt{\frac{\chi_f^2}{f}}\right) - \Phi\left(\frac{Z}{\sqrt{anR_0^*}} + k\sqrt{\frac{\chi_f^2}{f}}\right) \geq p\right] = 1 - \alpha.$$

This equation is the same as the one in (1.2.3), and so applying Result 1.2.1 (iii), we can get an approximate value of k, say k_M, given by

$$k_M \simeq \left(\frac{f^* \chi_{1;p}^2((anR_0^*)^{-1})}{\chi_{f^*;\alpha}^2}\right)^{\frac{1}{2}}, \qquad (4.5.3)$$

where f^* is as defined for (4.3.7). This factor leads to an approximate two-sided tolerance interval for $N(\mu, \sigma_\tau^2 + \sigma_e^2)$, given by

$$\bar{Y}_{..} \pm k_M \sqrt{\hat{\sigma}_\tau^2 + \hat{\sigma}_e^2}. \qquad (4.5.4)$$

4.5.2 The Liao-Lin-Iyer Approach

Note that in order to derive a two-sided tolerance interval, we have to obtain a margin of error statistic D_1, a function of $\hat{\sigma}_\tau^2$ and $\hat{\sigma}_e^2$, or equivalently a function of SS_τ and SS_e, so that

$$P_{\bar{Y}_{..}, SS_\tau, SS_e}\left[P_Y\{\bar{Y}_{..} - D_1 \leq Y \leq \bar{Y}_{..} + D_1 | \bar{Y}_{..}, SS_\tau, SS_e,\} \geq p\right] = 1 - \alpha, \qquad (4.5.5)$$

where $Y \sim N(\mu, \sigma_\tau^2 + \sigma_e^2)$. Once D_1 is obtained, the tolerance interval is given by $\bar{Y}_{..} \pm D_1$. Write

$$\sigma_1^2 = \sigma_\tau^2 + \sigma_e^2, \quad \text{and} \quad \sigma_2^2 = (n\sigma_\tau^2 + \sigma_e^2)/(an), \qquad (4.5.6)$$

so that $Z = \frac{\bar{Y}_{..} - \mu}{\sigma_2} \sim N(0,1)$; see (4.1.3). Similar to (4.5.1), the condition (4.5.5) can equivalently be expressed as

$$P_{\bar{Y}_{..}, SS_\tau, SS_e}\left[\Phi\left(\frac{\sigma_2 Z + D_1}{\sigma_1}\right) - \Phi\left(\frac{\sigma_2 Z - D_1}{\sigma_1}\right) \geq p\right] = 1 - \alpha. \qquad (4.5.7)$$

For a fixed u, consider the equation

$$\Phi(u+k) - \Phi(u-k) = p,$$

and suppose we want to obtain a solution to k. The following relation has been obtained by Howe (1969):

$$k^2 = z_{\frac{1+p}{2}}^2 \left[1 + u^2 + \frac{3 - z_{\frac{1+p}{2}}^2}{6} u^4 + ... \right].$$

Notice that (4.5.7) is the same as equation (1.2.3). In the proof of Result 1.2.3 (ii), it was noted that $\Phi(u+k) - \Phi(u-k)$ is an increasing function of k. Thus the condition

$$\Phi \left(\frac{\sigma_2 Z + D_1}{\sigma_1} \right) - \Phi \left(\frac{\sigma_2 Z - D_1}{\sigma_1} \right) \geq p$$

is equivalent to

$$D_1^2 \geq \sigma_1^2 z_{\frac{1+p}{2}}^2 \left[1 + Z^2 \frac{\sigma_2^2}{\sigma_1^2} + \frac{\left(3 - z_{\frac{1+p}{2}}^2 \right)}{6} Z^4 \frac{\sigma_2^4}{\sigma_1^4} + ... \right].$$

Thus, (4.5.7) is equivalent to

$$P_{Z,D_1^2} \left\{ D_1^2 \geq \sigma_1^2 z_{\frac{1+p}{2}}^2 \left[1 + Z^2 \frac{\sigma_2^2}{\sigma_1^2} + \frac{\left(3 - z_{\frac{1+p}{2}}^2 \right)}{6} Z^4 \frac{\sigma_2^4}{\sigma_1^4} + ... \right] \right\} = 1 - \alpha. \quad (4.5.8)$$

Following Liao, Lin and Iyer (2005), we replace the powers of Z by their expected values, and keep only two terms. Thus, as an approximation, (4.5.8) can be expressed as

$$P_{D_1^2} \left\{ D_1^2 \geq z_{\frac{1+p}{2}}^2 (\sigma_1^2 + \sigma_2^2) \right\} = 1 - \alpha. \quad (4.5.9)$$

The above equation implies that D_1^2 is a $1 - \alpha$ upper confidence limit for $z_{\frac{1+p}{2}}^2 (\sigma_1^2 + \sigma_2^2)$.

¿From the definition of σ_1^2 and σ_2^2 in (4.5.6), we note that

$$\begin{aligned} \sigma_1^2 + \sigma_2^2 &= \left(1 + \frac{1}{a} \right) \sigma_\tau^2 + \left(1 + \frac{1}{an} \right) \sigma_e^2 \\ &= \left(1 + \frac{1}{a} \right) \frac{n\sigma_\tau^2 + \sigma_e^2}{n} + \left(1 - \frac{1}{n} \right) \sigma_e^2. \end{aligned}$$

A $1 - \alpha$ upper confidence limit for $\sigma_1^2 + \sigma_2^2$ can be obtained using the generalized confidence interval idea. A GPQ for $\sigma_1^2 + \sigma_2^2$ is easily seen to be

$$\left(1 + \frac{1}{a} \right) \frac{1}{n} \frac{n\sigma_\tau^2 + \sigma_e^2}{SS_\tau} ss_\tau + \left(1 - \frac{1}{n} \right) \frac{\sigma_e^2}{SS_e} ss_e = \left(1 + \frac{1}{a} \right) \frac{1}{n} \frac{ss_\tau}{U_\tau^2} + \left(1 - \frac{1}{n} \right) \frac{ss_e}{U_e^2}, \quad (4.5.10)$$

where U_τ^2 and U_e^2 are the chi-square random variables defined in (4.1.3), and ss_τ and ss_e denote the observed values of SS_τ and SS_e, respectively. The $1 - \alpha$ quantile of the above GPQ will provide a $1 - \alpha$ upper confidence limit for $\sigma_1^2 + \sigma_2^2$. The square root of this upper confidence limit, multiplied by $z_{\frac{1+p}{2}}$, provides a margin of error statistic D_1 required for computing the approximate $(p, 1 - \alpha)$ two-sided tolerance interval $\bar{Y}_{..} \pm D_1$.

We note that it is also possible to use the Satterthwaite approximation to obtain a confidence limit for $\sigma_1^2 + \sigma_2^2$. The unbiased estimator of $\sigma_1^2 + \sigma_2^2$ is

$$
\begin{aligned}
\hat{\sigma}_1^2 + \hat{\sigma}_2^2 &= \left(1 + \frac{1}{a}\right)\hat{\sigma}_\tau^2 + \left(1 + \frac{1}{an}\right)\hat{\sigma}_e^2 \\
&= \frac{1}{n}\left(1 + \frac{1}{a}\right)MS_\tau + \left(1 - \frac{1}{n}\right)MS_e,
\end{aligned}
$$

where MS_τ and MS_e are the mean squares given in Table 4.1. Since the above estimator is always nonnegative, we can approximate the distribution of $(\hat{\sigma}_1^2 + \hat{\sigma}_2^2)/(\sigma_1^2 + \sigma_2^2)$ with a scalar multiple of a chi-square; see Section Result 1.2.2 in Chapter 1. Yet another option to compute a confidence limit for $\sigma_1^2 + \sigma_2^2$ is to use the modified large sample procedure described in Section 1.3 of Chapter 1.

Liao and Iyer (2004) have derived a margin of error statistic D_1, using a slightly different approximation than the one given above. However, based on coverage studies, Liao, Lin and Iyer (2005) conclude that the margin of error statistic derived in this section provides a tolerance interval that exhibits somewhat improved performance, compared to the one in Liao and Iyer (2004).

Example 4.1 (continued)

We shall now construct $(p, 1 - \alpha) = (0.90, 0.95)$ two-sided tolerance intervals for the examples in Section 4.1 using the different approaches. For Example 4.1, $a = 5$, $n = 5$, $ss_\tau = 4163.4$, $ss_e = 1578.4$.

Mee's Approach: Since $1 - \alpha = 0.95$, we choose $1 - \gamma = 0.905$. The F critical value to compute R_0^* is $F_{4,20;\gamma} = F_{4,20;.095} = 0.252042$, $R^* = 10.2671$, $R_0^* = 0.21528$. Furthermore, $f^* = 4.8482$, $(anR_0^*)^{-1} = 0.1858$, $\chi_{1;.90}^2(0.1858) = 3.20884$ and $\chi_{4.8482;.05}^2 = 1.0755$. Thus, the tolerance factor

$$
k_M = \left(\frac{4.8482 \times 3.20884}{1.0755}\right)^{\frac{1}{2}} = 3.8033,
$$

and the (0.90, 0.95) tolerance interval is $\bar{y}_{..} \pm k_M\sqrt{\sigma_\tau^2 + \sigma_e^2} = 388.36 \pm 3.8033 \times 16.4706 = 388.36 \pm 62.64$.

The Liao-Lin-Iyer Approach: To apply this approach, we first estimated a 95% generalized upper confidence limit for $\sigma_1^2+\sigma_2^2$ as 1471.57. Monte Carlo simulation consisting of 100,000 runs was applied to (4.5.10) to get this limit. Noting that $z_{\frac{1+p}{2}} = z_{.95} = 1.6449$, we found $D_1 = z_{.95} \times \sqrt{1471.57} = 63.1$. Thus, the (0.90, 0.95) tolerance interval is given by $\bar{y}_{..} \pm D_1 = 388.36 \pm 63.1$.

We note that the results based on both methods are in good agreement.

We can also use the Satterthwaite approximation, or the MLS methodology in order to compute an upper confidence limit for $\sigma_1^2+\sigma_2^2$. We shall now illustrate the application of the MLS methodology. Note that $n\hat{\sigma}_\tau^2 + \hat{\sigma}_e^2 = ss_\tau/(a-1) = 1040.85$ and $\hat{\sigma}_e^2 = ss_e/a(n-1) = 78.92$. Using the definitions of σ_1^2 and σ_2^2, we get

$$\hat{\sigma}_1^2 + \hat{\sigma}_2^2 = \left(1 + \frac{1}{a}\right)\frac{n\hat{\sigma}_\tau^2 + \hat{\sigma}_e^2}{n} + \left(1 - \frac{1}{n}\right)\hat{\sigma}_e^2 = 312.94.$$

We shall now apply formula (1.3.3) of Chapter 1 to get the 95% MLS upper confidence limit for $\sigma_1^2+\sigma_2^2$ (note that the notation σ_i^2 used in (1.3.3) is different from the notation σ_1^2 and σ_2^2 used above). The MLS upper confidence limit is given by

$$\left[\left(1 + \frac{1}{a}\right)\frac{n\hat{\sigma}_\tau^2 + \hat{\sigma}_e^2}{n} + \left(1 - \frac{1}{n}\right)\hat{\sigma}_e^2\right] + \left\{\left(1 + \frac{1}{a}\right)^2 \frac{(n\hat{\sigma}_\tau^2 + \hat{\sigma}_e^2)^2}{n^2}\left(\frac{a-1}{\chi_{a-1;\alpha}^2} - 1\right)^2\right.$$
$$\left. + \left(1 - \frac{1}{n}\right)^2 \hat{\sigma}_e^4\left(\frac{a(n-1)}{\chi_{a(n-1);\alpha}^2} - 1\right)^2\right\}^{1/2}.$$

For $1 - \alpha = 0.95$, $a = 5$ and $n = 5$, we have $\chi_{a-1;\alpha}^2 = 0.7107$ and $\chi_{a(n-1);\alpha}^2 = 10.8508$. The above expression for the 95% MLS upper confidence limit simplifies to 1474.94. We note that this upper limit is very close to the 95% generalized upper confidence limit given above, and hence the resulting tolerance limits are also very close.

Example 4.2 (continued)

Mee's Approach: We shall compute (0.90, 0.95) tolerance intervals for this example. Note that here $a = 9$, $n = 5$, $ss_\tau = 5037$ and $ss_e = 18918$. The F critical value to compute R_0^* is $F_{8,36;.095} = 0.413907$, $R^* = 0.37894$, $R_0^* = 0.47636$ and $f^* = 31.864$. Furthermore, $(anR_0^*)^{-1} = 0.04665$, $\chi_{1;.90}^2(0.04665) = 2.83198$ and $\chi_{31.864;.05}^2 = 19.9641$. Thus, the tolerance factor

$$k_M = \left(\frac{31.864 \times 2.83198}{19.9641}\right)^{\frac{1}{2}} = 2.1260,$$

and the (0.90, 0.95) tolerance interval is $\bar{y}_{..} \pm k_M \sqrt{\hat{\sigma}_\tau^2 + \hat{\sigma}_e^2} = 543.2 \pm 2.1260 \times 23.3838 = 543.2 \pm 49.71$.

Liao-Lin-Iyer Approach: A 95% generalized upper confidence limit for $\sigma_1^2 + \sigma_2^2$ is 932.497. Monte Carlo simulation consisting of 100,000 runs was applied to (4.5.10) to get this limit. Noting that $z_{\frac{1+p}{2}} = z_{.95} = 1.6449$, we found $D_1 = z_{.95} \times \sqrt{932.497} = 50.23$. Thus, the (0.90, 0.95) tolerance interval is given by $\bar{y}_{..} \pm D_1 = 543.2 \pm 50.23$.

Proceeding as with the previous example, a 95% MLS upper confidence limit for $\sigma_1^2 + \sigma_2^2$ comes out to be 986.63. This results in $D_1 = z_{0.95} \times \sqrt{986.63} = 51.67$. Thus, the (0.90, 0.95) tolerance interval is given by $\bar{y}_{..} \pm D_1 = 543.2 \pm 51.67$

We note again that the results are in good agreement.

4.6   Two-Sided Tolerance Intervals for $N(\mu, \sigma_\tau^2)$

We shall use the Liao, Lin and Iyer (2005) approach for deriving a tolerance interval for $N(\mu, \sigma_\tau^2)$. We shall take the tolerance interval to be of the form $\bar{Y}_{..} \pm D_2$, where D_2 is a margin of error statistic to be determined. D_2 will be a function of SS_τ and SS_e and satisfies

$$P_{\bar{Y}_{..},\,SS_\tau,\,SS_e}\left[P_Y \left\{ \bar{Y}_{..} - D_2 \leq \mu + \tau \leq \bar{Y}_{..} + D_2 | \bar{Y}_{..},\, SS_\tau,\, SS_e, \right\} \geq p \right] = 1 - \alpha, \tag{4.6.1}$$

where $\mu + \tau \sim N(\mu, \sigma_\tau^2)$. Following the derivations in the previous section, we conclude that an approximate condition to be satisfied by D_2 is

$$P_{D_2^2}\left\{ D_2^2 \geq z_{\frac{1+p}{2}}^2 (\sigma_\tau^2 + \sigma_2^2) \right\} = 1 - \alpha, \tag{4.6.2}$$

where σ_2^2 is defined in (4.5.6). The above equation implies that D_2^2 is a $1-\alpha$ upper confidence limit for $z_{\frac{1+p}{2}}^2 (\sigma_\tau^2 + \sigma_2^2)$.

¿From the definition of σ_2^2 in (4.5.6), we note that

$$\sigma_\tau^2 + \sigma_2^2 = \left(1 + \frac{1}{a}\right) \frac{n\sigma_\tau^2 + \sigma_e^2}{n} - \frac{\sigma_e^2}{n}.$$

A $1-\alpha$ upper confidence limit for $\sigma_\tau^2 + \sigma_2^2$ can be obtained using the generalized confidence interval idea. A GPQ for $\sigma_\tau^2 + \sigma_2^2$ is easily seen to be

$$\left[\left(1 + \frac{1}{a}\right) \frac{1}{n} \frac{n\sigma_\tau^2 + \sigma_e^2}{SS_\tau} ss_\tau - \frac{1}{n} \frac{\sigma_e^2}{SS_e} ss_e\right]_+ = \left[\left(1 + \frac{1}{a}\right) \frac{1}{n} \frac{ss_\tau}{U_\tau^2} - \frac{1}{n} \frac{ss_e}{U_e^2}\right]_+, \tag{4.6.3}$$

where for any scalar c, we define $c_+ = \max\{c, 0\}$. The $1 - \alpha$ quantile of the above GPQ will provide a $1 - \alpha$ upper confidence limit for $\sigma_\tau^2 + \sigma_2^2$. The square root of this upper confidence limit, multiplied by $z_{\frac{1+p}{2}}$, provides a margin of error statistic D_2 required for computing the approximate $(p, 1 - \alpha)$ two-sided tolerance interval $\bar{Y}_{..} \pm D_2$ for the distribution $N(\mu, \sigma_\tau^2)$.

Note that an unbiased estimator of $\sigma_\tau^2 + \sigma_2^2$ is given by $(1 + \frac{1}{a})\frac{1}{n}MS_\tau - \frac{1}{n}MS_e$, which can assume negative values. Hence its distribution cannot be approximated using a scalar multiple of a chi-square. In other words, it is not appropriate to use the Satterthwaite approximation to obtain an upper confidence limit for $\sigma_\tau^2 + \sigma_2^2$. However, the MLS procedure can be used to obtain such a confidence limit.

Example 4.1 (continued)

We shall illustrate the above method of constructing a tolerance interval for the distribution $N(\mu, \sigma_\tau^2)$ using Example 4.1. Recall that, for this example, $a = 5$, $n = 5$, $ss_\tau = 4163.4$, $ss_e = 1578.4$. To compute a $(0.90, 0.95)$ tolerance interval, we estimated the 95% generalized upper confidence limit for $\sigma_\tau^2 + \sigma_2^2$ as 1391.7; Monte Carlo simulation consisting of 100,000 runs was applied to (4.3.6) to get this limit. The value of D_2 can be obtained as $z_{.95}\sqrt{1391.7} = 61.4$. Thus, the desired tolerance interval is 388.36 ± 61.4.

Since $\hat{\sigma}_\tau^2 + \hat{\sigma}_2^2 = (1 + \frac{1}{a})\frac{n\hat{\sigma}_\tau^2 + \hat{\sigma}_e^2}{n} - \frac{\hat{\sigma}_e^2}{n}$ can assume negative values, it is not appropriate to use the Satterthwaite approximation to obtain an upper confidence bound for $\sigma_\tau^2 + \sigma_2^2$. However, the MLS procedure is applicable. Using the formula (1.3.4) of Chapter 1, the MLS upper bound for $\sigma_\tau^2 + \sigma_2^2$ is given by

$$\left[\left(1 + \frac{1}{a}\right)\frac{n\hat{\sigma}_\tau^2 + \hat{\sigma}_e^2}{n} - \frac{\hat{\sigma}_e^2}{n}\right] + \left\{\left(1 + \frac{1}{a}\right)^2 \frac{(n\hat{\sigma}_\tau^2 + \hat{\sigma}_e^2)^2}{n^2}\left(\frac{a-1}{\chi_{a-1;\alpha}^2} - 1\right)^2\right.$$

$$\left. + \frac{\sigma_e^4}{n^2}\left(\frac{a(n-1)}{\chi_{a(n-1);1-\alpha}^2} - 1\right)^2\right\}^{1/2}.$$

For $1 - \alpha = 0.95$, $a = 5$ and $n = 5$, we have $\chi_{a-1;\alpha}^2 = 0.7107$ and $\chi_{a(n-1);1-\alpha}^2 = 31.410$. Since $n\hat{\sigma}_\tau^2 + \hat{\sigma}_e^2 = ss_\tau/(a-1) = 1040.85$ and $\hat{\sigma}_e^2 = ss_e/a(n-1) = 78.92$, the above expression for the 95% MLS upper confidence limit simplifies to 1390.19, which is almost the same as the 95% generalized upper confidence limit given above. Thus the resulting tolerance limits are also practically the same.

Example 4.2 (continued)

For this example, $a = 9$, $n = 5$, $ss_\tau = 5037$, $ss_e = 18918$. To compute a $(0.90, 0.95)$ tolerance interval, we estimated the 95% generalized upper confidence limit for $\sigma_\tau^2 + \sigma_2^2$ as 302.1; Monte Carlo simulation consisting of 100,000 runs was applied to (4.3.6) to get this limit. The value of D_2 can be obtained as $z_{.95}\sqrt{302.1} = 28.6$. Thus, the desired tolerance interval is 543.2 ± 28.6. The MLS procedure gives the 95% upper confidence limit for $\sigma_\tau^2 + \sigma_2^2$ as 293.61, which is not very different from the above generalized upper confidence limit.

As already noted, in the context of computing two-sided tolerance intervals for $N(\mu, \sigma_\tau^2 + \sigma_e^2)$ and for $N(\mu, \sigma_\tau^2)$, for obtaining the margin of error statistic D_1 or D_2, the required upper confidence limit can be computed using the generalized confidence interval procedure, or using the MLS procedure. However, the numerical results in this section, and in the previous section show that the two upper confidence limits are nearly the same. Thus the MLS procedure can be recommended for practical use since the upper confidence limit has an explicit analytical form, and hence is easily computed, whereas the generalized upper confidence limit has to be estimated by Monte Carlo simulation.

4.7 Exercises

4.7.1. Consider the one-way random model with balanced data.

(a) Suppose it is desired to construct a confidence interval for $\sigma_\tau^2 + \sigma_e^2$, which is also the variance of Y_{ij}. Derive a generalized confidence interval for $\sigma_\tau^2 + \sigma_e^2$.

(b) Derive a generalized confidence interval for the ratio σ_τ^2/σ_e^2. Show that the interval reduces to the exact confidence interval that can be obtained using the F statistic $\frac{\sigma_e^2}{\sigma_\tau^2 + \sigma_e^2} \times \frac{MS_\tau}{MS_e}$.

4.7.2. Use the results in the previous problem to compute a 95% upper confidence limit for $\sigma_\tau^2 + \sigma_e^2$ based on the cement briquette breaking strength data given in Table 4.4. Use 10,000 simulated values of the GPQ.

4.7.3. For the one-way random model with balanced data, suppose the variance ratio $R = \sigma_\tau^2/\sigma_e^2$ is known (see Section 4.3.7). Use the relevant results in Chapter 2 to derive the following:

(a) a two-sided tolerance interval for $N(\mu, \sigma_\tau^2 + \sigma_e^2)$.

(b) one-sided and two-sided tolerance intervals for $N(\mu, \sigma_\tau^2)$.

(c) equal-tailed tolerance intervals for $N(\mu, \sigma_\tau^2 + \sigma_e^2)$ and $N(\mu, \sigma_\tau^2)$.

4.7.4. For the one-way random model with balanced data, where the variance ratio $R = \sigma_\tau^2/\sigma_e^2$ is known, explain how you will estimate the expected value of the upper tolerance limits for $N(\mu, \sigma_\tau^2 + \sigma_e^2)$ and $N(\mu, \sigma_\tau^2)$, by Monte Carlo simulation. For $p = 0.90$, $1 - \alpha = 0.95$, $a = 10$, $n = 5$, and $\sigma_e^2 = 1$, estimate the above expected values for various values of $R = \sigma_\tau^2/\sigma_e^2$. Are the expected values increasing in R? Should we expect this?

4.7.5. For the one-way random model with balanced data, derive an upper confidence bound for the exceedance probability concerning the true values, namely $P(\mu + \tau > L)$, where $\tau \sim N(0, \sigma_\tau^2)$ and L is a specified limit.

4.7.6. Consider the one-way random models $Y_{1ij} = \mu_1 + \tau_{1i} + e_{1ij}$, $j = 1, 2, ..., n_1$; $i = 1, 2, ..., a_1$ and $Y_{2ij} = \mu_2 + \tau_{2i} + e_{2ij}$, $j = 1, 2, ..., n_2$; $i = 1, 2, ..., a_2$, where $\tau_{1i} \sim N(0, \sigma_{\tau_1}^2)$, $e_{1ij} \sim N(0, \sigma_{e_1}^2)$, $\tau_{2i} \sim N(0, \sigma_{\tau_2}^2)$, $e_{2ij} \sim N(0, \sigma_{e_2}^2)$, and all the random variables are independently distributed.

(a) Describe procedures for computing one-sided and two-sided tolerance intervals for the distribution of $Y_1 - Y_2$, where Y_1 and Y_2 have the distributions of Y_{1ij} and Y_{2ij}, respectively.

(b) Describe procedures for computing one-sided and two-sided tolerance intervals for the distribution of the difference between the true values $(\mu_1 + \tau_1) - (\mu_2 + \tau_2)$, where $\tau_1 \sim N(0, \sigma_{\tau_1}^2)$ and $\tau_2 \sim N(0, \sigma_{\tau_2}^2)$.

(c) Explain how you will compute upper confidence limits for the probabilities $P(Y_1 > Y_2)$ and $P[(\mu_1 + \tau_1) > (\mu_2 + \tau_2)]$, where Y_1, Y_2, τ_1 and τ_2 have the distributions specified in part (i) and part (ii).

4.7.7. The following data is from a study to compare two musical programs (Program I and Program II) in terms of promoting efficiency among factory workers. Two groups of 8 workers each is selected for administering the programs. For the 8 workers who were administered Program I (labeled A, B, ..., H), the number of units produced were recorded for 10 days. For those administered Program II (labeled I, J, ..., P), the number of units produced were recorded for 8 days. The following data give the means and variances of the observations for each worker; for Program I, the means and variances are for the 10 observations per worker, and for Program II, the means and variances are for the 8 observations per worker.

Means and variances per worker in Program I							
Worker	A	B	C	D	E	F	G	H
Mean	93.2	98.1	89.6	88.4	96.2	95.0	99.6	97.9
Variance	23.4	27.6	18.6	22.1	15.4	26.2	33.1	29.8
Means and variances per worker in Program II							
Worker	I	J	K	L	M	N	O	P
Mean	90.3	85.1	99.4	98.4	86.2	82.5	103.9	96.7
Variance	32.4	26.3	16.8	23.7	18.4	25.6	34.1	28.3

For the workers who were administered Program I, let Y_{1ij}, denote the number of units produced by ith worker on the jth day, $j = 1, 2, ..., 10$; $i = 1, 2, ..., 8$. Similarly, let Y_{2ij} denote the observations for the workers administered Program II, $j = 1, 2, ..., 8$; $i = 1, 2, ..., 8$. Suppose the data can be modeled using the one-way random effects models in the previous problem, where the τ_{1i}'s and the τ_{2i}'s denote random effects due to the workers (here we assume that there is no significant variability among the different days).

(a) Using the given data, compute the two sums of squares, say SS_{τ_1} and SS_{τ_2}, due to the workers in Program I and Program II.

(b) By pooling the given sample variances, compute the two error sums of squares for the two groups, say SS_{e_1} and SS_{e_2}.

(c) In order to compare the two programs, it is decided to compute a $(0.90, 0.95)$ two-sided tolerance interval for the distribution of $Y_1 - Y_2$, where Y_1 and Y_2 have the distributions of Y_{1ij} and Y_{2ij}, respectively. Apply the procedure developed in the previous problem, and obtain such a tolerance interval.

(d) Suppose we want to compute a $(0.90, 0.95)$ two-sided tolerance interval for the distribution of the difference between the true values $(\mu_1 + \tau_1) - (\mu_2 + \tau_2)$, where $\tau_1 \sim N(0, \sigma_{\tau_1}^2)$ and $\tau_2 \sim N(0, \sigma_{\tau_2}^2)$. Apply the procedure developed in the previous problem, and obtain such a tolerance interval.

4.7.8. Consider the Styrene exposure data in Table 4.7, where the one-way random model is appropriate.

(a) Compute two-sided $(0.90, 0.95)$ tolerance intervals for the distribution $N(\mu, \sigma_\tau^2 + \sigma_e^2)$ using Mee's approach and the Liao-Lin-Iyer approach described in Section 4.5.1 and Section 4.5.2. For the Liao-Lin-Iyer approach, compute the required upper confidence limit for $\sigma_1^2 + \sigma_2^2$ using the generalized confidence interval method, Satterthwaite approximate method and the MLS method. How different are the resulting tolerance intervals?

(b) Compute two-sided $(0.90, 0.95)$ tolerance intervals for the distribution $N(\mu, \sigma_T^2)$ using the Liao-Lin-Iyer approach described in Section 4.6. Compute the required upper confidence limit for $\sigma_T^2 + \sigma_2^2$ using the generalized confidence interval method and the MLS method. How different are the resulting tolerance intervals?

Chapter 5

The One-Way Random Model with Unbalanced Data

5.1 Notations and Preliminaries

The model considered in this chapter is the same as that in the previous chapter, corresponding to a single factor experiment, except that now we have n_i observations corresponding to the ith level of the factor, where not all the n_i's are equal. Thus we have

$$Y_{ij} = \mu + \tau_i + e_{ij}, \ j = 1, 2, ..., n_i, \ i = 1, 2, ..., a, \tag{5.1.1}$$

where μ is an unknown general mean, τ_i's represent random effects and e_{ij}'s represent error terms. It is assumed that τ_i's and e_{ij}'s are all independent having the distributions $\tau_i \sim N(0, \sigma_\tau^2)$ and $e_{ij} \sim N(0, \sigma_e^2)$, so that $Y_{ij} \sim N(\mu, \sigma_\tau^2 + \sigma_e^2)$. The problems to be addressed are once again on the computation of tolerance intervals for the distribution $N(\mu, \sigma_\tau^2 + \sigma_e^2)$, corresponding to the observable random variables (i.e., the Y_{ij}'s), and the distribution $N(\mu, \sigma_\tau^2)$ corresponding to the unobservable true values (i.e., the $(\mu + \tau_i)$'s).

If we define $\bar{Y}_{i.} = \sum_{j=1}^{n_i} Y_{ij}/n_i$, and $SS_e = \sum_{i=1}^{a} \sum_{j=1}^{n_i} (Y_{ij} - \bar{Y}_{i.})^2$, then

$$U_e^2 = SS_e/\sigma_e^2 \sim \chi_{N-a}^2, \quad \text{where} \quad N = \sum_{i=1}^{a} n_i. \tag{5.1.2}$$

However, if we define $SS_\tau = \sum_{i=1}^{a} n_i (\bar{Y}_{i.} - \bar{Y}_{..})^2$, it is not true that SS_τ is distributed as a scalar multiple of a chi-square random variable, unlike in the

117

balanced case. Also, in the balanced case, the average of all the observations, namely $\bar{Y}_{..}$, is known to be the uniformly minimum variance unbiased estimator (umvue) of μ. Such a umvue does not exist in the unbalanced case. All these aspects complicate the computation of tolerance intervals in the unbalanced case.

For the distributions $N(\mu, \sigma_\tau^2 + \sigma_e^2)$ and $N(\mu, \sigma_\tau^2)$, Krishnamoorthy and Mathew (2004) have investigated the computation of one-sided tolerance limits in the unbalanced case, using $\frac{1}{a}\sum_{i=1}^{a} \bar{Y}_{i.}$ as an estimator of μ, and using SS_e along with another sum of squares (different from SS_τ given above) that has an approximate chi-square distribution. The generalized confidence interval idea was once again used and the limited simulation results showed good performance. Later, Liao, Lin and Iyer (2005) have obtained another solution, once again based on the generalized confidence interval idea. These authors have succeeded in deriving one-sided as well as two-sided tolerance intervals. Computationally, the Liao, Lin and Iyer (2005) approach is slightly more involved. Also, the numerical results reported by these authors show that their solution for the one-sided tolerance limit, and the solution due to Krishnamoorthy and Mathew (2004) show nearly identical performance. Earlier, Bhaumik and Kulkarni (1991) also addressed the computation of a one-sided tolerance interval for $N(\mu, \sigma_\tau^2 + \sigma_e^2)$ in the unbalanced case. Their approach consists of obtaining the tolerance limit for a known variance ratio R, and then replacing R with an upper confidence limit, similar to the work of Mee and Owen (1983) and Mee (1984b) in the balanced case. However, no guidelines are given in Bhaumik and Kulkarni (1991) regarding the choice of the confidence level.

In this chapter, we shall describe the procedures due to Krishnamoorthy and Mathew (2004), and that due to Liao, Lin and Iyer (2005) for obtaining lower or upper tolerance limits for $N(\mu, \sigma_\tau^2 + \sigma_e^2)$ and $N(\mu, \sigma_\tau^2)$. For obtaining two-sided intervals for both of these distributions, we shall present the Liao, Lin and Iyer (2005) approach. All of the procedures use the generalized confidence interval idea.

5.2   Two Examples

Example 5.1 (Moisture content in white pine lumber)

This example taken from Ostle and Mensing (1975) is on a study of the effect of storage conditions on the moisture content of white pine lumber, and the relevant data are given in Ostle and Mensing (1975, p. 296). Five different storage conditions ($a = 5$) were studied with varying number of sample boards

stored under each condition. The example is also considered in Bhaumik and
Kulkarni (1991), and the data are given in Table II of their paper. The data are
reproduced below.

Table 5.1: The moisture content data

Storage condition	Moisture content (%)
1	7.3, 8.3, 7.6, 8.4, 8.3
2	5.4, 7.4, 7.1
3	8.1, 6.4
4	7.9, 9.5, 10.0
5	7.1

Reproduced with permission from Taylor and
Francis, Ltd.; http://www.informaworld.com

Assuming a one-way random model, Bhaumik and Kulkarni have considered
the problem of obtaining an upper tolerance limit for $N(\mu, \sigma_\tau^2 + \sigma_e^2)$. One can
also think of constructing an upper tolerance limit for the distribution of the
true moisture content, i.e., the distribution $N(\mu, \sigma_\tau^2)$.

Example 5.2 (Sulfur content in bottles of coal)

In this example taken from Liao, Lin and Iyer (2005), measurements were
obtained on the sulfur content of each of 6 bottles of coal, so that $a = 6$. A
balanced version of this example is given in Wang and Iyer (1994), and the
unbalanced case is discussed in Liao, Lin and Iyer (2005). The observed value
of the Y_{ij}'s, i.e., the sulfur contents (weight %), are given in Table 5.2; the data
are taken from Table 2 of Liao, Lin and Iyer (2005). The data are unbalanced
since on four bottles, two observations each were obtained, and on two bottles
only one observation each was obtained.

Table 5.2: The sulfur content data

Bottle	Observations (weight %)
1	4.66, 4.68
2	4.65, 4.72
3	4.50, 4.78
4	4.70, 4.69
5	4.57
6	4.74

Reprinted with permission from *Technometrics*.
Copyright [2005] by the American Statistical Association.

In Liao, Lin and Iyer (2005), a one-way random model was used for the above data, and a two-sided tolerance interval was constructed for $N(\mu, \sigma_\tau^2)$, i.e., for the distribution of $\mu + \tau_i$. Here the τ_i's represent random effects due to the bottles. In the balanced version of this example considered in Wang and Iyer (1994), a two-sided tolerance interval was once again constructed for the same distribution. The same balanced version was also considered by Bhaumik and Kulkarni (1996), who computed a lower tolerance limit for $N(\mu, \sigma_\tau^2 + \sigma_e^2)$.

5.3 One-Sided Tolerance Limits for $N(\mu, \sigma_\tau^2 + \sigma_e^2)$

In order to compute a $(p, 1 - \alpha)$ lower tolerance limit for $N(\mu, \sigma_\tau^2 + \sigma_e^2)$, we shall obtain a $1 - \alpha$ lower confidence limit for $\mu - z_p\sqrt{\sigma_\tau^2 + \sigma_e^2}$. Here we shall describe the approaches due to Krishnamoorthy and Mathew (2004) and Liao, Lin and Iyer (2005). Krishnamoorthy and Mathew (2004) use approximations that permit the direct use of the results in the balanced case. The generalized confidence interval idea is used in both of the above articles; the generalized pivotal quantity (GPQ) constructed by Liao, Lin and Iyer (2005) is based on ideas similar to that in Iyer, Wang and Mathew (2004). Recall that in the case of balanced data, GPQs for μ and σ_τ^2 can be easily constructed using the distributions of $\bar{Y}_{..}$, SS_τ and SS_e, and the corresponding observed values. In the unbalanced case, the construction of GPQs for μ and σ_τ^2 are not very clear; in fact, it is possible to come up with several such constructions. Krishnamoorthy and Mathew (2004) use an approximate distribution to construct the GPQs; Liao, Lin and Iyer (2005) construct GPQs in an implicit manner.

5.3.1 The Krishnamoorthy and Mathew Approach

Recall that for $SS_e = \sum_{i=1}^{a} \sum_{j=1}^{n_i} (Y_{ij} - \bar{Y}_{i.})^2$, we have the chi-square distribution given in (5.1.2). Define

$$\tilde{n} = \frac{1}{a}\sum_{i=1}^{a} n_i^{-1}, \quad \bar{\bar{Y}} = \frac{1}{a}\sum_{i=1}^{a} \bar{Y}_{i.}, \quad \text{and} \quad SS_{\bar{y}} = \sum_{i=1}^{a}(\bar{Y}_{i.} - \bar{\bar{Y}})^2. \tag{5.3.1}$$

Then

$$\bar{\bar{Y}} \sim N\left(\mu, \frac{\sigma_\tau^2 + \tilde{n}\sigma_e^2}{a}\right).$$

By direct calculation, it can be verified that $E(SS_{\bar{y}}) = (a - 1)(\sigma_\tau^2 + \tilde{n}\sigma_e^2)$. An approximate chi-square distribution associated with $SS_{\bar{y}}$ has been obtained by

Thomas and Hultquist (1978); the result states that

$$U_{\bar{y}}^2 = \frac{SS_{\bar{y}}}{\sigma_\tau^2 + \tilde{n}\sigma_e^2} \sim \chi_{a-1}^2, \quad \text{approximately.} \tag{5.3.2}$$

It can also be verified that $\bar{\bar{Y}}$ and SS_e are independently distributed, and $SS_{\bar{y}}$ and SS_e are independently distributed. However, $\bar{\bar{Y}}$ and $SS_{\bar{y}}$ are not independent (except in the balanced case). We shall ignore this dependence while developing the tolerance limit. The performance of the tolerance limit so developed can be assessed based on numerical results.

Using $\bar{\bar{Y}}$, $SS_{\bar{y}}$ and SS_e, we shall now imitate the derivation for the balanced case (given in Section 4.3.3) to obtain a lower tolerance limit for $N(\mu, \sigma_\tau^2 + \sigma_e^2)$. Let $\bar{\bar{y}}$, $ss_{\bar{y}}$ and ss_e denote the observed values of $\bar{\bar{Y}}$, $SS_{\bar{y}}$ and SS_e, respectively. Following equation (4.3.16) in the balanced case, let

$$G_\mu = \bar{\bar{y}} + \frac{\sqrt{a}(\bar{\bar{Y}} - \mu)}{\sqrt{SS_{\bar{y}}}}\sqrt{\frac{ss_{\bar{y}}}{a}} = \bar{\bar{y}} + \frac{Z}{\sqrt{U_{\bar{y}}^2}}\sqrt{\frac{ss_{\bar{y}}}{a}}, \quad G_{\sigma_e^2} = \frac{\sigma_e^2}{SS_e}ss_e = \frac{ss_e}{U_e^2}, \tag{5.3.3}$$

and

$$G_{\sigma_\tau^2} = \frac{\sigma_\tau^2 + \tilde{n}\sigma_e^2}{SS_{\bar{y}}}ss_{\bar{y}} - \frac{\tilde{n}\sigma_e^2}{SS_e}ss_e = \frac{ss_{\bar{y}}}{U_{\bar{y}}^2} - \tilde{n}\frac{ss_e}{U_e^2} \overset{d}{\sim} \frac{ss_{\bar{y}}}{\chi_{a-1}^2} - \tilde{n}\frac{ss_e}{\chi_{N-a}^2}, \tag{5.3.4}$$

where $\overset{d}{\sim}$ means 'approximately distributed as', χ_{a-1}^2 denotes a chi-square random variable with $a - 1$ degrees of freedom, U_e^2 is given in (5.1.2), $U_{\bar{y}}^2$ is given in (5.3.2), $N = \sum_{i=1}^{a} n_i$, and $Z = \frac{\sqrt{a}(\bar{\bar{Y}} - \mu)}{\sqrt{\sigma_\tau^2 + \tilde{n}\sigma_e^2}} \sim N(0, 1)$. The quantity $G_{\sigma_e^2}$ is a GPQ for σ_e^2, and G_μ and $G_{\sigma_\tau^2}$ are approximate GPQs for μ and σ_τ^2, respectively. Define

$$\begin{aligned} G_3 &= G_\mu - z_p\sqrt{G_{\sigma_\tau^2} + G_{\sigma_e^2}} \\ &\overset{d}{\sim} \bar{\bar{y}} + \frac{Z}{\sqrt{\chi_{a-1}^2}}\sqrt{\frac{ss_{\bar{y}}}{a}} - z_p\left[\frac{ss_{\bar{y}}}{\chi_{a-1}^2} + (1 - \tilde{n})\frac{ss_e}{U_e^2}\right]^{1/2}. \end{aligned} \tag{5.3.5}$$

Recall once again that the actual distribution of $U_{\bar{y}} = \frac{SS_{\bar{y}}}{\sigma_\tau^2 + \tilde{n}\sigma_e^2}$ will depend on the unknown variance components, and the chi-square distribution is only an approximation. Furthermore, Z and $U_{\bar{y}}$ are not independent. Thus the actual distribution of G_3 does depend on unknown parameters, even though the second expression in (5.3.5) has a distribution free of any unknown parameters. However, as an approximation, we shall use the second expression in (5.3.5). The α

quantile of the random variable given in the second expression in (5.3.5) is our $(p, 1 - \alpha)$ lower tolerance limit for $N(\mu, \sigma_\tau^2 + \sigma_e^2)$ when the data are unbalanced. We shall denote this percentile as $G_{3,\alpha}$. Thus $G_{3,\alpha}$ is an approximation for the α quantile of the first expression in (5.3.5). An algorithm similar to Algorithm 4.1 can be used to compute $G_{3,\alpha}$.

An approximation similar to that in equation (4.3.20) can be developed for the unbalanced case also; the derivation is similar to that of equation (4.3.20). The approximate lower tolerance limit is given by

$$\bar{\bar{y}} - t_{a-1;1-\alpha}(\delta_3)\sqrt{\frac{ss_{\bar{y}}}{a(a-1)}},$$

$$\text{where} \quad \delta_3 \quad = \quad z_p\left(a + \frac{a(a-1)(1-\tilde{n})}{N-a}\frac{ss_e}{ss_{\bar{y}}}F_{a-1,N-a;\alpha}\right)^{1/2}. \quad (5.3.6)$$

It is readily verified that (5.3.6) reduces to the approximation in equation (4.3.20) when we have balanced data; simply use the relations that in the balanced case, $N = an$, $\tilde{n} = \frac{1}{n}$ and $ss_{\bar{y}} = \frac{ss_\tau}{n}$.

A $(p, 1 - \alpha)$ upper tolerance limit for $N(\mu, \sigma_\tau^2 + \sigma_e^2)$ can be similarly obtained as the $1 - \alpha$ quantile of

$$\bar{\bar{y}} + \frac{Z}{\sqrt{\chi_{a-1}^2}}\sqrt{\frac{ss_{\bar{y}}}{a}} + z_p\left[\frac{ss_{\bar{y}}}{\chi_{a-1}^2} + (1-\tilde{n})\frac{ss_e}{U_e^2}\right]^{1/2},$$

and has the approximate expression $\bar{\bar{y}} + t_{a-1;1-\alpha}(\delta_3)\sqrt{\frac{ss_{\bar{y}}}{a(a-1)}}$.

In the work of Bhaumik and Kulkarni (1991), the authors derive the tolerance limit when the variance ratio $R = \sigma_\tau^2/\sigma_e^2$ is known, and then replace the unknown variance ratio by a confidence limit. Their suggestion is to use an upper confidence limit for R when $\tilde{n} < 1$ and a lower confidence limit for R when $\tilde{n} > 1$. However, note that $\tilde{n}$ is always less than 1, and hence R should be replaced by an upper confidence limit. No guideline is provided in Bhaumik and Kulkarni (1991) regarding the choice of the confidence level to be used. However, if we replace R by its $100\alpha\%$ upper confidence limit constructed in Thomas and Hultquist (1978), the Bhaumik and Kulkarni (1991) tolerance limit actually coincides with the approximation (5.3.6). In order to see this, we note that if $R = \sigma_\tau^2/\sigma_e^2$ is known, the $(p, 1 - \alpha)$ lower tolerance limit derived by Bhaumik and Kulkarni (1991) is given by

$$\bar{\bar{y}} - t_{a-1;1-\alpha}(\delta_{BK})\sqrt{\frac{ss_{\bar{y}}}{a(a-1)}}, \quad \text{where} \quad \delta_{BK} = z_p\sqrt{\frac{a(R+1)}{R+\tilde{n}}}.$$

The $100\alpha\%$ upper confidence limit for R due to Thomas and Hultquist (1978) is

$$\tilde{n}\left[\frac{N-a}{\tilde{n}(a-1)}\frac{ss_{\bar{y}}}{ss_e}\frac{1}{F_{a-1,N-a;\,\alpha}}-1\right].$$

Let $\hat{\delta}_{BK}$ denote δ_{BK} with R replaced by the above upper limit. Then, it can be readily verified that $\hat{\delta}_{BK}$ coincides with δ_3 given in (5.3.6), and hence the tolerance limit $\bar{\bar{y}} - t_{a-1;1-\alpha}(\hat{\delta}_{BK})\sqrt{\frac{ss_{\bar{y}}}{a(a-1)}}$ coincides with the limit in (5.3.6).

5.3.2 The Liao, Lin and Iyer Approach

Note that

$$\bar{Y}_{i.} = \frac{1}{n_i}\sum_{j=1}^{n_i} Y_{ij} \sim N(\mu, v_i), \quad \text{where } v_i = \sigma_\tau^2 + \frac{\sigma_e^2}{n_i}. \tag{5.3.7}$$

Write

$$\bar{\boldsymbol{Y}} = (\bar{Y}_{1.}, \bar{Y}_{2.}, ..., \bar{Y}_{a.})' \quad \text{and} \quad \mathbf{V} = \mathrm{diag}(v_1, v_2, ..., v_a). \tag{5.3.8}$$

Then we have the distribution

$$\bar{\boldsymbol{Y}} \sim N(\mu \mathbf{1}_a, \mathbf{V}), \tag{5.3.9}$$

where $\mathbf{1}_a$ is an $a \times 1$ vector of ones. If the variance components σ_τ^2 and σ_e^2 are known, so that the v_i's are known, the umvue of μ, say $\bar{Y}_v$ is

$$
\begin{aligned}
\bar{Y}_v &= \left(\mathbf{1}_a'\mathbf{V}^{-1}\mathbf{1}_a\right)^{-1}\mathbf{1}_a'\mathbf{V}^{-1}\bar{\boldsymbol{Y}} \\
&= \frac{\sum_{i=1}^a \frac{1}{v_i}\bar{Y}_{i.}}{\sum_{i=1}^a \frac{1}{v_i}} \sim N\left(\mu, \left(\sum_{i=1}^a v_i^{-1}\right)^{-1}\right).
\end{aligned}
\tag{5.3.10}
$$

The second expression for $\bar{Y}_v$ in (5.3.10) is obtained by direct simplification. The residual sum of squares SS_0 under the model (5.3.9) is given by

$$
\begin{aligned}
SS_0 &= (\bar{\boldsymbol{Y}} - \bar{Y}_v\mathbf{1}_a)'\mathbf{V}^{-1}(\bar{\boldsymbol{Y}} - \bar{Y}_v\mathbf{1}_a) \\
&= \sum_{i=1}^a \frac{1}{v_i}(\bar{Y}_{i.} - \bar{Y}_v)^2 \\
&= \sum_{i=1}^a \frac{1}{v_i}\bar{Y}_{i.}^2 - \frac{1}{\sum_{i=1}^a \frac{1}{v_i}}\left(\sum_{i=1}^a \frac{1}{v_i}\bar{Y}_{i.}\right)^2.
\end{aligned}
\tag{5.3.11}
$$

The second expression in (5.3.11) is obtained by direct simplification using the expression for $\bar{Y}_v$, given in (5.3.10). Note that $SS_0 \sim \chi_{a-1}^2$.

Recall that $G_{\sigma_e^2}$ in (5.3.3) is an exact GPQ for σ_e^2, and G_μ and $G_{\sigma_\tau^2}$ in (5.3.3) and (5.3.4) are approximate GPQs for μ and σ_τ^2, respectively. We shall now derive exact GPQs for μ and σ_τ^2, where we shall continue to use the notation G_μ and $G_{\sigma_\tau^2}$ for the exact GPQs to be derived. When the notation G_μ and $G_{\sigma_\tau^2}$ is used in this chapter, it will be clear from the context whether we are referring to the exact GPQs, or the approximate ones in (5.3.3) and (5.3.4). For deriving the exact GPQs for μ and σ_τ^2, let $\bar{y}_{1.}$, $\bar{y}_{2.}$, ... $\bar{y}_{a.}$, and ss_e denote the observed values of $\bar{Y}_{1.}$, $\bar{Y}_{2.}$, ... $\bar{Y}_{a.}$, and SS_e, respectively. Once $G_{\sigma_\tau^2}$ is obtained, a GPQ, say G_{v_i}, for v_i (defined in (5.3.7)) is given by

$$G_{v_i} = G_{\sigma_\tau^2} + \frac{G_{\sigma_e^2}}{n_i}. \tag{5.3.12}$$

Motivated by (5.3.11), an exact GPQ $G_{\sigma_\tau^2}$ is obtained as the solution of the equation

$$U_0^2 = \sum_{i=1}^a \frac{1}{G_{v_i}} \bar{y}_{i.}^2 - \frac{1}{\sum_{i=1}^a \frac{1}{G_{v_i}}} \left(\sum_{i=1}^a \frac{1}{G_{v_i}} \bar{y}_{i.} \right)^2, \tag{5.3.13}$$

where $U_0^2 \sim \chi_{a-1}^2$. Note that the right hand side of (5.3.13) is a function of $G_{\sigma_\tau^2}$, since G_{v_i} (given in (5.3.12)) is a function of $G_{\sigma_\tau^2}$. Solving (5.3.13) for $G_{\sigma_\tau^2}$, we get the GPQ for σ_τ^2. It can be shown that the right hand side of (5.3.13) is a monotone decreasing function of $G_{\sigma_\tau^2}$, with limiting value 0 as $G_{\sigma_\tau^2} \to \infty$. For this, it is enough to show that SS_0 given in (5.3.11) satisfies a similar property as a function of σ_τ^2, since the right hand side of (5.3.13), as a function of $G_{\sigma_\tau^2}$, is similar to the right hand side of SS_0 in (5.3.11), as a function of σ_τ^2. A proof of this monotonicity property of SS_0 is given following equation (5.3.16) below. In view of this monotonicity property, there exists a unique $G_{\sigma_\tau^2}$ that solves (5.3.13), provided $G_{\sigma_\tau^2}$ is not restricted to be nonnegative.

Thus we have constructed GPQs $G_{\sigma_e^2}$, G_{v_i} and $G_{\sigma_\tau^2}$ with observed values σ_e^2, v_i and σ_τ^2, respectively, where $v_i = \sigma_\tau^2 + (\sigma_e^2/n_i)$. Furthermore, given the observed data, the distribution of $(G_{\sigma_e^2}, G_{v_i}, G_{\sigma_\tau^2})$ is free of unknown parameters. Now we shall construct a GPQ G_μ for μ. This is given by

$$\begin{aligned}
G_\mu &= \frac{\sum_{i=1}^a \frac{1}{G_{v_i}} \bar{y}_{i.}}{\sum_{i=1}^a \frac{1}{G_{v_i}}} - \frac{\bar{Y}_v - \mu}{\sqrt{\left(\sum_{i=1}^a v_i^{-1}\right)^{-1}}} \sqrt{\left[\left(\sum_{i=1}^a G_{v_i}^{-1}\right)^{-1}\right]_+} \\
&= \frac{\sum_{i=1}^a \frac{1}{G_{v_i}} \bar{y}_{i.}}{\sum_{i=1}^a \frac{1}{G_{v_i}}} - Z \sqrt{\left[\left(\sum_{i=1}^a G_{v_i}^{-1}\right)^{-1}\right]_+}, \tag{5.3.14}
\end{aligned}$$

where $x_+ = \max\{0, x\}$ and $Z = \dfrac{\bar{Y}_v - \mu}{\sqrt{\left(\sum_{i=1}^a v_i^{-1}\right)^{-1}}} \sim N(0,1)$; see (5.3.10). Because we allow $G_{\sigma_\tau^2}$ to be negative, $\sum_{i=1}^a G_{v_i}^{-1}$ could be negative, and so we take $\left[\left(\sum_{i=1}^a G_{v_i}^{-1}\right)^{-1}\right]_+$ under the radical sign.

Now we are ready to compute a $(p, 1 - \alpha)$ lower tolerance limit for $N(\mu, \sigma_\tau^2 + \sigma_e^2)$, i.e., $1 - \alpha$ lower confidence limit for $\mu - z_p\sqrt{\sigma_\tau^2 + \sigma_e^2}$. A GPQ for $\mu - z_p\sqrt{\sigma_\tau^2 + \sigma_e^2}$, say G_4, is given by

$$G_4 = G_\mu - z_p\sqrt{[G_{\sigma_\tau^2} + G_{\sigma_e^2}]_+}. \tag{5.3.15}$$

$G_{4,\alpha}$, the α quantile of G_4, gives a $1 - \alpha$ generalized lower confidence limit for $\mu - z_p\sqrt{\sigma_\tau^2 + \sigma_e^2}$, and hence a $(p, 1 - \alpha)$ lower tolerance limit for $N(\mu, \sigma_\tau^2 + \sigma_e^2)$.

Remark 5.1 Note that an iterative procedure is necessary to solve the equation (5.3.13) to find the exact GPQ $G_{\sigma_\tau^2}$. Instead, we could use the approximate GPQ $G_{\sigma_\tau^2}$ defined in (5.3.4). This way we can eliminate the iterative procedure for root finding. As will be seen in the sequel, our examples show that both methods of obtaining $G_{\sigma_\tau^2}$ produce practically the same result.

The following algorithm can be used for estimating the α quantile of G_4.

Algorithm 5.1

1. Once the data are obtained, compute the observed values $\bar{y}_1., \bar{y}_2., ..., \bar{y}_a.,$ and ss_e.

2. Let K denote the simulation sample size. For $k = 1, 2, ..., K$, perform the following steps.

3. Generate independent random variables $U_{e,k}^2 \sim \chi_{N-a}^2$, $U_{0,k}^2 \sim \chi_{a-1}^2$, and $Z_k \sim N(0,1)$.

4. Compute $G_{\sigma_e^2,k} = ss_e/U_{e,k}^2$.

5. (Use either Step a or b.)

   a. Solve the following equation to get $G_{\sigma_\tau^2,k}$:

   $$U_{0,k}^2 = \sum_{i=1}^a \frac{1}{G_{v_i,k}} \bar{y}_i.^2 - \frac{1}{\sum_{i=1}^a \frac{1}{G_{v_i,k}}} \left(\sum_{i=1}^a \frac{1}{G_{v_i,k}} \bar{y}_i.\right)^2,$$

   where $G_{v_i,k} = G_{\sigma_\tau^2,k} + \dfrac{G_{\sigma_e^2,k}}{n_i}$.

b. Compute $G_{\sigma_\tau^2,k} = \frac{ss_{\bar{y}}}{\chi_{a-1}^2} - \tilde{n}\frac{ss_e}{U_e^2}$, where $\tilde{n}$ is defined in (5.3.1) and $ss_{\bar{y}}$ is the observed value of $SS_{\bar{y}}$ in (5.3.1).

6. Let $G_{v_i,k} = G_{\sigma_\tau^2,k} + \frac{G_{\sigma_e^2,k}}{n_i}$. Compute

$$G_{\mu,k} = \frac{\sum_{i=1}^{a} \frac{1}{G_{v_i,k}} \bar{y}_{i.}}{\sum_{i=1}^{a} \frac{1}{G_{v_i,k}}} - Z_k \sqrt{\left[\left(\sum_{i=1}^{a} G_{v_i,k}^{-1}\right)^{-1}\right]_+}.$$

7. Compute

$$G_{4,k} = G_{\mu,k} - z_p\sqrt{\left[G_{\sigma_\tau^2,k} + G_{\sigma_e^2,k}\right]_+}.$$

The 100αth percentile of the $G_{4,k}$ values ($k = 1, 2, ..., K$) gives an estimate of the α quantile $G_{4,\alpha}$ of G_4, which in turn gives an estimate of the required $(p, 1 - \alpha)$ lower confidence limit for $N(\mu, \sigma_\tau^2 + \sigma_e^2)$.

If a $(p, 1 - \alpha)$ upper tolerance limit is required, we consider

$$G_5 = G_\mu + z_p\sqrt{[G_{\sigma_\tau^2,k} + G_{\sigma_e^2,k}]_+}, \qquad (5.3.16)$$

and the $1 - \alpha$ quantile $G_{5,1-\alpha}$ of G_5 gives the $(p, 1 - \alpha)$ upper tolerance limit. The limit can obviously be estimated as before.

Proof that SS_0 in (5.3.11) is a decreasing function of σ_τ^2

We shall give an alternative representation for SS_0 given in the first expression in (5.3.11). Towards this, we first note that

$$\begin{aligned}
SS_0 &= (\bar{Y} - \bar{Y}_v\mathbf{1}_a)'\mathbf{V}^{-1}(\bar{Y} - \bar{Y}_v\mathbf{1}_a) \\
&= \bar{Y}'\left[\mathbf{V}^{-1} - \frac{\mathbf{V}^{-1}\mathbf{1}_a\mathbf{1}_a'\mathbf{V}^{-1}}{\mathbf{1}_a'\mathbf{V}^{-1}\mathbf{1}_a}\right]\bar{Y},
\end{aligned}$$

where the second expression above is obtained from the first by using the expression for $\bar{Y}_v$, given in (5.3.10). Now let $\mathbf{H}$ be an $a \times (a - 1)$ matrix of rank $a - 1$ satisfying $\mathbf{H}'\mathbf{1}_a = 0$. Using a theorem in Searle, Casella and McCulloch (1992, p. 451), we can rewrite SS_0 as

$$SS_0 = \bar{Y}'\mathbf{H}(\mathbf{H}'\mathbf{VH})^{-1}\mathbf{H}'\bar{Y} = \mathbf{Y}_0'(\mathbf{H}'\mathbf{VH})^{-1}\mathbf{Y}_0, \qquad (5.3.17)$$

where $\mathbf{Y}_0 = \mathbf{H}'\bar{Y} \sim N(0, \mathbf{H}'\mathbf{VH})$, using (5.3.9). Note that $\mathbf{H}'\mathbf{VH}$ is an $(a - 1) \times (a - 1)$ positive definite matrix.

Now suppose $\sigma_{\tau,1}^2 > \sigma_{\tau,2}^2$. Furthermore, let SS_{01} and SS_{02} denote the expressions for SS_0, when $\sigma_\tau^2 = \sigma_{\tau,1}^2$ and $\sigma_{\tau,2}^2$, respectively. We shall show that $SS_{01} < SS_{02}$.

Let

$$v_{i1} = \sigma_{\tau,1}^2 + (\sigma_e^2/n_i), \;\; v_{i2} = \sigma_{\tau,2}^2 + (\sigma_e^2/n_i)$$
$$\mathbf{V}_1 = \operatorname{diag}(v_{11}, v_{21}, ..., v_{a1}), \;\; \text{and} \;\; \mathbf{V}_2 = \operatorname{diag}(v_{12}, v_{22}, ..., v_{a2})$$

Since $\sigma_{\tau,1}^2 > \sigma_{\tau,2}^2$, we have $v_{i1} > v_{i2}$ for $i = 1, 2, ..., a$. Thus the matrix $\mathbf{V}_1 - \mathbf{V}_2$ is positive definite. Hence $\mathbf{H}'\mathbf{V}_1\mathbf{H} - \mathbf{H}'\mathbf{V}_2\mathbf{H}$ is positive definite; consequently, $(\mathbf{H}'\mathbf{V}_2\mathbf{H})^{-1} - (\mathbf{H}'\mathbf{V}_1\mathbf{H})^{-1}$ is positive definite. It now follows from (5.3.17) that $SS_{02} > SS_{01}$. This establishes the assertion that SS_0 is a decreasing function of σ_τ^2. Also, since $\mathbf{H}'\mathbf{V}\mathbf{H} \to \infty$ as $\sigma_\tau^2 \to \infty$, we conclude that $SS_0 \to 0$ as $\sigma_\tau^2 \to \infty$. This proof also appears in Lin, Liao and Iyer (2008, Appendix B).

Numerical results on the performance of the lower tolerance limits $G_{3,\alpha}$ and $G_{4,\alpha}$, as well as the approximate lower tolerance limit (5.3.6), are reported in Krishnamoorthy and Mathew (2004) and in Liao, Lin and Iyer (2005). These authors have reported the simulated confidence levels corresponding to the different tolerance limits for various sample size and parameter configurations. The overall conclusion in Krishnamoorthy and Mathew (2004) is that the approximate tolerance limit (5.3.6) is unsatisfactory when the intraclass correlation $\sigma_\tau^2/(\sigma_\tau^2 + \sigma_e^2)$ is small; in this case the actual confidence levels are smaller than the assumed nominal level $1 - \alpha$. In general, the approximate tolerance limit (5.3.6) is satisfactory when the intraclass correlation is at least 0.5. Recall that we had the same conclusion in the balanced case. The tolerance limit $G_{3,\alpha}$ turns out to be quite satisfactory in maintaining the confidence level, even though it can be slightly conservative in some cases. The numerical results in Liao, Lin and Iyer (2005) show that the tolerance limit $G_{4,\alpha}$ resulting from Algorithm 5.1 exhibits performance very close to that of $G_{3,\alpha}$. Our recommendation is to use the approximate tolerance limit (5.3.6) if there is reason to believe that the intra-class correlation is at least 0.5 (i.e., if $\sigma_\tau^2 \geq \sigma_e^2$). If no such information is available, one should either use the limit $G_{3,\alpha}$, or the limit $G_{4,\alpha}$.

Example 5.1 (continued)

Based on the moisture content data in Table 5.1, let us compute a (0.90, 0.95) upper tolerance limit for the moisture content. Here $a = 5$ and $N = 14$. The observed values are:

$\bar{y}_{1.} = 7.98, \;\; \bar{y}_{2.} = 6.63, \;\; \bar{y}_{3.} = 7.25, \;\; \bar{y}_{4.} = 9.13, \;\; \bar{y}_{5.} = 7.10, \;\; \bar{\bar{y}} = 7.62, \;\; ss_{\bar{y}} = 3.80$, and $ss_e = 7.17$.

For $p = 0.90$ and $1 - \alpha = 0.95$, we have $z_p = 1.282$ and $F_{a-1,N-a;\alpha} = F_{4,9;0.05} = 0.1667$. Since $\tilde{n} = 0.4733$, we have

$$\delta_3 = z_p \left(a + \frac{a(a-1)(1-\tilde{n})}{N-a} \frac{ss_e}{ss_{\bar{y}}} F_{a-1,N-a;\alpha} \right)^{1/2} = 2.9703$$

Using the approximation developed in Section 5.3.1, a $(0.90, 0.95)$ upper toler-ance limit for the moisture content distribution $N(\mu, \sigma_\tau^2 + \sigma_e^2)$ is given by

$$\bar{\bar{y}} + t_{a-1;1-\alpha}(\delta_3)\sqrt{\frac{ss_{\bar{y}}}{a(a-1)}} = 7.62 + 7.8437 \times \sqrt{\frac{3.80}{20}} = 11.04.$$

Instead of using the approximation, let us compute the upper tolerance limit as the 95th percentile of

$$\bar{\bar{y}} + \frac{Z}{\sqrt{\chi_{a-1}^2}}\sqrt{\frac{ss_{\bar{y}}}{a}} + z_p \left[\frac{ss_{\bar{y}}}{\chi_{a-1}^2} + (1-\tilde{n})\frac{ss_e}{U_e} \right]^{1/2}$$

$$= 7.62 + \frac{Z}{\sqrt{\chi_4^2}}\sqrt{\frac{3.80}{5}} + 1.282 \left[\frac{3.80}{\chi_4^2} + (1 - 0.4733)\frac{7.17}{U_e} \right]^{1/2}.$$

The estimate of the 95th percentile, based on 10,000 simulations, turned out to be 11.12. We note that this is very close to the approximation obtained above.

We shall now use the Liao, Lin and Iyer (2005) approach as described in Algorithm 5.1. Using Monte Carlo simulation with $K = 10,000$ runs and Step 5a, we estimated the $(0.90, 0.95)$ upper tolerance limit as 11.35; using Step 5b, we obtained 11.31. As noted in Remark 5.1, both Steps 5a and 5b produced practically the same limit.

5.3.3 One-Sided Confidence Limits for Exceedance Probabilities

In this section, we shall extend the results in Section 4.3.6 to the case of unbal-anced data. Thus the problem is the computation of an upper or lower confidence limit for $\eta = P(Y > L)$, where $Y \sim N(\mu, \sigma_\tau^2 + \sigma_e^2)$, and L is a specified limit, and unbalanced data are available on Y, as specified in (5.1.1). Following the arguments in Section 4.3.6, we need to equate L to an upper tolerance limit for $N(\mu, \sigma_\tau^2 + \sigma_e^2)$, having content $1 - A$ and confidence level $1 - \alpha$, and then solve for A. The solution so obtained is a $1 - \alpha$ upper confidence limit for η.

If we use the approximation to the upper tolerance limit given in Section

5.3.1, then the solution to A is to be obtained from the equation

$$\bar{\bar{y}} + t_{a-1;1-\alpha}(\delta_3)\sqrt{\frac{ss_{\bar{y}}}{a(a-1)}} = L,$$

$$\text{where} \quad \delta_3 \;=\; z_{1-A}\left(a + \frac{a(a-1)(1-\tilde{n})}{N-a}\frac{ss_e}{ss_{\bar{y}}}F_{a-1,N-a;\alpha}\right)^{1/2}.$$

If the approximation to the upper tolerance limit is not used, then trial and error is necessary in order to numerically obtain the value of A for which the $(1 - A, 1 - \alpha)$ upper tolerance limit is equal to L. The value of A obtained based on the approximate upper tolerance limit can be used as a starting value for this purpose, as noted in the balanced case. The value can be adjusted and the $(1 - A, 1 - \alpha)$ upper tolerance limit can be recomputed until the tolerance limit is approximately equal to the value of L.

A $100(1 - \alpha)\%$ lower confidence bound for η can be obtained as the value of A for which a lower tolerance limit for $N(\mu, \sigma_\tau^2 + \sigma_e^2)$, having content A and confidence level $1 - \alpha$, is equal to L.

Example 5.3 (Nickel dust exposures at a nickel-producing complex)

This example and the relevant data are taken from Lyles, Kupper and Rappaport (1997b, Table D2). The data are measurements (mg/m^3) on nickel dust exposure on maintenance mechanics at a nickel producing complex. The log-transformed data, reproduced in Table 5.3, consist of measurements obtained from 23 mechanics. The data are obviously unbalanced, and the total number of observations is $N = 34$. In Lyles, Kupper and Rappaport (1997b) as well as in Krishnamoorthy, Mathew and Ramachandran (2007), the data were analyzed using a one-way random effects model, where the random effect represents the effect due to the mechanic. The problem of interest is the estimation of the proportion of exposure measurements that exceed the occupation exposure limit (OEL) of 1 mg/m^3 for nickel exposure. Thus on the logarithmic scale, we are interested in estimating $P(Y > 0)$, where $Y \sim N(\mu, \sigma_\tau^2 + \sigma_e^2)$.

Computations based on the data in Table 5.3 resulted in $\bar{\bar{y}} = -3.683$, $\tilde{n} = 0.855$, $ss_{\bar{y}} = 16.081$ and $ss_e = 2.699$. Thus

$$\left(a + \frac{a(a-1)(1-\tilde{n})}{N-a}\frac{ss_e}{ss_{\bar{y}}}F_{a-1,N-a;\alpha}\right)^{1/2} = \left(23 + \frac{73.37}{34-23}\frac{2.699}{16.081}(0.4428)\right)^{1/2}$$

$$= 4.9112,$$

yielding $\delta_3 = 4.9112 \times z_{1-A}$. Since $L = 0$, the equation to be solved to get a

Table 5.3: Log-transformed nickel dust exposures on mechanics at a nickel-producing complex

Worker	Observations	Worker	Observations
1	-2.900	13	-3.244
2	-6.215	14	$-2.617, -1.772,$
3	$-4.423, -4.906, -4.828$		$-3.352, -3.026$
4	-3.907	15	-3.037
5	$-3.772, -3.540$	16	-2.375
6	-4.320	17	$-3.817, -4.510, -3.689$
7	-4.423	18	-3.863
8	$-4.343, -4.962, -3.772, -4.129$	19	-4.465
9	-2.937	20	-3.378
10	-3.817	21	-3.576
11	-2.882	22	-2.577
12	-3.689	23	-3.730

95% upper confidence limit for $P(Y > 0)$ is

$$\bar{\bar{y}} + t_{a-1;0.95}(4.9112 \times z_{1-A})\sqrt{\frac{ss_{\bar{y}}}{a(a-1)}} = 0,$$

where we need to solve for A. That is

$$t_{22;0.95}(4.9112 \times z_{1-A}) = -\bar{\bar{y}} \times \sqrt{\frac{a(a-1)}{ss_{\bar{y}}}} = 3.863 \times \sqrt{\frac{23 \times 22}{16.081}} = 20.6595.$$

Using the PC calculator *StatCalc* by Krishnamoorthy (2006), we get $4.9112 \times z_{1-A} = 15.1994$, i.e., $z_{1-A} = 3.0948$. The last equation implies that A = 0.0010. Thus, a 95% upper confidence limit for $P(Y > 0)$ is 0.001. This means that less than 0.1% of exposure measurements exceed the OEL, with 95% confidence.

Inference concerning exceedance probabilities have important applications in industrial hygiene. Such problems in the context of a one-way random model with unbalanced data are addressed in Krishnamoorthy and Guo (2005).

5.4   One-Sided Tolerance Limits for $N(\mu, \sigma_\tau^2)$

The procedure developed by Krishnamoorthy and Mathew (2004), and the one due to Liao, Lin and Iyer (2005) described in the previous section, can be easily adapted to derive one-sided tolerance limits for $N(\mu, \sigma_\tau^2)$. What follows is a brief development of this. We once again use the fact that a $(p, 1-\alpha)$ lower tolerance limit for $N(\mu, \sigma_\tau^2)$ is a $1-\alpha$ lower confidence limit for $\mu - z_p\sigma_\tau$ and a $(p, 1-\alpha)$ upper tolerance limit for $N(\mu, \sigma_\tau^2)$ is a $1-\alpha$ upper confidence limit for $\mu + z_p\sigma_\tau$.

5.4.1 The Krishnamoorthy and Mathew Approach

Similar to G_3 in (5.3.5), define

$$
\begin{aligned}
G_6 &= \bar{\bar{y}} + \frac{\sqrt{a}(\bar{\bar{Y}} - \mu)}{\sqrt{SS_{\bar{y}}}}\sqrt{\frac{SS_{\bar{y}}}{a}} - z_p\left[\left(\frac{\sigma_\tau^2 + \tilde{n}\sigma_e^2}{SS_{\bar{y}}}ss_{\bar{y}} - \frac{\tilde{n}\sigma_e^2}{SS_e}ss_e\right)\right]_+^{1/2} \\
&\overset{d}{\sim} \bar{\bar{y}} + \frac{Z}{\sqrt{\chi_{a-1}^2}}\sqrt{\frac{ss_{\bar{y}}}{a}} - z_p\left[\frac{ss_{\bar{y}}}{\chi_{a-1}^2} - \tilde{n}\frac{ss_e}{U_e^2}\right]_+^{1/2},
\end{aligned}
\tag{5.4.1}
$$

where for any real number c, $c_+ = \max\{c, 0\}$, and the other notations in (5.4.1) are the same as those in (5.3.5). Let $G_{6,\alpha}$ denote the 100αth percentile of the second expression in (5.4.1). Then $G_{6,\alpha}$ gives a $(p, 1-\alpha)$ lower tolerance limit for $N(\mu, \sigma_\tau^2)$. Similar to (5.3.6), an approximation for this percentile is given by

$$
\bar{\bar{y}} - t_{a-1;1-\alpha}(\delta_4)\sqrt{\frac{ss_{\bar{y}}}{a(a-1)}},
$$

$$
\text{where } \delta_4 = z_p\left(a - \frac{a(a-1)\tilde{n}}{N-a}\frac{ss_e}{ss_{\bar{y}}}F_{a-1,N-a;\alpha}\right)_+^{1/2}.
\tag{5.4.2}
$$

A $(p, 1-\alpha)$ upper tolerance limit for $N(\mu, \sigma_\tau^2)$ can be similarly obtained, and has the approximate expression $\bar{\bar{y}} + t_{a-1;1-\alpha}(\delta_4)\sqrt{\frac{ss_{\bar{y}}}{a(a-1)}}$, where δ_4 is as given above.

The numerical results in Krishnamoorthy and Mathew (2004) show that the above approximations are quite satisfactory, even though the approximate limit is somewhat conservative for small values of the intra-class correlation $\sigma_\tau^2/(\sigma_\tau^2 + \sigma_e^2)$.

5.4.2 The Liao, Lin and Iyer Approach

Similar to G_4 in (5.3.15), define

$$
G_7 = G_\mu - z_p\sqrt{[G_{\sigma^2}]_+},
\tag{5.4.3}
$$

where the notations are as in Section 5.3.2. Note that G_7 is a GPQ for $\mu - z_p\sigma_\tau$. The 100αth percentile of G_7, say $G_{7,\alpha}$, gives a $(p, 1-\alpha)$ lower tolerance limit for $N(\mu, \sigma_\tau^2)$. Furthermore, $G_{7,\alpha}$ can be estimated using an algorithm similar to the one given in Section 5.3.2. We also note that for G_{σ^2}, we can use either the approximate GPQ or the exact GPQ.

Liao, Lin and Iyer (2005) have compared their limit $G_{7,\alpha}$ with $G_{6,\alpha}$, and have noted very similar performance. It appears that for obtaining a lower tolerance limit for $N(\mu, \sigma_\tau^2)$, the simple approximation (5.4.2) can be recommended. The similar approximation given below equation (5.4.2) can be recommended for computing an upper tolerance limit for $N(\mu, \sigma_\tau^2)$.

Example 5.1 (continued)

Based the moisture content data in Table 5.1, let us compute a (0.90, 0.95) upper tolerance limit for the distribution $N(\mu, \sigma_\tau^2)$ of the "true" moisture content. The summary statistics are given in Section 5.3.2.

We now have

$$\delta_4 = z_p \left(a - \frac{a(a-1)\tilde{n}}{N-a} \frac{ss_e}{ss_{\bar{y}}} F_{a-1,N-a;\alpha} \right)_+^{1/2} = 2.7702$$

Using the approximation developed in Section 5.4.1, a (0.90,0.95) upper tolerance limit for the "true" moisture content distribution $N(\mu, \sigma_\tau^2)$ is given by

$$\bar{\bar{y}} + t_{a-1;1-\alpha}(\delta_4)\sqrt{\frac{ss_{\bar{y}}}{a(a-1)}} = 7.62 + 7.4121 \times \sqrt{\frac{3.80}{20}} = 10.85.$$

Instead of using the approximation, suppose we compute the upper tolerance limit as the 95th percentile of

$$\bar{\bar{y}} + \frac{Z}{\sqrt{\chi_{a-1}^2}}\sqrt{\frac{ss_{\bar{y}}}{a}} + z_p \left[\frac{ss_{\bar{y}}}{\chi_{a-1}^2} - \tilde{n}\frac{ss_e}{U_e} \right]_+^{1/2}$$

$$= 7.62 + \frac{Z}{\sqrt{\chi_4^2}}\sqrt{\frac{3.80}{5}} + 1.282 \left[\frac{3.80}{\chi_4^2} - 0.4733\frac{7.17}{U_e} \right]_+^{1/2}.$$

The estimate of the 95th percentile, based on 10,000 simulations, turned out to be 10.94, very close to the approximation obtained above.

As mentioned earlier, to obtain an upper tolerance limit using the Liao, Lin and Iyer (2005) approach, we can use Algorithm 5.1 without G_{σ^2}. In particular, using Algorithm 5.1 with $K = 10,000$ and Step 5a, we found (0.90, 0.95) upper tolerance limitas 10.98; the same limit was also obtained using Step 5b.

5.5   Two-Sided Tolerance Intervals

In this section, we shall describe the computation of a two-sided $(p, 1 - \alpha)$ tolerance interval for $N(\mu, \sigma_\tau^2 + \sigma_e^2)$ and for $N(\mu, \sigma_\tau^2)$. The procedures described in this section are due to Liao, Lin and Iyer (2005).

5.5.1   A Two-Sided Tolerance Interval for $N(\mu, \sigma_\tau^2 + \sigma_e^2)$

Let $\tilde{\mu}$ denote an estimator of μ (to be determined), where $\tilde{\mu}$ follows a normal distribution with mean μ. As in the balanced case discussed in Section 4.4.1, we need to determine a margin of error statistic D_3 so that $\tilde{\mu} \pm D_3$ is a $(p, 1 - \alpha)$ tolerance interval for $N(\mu, \sigma_\tau^2 + \sigma_e^2)$. The quantities $\tilde{\mu}$ and D_3 are functions of $\bar{Y}$ (defined in (5.3.8)) and SS_e, and are required to satisfy the tolerance interval condition

$$P_{\bar{Y}, \, SS_e} \left[P_Y \left\{ \tilde{\mu} - D_3 \leq Y \leq \tilde{\mu} + D_3 | \bar{Y}, \, SS_e, \right\} \geq p \right] = 1 - \alpha, \qquad (5.5.1)$$

where $Y \sim N(\mu, \sigma_\tau^2 + \sigma_e^2)$. Suppose we take $\tilde{\mu} = \bar{Y}_v$, defined in (5.3.10). Obviously we cannot use this as an estimator of μ since $\bar{Y}_v$ depends on unknown parameters; we shall address this issue shortly. Following a derivation similar to that in Section 4.4.1, we can conclude that D_3 can be taken to be a $1 - \alpha$ upper confidence limit of the parametric function

$$z_{\frac{1+p}{2}} \sqrt{\sigma_\tau^2 + \sigma_e^2 + \left(\sum_{i=1}^{a} 1/v_i \right)^{-1}}.$$

Such a confidence limit can be obtained as the $1 - \alpha$ quantile of

$$z_{\frac{1+p}{2}} \sqrt{\left[G_{\sigma_\tau^2} + G_{\sigma_e^2} + \left(\sum_{i=1}^{a} 1/G_{v_i} \right)^{-1} \right]_+},$$

where the quantities $G_{\sigma_\tau^2}$, $G_{\sigma_e^2}$, and G_{v_i} are defined in Section 5.3.2, and $x_+ = \max\{0, x\}$. Once the confidence limit D_3 is obtained, a $(p, 1 - \alpha)$ tolerance interval for $N(\mu, \sigma_\tau^2 + \sigma_e^2)$ is given by $\tilde{\mu} \pm D_3$.

We now discuss the choice of $\tilde{\mu}$. Since $\bar{Y}_v$ depends on unknown parameters, one possibility is to consider the GPQ G_μ given in (5.3.14), and take $\tilde{\mu}$ to be the mean (or the median) of G_μ. This mean (or median) can be easily estimated based on the realizations $G_{\mu,k}$ of G_μ; see Step 6 of the Algorithm 5.1.

The following algorithm can be used to implement the above procedure to compute a $(p, 1 - \alpha)$ tolerance interval for $N(\mu, \sigma_\tau^2 + \sigma_e^2)$:

Algorithm 5.2

1. Once the data are obtained, compute the observed values $\bar{y}_{1.}, \bar{y}_{2.}, \ldots, \bar{y}_{a.}$, and ss_e.

2. Let K denote the simulation sample size. For $k = 1, 2, \ldots, K$, perform Steps 3–6 of Algorithm 5.1 in Section 5.3.2.

3. Compute $\left[G_{\sigma_\tau^2, k} + G_{\sigma_e^2, k} + \left(\sum_{i=1}^a G_{v_i, k}^{-1}\right)^{-1}\right]_+$.

4. Compute $\tilde{\mu}$, the mean of the $G_{\mu, k}$ values.

5. Compute D_3, the $1-\alpha$ quantile of $z_{\frac{1+p}{2}} \left[G_{\sigma_\tau^2, k} + G_{\sigma_e^2, k} + \left(\sum_{i=1}^a G_{v_i, k}^{-1}\right)^{-1}\right]_+^{\frac{1}{2}}$, where $x_+ = \max\{0, x\}$.

6. A $(p, 1-\alpha)$ tolerance interval for $N(\mu, \sigma_\tau^2 + \sigma_e^2)$ is given by $\tilde{\mu} \pm D_3$.

5.5.2   A Two-Sided Tolerance Interval for $N(\mu, \sigma_\tau^2)$

A $(p, 1-\alpha)$ tolerance interval for $N(\mu, \sigma_\tau^2)$ is given by $\tilde{\mu} \pm D_4$, where $\tilde{\mu}$ is as defined before and the margin of error statistic D_4 is such that D_4 is a $1-\alpha$ upper confidence limit of the parametric function $z_{\frac{1+p}{2}} \left(\sigma_\tau^2 + \left(\sum_{i=1}^a v_i^{-1}\right)^{-1}\right)^{\frac{1}{2}}$. The quantity $\tilde{\mu}$ can be computed as in Step 4 of Algorithm 5.2, and D_4 can be obtained proceeding similar to Step 5 of Algorithm 5.2.

Extensive numerical results are reported in Liao, Lin and Iyer (2005) regarding the performance of the two-sided tolerance intervals in this section for the distributions $N(\mu, \sigma_\tau^2 + \sigma_e^2)$ and $N(\mu, \sigma_\tau^2)$. Slight conservatism has been noted when the intra-class correlation $\frac{\sigma_\tau^2}{\sigma_\tau^2 + \sigma_e^2}$ is small. Otherwise, the tolerance intervals maintain the confidence level quite satisfactorily.

Example 5.2 (continued)

Based on the sulfur content data in Table 5.2, let us compute a $(0.99, 0.95)$ two-sided tolerance interval for the sulfur content distribution $N(\mu, \sigma_\tau^2 + \sigma_e^2)$, and for the "true" sulfur content distribution $N(\mu, \sigma_\tau^2)$. Here $a = 6$ and $N = 10$. The observed values are: $\bar{y}_{1.} = 4.670$, $\bar{y}_{2.} = 4.685$, $\bar{y}_{3.} = 4.640$, $\bar{y}_{4.} = 4.695$, $\bar{y}_{5.} = 4.570$, $\bar{y}_{6.} = 4.740$, and $ss_e = 0.0419$. Using simulation consisting of 100,000 runs, $\tilde{\mu}$ is computed as 4.67, and the tolerance interval for the sulfur content distribution is obtained as $(4.22, 5.12)$. The tolerance interval for the true concentration is obtained as $(4.43, 4.91)$. The tolerance intervals (for both distributions) were obtained using Step 5b, that is, using $G_{\sigma_\tau^2}$ in $(5.3.4)$.

5.6 Exercises

5.6.1. For the one-way random model with unbalanced data, let $\bar{Y}_{i.} = \sum_{j=1}^{n_i} Y_{ij}/n_i$, $i = 1, 2, ..., a$, and define $\bar{Y} = (\bar{Y}_{1.}, \bar{Y}_{2.}, ..., \bar{Y}_{a.})'$, so that $\bar{Y}$ has a multivariate normal distribution.

  (a) Derive the mean vector and the variance-covariance matrix of $\bar{Y}$.

  (b) Express $U_{\bar{y}}^2$ in $(5.3.2)$ as a quadratic form in $\bar{Y}$.

  (c) Obtain a condition under which $U_{\bar{y}}^2$ has a chi-square distribution (use the results on the distribution of quadratic forms; see Rao (1973) or Searle (1971)).

  (d) Show that the condition in (c) will always hold in the case of balanced data.

  (e) Show that $U_{\bar{y}}^2$ in $(5.3.2)$ reduces to U_τ^2 in $(4.1.3)$ in the case of balanced data.

5.6.2. When the data are balanced, show that

  (a) the GPQ in $(5.3.5)$ reduces to the GPQ in $(4.3.16)$, and the approximate tolerance limit in $(5.3.6)$ reduces to that in $(4.3.20)$.

  (b) The GPQ in $(5.4.1)$ and the approximate tolerance limit in $(5.4.2)$ reduce to the corresponding quantities in $(4.4.1)$ and $(4.4.2)$, respectively.

5.6.3. For the one-way random model with unbalanced data, show that $\bar{\bar{Y}}$ and SS_e are independently distributed, and $SS_{\bar{y}}$ and SS_e are also independently distributed. Show also that $\bar{\bar{Y}}$ and $SS_{\bar{y}}$ are not independently distributed, unless the data are balanced.

5.6.4. Consider the log-transformed nickel dust exposure data in Table 5.3.

  (a) Compute $(0.95, 0.95)$ upper tolerance limits for the distributions $N(\mu, \sigma_\tau^2 + \sigma_e^2)$ and $N(\mu, \sigma_\tau^2)$, by estimating the percentiles of the respective approximate GPQs. Also compute the upper tolerance limits using approximations similar to those in $(5.3.6)$ and $(5.4.2)$. Do the results based on the approximations agree with that obtained by estimating the percentiles of the respective approximate GPQs?

(b) Compute (0.95, 0.95) two sided tolerance intervals for the distributions $N(\mu, \sigma_\tau^2 + \sigma_e^2)$ and $N(\mu, \sigma_\tau^2)$. Use Algorithm 5.2.

5.6.5. Modify the Mee-Owen approach described in Section 4.3.1 to the case of unbalanced data by using the approximate chi-square distribution of $U_{\bar{y}}^2$ in the place of the exact chi-square distribution of U_τ^2 (given in Chapter 4). Follow the derivations in Section 4.3.1 to arrive at an approximate tolerance factor for the case of unbalanced data, similar to the one in (4.3.5) for the case of balanced data. Using the solution so obtained, compute a (0.90, 0.95) upper tolerance limit for the distribution $N(\mu, \sigma_\tau^2 + \sigma_e^2)$ of the moisture content, in the context of Example 5.1. Does the result agree with those obtained in Section 5.3.2?

5.6.6. Follow the suggestion in the previous problem to derive a two-sided tolerance interval for $N(\mu, \sigma_\tau^2 + \sigma_e^2)$ in the case of unbalanced data , by imitating Mee's approach given in Section 4.5.1. Using the solution so obtained, compute a (0.99, 0.95) two-sided tolerance interval for the sulfur content distribution $N(\mu, \sigma_\tau^2 + \sigma_e^2)$, in the context of Example 5.2. Does the result agree with those obtained in Section 5.5.2?

5.6.7. For the one-way random model with unbalanced data, proceed as in Section 5.3.3 to derive an upper confidence bound for the exceedance probability concerning the true values, namely $P(\mu + \tau > L)$, where $\tau \sim N(0, \sigma_\tau^2)$ and L is a specified limit. Apply the procedure to derive a 95% upper confidence limit for $P(\mu + \tau > 0)$ using the log-transformed nickel dust exposure data in Table 5.3.

5.6.8. For the one-way random model with unbalanced data, suppose the variance ratio, $R = \sigma_\tau^2 / \sigma_e^2$ is known. Use the relevant results in Chapter 2 to derive the following:

(a) one-sided and two-sided tolerance intervals for $N(\mu, \sigma_\tau^2 + \sigma_e^2)$.

(b) one-sided and two-sided tolerance intervals for $N(\mu, \sigma_\tau^2)$.

(c) equal-tailed tolerance intervals for $N(\mu, \sigma_\tau^2 + \sigma_e^2)$ and $N(\mu, \sigma_\tau^2)$.

5.6.9. Explain how you will address parts (a)–(c) of Problem 4.7.6 of Chapter 4, in the case of two one-way random models with unbalanced data.

Chapter 6

Some General Mixed Models

6.1 Notations and Preliminaries

When we have a general mixed effects model with the normality assumption and balanced data, it is possible to derive tolerance intervals (one-sided or two-sided) by suitably applying the generalized confidence interval idea. The problem becomes tractable in the case of balanced data since a set of sufficient statistics can be found consisting of independent normal and scaled chi-square random variables. In the case of one-sided tolerance limits, the generalized confidence interval procedure can be directly applied, since the problem reduces to that of computing a confidence limit for a percentile. For computing a two-sided tolerance interval, an approximate margin of error statistic can be found as the square root of a generalized upper confidence limit for a linear combination of the variance components. In other words, the generalized confidence interval based procedures in Chapter 4, developed for the one-way random model with balanced data, can be easily extended for a general mixed effects model with balanced data; this has been carried out by Liao, Lin and Iyer (2005).

For a random vector Y, a general mixed effects model can be written as

$$
\begin{aligned}
Y &= \sum_{i=1}^{\nu_1} \mathbf{X}_i \boldsymbol{\beta}_i + \sum_{i=1}^{\nu_2} \mathbf{Z}_i \boldsymbol{\delta}_i + e \\
&= \mathbf{X}\boldsymbol{\beta} + \sum_{i=1}^{\nu_2} \mathbf{Z}_i \boldsymbol{\delta}_i + e,
\end{aligned} \tag{6.1.1}
$$

where the vectors $\boldsymbol{\beta}_i$'s represent fixed effects, $\boldsymbol{\delta}_i$'s are the random effects, $\mathbf{X}_i$'s and $\mathbf{Z}_i$'s are design matrices, e is the error vector, $\mathbf{X} = (\mathbf{X}_1, \mathbf{X}_2, ..., \mathbf{X}_{\nu_1})$, and

$\boldsymbol{\beta} = (\boldsymbol{\beta}_1', \boldsymbol{\beta}_2', ..., \boldsymbol{\beta}_{\nu_1}')'$. In (6.1.1), it is assumed that there are ν_1 factors whose effects are fixed, and ν_2 factors whose effects are random. The dimensions of $\boldsymbol{\beta}_i$ and $\boldsymbol{\delta}_i$ are simply the number of levels of the ith fixed factor, and the ith random factor, respectively. We make the usual normality assumptions: $\boldsymbol{\delta}_i \sim N(0, \sigma_{\delta_i}^2 \mathbf{I})$, $\boldsymbol{e} \sim N(0, \sigma_e^2 \mathbf{I})$, and all the random variables are independent. If the above model is an ANOVA model with mixed effects and balanced data, a set of complete sufficient statistics consists of the ordinary least squares estimators of the fixed effect parameters, along with the $\nu_2 + 1$ sums of squares corresponding to the ν_2 random effects, and that corresponding to the error. Furthermore, these quantities are all independent, and each sum of squares is distributed as a scaled chi-square. In particular, if SS_e denotes the error sum of squares, then SS_e/σ_e^2 is distributed as a chi-square. We shall not elaborate on these properties in a general balanced data setting; the interested reader may refer to Searle (1971), Khuri (1982), Khuri, Mathew and Sinha (1998) and Montgomery (2009). For illustration, we shall provide the details for just one special model.

A Two-Way Nested Model

Consider two factors A and B with the levels of B nested within the levels of A. Suppose A has a levels, and b levels of B are nested within each level of A, with n observations obtained on each level combination. Let Y_{ijl} denote the lth observation corresponding to the jth level of B nested within the ith level of A. We then have the model

$$Y_{ijl} = \mu + \tau_i + \beta_{j(i)} + e_{ijl}, \tag{6.1.2}$$

$i = 1, 2, ..., a$; $j = 1, 2, ..., b$; $l = 1, 2, ..., n$, where μ is a general mean, τ_i is the main effect due to the ith level of A, $\beta_{j(i)}$ is the effect due to the jth level of B nested within the ith level of A, and e_{ijl}'s are the error terms with $e_{ijl} \sim N(0, \sigma_e^2)$. If the levels of A are fixed, and the levels of B are randomly selected, we are in the mixed model set up, and we assume that $\sum_{i=1}^a \tau_i = 0$, $\beta_{j(i)} \sim N(0, \sigma_\beta^2)$, where the $\beta_{j(i)}$'s and e_{ijl}'s are assumed to be independent. If the levels of A are also randomly selected, then we have a random effects model, and in addition to the above distributional assumptions, we also assume $\tau_i \sim N(0, \sigma_\tau^2)$, independent of the $\beta_{j(i)}$'s and e_{ijl}'s. Writing $\boldsymbol{Y} = (Y_{111}, Y_{112}, ..., Y_{11n}, Y_{121}, Y_{122}, ..., Y_{12n}, ..., Y_{ab1}, Y_{ab2}, ..., Y_{abn})'$, $\boldsymbol{e}$ defined similarly, $\boldsymbol{\tau} = (\tau_1, \tau_2, ..., \tau_a)'$, and $\boldsymbol{\beta} = (\beta_{1(1)}, \beta_{2(1)}, ..., \beta_{b(1)}, ..., \beta_{1(a)}, \beta_{2(a)}, ..., \beta_{b(a)})'$, the model (6.1.2) can be written as

$$\boldsymbol{Y} = \mathbf{1}_{abn}\mu + (\mathbf{I}_a \otimes \mathbf{1}_b \otimes \mathbf{1}_n)\boldsymbol{\tau} + (\mathbf{I}_a \otimes \mathbf{I}_b \otimes \mathbf{1}_n)\boldsymbol{\beta} + \boldsymbol{e}, \tag{6.1.3}$$

where $\otimes$ denotes the Kronecker product, and $\mathbf{1}_r$ denotes an $r \times 1$ vector of ones. Let $\bar{Y}_{ij.} = \frac{1}{n}\sum_{l=1}^{n} Y_{ijl}$, $\bar{Y}_{i..} = \frac{1}{bn}\sum_{j=1}^{b}\sum_{l=1}^{n} Y_{ijl}$, $\bar{Y}_{...} = \frac{1}{abn}\sum_{i=1}^{a}\sum_{j=1}^{b}\sum_{l=1}^{n} Y_{ijl}$, and let

$$SS_\tau = bn \sum_{i=1}^{a}(\bar{Y}_{i..} - \bar{Y}_{...})^2$$

$$SS_\beta = n \sum_{i=1}^{a}\sum_{j=1}^{b}(\bar{Y}_{ij.} - \bar{Y}_{i..})^2$$

$$SS_e = \sum_{i=1}^{a}\sum_{j=1}^{b}\sum_{l=1}^{n}(Y_{ijl} - \bar{Y}_{ij.})^2. \qquad (6.1.4)$$

When the $\beta_{j(i)}$'s are random, having the normal distribution mentioned above, we have

$$U_\beta^2 = \frac{SS_\beta}{n\sigma_\beta^2 + \sigma_e^2} \sim \chi_{a(b-1)}^2 \quad \text{and} \quad U_e^2 = \frac{SS_e}{\sigma_e^2 \sim \chi_{ab(n-1)}^2}, \qquad (6.1.5)$$

where the above chi-squares are also independent. When the τ_i's are fixed effects, define $\mu_i = \mu + \tau_i$. Then an estimator of μ_i is

$$\hat{\mu}_i = \bar{Y}_{i..} \sim N\left(\mu_i, \frac{n\sigma_\beta^2 + \sigma_e^2}{bn}\right), \qquad (6.1.6)$$

independent of the chi-square random variables in (6.1.5). A set of sufficient statistics for the mixed model (6.1.2) consists of the $\bar{Y}_{i..}$'s ($i = 1, 2, \ldots, a$), along with SS_β and SS_e. Suppose a tolerance interval is required for an observation corresponding to the ith level of A, i.e., for the distribution $N(\mu_i, \sigma_\beta^2 + \sigma_e^2)$. In Section 6.3, we shall show how such a tolerance interval can be computed. We shall also develop a procedure for computing a tolerance interval for the true value $\mu_i + \beta_{j(i)}$ corresponding to the ith level of A, i.e., for the distribution $N(\mu_i, \sigma_\beta^2)$.

If the τ_i's are also random in the model (6.1.2), having the normal distribution mentioned above, we also have

$$U_\tau^2 = \frac{SS_\tau}{(bn\sigma_\tau^2 + n\sigma_\beta^2 + \sigma_e^2)} \sim \chi_{a-1}^2, \qquad (6.1.7)$$

independent of the chi-square distributions in (6.1.5), where SS_τ is defined in (6.1.4). Now the only fixed parameter in the model is μ, and we have the estimator

$$\hat{\mu} = \bar{Y}_{...} \sim N\left(\mu, \frac{bn\sigma_\tau^2 + n\sigma_\beta^2 + \sigma_e^2}{abn}\right), \qquad (6.1.8)$$

independent of the chi-square random variables in (6.1.5) and (6.1.7). A set of sufficient statistics for the random effects model (6.1.2) consists of $\bar{Y}_{...}$ along with SS_τ, SS_β and SS_e. Now a tolerance interval for an observation corresponding to any level combination of A and B is a tolerance interval for the distribution $N(\mu, \sigma_\tau^2 + \sigma_\beta^2 + \sigma_e^2)$. Furthermore, a tolerance interval for the true value $\mu + \tau_i + \beta_{j(i)}$ is a tolerance interval for the distribution $N(\mu, \sigma_\tau^2 + \sigma_\beta^2)$. The results in Section 6.3 can once again be applied to derive such tolerance intervals. Further details concerning the distributional results mentioned above for the nested model can be found in Montgomery (2009) or Searle (1971). Also, the derivation of tolerance intervals under the model (6.1.2) is addressed in Fonseca et al. (2007), in the random effects as well as in the mixed effects scenarios.

In Section 6.3, we shall derive one-sided and two-sided tolerance intervals in a very general setting applicable to mixed and random models with balanced data. In our general setting, the problem consists of constructing a tolerance interval for the distribution $N(\theta, \sum_{i=1}^{q} c_i \sigma_i^2)$, given the independent statistics $\hat{\theta}$ and SS_i ($i = 1, 2, ..., q$), having the distributions $\hat{\theta} \sim N(\theta, \sum_{i=1}^{q} d_i \sigma_i^2)$, and $SS_i/\sigma_i^2 \sim \chi_{f_i}^2$. Here the c_i's and d_i's are known constants. For example, in the context of the model (6.1.2) where the τ_i's are fixed and the $\beta_{j(i)}$'s are random (having the normal distribution mentioned earlier), suppose we want a tolerance interval for the distribution of the Y_{ijl} for a fixed i, i.e., for the distribution $N(\mu_i, \sigma_\beta^2 + \sigma_e^2)$, where $\mu_i = \mu + \tau_i$. Define $SS_1 = SS_\beta$, $SS_2 = SS_e$, $\sigma_1^2 = n\sigma_\beta^2 + \sigma_e^2$, $\sigma_2^2 = \sigma_e^2$, $f_1 = a(b-1)$, and $f_2 = ab(n-1)$. Then $SS_i/\sigma_i^2 \sim \chi_{f_i}^2$, $i = 1, 2$; see (6.1.5). Furthermore, taking $\theta = \mu_i$, $\hat{\theta} = \hat{\mu}_i$ follows a normal distribution; see (6.1.6). From (6.1.6), it is readily verified that the variance of $\hat{\mu}_i$ is a linear combination of the σ_i^2's; in fact, it is just a multiple of σ_1^2. Furthermore, in the distribution $N(\mu_i, \sigma_\beta^2 + \sigma_e^2)$, for which a tolerance interval is required, the variance $\sigma_\beta^2 + \sigma_e^2$ is also a linear combination of the σ_i^2's. In other words, this tolerance interval problem is a special case of the general problem of constructing a tolerance interval for the distribution $N(\theta, \sum_{i=1}^{q} c_i \sigma_i^2)$, mentioned above. This is also true for computing a tolerance interval for the distribution $N(\mu_i, \sigma_\beta^2)$, and also for computing tolerance intervals for the distributions $N(\mu, \sigma_\tau^2 + \sigma_\beta^2 + \sigma_e^2)$ and $N(\mu, \sigma_\tau^2 + \sigma_\beta^2)$ in the context of the model (6.1.2) with all effects random. Our results on this general tolerance interval problem, described in Section 6.3, are due to Liao, Lin and Iyer (2005), and are generalizations of several results in Chapter 4.

In Section 6.4, we shall consider a general model with exactly one random effect, and hence two variance components: the random effect variance component and the error variance component. The model is thus a special case of

(6.1.1) with $\nu_2 = 1$, and can be written as

$$Y = \mathbf{X}\beta + \mathbf{Z}\delta + e, \qquad (6.1.9)$$

where β is a fixed unknown parameter vector, δ is a vector of random effects having the distribution $\delta \sim N(0, \sigma_\delta^2 \mathbf{I})$, independent of $e \sim N(0, \sigma_e^2 \mathbf{I})$. The one-way random model with balanced or unbalanced data is a special case of such a model with two variance components. We also note that the model (6.1.2) with mixed effects is also a special case of such a model, even if the data are unbalanced. Another special case of (6.1.9) is a two-way crossed classification model without interaction and with mixed effects. For the model (6.1.9) with exactly one random effect, one-sided and two-sided tolerance intervals are derived in Section 6.4, generalizing the results in Section 5.3.2, Section 5.4.2, and Section 5.5. The examples in the next section will be used to illustrate our results.

Bagui, Bhaumik and Parnes (1996) have made an attempt at constructing one-sided tolerance limits in a general unbalanced random effects model with more than two variance components. They have computed an upper tolerance limit assuming that the variance components are known, and then replaced the unknown variance components by estimates. The actual confidence levels of the resulting tolerance limit can be quite different from the assumed nominal level. This is also noted in Smith (2002), who has addressed the computation of one-sided tolerance limits under a two-way crossed classification model with interaction, with random effects, and with unbalanced data, relevant to some applications in environmental monitoring and regulation. Our results in Section 6.4 deal with general models with two variance components only. In Section 6.5, we shall describe the computation of tolerance intervals for a one-way random model with covariates and heteroscedastic error variances. The topic of bioequivalence testing is discussed in Section 6.6, and tolerance intervals are proposed for testing individual bioequivalence. For general mixed effects or random effects models with unbalanced data, a unified methodology is currently not available for computing tolerance intervals.

6.2 Some Examples

Here we shall give a couple of motivating examples. These and a few other examples will be taken up later in this chapter.

Example 6.1 (A glucose monitoring meter experiment)

This example is on a glucose meter experiment and was first considered by Liao and Iyer (2001) and later by Liao and Iyer (2004) and Liao, Lin and Iyer (2005). The application deals with a gauge study for comparing the quality between a test meter and a reference meter. The test meter is a newly developed glucose monitoring meter for in-home use by diabetes patients, and the reference meter is an already marketed glucose monitoring meter. Suppose n replicate measurements each are obtained with strips that come from L lots, on B blood samples, using m_1 test meters and m_2 reference meters. In the models introduced below, the effects due to the lots, blood samples and meters will all be random effects.

In order to describe the model and the tolerance interval problem, let W_{ijlr} denote the rth measurement using a strip from the lth lot, on the jth blood sample, using the ith test meter, $r = 1, 2, ..., E$; $l = 1, 2, ..., L$; $j = 1, 2, ..., B$; $i = 1, 2, ..., m_1$. Also let Y_{ijlr} denote the corresponding observation for the ith reference meter; $r = 1, 2, ..., E$; $l = 1, 2, ..., L$; $j = 1, 2, ..., B$; $i = 1, 2, ..., m_2$. The models are given by

$$
\begin{aligned}
W_{ijlr} &= \mu_T + M_{Ti} + B_j + L_l + e_{ijlr} \\
Y_{ijlr} &= \mu_R + M_{Ri} + B_j + L_l + e'_{ijlr},
\end{aligned}
\tag{6.2.1}
$$

where μ_T and μ_R represent expected readings when using a test meter and a reference meter, respectively, M_{Ti} and M_{Ri} represent the effects due to the ith test meter and the ith reference meter, respectively, B_j denotes the effect due to the jth blood sample, L_l denotes the effect due to the lth strip lot, and e_{ijlr} and e'_{ijlr} denote the error terms. Here μ_T and μ_R are fixed unknown parameters, and the rest are all random effects. We assume $M_{Ti} \sim N(0, \sigma_T^2)$, $M_{Ri} \sim N(0, \sigma_R^2)$, $B_j \sim N(0, \sigma_B^2)$, $L_l \sim N(0, \sigma_L^2)$, $e_{ijlr} \sim N(0, \sigma_e^2)$ and $e'_{ijlr} \sim N(0, \sigma_e^2)$, where all the random variables are assumed to be independent. Note that e_{ijlr} and e'_{ijlr} are assumed to have the same variance. The true average reading for the ith test meter, based on the jth blood sample and lth strip lot is $\mu_T + M_{Ti} + B_j + L_l$. The average reference reading for the same blood sample and strip lot, averaged over all the reference meters, is $\mu_R + B_j + L_l$. In order to assess the quality of individual test meters, the mean value $\mu_T + M_{Ti} + B_j + L_l$ is to be compared against the reference value $\mu_R + B_j + L_l$. This can be accomplished by computing a tolerance interval for the difference $(\mu_T + M_{Ti} + B_j + L_l) - (\mu_R + B_j + L_l) = \mu_T - \mu_R + M_{Ti} \sim N(\mu_T - \mu_R, \sigma_T^2)$. In other words, what is required is a tolerance interval for the distribution $N(\mu_T - \mu_R, \sigma_T^2)$. Later we shall see that this is a special case of the general set up considered in the next section. If a (0.95, 0.90) tolerance interval for $N(\mu_T - \mu_R, \sigma_T^2)$ is within [-5, 5], we conclude that the quality requirements have been met by a batch of test meters; see Liao and Iyer (2001, 2004).

The relevant data for this example are based on blood samples that were prepared so that the true glucose was 50. Furthermore, the data corresponded to $m_1 = 44$, $m_2 = 20$, and $B = L = E = 3$. Part of the data are reported in Table A1, Appendix A. The full data can be accessed at the Wiley ftp site provided at the bottom of Table A1.

Example 6.2 (Strength measurements from 8 batches)

This is one of the example data sets distributed along with the FORTRAN program RECIPE (due to Mark Vangel) available at the National Institutes of Standards and Technology (NIST) website*. The data set that we shall use is from the file ex4.dat posted at the above site. This example is also discussed in Liao, Lin and Iyer (2005). The example is on strength measurements from 8 batches obtained at two fixed temperatures within each batch: 75°F and −67°F. Let Y_{ijl} denote the lth strength measurement obtained at the jth temperature for the ith batch; $l = 1, 2, ..., n_{ij}$; $j = 1, 2$, and $i = 1, 2, ..., 8$. The data are given in Table 6.1 with the temperature labeled as 1 and 2, corresponding to 75°F and −67°F, respectively.

From the data, we see that $n_{11} = 5$, $n_{21} = 6$, $n_{31} = 5$, $n_{41} = 5$, $n_{51} = 5$, $n_{61} = 5$, $n_{71} = 0$, $n_{81} = 0$, $n_{12} = 6$, $n_{22} = 6$, $n_{32} = 6$, $n_{42} = 6$, $n_{52} = 6$, $n_{62} = 0$, $n_{72} = 6$, $n_{82} = 5$. We shall model the above data using the following simple linear regression model with a random effect; see the above NIST site, and see also Liao, Lin and Iyer (2005):

$$Y_{ijl} = \beta_0 + \beta_1 x + \tau_i + e_{ijl}, \tag{6.2.2}$$

where x represents the temperature (taking only two values 75°F and −67°F, labeled as 1 and 2 in Table 6.1), β_0 and β_1 are the slope and intercept parameters, respectively, τ_i's represent the random batch effects, and e_{ijl}'s are the error terms. We assume $\tau_i \sim N(0, \sigma_\tau^2)$ and $e_{ijl} \sim N(0, \sigma_e^2)$, where the τ_i's and e_{ijl}'s are all assumed to be independent. We shall now exhibit the model (6.2.2) as a special case of the general model (6.1.9). Note that $\sum_{i=1}^{8} n_{i1} = 31$ and $\sum_{i=1}^{8} n_{i2} = 41$. Write

$$\mathbf{Y} = (Y_{111}, Y_{112}, ..., Y_{11n_{11}}, Y_{211}, Y_{212}, ..., Y_{21n_{21}}, ..., Y_{821}, Y_{822}, ..., Y_{82n_{82}})',$$

with $\mathbf{e}$ defined similarly consisting of the e_{ijl}'s, $\boldsymbol{\tau} = (\tau_1, \tau_2, ..., \tau_8)'$, $\boldsymbol{\beta} = (\beta_0, \beta_1)'$, and

$$\mathbf{X} = \left(\begin{array}{cc} \mathbf{1}_{31} & 75 \times \mathbf{1}_{31} \\ \mathbf{1}_{41} & -67 \times \mathbf{1}_{41} \end{array} \right),$$

where $\mathbf{1}_r$ denotes an $r \times 1$ vector of ones, and we have also used the fact that the temperature x can take the two values 75 and −67, labeled as 1 and 2 in

---

*http://www.itl.nist.gov/div898/software/recipe/homepage.html.

Table 6.1. In order to define the matrix $\mathbf{Z}$, let us denote by $\mathbf{0}_r$ an $r \times 1$ vector of zeros. Now define

$$
\mathbf{Z} = \begin{pmatrix}
\text{diag}(\mathbf{1}_5, \mathbf{1}_6, \mathbf{1}_5, \mathbf{1}_5, \mathbf{1}_5, \mathbf{1}_5), \mathbf{0}_{31}, \mathbf{0}_{31} \\[2mm]
\text{diag}(\mathbf{1}_6, \mathbf{1}_6, \mathbf{1}_6, \mathbf{1}_6, \mathbf{1}_6), \mathbf{0}_{30}, \mathbf{0}_{30}, \mathbf{0}_{30} \\[2mm]
\mathbf{0}_{11}, \mathbf{0}_{11}, \mathbf{0}_{11}, \mathbf{0}_{11}, \mathbf{0}_{11}, \mathbf{0}_{11}, \text{diag}(\mathbf{1}_6, \mathbf{1}_5)
\end{pmatrix}.
$$

With the above notation, the model (6.2.2) is a special case of the general model (6.1.9) with $\boldsymbol{\delta} = \boldsymbol{\tau}$.

If we are interested in computing a tolerance interval for the distribution of the strength measurements at a specific temperature x_0, the problem reduces to computing a tolerance interval for the distribution $N(\mu, \sigma_\tau^2 + \sigma_e^2)$, where $\mu = \beta_0 + \beta_1 x_0$.

We note that a general model of the type (6.1.9) can also come up in the context of certain mixed or random effects models involving fixed covariates. A one-way random model with a fixed covariate is discussed in Limam and Thomas (1988b), and is a special case of the model (6.1.9). The model (6.2.2) corresponding to Example 6.2 is in fact a one-way random model with a fixed covariate. Such models will be taken up again in Section 6.5 when the errors are heteroscedastic.

6.3 Tolerance Intervals in a General Setting Applicable to Models with Balanced Data

As pointed out in Section 6.1, we shall now describe procedures for constructing a tolerance interval for the distribution $N(\theta, \sum_{i=1}^{q} c_i \sigma_i^2)$, based on the independent statistics $\hat{\theta}$ and SS_i ($i = 1, 2, ..., q$), with

$$
Z = \frac{\hat{\theta} - \theta}{\sqrt{\sum_{i=1}^{q} d_i \sigma_i^2}} \sim N(0,1) \quad \text{and} \quad U_i^2 = \frac{SS_i}{\sigma_i^2} \sim \chi_{f_i}^2. \tag{6.3.1}
$$

Here the c_i's and d_i's are known constants, not necessarily nonnegative. We shall first consider the one-sided tolerance limit problem, followed by the construction of two-sided tolerance intervals. Generalized confidence intervals will be used in both cases.

Table 6.1: The strength measurement data

Temperature	Batch	Strength measurement	Temperature	Batch	Strength measurement
1	1	328.1174	2	1	344.1524
1	1	334.7674	2	2	308.6256
1	1	347.7833	2	2	315.1819
1	1	346.2661	2	2	317.6867
1	1	338.7314	2	2	313.9832
1	2	297.0387	2	2	309.3132
1	2	293.4595	2	2	275.1758
1	2	308.0419	2	3	321.4128
1	2	326.4864	2	3	316.4652
1	2	318.1297	2	3	331.3724
1	2	309.0487	2	3	304.8643
1	3	337.0930	2	3	309.6249
1	3	317.7319	2	3	347.8449
1	3	321.4292	2	4	331.5487
1	3	317.2652	2	4	316.5891
1	3	291.8881	2	4	303.7171
1	4	297.6943	2	4	320.3625
1	4	327.3973	2	4	315.2963
1	4	303.8629	2	4	322.8280
1	4	313.0984	2	5	340.0990
1	4	323.2769	2	5	348.9354
1	5	312.9743	2	5	331.2500
1	5	324.5192	2	5	330.0000
1	5	334.5965	2	5	340.9836
1	5	314.9458	2	5	329.4393
1	5	322.7194	2	7	330.9309
1	6	291.1215	2	7	328.4553
1	6	309.7852	2	7	344.1026
1	6	304.8499	2	7	343.3584
1	6	288.0184	2	7	344.4717
1	6	294.1995	2	7	351.2776
2	1	340.8146	2	8	331.0259
2	1	343.5855	2	8	322.4052
2	1	334.1746	2	8	327.6699
2	1	348.6610	2	8	296.8215
2	1	356.3232	2	8	338.1995

6.3.1 One-Sided Tolerance Intervals

Suppose a $(p, 1-\alpha)$ lower tolerance limit is required for $N(\theta, \sum_{i=1}^{q} c_i \sigma_i^2)$. That is, we need to compute a $100(1 - \alpha)\%$ lower confidence limit for $\theta - z_p \sqrt{\sum_{i=1}^{q} c_i \sigma_i^2}$, where z_p denotes the p percentile of the standard normal distribution. Let $\hat{\theta}_{\text{obs}}$ and ss_i, respectively, denote the observed values of $\hat{\theta}$ and SS_i ($i = 1, 2, ..., q$), and define

$$G_i = \frac{\sigma_i^2}{SS_i} ss_i = \frac{ss_i}{U_i^2}$$

$$G_\theta = \hat{\theta}_{\text{obs}} - \frac{\hat{\theta} - \theta}{\sqrt{\sum_{i=1}^q d_i \sigma_i^2}} \sqrt{\sum_{i=1}^q d_i G_i}$$

$$= \hat{\theta}_{\text{obs}} - Z \times \sqrt{\sum_{i=1}^q d_i G_i}, \tag{6.3.2}$$

where Z and the U_i^2's are defined in (6.3.1). We note that given the observed data, the distribution of $(G_\theta, G_1, G_2, ..., G_q)$ is free of unknown parameters. Furthermore, the observed value of G_i is σ_i^2 and that of G_θ is θ. In other words G_i is a generalized pivotal quantity (GPQ) for σ_i^2, and G_θ is a GPQ for θ. Thus a GPQ for $\theta - z_p\sqrt{\sum_{i=1}^q c_i \sigma_i^2}$ is given by

$$G_8 = G_\theta - z_p \sqrt{\sum_{i=1}^q c_i G_i}, \tag{6.3.3}$$

and the 100αth percentile of G_8 gives a $100(1-\alpha)\%$ generalized lower confidence limit for $\theta - z_p\sqrt{\sum_{i=1}^q c_i \sigma_i^2}$, which in turn gives a $(p, 1-\alpha)$ lower tolerance limit for $N(\theta, \sum_{i=1}^q c_i \sigma_i^2)$. The following algorithm can be used for estimating the 100αth percentile of G_8:

Algorithm 6.1

1. Once the data are obtained, compute the observed values $\hat{\theta}_{\text{obs}}$, ss_1, ss_2, ..., ss_q.

2. Let K denote the simulation sample size. For $k = 1, 2, ..., K$, perform the following steps.

3. Generate independent random variables $Z_k \sim N(0, 1)$, and $U_{i,k}^2 \sim \chi_{f_i}^2$, $i = 1, 2, ..., q$.

4. Compute $G_{i,k} = \frac{ss_i}{U_{i,k}^2}$ and $G_{\theta,k} = \hat{\theta}_{\text{obs}} - Z_k \times \sqrt{\sum_{i=1}^q d_i G_{i,k}}$.

5. Compute

$$G_{8,k} = G_{\theta,k} - z_p \sqrt{\sum_{i=1}^{q} c_i G_{i,k}}.$$

The 100αth percentile of the $G_{8,k}$ values ($k = 1, 2, ..., K$) gives an estimate of the 100αth percentile of G_8, which in turn gives an estimate of the required $(p, 1 - \alpha)$ lower confidence limit.

In order to compute a $(p, 1 - \alpha)$ upper tolerance limit for $N(\theta, \sum_{i=1}^{q} c_i \sigma_i^2)$, consider

$$G_9 = G_\theta + z_p \sqrt{\sum_{i=1}^{q} c_i G_i}. \tag{6.3.4}$$

The $100(1 - \alpha)$th percentile of G_9 gives a $100(1 - \alpha)\%$ generalized upper confidence limit for $\theta + z_p \sqrt{\sum_{i=1}^{q} c_i \sigma_i^2}$, which in turn gives a $(p, 1 - \alpha)$ upper tolerance limit for $N(\theta, \sum_{i=1}^{q} c_i \sigma_i^2)$. This limit can obviously be estimated using an algorithm similar to the one given above.

6.3.2 Two-Sided Tolerance Intervals

The procedure to be described here, due to Liao, Lin and Iyer (2005), is an immediate extension of the approach described in Section 4.5.2 and Section 4.6. Thus we shall obtain a margin of error statistic D_5 that is a function of the SS_i's, so that

$$P_{\hat{\theta}, \, SS_1, ..., SS_q} \left[P_Y \left\{ \hat{\theta} - D_5 \leq Y \leq \hat{\theta} + D_5 | \hat{\theta}, \, SS_1, ..., SS_q \right\} \geq p \right] = 1 - \alpha, \tag{6.3.5}$$

where $Y \sim N(\theta, \sum_{i=1}^{q} c_i \sigma_i^2)$. Thus the tolerance interval is $\hat{\theta} \pm D_5$. Write

$$\sigma_c^2 = \sum_{i=1}^{q} c_i \sigma_i^2 \quad \text{and} \quad \sigma_d^2 = \sum_{i=1}^{q} d_i \sigma_i^2, \tag{6.3.6}$$

so that $Z = (\hat{\theta} - \theta)/\sigma_d \sim N(0, 1)$; see (6.3.1). Similar to (4.5.7), the condition (6.3.5) can equivalently be expressed as

$$P_{\hat{\theta}, \, SS_1, ..., SS_q} \left[\Phi \left(\frac{\sigma_d Z + D_5}{\sigma_c} \right) - \Phi \left(\frac{\sigma_d Z - D_5}{\sigma_c} \right) \geq p \right] = 1 - \alpha. \tag{6.3.7}$$

Arguing as in Section 4.5.2, we conclude that as an approximation, D_5 should satisfy

$$P_{D_5^2} \left\{ D_5^2 \geq z_{\frac{1+p}{2}}^2 (\sigma_c^2 + \sigma_d^2) \right\} = 1 - \alpha. \tag{6.3.8}$$

The above equation implies that D_5^2 is an upper $100(1-\alpha)\%$ confidence limit for $z_{\frac{1+p}{2}}^2(\sigma_c^2 + \sigma_d^2)$.

From the definition of σ_c^2 and σ_d^2 in (6.3.6), we note that

$$\sigma_c^2 + \sigma_d^2 = \sum_{i=1}^{q}(c_i + d_i)\sigma_i^2.$$

A $100(1-\alpha)\%$ upper confidence limit for $\sigma_c^2 + \sigma_d^2$ can be obtained using the generalized confidence interval idea. A GPQ for $\sigma_c^2 + \sigma_d^2$ is easily seen to be

$$\sum_{i=1}^{q}(c_i + d_i)G_i = \sum_{i=1}^{q}(c_i + d_i)\frac{ss_i}{U_i^2}, \qquad (6.3.9)$$

where U_i^2's are the chi-square random variables defined in (6.3.1) and G_i is the GPQ defined in (6.3.2). The $100(1-\alpha)$th percentile of the above GPQ will provide a $100(1-\alpha)\%$ upper confidence limit for $\sigma_c^2 + \sigma_d^2$. The required margin of error statistic D_5 is simply the product of $z_{\frac{1+p}{2}}$ and the square root of the $100(1-\alpha)\%$ upper confidence limit for $\sigma_c^2 + \sigma_d^2$. A $(p, 1-\alpha)$ two-sided tolerance interval for $N(\theta, \sum_{i=1}^{q}c_i\sigma_i^2)$ is finally obtained as $\hat{\theta} \pm D_5$. An algorithm for the required computations is given below.

Algorithm 6.2

1. Once the data are obtained, compute the observed values $\hat{\theta}_{\text{obs}}$, ss_1, ss_2, ..., ss_q.

2. Let K denote the simulation sample size. For $k = 1, 2, ..., K$, perform the following steps.

3. Generate independent random variables $U_{i,k}^2 \sim \chi_{f_i}^2$, $i = 1, 2, ..., q$.

4. Compute $G_{i,k} = ss_i/U_{i,k}^2$.

5. Compute $\sum_{i=1}^{q}(c_i + d_i)G_{i,k}$

6. Compute the $100(1-\alpha)$th percentile of the values $\sum_{i=1}^{q}(c_i + d_i)G_{i,k}$, $k = 1, 2, ..., K$; take the square root of the percentile so obtained and multiply with $z_{\frac{1+p}{2}}$ to obtain D_5.

A $(p, 1-\alpha)$ two-sided tolerance interval for $N(\theta, \sum_{i=1}^{q}c_i\sigma_i^2)$ can then be computed as $\hat{\theta}_{\text{obs}} \pm D_5$.

A $1-\alpha$ upper confidence limit for $\sigma_c^2 + \sigma_d^2$ can also be computed using the Satterthwaite approximation (Result 1.2.2 in Chapter 1), provided $c_i + d_i \geq 0$, so that $\sigma_c^2 + \sigma_d^2$ is a linear combination of the σ_i^2's with nonnegative coefficients. The modified large sample (MLS) procedure in Section 1.3 of Chapter 1 provides another easy and convenient way to compute an upper confidence limit for $\sigma_c^2 + \sigma_d^2$ without requiring the condition $c_i + d_i \geq 0$.

Let's now consider the two-way nested model (6.1.2) and assume that all the effects are random. Suppose a tolerance interval is required for $N(\mu, \sigma_\tau^2 + \sigma_\delta^2 + \sigma_e^2)$. Define $SS_1 = SS_\tau$, $SS_2 = SS_\delta$, $SS_3 = SS_e$, $\sigma_1^2 = bn\sigma_\tau^2 + n\sigma_\delta^2 + \sigma_e^2$, $\sigma_2^2 = n\sigma_\delta^2 + \sigma_e^2$, $\sigma_3^2 = \sigma_e^2$, $f_1 = a - 1$, $f_2 = a(b-1)$, and $f_3 = ab(n-1)$. Then $SS_i/\sigma_i^2 \sim \chi_{f_i}^2$, $i = 1, 2, 3$; see (6.1.5) and (6.1.7). With $\hat{\mu}$ in (6.1.8) taking the place of $\hat{\theta}$, we thus have a special case of (6.3.1). Furthermore,

$$
\begin{aligned}
\sigma_c^2 &= \sigma_\tau^2 + \sigma_\delta^2 + \sigma_e^2 = \frac{1}{bn}\sigma_1^2 + \frac{1}{n}\left(1 - \frac{1}{b}\right)\sigma_2^2 + \left(1 - \frac{1}{n}\right)\sigma_3^2 \\
\sigma_d^2 &= \frac{bn\sigma_\tau^2 + n\sigma_\delta^2 + \sigma_e^2}{abn} = \frac{\sigma_1^2}{abn},
\end{aligned}
\tag{6.3.10}
$$

where the expression for σ_d^2 is the expression for the variance given in (6.1.8). We note that $\sigma_c^2 + \sigma_d^2$ is a linear combination of σ_1^2, σ_2^2 and σ_3^2 with positive coefficients; thus the generalized confidence interval approach, the Satterthwaite approximation method, or the MLS procedure can be used to compute an upper confidence limit for $\sigma_c^2 + \sigma_d^2$. A $(p, 1-\alpha)$ tolerance interval for $N(\mu, \sigma_c^2 + \sigma_\delta^2 + \sigma_e^2)$ is $\bar{y}_{...} \pm D_5$, where D_5 is the product of $z_{\frac{1+p}{2}}$ and the square root of the $100(1-\alpha)\%$ upper confidence limit for $\sigma_c^2 + \sigma_d^2$.

Now suppose we require a tolerance interval for the distribution of the true value $\mu + \tau_i + \delta_{j(i)}$ corresponding to the jth level of B nested within the ith level of A; i.e., for the distribution $N(\mu, \sigma_\tau^2 + \sigma_\delta^2)$. Now

$$
\sigma_c^2 = \sigma_\tau^2 + \sigma_\delta^2 = \frac{1}{bn}\sigma_1^2 + \frac{1}{n}\left(1 - \frac{1}{b}\right)\sigma_2^2 - \frac{1}{n}\sigma_3^2,
\tag{6.3.11}
$$

and σ_d^2 is as given in (6.3.10). Now $\sigma_c^2 + \sigma_d^2$ is a linear combination of σ_1^2, σ_2^2 and σ_3^2, but the coefficient of σ_3^2 is negative. Thus the generalized confidence interval approach or the MLS procedure (but not the Satterthwaite approximation) can be used to compute an upper confidence limit for $\sigma_c^2 + \sigma_d^2$, which can then be used to compute a $(p, 1-\alpha)$ tolerance interval for $N(\mu, \sigma_\tau^2 + \sigma_\delta^2)$.

If (6.1.2) has mixed effects, we can similarly derive tolerance intervals for $N(\mu_i, \sigma_\delta^2 + \sigma_e^2)$ and $N(\mu_i, \sigma_\delta^2)$, where $\mu_i = \mu + \tau_i$.

Example 6.1 (continued)

We now consider the glucose monitoring meter experiment described in Example 6.1, where we have the models in (6.2.1) for the observations W_{ijlr} made on the test meters ($r = 1, 2, ..., E$; $l = 1, 2, ..., L$; $j = 1, 2, ..., B$; $i = 1, 2, ..., m_1$) and the observations Y_{ijlr} made on the reference meters ($r = 1, 2, ..., E$; $l = 1, 2, ..., L$; $j = 1, 2, ..., B$; $i = 1, 2, ..., m_2$), where $m_1 = 44$, $m_2 = 20$, and $B = L = E = 3$. Under the model (6.2.1), a (0.95, 0.90) tolerance interval is required for $N(\mu_T - \mu_R, \sigma_T^2)$. If the tolerance interval is within [-5, 5], we conclude that the quality requirements are met by the test meters.

Let

$$\bar{W}_{i..} = \frac{1}{BLE}\sum_{j=1}^{3}\sum_{l=1}^{3}\sum_{r=1}^{3}W_{ijlr}, \quad \bar{W} = \frac{1}{m_1 BLE}\sum_{i=1}^{m_1}\sum_{j=1}^{3}\sum_{l=1}^{3}\sum_{r=1}^{3}W_{ijlr},$$

and define $\bar{Y}_{i..}$ and $\bar{Y}$ similarly. Also define

$$SS_T = BLE\sum_{i=1}^{m_1}(\bar{W}_{i..} - \bar{W})^2, \quad SS_R = BLE\sum_{i=1}^{m_2}(\bar{Y}_{i..} - \bar{Y})^2,$$

$$\sigma_1^2 = BLE\sigma_T^2 + \sigma_e^2, \quad \sigma_2^2 = BLE\sigma_R^2 + \sigma_e^2 \quad \text{and} \quad \sigma_3^2 = \sigma_e^2.$$

Furthermore, let SS_{e1} and SS_{e2} be the error sums of squares based on the models for W_{ijlr} and Y_{ijlr}, respectively, and define $SS_e = SS_{e1} + SS_{e2}$. Thus SS_e is the pooled error sum of squares from both the models (since we are assuming that the errors have a common variance). Then we have the independent distributions

$$\bar{W} - \bar{Y} \sim N\left(\mu_T - \mu_R, \frac{\sigma_1^2}{m_1 BLE} + \frac{\sigma_2^2}{m_2 BLE}\right), \quad SS_T/\sigma_1^2 \sim \chi_{m_1-1}^2,$$

$$SS_R/\sigma_2^2 \sim \chi_{m_2-1}^2, \quad \text{and} \quad SS_e/\sigma_3^2 \sim \chi_v^2,$$

$$\text{where } v = (m_1 + m_2)(BLE - 1) - 2B - 2L + 4.$$

Based on the data for Example 6.1, the above quantities have the observed values

$$\bar{W} - \bar{Y} = -1.3791, \ SS_T = 718.9790, \ SS_R = 280.3167, \ \text{and} \ SS_e = 8376.4525.$$

Also, since $m_1 = 44$, $m_2 = 20$, and $B = L = E = 3$, the df for SS_T, SS_R, and SS_e are $m_1 - 1 = 43$, $m_2 - 1 = 19$ and $v = 1656$, respectively. Recall that a (0.95, 0.90) tolerance interval is required for $N(\mu_T - \mu_R, \sigma_T^2)$. We can use the definitions of σ_1^2 and σ_3^2 given above to write

$$\sigma_T^2 = \frac{1}{BLE}\left(\sigma_1^2 - \sigma_3^2\right)$$

Thus our tolerance interval problem is in the set up introduced in the beginning of this section, where $\hat{\theta} = \bar{W} - \bar{Y}$ has the observed value $\hat{\theta}_{\text{obs}} = -1.3791$, and the quantities in (6.3.6) are

$$\sigma_c^2 = \sigma_T^2 = \frac{1}{BLE}\left(\sigma_1^2 - \sigma_3^2\right) \text{ and } \sigma_d^2 = \frac{\sigma_1^2}{m_1 BLE} + \frac{\sigma_2^2}{m_2 BLE},$$

so that

$$\sigma_c^2 + \sigma_d^2 = \left(1 + \frac{1}{m_1}\right)\frac{\sigma_1^2}{BLE} + \frac{\sigma_2^2}{m_2 BLE} - \frac{\sigma_3^2}{BLE}.$$

The required (0.95, 0.90) tolerance interval for $N(\mu_T - \mu_R, \sigma_T^2)$ is then given by $\hat{\theta}_{\text{obs}} \pm D_5$, where D_5 is the product of $z_{.975}$ and the square root of a 90% upper confidence limit for $\sigma_c^2 + \sigma_d^2$ (given above).

We used the MLS procedure to compute a 90% upper confidence limit for $\sigma_c^2 + \sigma_d^2$, and the limit is 0.70191. Since $z_{.975} = 1.96$, $D_5 = 1.96 \times \sqrt{0.70191} = 1.6505$. Since $\hat{\theta}_{\text{obs}} = -1.3791$, the tolerance interval simplifies to $(-3.0296, 0.2713)$. Since the tolerance interval is within [-5, 5], we conclude that the quality requirements have been met by the test meters.

6.4   A General Model with Two Variance Components

In this section, we shall consider the model (6.1.9) involving just two variance components, and derive one-sided and two-sided tolerance intervals. Since we are considering a general set up of such a model, the derivation of the tolerance intervals in this section is a bit more involved compared to what we have seen so far. The model is given by

$$\mathbf{Y} = \mathbf{X}\boldsymbol{\beta} + \mathbf{Z}\boldsymbol{\delta} + \mathbf{e}, \tag{6.4.1}$$

where $\mathbf{Y}$ is an $N \times 1$ vector of observations, $\boldsymbol{\beta}$ is a fixed unknown parameter vector, $\boldsymbol{\delta}$ is a vector of random effects having the distribution $\boldsymbol{\delta} \sim N(0, \sigma_\delta^2 \mathbf{I})$, independent of $\mathbf{e} \sim N(0, \sigma_e^2 \mathbf{I})$. We shall assume that $\text{rank}(\mathbf{X}, \mathbf{Z}) = r$. We shall also assume that $\mathbf{X}$ is an $N \times b$ matrix of full column rank; otherwise a reparametrization can be carried out so that $\mathbf{X}$ becomes a matrix of full column rank.

Let $\mathbf{x}_0'$ and $\mathbf{z}_0'$ be fixed design vectors corresponding to the fixed effect and random effect, respectively, and let Y_0 be the corresponding observation. In view of the above assumptions concerning the model (6.4.1), we get the distribution

$$Y_0 \sim N\left(\mathbf{x}_0'\boldsymbol{\beta}, (\mathbf{z}_0'\mathbf{z}_0)\sigma_\delta^2 + \sigma_e^2\right). \tag{6.4.2}$$

In this section, we shall consider the problem of computing one-sided and two-sided tolerance intervals for Y_0, i.e., for the distribution $N(\mathbf{x}_0'\boldsymbol{\beta}, (\mathbf{z}_0'\mathbf{z}_0)\sigma_\delta^2 + \sigma_e^2)$. We shall also consider the problem of computing tolerance limits for the true value $\mathbf{x}_0'\boldsymbol{\beta} + \mathbf{z}_0'\boldsymbol{\delta}$, i.e., for the distribution $N(\mathbf{x}_0'\boldsymbol{\beta}, (\mathbf{z}_0'\mathbf{z}_0)\sigma_\delta^2)$. The results in this section are generalizations of those in Section 5.3.2, Section 5.4.2 and Section 5.5.

Let

$$P_{(\mathbf{X},\mathbf{Z})} = (\mathbf{X}, \mathbf{Z}) \left[(\mathbf{X}, \mathbf{Z})'(\mathbf{X}, \mathbf{Z}) \right]^- (\mathbf{X}, \mathbf{Z})',$$

where, for any matrix $\mathbf{A}$, $\mathbf{A}^-$ denotes a generalized inverse. Thus $P_{(\mathbf{X},\mathbf{Z})}$ is the orthogonal projection matrix onto the $r-$dimensional vector space spanned by the columns of $(\mathbf{X}, \mathbf{Z})$. The error sum of squares SS_e, under the model (6.4.1), is then given by

$$
\begin{aligned}
SS_e &= \mathbf{Y}' \left[\mathbf{I}_N - P_{(\mathbf{X},\mathbf{Z})} \right] \mathbf{Y} \\
\text{with } U_e^2 &= \frac{SS_e}{\sigma_e^2} \sim \chi_{N-r}^2.
\end{aligned}
\tag{6.4.3}
$$

Let $\mathbf{F}$ be an $N \times r$ matrix such that

$$P_{(\mathbf{X},\mathbf{Z})} = \mathbf{F}\mathbf{F}' \quad \text{and} \quad \mathbf{F}'\mathbf{F} = \mathbf{I}_r.$$

In other words, the columns of $\mathbf{F}$ form an orthonormal basis for the vector space spanned by the columns of $(\mathbf{X}, \mathbf{Z})$. Now consider the model for $\mathbf{F}'\mathbf{Y}$, obtained from (6.4.1):

$$
\begin{aligned}
\mathbf{F}'\mathbf{Y} &\sim N(\mathbf{F}'\mathbf{X}\boldsymbol{\beta}, \mathbf{V}), \\
\text{where } \mathbf{V} &= \sigma_\delta^2 \mathbf{F}'\mathbf{Z}\mathbf{Z}'\mathbf{F} + \sigma_e^2 \mathbf{I}_r.
\end{aligned}
\tag{6.4.4}
$$

We note that since the columns of $\mathbf{F}$ span the vector space generated by the columns of $(\mathbf{X}, \mathbf{Z})$, and since $\mathbf{X}$ is of full column rank, so is $\mathbf{F}'\mathbf{X}$, with rank equal to the rank of $\mathbf{X}$. It can also be shown that SS_e and $\mathbf{F}'\mathbf{Y}$ are independently distributed. As an illustration of the model obtained in (6.4.4), consider the one-way random model with unbalanced data; the model (5.1.1) in Chapter 5. After writing the model using vector-matrix notation, it can be checked that the matrix $\mathbf{F}$ can be taken to be $\text{diag}\left(\frac{1}{\sqrt{n_1}}\mathbf{1}_{n_1}, \frac{1}{\sqrt{n_2}}\mathbf{1}_{n_2}, ..., \frac{1}{\sqrt{n_a}}\mathbf{1}_{n_a}\right)$. With this choice of $\mathbf{F}$, $\mathbf{F}'\mathbf{Y}$ and $\mathbf{V}$ are equivalent to the quantities given in (5.3.8).

If σ_δ^2 and σ_e^2 are known, so that $\mathbf{V}$ is known, the best linear unbiased estimator of $\boldsymbol{\beta}$, say $\hat{\boldsymbol{\beta}}_{\mathbf{V}}$ is given by

$$
\begin{aligned}
\hat{\boldsymbol{\beta}}_{\mathbf{V}} &= (\mathbf{X}'\mathbf{F}\mathbf{V}^{-1}\mathbf{F}'\mathbf{X})^{-1}\mathbf{X}'\mathbf{F}\mathbf{V}^{-1}\mathbf{F}'\mathbf{Y} \\
\text{and } \hat{\boldsymbol{\beta}}_{\mathbf{V}} &\sim N(\boldsymbol{\beta}, (\mathbf{X}'\mathbf{F}\mathbf{V}^{-1}\mathbf{F}'\mathbf{X})^{-1}).
\end{aligned}
\tag{6.4.5}
$$

Also, the residual sum of squares, say SS_0, under the model (6.4.4) is given by

$$
\begin{aligned}
SS_0 &= (\mathbf{F}'\mathbf{Y} - \mathbf{F}'\mathbf{X}\hat{\boldsymbol{\beta}}_{\mathbf{V}})'\mathbf{V}^{-1}(\mathbf{F}'\mathbf{Y} - \mathbf{F}'\mathbf{X}\hat{\boldsymbol{\beta}}_{\mathbf{V}}) \\
&= \mathbf{Y}'\mathbf{F}\left[\mathbf{V}^{-1} - \mathbf{V}^{-1}\mathbf{F}'\mathbf{X}(\mathbf{X}'\mathbf{F}\mathbf{V}^{-1}\mathbf{F}'\mathbf{X})^{-1}\mathbf{X}'\mathbf{F}\mathbf{V}^{-1}\right]\mathbf{F}'\mathbf{Y}. \quad (6.4.6)
\end{aligned}
$$

Note that $SS_0 \sim \chi^2_{r-b}$; the degrees of freedom is obtained using the observations that $\mathbf{F}'\mathbf{Y}$ is an $r \times 1$ vector, and the matrix $\mathbf{X}$ (and hence $\mathbf{F}'\mathbf{X}$) is assumed to be of full column rank equal to b. We would like to emphasize that $\hat{\boldsymbol{\beta}}_{\mathbf{V}}$ cannot be used to estimate $\boldsymbol{\beta}$, since it depends on the unknown variance components; see the expression for $\mathbf{V}$ in (6.4.4). In fact both $\hat{\boldsymbol{\beta}}_{\mathbf{V}}$ and SS_0 depend on the unknown variance components. These quantities have been defined only to facilitate the development of the required GPQs.

We shall now derive GPQs for $\mathbf{x}_0'\boldsymbol{\beta}$, σ_δ^2 and σ_e^2, to be denoted by $G_{\mathbf{x}_0'\boldsymbol{\beta}}$, $G_{\sigma_\delta^2}$ and $G_{\sigma_e^2}$, respectively. Let $\tilde{\mathbf{Y}}$ and ss_e denote the observed values of $\mathbf{Y}$ and SS_e, respectively. Let

$$
\begin{aligned}
G_{\sigma_e^2} &= \frac{\sigma_e^2}{SS_e}ss_e = \frac{ss_e}{U_e^2} \\
\mathbf{V}_G &= G_{\sigma_\delta^2}\mathbf{F}'\mathbf{ZZ}'\mathbf{F} + G_{\sigma_e^2}\mathbf{I}_r, \quad (6.4.7)
\end{aligned}
$$

where U_e^2 is the chi-square random variable defined in (6.4.2), and $G_{\sigma_\delta^2}$ is the GPQ for σ_δ^2, to be obtained. It is clear that $G_{\sigma_e^2}$ is a GPQ for σ_e^2. Furthermore, once $G_{\sigma_\delta^2}$ is determined, $\mathbf{V}_G$ is a "matrix GPQ" for $\mathbf{V}$. Motivated by (6.4.6), consider the equation

$$
U_0^2 = \tilde{\mathbf{Y}}'\mathbf{F}\left[\mathbf{V}_G^{-1} - \mathbf{V}_G^{-1}\mathbf{F}'\mathbf{X}(\mathbf{X}'\mathbf{F}\mathbf{V}_G^{-1}\mathbf{F}'\mathbf{X})^{-1}\mathbf{X}'\mathbf{F}\mathbf{V}_G^{-1}\right]\mathbf{F}'\tilde{\mathbf{Y}}, \quad (6.4.8)
$$

where $U_0^2 \sim \chi^2_{r-b}$, $\tilde{\mathbf{Y}}$ is the observed value of $\mathbf{Y}$, and $\mathbf{V}_G$ is defined in (6.4.7). The right hand side of (6.4.8) is a function of $G_{\sigma_\delta^2}$. It can be shown that the right hand side of (6.4.8) is a monotone decreasing function of $G_{\sigma_\delta^2}$, and the limit is zero as $G_{\sigma_\delta^2} \to \infty$. For this, it is enough to show that SS_0 given in (6.4.6) is a decreasing function of σ_δ^2, and the limit is zero as $\sigma_\delta^2 \to \infty$. A proof of this monotonicity is given following equation (6.4.14), and is similar to the corresponding proof in Section 5.3.2. Once we show that the right hand side of (6.4.8) is a monotone decreasing function of $G_{\sigma_\delta^2}$, we can conclude that the solution to (6.4.8) exists uniquely, even though the solution $G_{\sigma_\delta^2}$ so obtained need not be nonnegative. Note that even though the matrix $\mathbf{V}$ in (6.4.4) is positive definite, the matrix GPQ $\mathbf{V}_G$ in (6.4.7) need not be so, when $G_{\sigma_\delta^2}$ takes a negative value.

In order to define $G_{\mathbf{x}_0'\boldsymbol{\beta}}$, note from (6.4.5) that

$$Z = \frac{\mathbf{x}_0'\hat{\boldsymbol{\beta}}_V - \mathbf{x}_0'\boldsymbol{\beta}}{\sqrt{\mathbf{x}_0'(\mathbf{X}'\mathbf{F}\mathbf{V}^{-1}\mathbf{F}'\mathbf{X})^{-1}\mathbf{x}_0}} \sim N(0,1). \tag{6.4.9}$$

Now define

$$
\begin{aligned}
G_{\mathbf{x}_0'\boldsymbol{\beta}} &= \mathbf{x}_0'(\mathbf{X}'\mathbf{F}\mathbf{V}_G^{-1}\mathbf{F}'\mathbf{X})^{-1}\mathbf{X}'\mathbf{F}\mathbf{V}_G^{-1}\mathbf{F}'\tilde{\mathbf{Y}} \\
&\quad - \frac{\mathbf{x}_0'\hat{\boldsymbol{\beta}}_V - \mathbf{x}_0'\boldsymbol{\beta}}{\sqrt{\mathbf{x}_0'(\mathbf{X}'\mathbf{F}\mathbf{V}^{-1}\mathbf{F}'\mathbf{X})^{-1}\mathbf{x}_0}} \times \sqrt{\left[\mathbf{x}_0'(\mathbf{X}'\mathbf{F}\mathbf{V}_G^{-1}\mathbf{F}'\mathbf{X})^{-1}\mathbf{x}_0\right]_+} \\
&= \mathbf{x}_0'(\mathbf{X}'\mathbf{F}\mathbf{V}_G^{-1}\mathbf{F}'\mathbf{X})^{-1}\mathbf{X}'\mathbf{F}\mathbf{V}_G^{-1}\mathbf{F}'\tilde{\mathbf{Y}} - Z\sqrt{\left[\mathbf{x}_0'(\mathbf{X}'\mathbf{F}\mathbf{V}_G^{-1}\mathbf{F}'\mathbf{X})^{-1}\mathbf{x}_0\right]_+},
\end{aligned}
\tag{6.4.10}
$$

where $\mathbf{V}_G$ is defined in (6.4.7), $\tilde{\mathbf{Y}}$ is the observed value of $\mathbf{Y}$, and for any real number x, we define $x_+ = \max\{x, 0\}$. It can be directly verified that given the observed data, the distribution of $G_{\mathbf{x}_0'\boldsymbol{\beta}}$ is free of any unknown parameters, and the observed value of $G_{\mathbf{x}_0'\boldsymbol{\beta}}$ is $\mathbf{x}_0'\boldsymbol{\beta}$. In other words $G_{\mathbf{x}_0'\boldsymbol{\beta}}$ is a GPQ for $\mathbf{x}_0'\boldsymbol{\beta}$.

We are now ready to derive one-sided and two-sided tolerance intervals for $N(\mathbf{x}_0'\boldsymbol{\beta}, (\mathbf{z}_0'\mathbf{z}_0)\sigma_\delta^2 + \sigma_e^2)$, and for $N(\mathbf{x}_0'\boldsymbol{\beta}, (\mathbf{z}_0'\mathbf{z}_0)\sigma_\delta^2)$.

6.4.1   One-Sided Tolerance Limits

Recalling that a $(p, 1 - \alpha)$ lower tolerance limit for $N(\mathbf{x}_0'\boldsymbol{\beta}, (\mathbf{z}_0'\mathbf{z}_0)\sigma_\delta^2 + \sigma_e^2)$ is a $100(1 - \alpha)\%$ lower confidence limit for $\mathbf{x}_0'\boldsymbol{\beta} - z_p\sqrt{(\mathbf{z}_0'\mathbf{z}_0)\sigma_\delta^2 + \sigma_e^2}$, a GPQ for this is given by

$$G_{10} = G_{\mathbf{x}_0'\boldsymbol{\beta}} - z_p\sqrt{\left[(\mathbf{z}_0'\mathbf{z}_0)G_{\sigma_\delta^2} + G_{\sigma_e^2}\right]_+}. \tag{6.4.11}$$

The 100αth percentile of G_{10} gives the required $(p, 1 - \alpha)$ lower tolerance limit for $N(\mathbf{x}_0'\boldsymbol{\beta}, (\mathbf{z}_0'\mathbf{z}_0)\sigma_\delta^2 + \sigma_e^2)$. Here is the algorithm for performing the relevant calculations:

Algorithm 6.3

1. Once the $N \times 1$ data vector $\tilde{\mathbf{Y}}$ is observed, compute the observed value ss_e.

2. Let $b = \text{rank}(\mathbf{X})$ and $r = \text{rank}(\mathbf{X}, \mathbf{Z})$. Compute an $N \times r$ matrix $\mathbf{F}$ whose columns are orthonormal and form a basis for the vector space spanned by the columns of $(\mathbf{X}, \mathbf{Z})$.

3. Let K denote the simulation sample size. For $k = 1, 2, ..., K$, perform the following steps.

4. Generate independent random variables $U_{e,k}^2 \sim \chi_{N-r}^2$, $U_{0,k}^2 \sim \chi_{r-b}^2$, and $Z_k \sim N(0,1)$.

5. Compute $G_{\sigma_e^2,k} = ss_e/U_{e,k}^2$.

6. Let $\mathbf{V}_{G,k} = G_{\sigma_\delta^2,k}\mathbf{F}'\mathbf{Z}\mathbf{Z}'\mathbf{F} + G_{\sigma_e^2,k}\mathbf{I}_r$. Now solve the following equation to get $G_{\sigma_\delta^2,k}$:

$$U_{0,k}^2 = \tilde{\mathbf{Y}}'\mathbf{F}\left[\mathbf{V}_{G,k}^{-1} - \mathbf{V}_{G,k}^{-1}\mathbf{F}'\mathbf{X}(\mathbf{X}'\mathbf{F}\mathbf{V}_{G,k}^{-1}\mathbf{F}'\mathbf{X})^{-1}\mathbf{X}'\mathbf{F}\mathbf{V}_{G,k}^{-1}\right]\mathbf{F}'\tilde{\mathbf{Y}}.$$

7. Compute

$$G_{\mathbf{x}_0'\boldsymbol{\beta},k} = \mathbf{x}_0'(\mathbf{X}'\mathbf{F}\mathbf{V}_{G,k}^{-1}\mathbf{F}'\mathbf{X})^{-1}\mathbf{X}'\mathbf{F}\mathbf{V}_{G,k}^{-1}\mathbf{F}'\tilde{\mathbf{Y}} - Z_k\sqrt{\left[\mathbf{x}_0'(\mathbf{X}'\mathbf{F}\mathbf{V}_{G,k}^{-1}\mathbf{F}'\mathbf{X})^{-1}\mathbf{x}_0\right]_+}$$

8. Compute

$$G_{10,k} = G_{\mathbf{x}_0'\boldsymbol{\beta},k} - z_p\sqrt{\left[(\mathbf{z}_0'\mathbf{z}_0)G_{\sigma_\delta^2,k} + G_{\sigma_e^2,k}\right]_+}.$$

The 100αth percentile of the $G_{10,k}$ values ($k = 1, 2, ..., K$) gives an estimate of the 100αth percentile of G_{10}, which in turn gives an estimate of the required $(p, 1 - \alpha)$ lower confidence limit for $N(\mathbf{x}_0'\boldsymbol{\beta}, (\mathbf{z}_0'\mathbf{z}_0)\sigma_\delta^2 + \sigma_e^2)$.

A $(p, 1 - \alpha)$ upper tolerance limit for $N(\mathbf{x}_0'\boldsymbol{\beta}, (\mathbf{z}_0'\mathbf{z}_0)\sigma_\delta^2 + \sigma_e^2)$ can be obtained as the $100(1 - \alpha)$th percentile of the following GPQ:

$$G_{11} = G_{\mathbf{x}_0'\boldsymbol{\beta}} + z_p\sqrt{\left[(\mathbf{z}_0'\mathbf{z}_0)G_{\sigma_\delta^2} + G_{\sigma_e^2}\right]_+}. \qquad (6.4.12)$$

Similarly, a $(p, 1-\alpha)$ lower tolerance limit for $N(\mathbf{x}_0'\boldsymbol{\beta}, (\mathbf{z}_0'\mathbf{z}_0)\sigma_\delta^2)$ is the 100αth percentile of the GPQ

$$G_{12} = G_{\mathbf{x}_0'\boldsymbol{\beta}} - z_p\sqrt{\left[(\mathbf{z}_0'\mathbf{z}_0)G_{\sigma_\delta^2}\right]_+}. \qquad (6.4.13)$$

Finally, $(p, 1-\alpha)$ upper tolerance limit for $N(\mathbf{x}_0'\boldsymbol{\beta}, (\mathbf{z}_0'\mathbf{z}_0)\sigma_\delta^2)$ can be obtained as the $100(1-\alpha)$th percentile of the following GPQ:

$$G_{13} = G_{\mathbf{x}_0'\boldsymbol{\beta}} + z_p\sqrt{\left[(\mathbf{z}_0'\mathbf{z}_0)G_{\sigma_\delta^2}\right]_+}. \qquad (6.4.14)$$

Proof that SS_0 in (6.4.6) is a decreasing function of σ_δ^2

The proof is similar to the corresponding proof in Section 5.3.2. We shall first give an alternative representation for SS_0. Recall that $\mathbf{F}'\mathbf{X}$ is an $r \times b$ matrix of rank b. Let $\mathbf{H}$ be an $r \times (r-b)$ matrix of rank $r-b$ satisfying $\mathbf{H}'\mathbf{F}'\mathbf{X} = 0$. Using a theorem in Searle, Casella and McCulloch (1992, p. 451), we can rewrite SS_0 in (6.4.6) as

$$SS_0 = \mathbf{Y}_0'(\mathbf{H}'\mathbf{V}\mathbf{H})^{-1}\mathbf{Y}_0,$$

where $\mathbf{Y}_0 = \mathbf{H}'\mathbf{F}'\mathbf{Y}$. The proof can now be completed following the arguments in the corresponding proof in Section 5.3.2.

6.4.2 Two-Sided Tolerance Intervals

Following the arguments in Sections 4.4.1 and 5.5.1, a $(p, 1-\alpha)$ two-sided tolerance interval for $N(\mathbf{x}_0'\boldsymbol{\beta}, (\mathbf{z}_0'\mathbf{z}_0)\sigma_\delta^2 + \sigma_e^2)$ can be taken to be of the form $\mathbf{x}_0'\tilde{\boldsymbol{\beta}} \pm D_6$, where $\mathbf{x}_0'\tilde{\boldsymbol{\beta}}$ is an estimator of $\mathbf{x}_0'\boldsymbol{\beta}$, to be determined, and D_6 is a margin of error statistic. The quantity D_6^2 can be taken to be a $100(1-\alpha)\%$ upper confidence limit of $z_{\frac{1+p}{2}}^2\left((\mathbf{z}_0'\mathbf{z}_0)\sigma_\delta^2 + \sigma_e^2 + \mathbf{x}_0'(\mathbf{X}'\mathbf{F}\mathbf{V}^{-1}\mathbf{F}'\mathbf{X})^{-1}\mathbf{x}_0\right)$. Such a confidence limit can be obtained as the $100(1-\alpha)$th percentile of

$$z_{\frac{1+p}{2}}^2 \times \left[(\mathbf{z}_0'\mathbf{z}_0)G_{\sigma_\delta^2} + G_{\sigma_e^2} + \mathbf{x}_0'(\mathbf{X}'\mathbf{F}\mathbf{V}_G^{-1}\mathbf{F}'\mathbf{X})^{-1}\mathbf{x}_0\right]_+,$$

where $\mathbf{V}_G$ is given in (6.4.7).

Regarding the choice of $\mathbf{x}_0'\tilde{\boldsymbol{\beta}}$, we can consider the GPQ $G_{\mathbf{x}_0'\boldsymbol{\beta}}$ given in (6.4.10), and take $\mathbf{x}_0'\tilde{\boldsymbol{\beta}}$ to be the mean (or the median) of $G_{\mathbf{x}_0'\boldsymbol{\beta}}$. This mean (or median) can be easily estimated based on the realizations $G_{\mathbf{x}_0'\boldsymbol{\beta},k}$ of $G_{\mathbf{x}_0'\boldsymbol{\beta}}$; see Step 7 of Algorithm 6.3 given above. Note that $\mathbf{x}_0'\hat{\boldsymbol{\beta}}_\mathbf{V}$ defined in (6.4.5) cannot be used as an estimator of $\mathbf{x}_0'\boldsymbol{\beta}$ since $\mathbf{V}$ (and hence $\mathbf{x}_0'\hat{\boldsymbol{\beta}}_\mathbf{V}$) depends on the unknown variance components.

Similar to Algorithm 5.2, the following algorithm can be used to compute a $(p, 1-\alpha)$ two-sided tolerance interval for $N(\mathbf{x}_0'\boldsymbol{\beta}, (\mathbf{z}_0'\mathbf{z}_0)\sigma_\delta^2 + \sigma_e^2)$.

Algorithm 6.4

Step 1. Once the $N \times 1$ data vector $\tilde{Y}$ is observed, compute the observed value ss_e.

Step 2. Let K denote the simulation sample size. For $k = 1, 2, ..., K$, perform steps 2–7 of Algorithm 6.3.

Step 3. Compute $(\mathbf{z}_0'\mathbf{z}_0)G_{\sigma_\delta^2} + G_{\sigma_e^2} + \mathbf{x}_0'(\mathbf{X}'\mathbf{F}\mathbf{V}_G^{-1}\mathbf{F}'\mathbf{X})^{-1}\mathbf{x}_0$, where $\mathbf{V}_G$ is given in (6.4.7).

Step 4. Compute $\mathbf{x}_0'\tilde{\boldsymbol{\beta}}$, the mean of the $G_{\mathbf{x}_0'\boldsymbol{\beta},k}$ values.

Step 5. Compute $D_6^2 = z_{\frac{1+p}{2}}^2 \times$ the $100(1 - \alpha)$th percentile of

$$\left[(\mathbf{z}_0'\mathbf{z}_0)G_{\sigma_\delta^2} + G_{\sigma_e^2} + \mathbf{x}_0'(\mathbf{X}'\mathbf{F}\mathbf{V}_G^{-1}\mathbf{F}'\mathbf{X})^{-1}\mathbf{x}_0 \right]_+$$

Step 6. A $(p, 1 - \alpha)$ two-sided tolerance interval for $N(\mathbf{x}_0'\boldsymbol{\beta}, (\mathbf{z}_0'\mathbf{z}_0)\sigma_\delta^2 + \sigma_e^2)$ is then given by $\mathbf{x}_0'\tilde{\boldsymbol{\beta}} \pm D_6$.

In order to obtain a $(p, 1-\alpha)$ two-sided tolerance interval for $N(\mathbf{x}_0'\boldsymbol{\beta}, (\mathbf{z}_0'\mathbf{z}_0)\sigma_\delta^2)$, we compute $D_7^2 = z_{\frac{1+p}{2}}^2 \times$ the $100(1 - \alpha)$th percentile of

$$\left[(\mathbf{z}_0'\mathbf{z}_0)G_{\sigma_\delta^2} + \mathbf{x}_0'(\mathbf{X}'\mathbf{F}\mathbf{V}_G^{-1}\mathbf{F}'\mathbf{X})^{-1}\mathbf{x}_0 \right]_+ .$$

The interval is then given by $\mathbf{x}_0'\tilde{\boldsymbol{\beta}} \pm D_7$.

Example 6.2 (continued)

We already noted that for the data in Table 6.1, the model (6.2.2) can be expressed in the form (6.4.1) ; see the discussion below (6.2.2). With $\mathbf{X}$ and $\mathbf{Z}$ as given below (6.2.2), it can be verified that the matrix $\mathbf{F}$ consists of the columns of $\mathbf{Z}$ along with the second column of $\mathbf{X}$, namely the column $\begin{pmatrix} 75 \times \mathbf{1}_{31} \\ -67 \times \mathbf{1}_{41} \end{pmatrix}$, after an orthonormalization of these columns. Thus $\mathbf{F}$ is a 72×9 matrix of rank 9. For the data in Table 6.1, we also have the observed value $ss_e = 8663.776$.

Let us now compute a $(0.90, 0.95)$ lower tolerance limit for the strength distribution corresponding to the temperatures -50 degrees and 50 degrees. In other words, we need a lower tolerance limit for the distribution $N(\beta_0 + \beta_1 x_0)$ when $x_0 = -50$ and when $x_0 = 50$. Applying Algorithm 6.3, the (0.90, 0.95) lower tolerance limits turned out to be 312.216 and 302.392, corresponding to the temperatures -50 degrees and 50 degrees, respectively.

6.5 A One-Way Random Model with Covariates and Unequal Variances

A one-way random model with heteroscedastic variances is very common in applications, most notably in the context of inter-laboratory studies; for details and further references, see Rukhin and Vangel (1998), Vangel and Rukhin (1999) and Iyer, Wang and Mathew (2004). If covariates are present in a study, obviously they have to be incorporated into the model. In fact the model (6.2.2) corresponding to Example 6.2 is a one-way random model that also includes a covariate; however, the errors are assumed to be homoscedastic. Such a model was also considered by Limam and Thomas (1988b), once again assuming homoscedasticity. A one-way random model that contains several covariates and having heteroscedastic error variances has recently been considered by Lin, Liao and Iyer (2008); they have addressed the problem of constructing one-sided and two-sided tolerance intervals. These problems are addressed in this section based on the work of Lin, Liao and Iyer (2008). In the presence of b covariates and heteroscedasticity, the one-way random model can be written as

$$Y_{ij} = \mathbf{x}_i'\boldsymbol{\beta} + \tau_i + e_{ij}, \tag{6.5.1}$$

$j = 1, 2, ..., n_i; i = 1, 2, ..., a$, where the $\mathbf{x}_i$'s ($i = 1, 2, ..., a$) are $b \times 1$ vectors of covariates (whose values are known), $\tau_i \sim N(0, \sigma_\tau^2)$, $e_{ij} \sim N(0, \sigma_{ei}^2)$, and all the random variables are assumed to be independent. In this section, we shall assume that $b < a$; the need to have this assumption will become clear later. The $b \times 1$ vector $\boldsymbol{\beta}$ and the variances σ_τ^2 and the σ_{ei}^2's are all unknown parameters. Note that we are allowing the variances σ_{ei}^2's to be unequal. Now consider an observation Y_0 that corresponds to the covariate vector $\mathbf{x}_0$, and suppose the observation is also based on the same process or phenomenon that generated $Y_{i1}, Y_{i2}, ..., Y_{in_i}$, for a fixed i. Thus we have the distribution

$$Y_0 \sim N(\mathbf{x}_0'\boldsymbol{\beta}, \sigma_\tau^2 + \sigma_{ei}^2). \tag{6.5.2}$$

We shall address the problem of computing tolerance intervals for the distribution $N(\mathbf{x}_0'\boldsymbol{\beta}, \sigma_\tau^2 + \sigma_{ei}^2)$, and also for the distribution $N(\mathbf{x}_0'\boldsymbol{\beta}, \sigma_\tau^2)$, where the latter distribution corresponds to the true values, free of the error term.

Define

$$\bar{Y}_{i.} = \sum_{j=1}^{n_i} Y_{ij}/n_i, \quad \bar{\boldsymbol{Y}} = (\bar{Y}_{1.}, \bar{Y}_{2.}, ..., \bar{Y}_{a.})', \quad SS_{ei} = \sum_{j=1}^{n_i}(Y_{ij} - \bar{Y}_{i.})^2,$$

$$\mathbf{X} = (\mathbf{x}_1, \mathbf{x}_2, ..., \mathbf{x}_a)', \quad \text{and} \quad \mathbf{V} = \operatorname{diag}\left(\frac{1}{\sigma_\tau^2 + \frac{\sigma_{e1}^2}{n_1}}, \frac{1}{\sigma_\tau^2 + \frac{\sigma_{e2}^2}{n_2}}, ..., \frac{1}{\sigma_\tau^2 + \frac{\sigma_{ea}^2}{n_a}}\right).$$

$$(6.5.3)$$

We shall assume that the $a \times b$ matrix $\mathbf{X}$ has rank b (recall our assumption $b < a$). Then

$$\bar{\boldsymbol{Y}} \sim N(\mathbf{X}\boldsymbol{\beta}, \mathbf{V}), \quad \text{and} \quad U_{ei}^2 = SS_{ei}/\sigma_{ei}^2 \sim \chi_{n_i-1}^2, \qquad (6.5.4)$$

$i = 1, 2, ..., a$. The best linear unbiased estimator of $\boldsymbol{\beta}$, say $\hat{\boldsymbol{\beta}}_{\mathbf{V}}$ is given by

$$\hat{\boldsymbol{\beta}}_{\mathbf{V}} = (\mathbf{X}'\mathbf{V}^{-1}\mathbf{X})^{-1}\mathbf{X}'\mathbf{V}^{-1}\bar{\boldsymbol{Y}} \sim N(\boldsymbol{\beta}, (\mathbf{X}'\mathbf{V}^{-1}\mathbf{X})^{-1}). \qquad (6.5.5)$$

Also, the residual sum of squares, say SS_0, under the model (6.5.4) for $\bar{\boldsymbol{Y}}$ is given by

$$
\begin{aligned}
SS_0 &= (\bar{\boldsymbol{Y}} - \mathbf{X}\hat{\boldsymbol{\beta}}_{\mathbf{V}})'\mathbf{V}^{-1}(\bar{\boldsymbol{Y}} - \mathbf{X}\hat{\boldsymbol{\beta}}_{\mathbf{V}}) \\
&= \bar{\boldsymbol{Y}}'\left[\mathbf{V}^{-1} - \mathbf{V}^{-1}\mathbf{X}(\mathbf{X}'\mathbf{V}^{-1}\mathbf{X})^{-1}\mathbf{X}'\mathbf{V}^{-1}\right]\bar{\boldsymbol{Y}}. \qquad (6.5.6)
\end{aligned}
$$

Note that $SS_0 \sim \chi_{a-b}^2$.

Since upper and lower tolerance limits for $N(\mathbf{x}_0'\boldsymbol{\beta}, \sigma_\tau^2 + \sigma_{ei}^2)$, and for $N(\mathbf{x}_0'\boldsymbol{\beta}, \sigma_\tau^2)$, are confidence limits for the appropriate percentiles, we shall follow the approach that should be familiar by now: first derive GPQs for the percentiles, which will permit the computation of generalized confidence limits, which in turn will provide the required tolerance limits. Towards this, we shall now derive GPQs for $\mathbf{x}_0'\boldsymbol{\beta}$, σ_τ^2 and σ_{ei}^2, to be denoted by $G_{\mathbf{x}_0'\boldsymbol{\beta}}$, $G_{\sigma_\tau^2}$ and $G_{\sigma_{ei}^2}$, respectively ($i = 1, 2, ..., a$). Let $\tilde{\bar{\boldsymbol{Y}}}$ and ss_i denote the observed values of $\bar{\boldsymbol{Y}}$ and SS_i, respectively. Let

$$G_{\sigma_{ei}^2} = \frac{\sigma_{ei}^2}{SS_{ei}}ss_{ei} = \frac{ss_{ei}}{U_{ei}^2}$$

$$\mathbf{V}_G = \operatorname{diag}\left(\frac{1}{G_{\sigma_\tau^2} + \frac{G_{\sigma_{e1}^2}}{n_1}}, \frac{1}{G_{\sigma_\tau^2} + \frac{G_{\sigma_{e2}^2}}{n_2}}, ..., \frac{1}{G_{\sigma_\tau^2} + \frac{G_{\sigma_{ea}^2}}{n_a}}\right), \quad (6.5.7)$$

where U_{ei} is the chi-square random variable defined in (6.5.4), and $G_{\sigma_\tau^2}$ is the GPQ for σ_τ^2, to be obtained. It is clear that $G_{\sigma_{ei}^2}$ is a GPQ for σ_{ei}^2. Motivated by (6.5.6), consider the equation

$$U_0^2 = \tilde{\bar{\boldsymbol{Y}}}'\left[\mathbf{V}_G^{-1} - \mathbf{V}_G^{-1}\mathbf{X}(\mathbf{X}'\mathbf{V}_G^{-1}\mathbf{X})^{-1}\mathbf{X}'\mathbf{V}_G^{-1}\right]\tilde{\bar{\boldsymbol{Y}}}, \qquad (6.5.8)$$

where $U_0^2 \sim \chi_{a-b}^2$, $\tilde{\bar{Y}}$ is the observed value of $\bar{Y}$, and $\mathbf{V}_G$ is defined in (6.5.7). The right hand side of (6.5.8) is a function of $G_{\sigma_\tau^2}$. It can be shown that right hand side of (6.5.8) is a monotone decreasing function of $G_{\sigma_\tau^2}$, and the limit is zero as $G_{\sigma_\tau^2} \to \infty$. The proof of this is similar to the corresponding proof in Section 6.4. Thus the solution $G_{\sigma_\tau^2}$ of the equation (6.5.8) exists uniquely, even though the solution need not be nonnegative.

Having obtained $G_{\sigma_{ei}^2}$ and $G_{\sigma_\tau^2}$, it now remains to construct a GPQ for $\mathbf{x}_0'\boldsymbol{\beta}$, say $G_{\mathbf{x}_0'\boldsymbol{\beta}}$. For this, note from (6.5.5) that

$$Z = \frac{\mathbf{x}_0'\hat{\boldsymbol{\beta}}_\mathbf{V} - \mathbf{x}_0'\boldsymbol{\beta}}{\sqrt{\mathbf{x}_0'(\mathbf{X}'\mathbf{V}^{-1}\mathbf{X})^{-1}\mathbf{x}_0}} \sim N(0,1). \tag{6.5.9}$$

Now define

$$
\begin{aligned}
G_{\mathbf{x}_0'\boldsymbol{\beta}} &= \mathbf{x}_0'(\mathbf{X}'\mathbf{V}_G^{-1}\mathbf{X})^{-1}\mathbf{X}'\mathbf{V}_G^{-1}\tilde{\bar{Y}} \\
&\quad - \frac{\mathbf{x}_0'\hat{\boldsymbol{\beta}}_\mathbf{V} - \mathbf{x}_0'\boldsymbol{\beta}}{\sqrt{\mathbf{x}_0'(\mathbf{X}'\mathbf{V}^{-1}\mathbf{X})^{-1}\mathbf{x}_0}} \times \sqrt{\left[\mathbf{x}_0'(\mathbf{X}'\mathbf{V}_G^{-1}\mathbf{X})^{-1}\mathbf{x}_0\right]_+} \\
&= \mathbf{x}_0'(\mathbf{X}'\mathbf{V}_G^{-1}\mathbf{X})^{-1}\mathbf{X}'\mathbf{V}_G^{-1}\tilde{\bar{Y}} - Z\sqrt{\left[\mathbf{x}_0'(\mathbf{X}'\mathbf{V}_G^{-1}\mathbf{X})^{-1}\mathbf{x}_0\right]_+},
\end{aligned}
$$
$$\tag{6.5.10}$$

where $\mathbf{V}_G$ is defined in (6.5.7) and $\tilde{\bar{Y}}$ is the observed value of $\bar{Y}$. It can be directly verified that given the observed data, the distribution of $G_{\mathbf{x}_0'\boldsymbol{\beta}}$ is free of any unknown parameters, and the observed value of $G_{\mathbf{x}_0'\boldsymbol{\beta}}$ is $\mathbf{x}_0'\boldsymbol{\beta}$. In other words $G_{\mathbf{x}_0'\boldsymbol{\beta}}$ is a GPQ for $\mathbf{x}_0'\boldsymbol{\beta}$.

Once the GPQs have been derived as given above, one-sided and two-sided tolerance intervals for the distributions $N(\mathbf{x}_0'\boldsymbol{\beta}, \sigma_\tau^2 + \sigma_{ei}^2)$ and $N(\mathbf{x}_0'\boldsymbol{\beta}, \sigma_\tau^2)$ can be easily constructed. The $100p$th percentiles of $N(\mathbf{x}_0'\boldsymbol{\beta}, \sigma_\tau^2 + \sigma_{ei}^2)$ and $N(\mathbf{x}_0'\boldsymbol{\beta}, \sigma_\tau^2)$ are $\mathbf{x}_0'\boldsymbol{\beta} + z_p\sqrt{\sigma_\tau^2 + \sigma_{ei}^2}$ and $\mathbf{x}_0'\boldsymbol{\beta} + z_p\sigma_\tau$, respectively. GPQs for these quantities are $G_{\mathbf{x}_0'\boldsymbol{\beta}} + z_p\sqrt{\left[G_{\sigma_\tau^2} + G_{\sigma_{ei}^2}\right]_+}$ and $G_{\mathbf{x}_0'\boldsymbol{\beta}} + z_p\sqrt{\left[G_{\sigma_\tau^2}\right]_+}$, respectively. The $100(1-\alpha)$th percentiles of these GPQs provide $(p, 1-\alpha)$ upper tolerance limits for $N(\mathbf{x}_0'\boldsymbol{\beta}, \sigma_\tau^2 + \sigma_{ei}^2)$ and $N(\mathbf{x}_0'\boldsymbol{\beta}, \sigma_\tau^2)$, respectively. An algorithm similar to Algorithm 5.1 or Algorithm 6.3 (with obvious modifications) can be used to compute these tolerance limits.

In order to obtain two-sided tolerance limits for $N(\mathbf{x}_0'\boldsymbol{\beta}, \sigma_\tau^2 + \sigma_{ei}^2)$ and $N(\mathbf{x}_0'\boldsymbol{\beta}, \sigma_\tau^2)$, we shall assume that the two sided $(p, 1-\alpha)$ tolerance limit for $N(\mathbf{x}_0'\boldsymbol{\beta}, \sigma_\tau^2 + \sigma_{ei}^2)$ is of the form $\mathbf{x}_0'\tilde{\boldsymbol{\beta}} \pm D_{10}$ and that for $N(\mathbf{x}_0'\boldsymbol{\beta}, \sigma_\tau^2)$ is of the form $\mathbf{x}_0'\tilde{\boldsymbol{\beta}} \pm D_{11}$, where $\mathbf{x}_0'\tilde{\boldsymbol{\beta}}$ is an estimator of $\mathbf{x}_0'\boldsymbol{\beta}$, to be determined, and D_{10} and D_{11} are margin of error statistics. The quantity D_{10}^2 can be taken to be a $100(1-\alpha)\%$ upper confidence limit of $z_{\frac{1+p}{2}}^2\left((\mathbf{z}_0'\mathbf{z}_0)\sigma_\tau^2 + \sigma_{ei}^2 + \mathbf{x}_0'(\mathbf{X}'\mathbf{V}^{-1}\mathbf{X})^{-1}\mathbf{x}_0\right)$. Such a confidence limit can be obtained as the $100(1-\alpha)$th percentile of

$$z_{\frac{1+p}{2}}^2 \times \left[(\mathbf{z}_0'\mathbf{z}_0)G_{\sigma_\tau^2} + G_{\sigma_{ei}^2} + \mathbf{x}_0'(\mathbf{X}'\mathbf{V}_G^{-1}\mathbf{X})^{-1}\mathbf{x}_0\right]_+,$$

where $\mathbf{V}_G$ is given in (6.5.7). Similarly, D_{11}^2 can be taken to be a $100(1-\alpha)\%$ upper confidence limit of $z_{\frac{1+p}{2}}^2\left((\mathbf{z}_0'\mathbf{z}_0)\sigma_\tau^2 + \mathbf{x}_0'(\mathbf{X}'\mathbf{V}^{-1}\mathbf{X})^{-1}\mathbf{x}_0\right)$. Such a confidence limit can be obtained as the $100(1-\alpha)$th percentile of

$$z_{\frac{1+p}{2}}^2 \times \left[(\mathbf{z}_0'\mathbf{z}_0)G_{\sigma_\tau^2} + \mathbf{x}_0'(\mathbf{X}'\mathbf{V}_G^{-1}\mathbf{X})^{-1}\mathbf{x}_0\right]_+,$$

The estimator $\mathbf{x}_0'\tilde{\boldsymbol{\beta}}$ can be taken to be the mean (or the median) of the GPQ $G_{\mathbf{x}_0'\boldsymbol{\beta}}$. This mean (or median) can be easily estimated based on the realizations $G_{\mathbf{x}_0'\boldsymbol{\beta},k}$ of $G_{\mathbf{x}_0'\boldsymbol{\beta}}$. An algorithm similar to Algorithm 5.2 or Algorithm 6.4 (with obvious modifications) can be used to compute the above two-sided tolerance intervals.

Example 6.4 (Sea urchin density and algae data)

This example is taken from Lin, Liao and Iyer (2008), who quote Andrew and Underwood (1993) for further details and the data. The data consist of the percentage cover of filamentous algae in 16 patches of reef in a shallow subtidal region in Australia. The sea urchin density was a covariate, taking four values: 0%, 33%, 66% and 100%. The data are reproduced in Table 6.2, where the mean and standard deviation are reported for the square root transformed data.

Lin, Liao and Iyer (2008) analyzed the data using the one-way random model for Y_{ij}, the square root transformed percentage cover of filamentous algae:

$$Y_{ij} = \beta_0 + \beta_1 x_i + \tau_i + e_{ij}, \qquad (6.5.11)$$

$j = 1, 2, \ldots, n_i;\ i = 1, 2, 3, \ldots, 16$, where x_i's represent the density, $\tau_i \sim N(0, \sigma_\tau^2)$, $e_{ij} \sim N(0, \sigma_{ei}^2)$. Following Lin, Liao and Iyer (2008), we assume that the σ_{ei}^2's corresponding to the same x_i are equal. Thus, $\sigma_{e1}^2 = \sigma_{e2}^2 = \sigma_{e3}^2 = \sigma_{e4}^2$ (since they correspond to $x_i = 1.00$, for $i = 1, 2, 3, 4$), and the next four σ_{ei}^2's are equal, and so on. In the notations introduced for the model (6.5.1), we have $a = 16$, $b = 2$, and all the n_i's are equal to 5, except $n_8 = 4$ and $n_{16} = 3$. Note that the observed values ss_{ei}'s are obtained by pooling the four variances corresponding to each density x_i, since there are only four heterogeneous variances to be estimated.

Table 6.2: Data on sea urchin density along with the sample mean and standard deviation of percentage cover of filamentous algae (square root transformed)

Patch	Sample Size	Density	Mean	SD
1	5	1.00	0.773	1.1196
2	5	1.00	0.000	0.0000
3	5	1.00	0.600	0.8944
4	5	1.00	0.721	1.6125
5	5	0.66	4.637	2.9366
6	5	0.66	5.275	3.3486
7	5	0.66	0.447	1.0000
8	4	0.66	3.500	4.0329
9	5	0.33	1.013	1.4027
10	5	0.33	0.000	0.0000
11	5	0.33	5.865	2.0023
12	5	0.33	4.985	3.7000
13	5	0.00	5.704	1.4419
14	5	0.00	7.861	0.4971
15	5	0.00	0.937	1.2854
16	3	0.00	8.800	0.4145

We shall denote the pooled sum of squares by ss_i, $i = 1, 2, 3, 4$. Their values are

$$ss_1 = 18.6144, \ ss_2 = 132.1386, \ ss_3 = 78.6652, \text{and } ss_4 = 16.2573.$$

Also note that in the matrix $\mathbf{V}$ defined in (6.5.3), the σ_{ei}^2's are equal for $i = 1$, 2, 3, 4, and they are also equal for the next groups of four each. Consequently, in the definition of $\mathbf{V}_G$ in (6.5.7), the GPQs $G_{\sigma_{ei}^2}$'s are equal for $i = 1, 2, 3, 4$, and they are also equal for the next groups of four each. The chi-square random variable U_0^2 has df $= a - b = 14$, and there are four independent chi-square random variables U_{ei}^2 in (6.5.4), to be denoted by U_i^2, $i = 1, 2, 3, 4$, having df 16, 15, 16 and 14, respectively.

The procedures discussed in this section can be applied to compute one-sided tolerance limits and two-sided tolerance intervals for the distributions $N(\beta_0 + \beta_1 x_0, \sigma_\tau^2 + \sigma_{ei}^2)$ and $N(\beta_0 + \beta_1 x_0, \sigma_\tau^2)$, for a specified value of the density x_0. For $x_0 = 50$, Lin, Liao and Iyer (2008) have reported a (0.90, 0.95) upper tolerance limit for the distribution $N(\beta_0 + \beta_1 x_0, \sigma_\tau^2)$, and the limit is 8.0055. Also, a (0.90, 0.95) two-sided tolerance interval for $N(\beta_0 + \beta_1 x_0, \sigma_\tau^2)$ is $(-2.8901, 8.8864)$. Since the random variable of interest is nonnegative, we shall take the lower limit to be zero. Thus the tolerance interval is $(0, 8.8864)$.

6.6   Testing Individual Bioequivalence

In this section, we shall address the role of tolerance intervals in the problem of bioequivalence testing. Bioequivalence trials are carried out in order to compare the bioavailabilities of two formulations of the same drug, or two drug products. The bioavailability of a drug is defined as the amount of the active drug ingredient that gets into the bloodstream, and becomes available at the site of drug action. Two drug products are said to be bioequivalent if they have similar bioavailabilities, where similarity is to be assessed based on appropriate criteria. Bioequivalence testing is the statistical procedure for assessing bioequivalence between two formulations of the same drug or between two drug products: usually a brand-name drug and a generic copy. The discussion that follows is very brief, and our goal in this section is to illustrate the application of tolerance intervals for the assessment of bioequivalence. For more information on the topic of bioequivalence, we refer to the books by Chow and Liu (2000), Patterson and Jones (2006) and Hauschke, Steinijans and Pigeot (2007).

The data for bioequivalence studies are generated using cross-over designs where the subjects in the study are healthy volunteers. Thus each individual in the study is administered both drug products, separated by a washout period that is long enough to guarantee that very little (or none) of the effect of the previous drug administration is left in the bloodstream. After each administration of the drug, the concentration of the active drug ingredient in the bloodstream is measured at several time points. Note that the blood concentration of the active drug ingredient will gradually increase over time, eventually reaching a peak, and will then gradually decline. A plot of the blood concentration versus time can then be obtained, and three responses are typically measured: (i) the area under the curve, denoted by AUC, (ii) the maximum blood concentration, denoted by C_{max}, and (iii) the time to reach the maximum concentration, denoted by T_{max}. Usually the AUC and C_{max} are the responses of interest, and these quantities typically follow a lognormal distribution. In our discussion in this section, we shall assume that the lognormality assumption is valid. Thus it is the log-transformed data that are modeled and analyzed for the assessment of bioequivalence. Average bioequivalence holds between two drug products if the difference between the average responses (corresponding to ln(AUC), say) is within a narrow interval, usually taken to be the interval $(-\ln(1.25), \ln(1.25))$. Two drug products are said to be individually bioequivalent if the responses from the same individual corresponding to the two drug products are close, according to a suitable criterion. For the assessment of individual bioequivalence, the use of tolerance intervals has been investigated by Esinhart and Chinchilli (1994), Brown, Iyer and Wang (1997) and Liao and Iyer (2004). A brief discussion of the application of tolerance intervals for testing individual bioequivalence also appears in Chow and Liu (2000, Section 15.4.3).

Here we shall consider only four period cross-over designs. Let R denote the reference drug or the brand-name drug, and T denote the test drug or the generic copy. Thus R and T denote the two formulations to be compared. A two-sequence and four-period cross-over design (or a 2×4 cross-over design) is given by

		Period			
		1	2	3	4
Sequence	1	T	R	T	R
	2	R	T	T	R

A four-sequence and four-period crossover design (or a 4×4 cross-over design) is given by

		Period			
		1	2	3	4
Sequence	1	R	R	T	T
	2	R	T	T	R
	3	T	T	R	R
	4	T	R	R	T

In the case of a 2×4 cross-over design, the subjects in the study are randomly assigned to the two sequences. In a 4×4 cross-over design, the subjects are randomly assigned to the four sequences. Thus each subject receives the reference drug and test drug twice each, separated by washout periods.

Conceptually, the tolerance interval idea proposed by Brown, Iyer and Wang (1997), and Liao and Iyer (2004) is as follows. Denote by μ_T and μ_R the population mean responses (based on the log-transformed data) corresponding to the two formulations T and R, respectively, and let η_{ijT} and η_{ijR} denote random subject effects corresponding to T and R, respectively, for the jth subject in the ith sequence. Thus the true responses based on T and R, for the jth subject in the ith sequence, are $\mu_T + \eta_{ijT}$ and $\mu_R + \eta_{ijR}$. We conclude individual bioequivalence if $\mu_T + \eta_{ijT}$ and $\mu_R + \eta_{ijR}$ are close, i.e., if a tolerance interval for $\mu_T - \mu_R + \eta_{ijT} - \eta_{ijR}$ is narrow. In their articles, Brown, Iyer and Wang (1997) and Liao and Iyer (2004) used the interval $(\ln(0.75), \ln(1.25))$ to decide if the tolerance interval is narrow enough to conclude individual bioequivalence. The interval is to be constructed in such a way that at least 75% of the individuals will have their true response differences, i.e., the quantities $\mu_T - \mu_R + \eta_{ijT} - \eta_{ijR}$, contained in the interval $(\ln(0.75), \ln(1.25))$ with 95% confidence. That is, conclude individual bioequivalence if a $(0.75, 0.95)$ two-sided tolerance interval for $\mu_T - \mu_R + \eta_{ijT} - \eta_{ijR}$ is a subset of $(\ln(0.75), \ln(1.25))$. A similar approach developed by Esinhart and Chinchilli (1994), and described in Chow and Liu (2000, Section 15.4.3), constructs a tolerance interval for the differences in the responses, including the within subject error terms as well. These authors use the interval $(-\ln(1.33), \ln(1.33))$ to decide whether the tolerance interval is narrow enough to conclude individual bioequivalence.

We shall now give the model for the responses, and then describe the procedure for constructing the required tolerance intervals. Consider an $s \times 4$ crossover design, and let n_i be the number of subjects in the ith sequence. Thus there are s sequences in the design; for example, s could be two or four. Furthermore, let Y_{ijrl} be the lth response (AUC or $C_{\max}$, log transformed) corresponding to drug formulation r for the jth subject in the ith sequence; $i = 1, 2, \ldots s$, $j = 1, 2, \ldots n_i$, $r = T, R$, $l = 1, 2$. A mixed effects model for Y_{ijrl} has been proposed by Chinchilli and Esinhart (1996), and is given by

$$Y_{ijrl} = \mu_r + \gamma_{irl} + \eta_{ijr} + \epsilon_{ijrl}, \qquad (6.6.1)$$

where μ_T and μ_R are the population mean responses, γ_{irl} is a fixed effect satisfying $\sum_{i=1}^{s} \sum_{l=1}^{2} \gamma_{irl} = 0$, for $r = T, R$, η_{ijr} is a random subject effect corresponding to formulation r for subject j in sequence i, and the ϵ_{ijrl}'s are within-subject errors. We assume that

$$\epsilon_{ijTl} \sim N(0, \sigma_{WT}^2), \quad \epsilon_{ijRl} \sim N(0, \sigma_{WR}^2).$$

Thus σ_{WT}^2 and σ_{WR}^2 are the within subject variances. We also assume that $(\eta_{ijT}, \eta_{ijR})'$ follows a bivariate normal distribution with zero means and variance-covariance matrix

$$\Sigma_B = \begin{pmatrix} \sigma_{BT}^2 & \sigma_{BTR} \\ \sigma_{BTR} & \sigma_{BR}^2 \end{pmatrix}.$$

Thus Σ_B is the between-subject covariance matrix. The above model is also given in the U. S. Food and Drug Administration's guidance document, U.S. FDA (2001). Note that the quantity $\mu_T - \mu_R + \eta_{ijT} - \eta_{ijR}$ for which a tolerance interval is required has the distribution

$$\mu_T - \mu_R + \eta_{ijT} - \eta_{ijR} \sim N(\mu_T - \mu_R, \sigma_D^2),$$
$$\text{where} \quad \sigma_D^2 = \text{Var}(\eta_{ijT} - \eta_{ijR}) = \sigma_{BT}^2 + \sigma_{BR}^2 - 2\sigma_{BTR}. \qquad (6.6.2)$$

The variance component σ_D^2 is referred to as the subject-by-formulation interaction. Instead of looking at the true difference $\mu_T - \mu_R + \eta_{ijT} - \eta_{ijR}$, if the within-subject errors are also to be included, then a tolerance interval is required for the quantity $\mu_T - \mu_R + \eta_{ijT} - \eta_{ijR} + \epsilon_{ijTl} - \epsilon_{ijRl'}$ having the distribution

$$\mu_T - \mu_R + \eta_{ijT} - \eta_{ijR} + \epsilon_{ijTl} - \epsilon_{ijRl'} \sim N(\mu_T - \mu_R, \sigma_D^2 + \sigma_{WT}^2 + \sigma_{WR}^2), \quad (6.6.3)$$

where l and l' take the values 1 or 2 (since each subject receives two administrations each, of T and R).

We shall now show that for obtaining tolerance intervals for the distributions in (6.6.2) and (6.6.3), we are in the set up described in Section 6.3. Towards this, define

$$\bar{Y}_{ijr} = (Y_{ijr1} + Y_{ijr2})/2, \quad \bar{Y}_{i.r} = \frac{1}{n_i} \sum_{j=1}^{n_i} \bar{Y}_{ijr}, \quad (r = T, R)$$

$$E_{ijr} = (Y_{ijr1} - Y_{ijr2}), \quad \bar{E}_{i.r} = \frac{1}{n_i} \sum_{j=1}^{n_i} E_{ijr}, \quad (r = T, R)$$

$$\bar{D} = \frac{1}{s} \sum_{i=1}^{s} \frac{1}{n_i} \sum_{j=1}^{n_i} (\bar{Y}_{ijT} - \bar{Y}_{ijR}),$$

$$c^2 = \frac{1}{s^2} \sum_{i=1}^{s} \frac{1}{n_i}, \quad v = \sum_{i=1}^{s} n_i - s,$$

$$SS_D = \sum_{i=1}^{s} \sum_{j=1}^{n_i} \left([\bar{Y}_{ijT} - \bar{Y}_{ijR}] - [\bar{Y}_{i.T} - \bar{Y}_{i.R}]\right)^2,$$

and $\quad SS_{Wr} = \frac{1}{2} \sum_{i=1}^{s} \sum_{j=1}^{n_i} (E_{ijr} - \bar{E}_{i.r})^2, \quad r = T, R. \quad (6.6.4)$

It can be verified that

$$\bar{D} \sim N\left(\mu_T - \mu_R, \; c^2[\sigma_D^2 + \frac{1}{2}(\sigma_{WT}^2 + \sigma_{WR}^2)]\right),$$

$$\frac{SS_D}{[\sigma_D^2 + \frac{1}{2}(\sigma_{WT}^2 + \sigma_{WR}^2)]} \sim \chi_v^2, \quad \frac{SS_{WT}}{\sigma_{WT}^2} \sim \chi_v^2, \quad \text{and} \quad \frac{SS_{WR}}{\sigma_{WR}^2} \sim \chi_v^2, \quad (6.6.5)$$

where the random variables $\bar{D}$, SS_D, SS_{WT} and SS_{WR} are also independently distributed. We note that SS_{WT} and SS_{WR}, respectively, are the error sum of squares obtained from an ANOVA based on the model for Y_{ijTl}, and the model for Y_{ijRl}.

It is clear that (6.6.5) is a special case of the set up (6.3.1). Let

$$\sigma_1^2 = \sigma_D^2 + \frac{1}{2}(\sigma_{WT}^2 + \sigma_{WR}^2), \quad \sigma_2^2 = \sigma_{WT}^2, \quad \text{and} \quad \sigma_3^2 = \sigma_{WR}^2.$$

Following the derivations in Section 6.3.2, we conclude that a $(p, 1 - \alpha)$ two-sided tolerance interval for the distribution $N(\mu_T - \mu_R, \sigma_D^2)$ is given by $\bar{D} \pm D_5$, where the margin of error statistic D_5 is such that D_5^2 is an upper $100(1 - \alpha)\%$ confidence limit for $z_{\frac{1+p}{2}}^2 \left[(1 + c^2)\sigma_1^2 - \frac{c^2}{2}(\sigma_2^2 + \sigma_3^2) \right]$. Similarly, a $(p, 1 - \alpha)$ two-sided tolerance interval for the distribution $N(\mu_T - \mu_R, \sigma_D^2 + \sigma_{WT}^2 + \sigma_{WR}^2)$ is given by $\bar{D} \pm \tilde{D}_5$, where the margin of error statistic $\tilde{D}_5$ is such that $\tilde{D}_5^2$ is an upper $100(1 - \alpha)\%$ confidence limit for $z_{\frac{1+p}{2}}^2 \left[(1 + c^2)\sigma_1^2 + (1 - \frac{c^2}{2})(\sigma_2^2 + \sigma_3^2) \right]$.

Example 6.5 (Bioequivalence study based on a 2×4 crossover design)

We shall now illustrate the above results with an example. The data for the example is taken from the U. S. FDA website.[1] The drug under consideration is Monamine Oxidase (MAO) inhibitor, a drug used for treating depression. Data on C_{max} were recorded for 38 subjects, obtained based on a 2×4 crossover design with 18 subjects assigned to sequence TRRT and 20 subjects assigned to sequence RTTR. The data are reproduced in Table A2, Appendix A.

The log-transformed C_{max} data were analyzed using the model (6.6.1), and gave the following observed values of $\bar{D}$, SS_D, SS_{WT} and SS_{WR}, denoted by $\bar{d}$, ss_D, ss_{WT} and ss_{WR}, respectively.

$$\bar{d} = -0.00501, \quad ss_D = 0.6804, \quad ss_{WT} = 1.1904, \quad ss_{WR} = 1.2956.$$

Note that the degrees of freedom v, given in (6.6.4) and appearing in (6.6.5), simplifies to $v = 36$. We shall use the MLS method to compute an upper confidence limit for $(1 + c^2)\sigma_1^2 - \frac{c^2}{2}(\sigma_2^2 + \sigma_3^2)$. This will permit the computation of the margin of error statistic D_5, which is such that D_5^2 is an upper $100(1-\alpha)\%$ confidence limit for $z_{\frac{1+p}{2}}^2 \left[(1 + c^2)\sigma_1^2 - \frac{c^2}{2}(\sigma_2^2 + \sigma_3^2) \right]$. ¿From the definition of c^2 in (6.6.4), we have

$$c^2 = \frac{1}{s^2} \sum_{i=1}^{s} \frac{1}{n_i} = \frac{1}{4}\left(\frac{1}{18} + \frac{1}{20} \right) = 0.026388.$$

Using the given values of ss_D, ss_{WT} and ss_{WR}, we also have the estimates

$$\hat{\sigma}_1^2 = \frac{ss_D}{36} = 0.0189, \quad \hat{\sigma}_2^2 = \frac{ss_{WT}}{36} = 0.033066, \quad \hat{\sigma}_3^2 = \frac{ss_{WR}}{36} = 0.035988.$$

For 36 df, we also have $\chi_{36;0.05}^2 = 23.27$ and $\chi_{36;0.95}^2 = 51$. Using the expression (1.3.4) in Section 1.3, the a 95% upper confidence limit for $(1+c^2)\sigma_1^2 - \frac{c^2}{2}(\sigma_2^2 + \sigma_3^2)$,

---

[1] www.fda.gov/cder/bioequivdata/drug14b.txt

based on the MLS procedure, simplifies to 0.0291. For $p = 0.75$, $z_{\frac{1+p}{2}} = 1.5035$. Hence $D_5^2 = 1.5035 \times 0.0291 = 0.043751$, and the margin of error statistic D_5 has the value 0.20916. The $(0.75, 0.95)$ two-sided tolerance interval $\bar{D} \pm D_5$ for the distribution $N(\mu_T - \mu_R, \sigma_D^2)$ now simplifies to the interval $(-0.2142, 0.2042)$. Since $(\ln(0.75), \ln(1.25)) = (-0.2877, 0.2231)$, it is clear that the interval $\bar{D} \pm D_5$ is a subset of $(\ln(0.75), \ln(1.25))$. In other words, we conclude individual bioequivalence.

Let us now construct a two-sided tolerance interval for the distribution $N(\mu_T - \mu_R, \sigma_D^2 + \sigma_{WT}^2 + \sigma_{WR}^2)$, corresponding to $\mu_T - \mu_R + \eta_{ijT} - \eta_{ijR} + \epsilon_{ijTl} - \epsilon_{ijRl'}$, and verify if it is contained in the interval $(-\ln(1.33), \ln(1.33))$, as recommended in Chow and Liu (2000, Section 15.4.3). As noted earlier, $\sigma_D^2 + \sigma_{WT}^2 + \sigma_{WR}^2 = (1 + c^2)\sigma_1^2 + (1 - \frac{c^2}{2})(\sigma_2^2 + \sigma_3^2)$, and our tolerance interval is of the form $\bar{D} \pm \tilde{D}_5$, where the margin of error statistic $\tilde{D}_5$ is such that $\tilde{D}_5^2$ is an upper $100(1 - \alpha)\%$ confidence limit for $z_{\frac{1+p}{2}}^2 \left[(1 + c^2)\sigma_1^2 + (1 - \frac{c^2}{2})(\sigma_2^2 + \sigma_3^2) \right]$. Applying the formula (1.3.3) of Section 1.3, the 95% MLS upper confidence limit for $(1 + c^2)\sigma_1^2 + (1 - \frac{c^2}{2})(\sigma_2^2 + \sigma_3^2)$ simplifies to 0.11597. Thus $\tilde{D}_5^2 = 1.5035 \times 0.11597 = 0.17436$. Consequently $\tilde{D}_5 = 0.4176$ and the $(0.75, 0.95)$ two-sided tolerance interval $\bar{D} \pm \tilde{D}_5$ for $N(\mu_T - \mu_R, \sigma_D^2 + \sigma_{WT}^2 + \sigma_{WR}^2)$ simplifies to $(-0.4226, 0.4126)$. Since $(-\ln(1.33), \ln(1.33)) = (-0.2852, 0.2852)$, it is clear that the interval $\bar{D} \pm D_5$ is not a subset of $(-\ln(1.33), \ln(1.33))$. In other words, now we cannot conclude individual bioequivalence.

Since the two approaches have resulted in opposite conclusions, what should we conclude regarding individual bioequivalence? In their article, Brown, Iyer and Wang (1997, p. 805) comment that "Any individual bioequivalence criterion should itself be stated independently of the magnitude of measurement error variances." If we accept this, then conclusion regarding individual bioequivalence should be based on a two-sided tolerance interval for the distribution $N(\mu_T - \mu_R, \sigma_D^2)$, and not for the distribution $N(\mu_T - \mu_R, \sigma_D^2 + \sigma_{WT}^2 + \sigma_{WR}^2)$. Thus our conclusion will be in favor of individual bioequivalence. We shall not comment on this further; for more information and alternative criteria for individual bioequivalence, we refer to Chow and Liu (2000, Chapter 15) and Patterson and Jones (2006, Chapter 6).

6.7 Exercises

6.7.1. Consider two independent two-way nested models with random effects given by

$$Y_{1ijk} = \mu_1 + \tau_{1i} + \beta_{1j(i)} + e_{1ijk}, \ k = 1, 2, ..., n_1, \ j = 1, 2, ..., b_1, \ i = 1, 2, ..., a_1$$
$$Y_{2ijk} = \mu_2 + \tau_{2i} + \beta_{2j(i)} + e_{2ijk}, \ k = 1, 2, ..., n_2, \ j = 1, 2, ..., b_2, \ i = 1, 2, ..., a_2$$

where for $l = 1, 2$, it is assumed that $\tau_{li} \sim N(0, \sigma_{\tau l}^2)$, $\beta_{lj(i)} \sim N(0, \sigma_{\beta l}^2)$, $e_{lijk} \sim N(0, \sigma_{el}^2)$, and all the random variables are independent.

(a) Explain how you will compute one-sided and two-sided tolerance intervals for the difference $Y_1 - Y_2$, where $Y_l \sim N(\mu_l, \sigma_{\tau l}^2 + \sigma_{\beta l}^2 + \sigma_{el}^2)$, $l = 1, 2$.

(b) Explain how you will compute one-sided and two-sided tolerance intervals for the difference between the true values $(\mu_1 + \tau_1 + \beta_1) - (\mu_2 + \tau_2 + \beta_2)$, where τ_l and β_l have the same distributions as those of τ_{li} and $\beta_{lj(i)}$, respectively, $l = 1, 2$.

6.7.2. Suppose the models given above are mixed effects models; that is, the τ_{1i}'s and the τ_{2i}'s are fixed effects, and the rest of the quantities are random having the distributions specified in the previous problem.

(a) Explain how you will compute one-sided and two-sided tolerance intervals for the difference $Y_{1i} - Y_{2i}$ for a fixed i, where $Y_{li} \sim N(\mu_l + \tau_{li}, \sigma_{\beta l}^2 + \sigma_{el}^2)$, $l = 1, 2$.

(b) Explain how you will compute one-sided and two-sided tolerance intervals for the difference between the true values $(\mu_1 + \tau_{1i} + \beta_1) - (\mu_2 + \tau_{2i} + \beta_2)$ for a fixed i, where β_l has the same distribution as that of $\beta_{lj(i)}$, $l = 1, 2$.

6.7.3. An experiment was carried out at the U.S. Army Ballistic Research Laboratory, Aberdeen Proving Ground, Maryland, to compare a new tube (NT) with a control tube (CT) to be used for firing ammunition from tanks. Twenty new tubes and twenty control tubes were randomly selected for the experiment with 4 tanks each for mounting the new tubes and the control tubes. Five new tubes were mounted on each of 4 tanks and 5 control tubes were mounted on each of the other 4 tanks. Three rounds were fired from each tube and the observations consisted of a miss distance (the unit used was 6400 mils per 365 degrees). Let CT_{ij} and NT_{ij}, respectively, denote the jth control tube and the jth new tube mounted on the ith tank ($j = 1, 2, 3, 4, 5; i = 1, 2, 3, 4$). The three measurements (the miss distances) corresponding to each CT_{ij} and NT_{ij} given below are reproduced from Zhou and Mathew (1994).

Tank	CT_{i1}	CT_{i2}	CT_{i3}	CT_{i4}	CT_{i5}	NT_{i1}	NT_{i2}	NT_{i3}	NT_{i4}	NT_{i5}
	2.76	1.83	1.60	1.53	2.20	1.92	1.98	2.28	1.52	1.28
1	2.10	1.65	1.56	2.29	2.59	1.77	1.56	1.90	1.82	1.61
	1.61	1.76	1.73	2.06	1.91	1.37	1.83	2.10	1.79	1.48
	1.35	1.15	1.68	1.70	1.34	1.70	1.61	1.78	1.60	1.69
2	1.64	1.83	1.71	1.26	1.26	1.82	1.71	2.31	1.65	1.72
	1.56	1.92	1.63	1.64	1.69	1.65	1.28	1.73	1.26	1.76
	1.33	1.65	1.94	1.72	1.81	1.79	1.64	1.84	1.80	1.73
3	1.28	1.76	1.86	1.56	2.13	1.39	1.88	1.67	1.49	1.83
	1.40	1.81	2.00	1.91	1.86	1.52	1.60	1.64	1.92	1.79
	1.64	1.77	1.01	1.04	1.27	1.49	1.88	1.77	1.46	2.10
4	1.80	1.63	1.63	1.78	1.38	1.60	1.60	1.56	1.29	1.46
	1.89	1.51	1.46	1.86	1.55	1.63	1.61	1.62	1.72	1.60

Let Y_{1ijk} and Y_{2ijk}, respectively, denote the kth observation corresponding to CT_{ij} and NT_{ij}, τ_{1i} denote the effect due to the ith tank on which a control tube was mounted, τ_{2i} denote the effect due to the ith tank on which a new tube was mounted, $\beta_{1j(i)}$ denote the effect due to CT_{ij} and $\beta_{2j(i)}$ denote the effect due to NT_{ij}. All these effects are assumed to be random, and the random effects models in Problem 6.7.1 will be used to analyze the data. In order to compare the miss distances resulting from the new tube and the control tube, it is decided to compute (0.95, 0.95) two-sided tolerance intervals for the differences mentioned in part (i) and part (ii) of Problem 6.7.1. Using the data given above, compute such tolerance intervals.

6.7.4. Consider the one-way random model with a single covariate given by

$$Y_{ij} = \mu + \tau_i + \beta x_i + e_{ij},$$

$j = 1, 2, ..., n$, $i = 1, 2, ..., a$, where β is an unknown parameter, and the x_i's denote the values of a covariate (assumed to be non-random). We also make the usual assumptions: $\tau_i \sim N(0, \sigma_\tau^2)$, $e_{ij} \sim N(0, \sigma_e^2)$, and all the random variables are independent.

(a) Express the model in the form (6.4.1).

(b) Explain how you will compute one-sided and two-sided tolerance intervals for the distribution $N(\mu + \beta x_0, \sigma_\tau^2 + \sigma_e^2)$, and for the distribution $N(\mu + \beta x_0, \sigma_\tau^2)$, for a fixed value x_0 of the covariate.

6.7.5. In a breeding experiment, genetically superior sires are bred for improved milk production in cows. Sires are evaluated based on their daughter's milking performance, where each sire has a large number of daughters through artificial insemination. Data are collected on the daughter's yearly milk production; however, each sire's breeding value is a covariate. The data given below are the yearly milk production (measured in kg) on each of five daughters for four sires, along with the breeding value (the x_i's) for four sires; the data are taken from Limam and Thomas (1988b).

x_i	Y_{ij}	x_i	Y_{ij}
780	8566	650	8240
	8586		8288
	8528		8328
	8540		8312
	8575		8271
920	8860	1170	9333
	8857		9313
	8876		9290
	8877		9310
	8836		9318

Reproduced with permission from Taylor and
Francis, Ltd.; http://www.informaworld.com

Assuming the model in the previous problem, compute $(0.95, 0.95)$ one-sided and two-sided tolerance intervals for the distribution $N(\mu + \beta x_0, \sigma_\tau^2 + \sigma_e^2)$, and for the distribution $N(\mu + \beta x_0, \sigma_\tau^2)$, for the breeding value $x_0 = 1000$.

6.7.6. In the previous problem, suppose it is decided to compare the daughters' milk yields from two different sires having the breeding values $x_{01} = 1100$ and $x_{02} = 1000$. Let Y_{01} and Y_{02} be random variables representing the corresponding milk yields.

(a) Suppose the comparison is to be made based on a $(0.90, 0.95)$ lower tolerance limit for the difference $Y_{01} - Y_{02}$. Explain the procedure for computing such a lower tolerance limit.

(b) Apply the procedure to the data in the previous problem, and compute the lower tolerance limit. Based on the lower tolerance limit, can you conclude that daughters' milk yields are expected to be higher for sires with breeding value $x_{01} = 1100$, compared to sires with breeding value $x_{02} = 1000$?

(c) Explain how you will compute a lower confidence limit for $P(Y_{01} \geq Y_{02})$.

6.7.7. In the bioequivalence problem, a large value of the subject-by-formulation interaction σ_D^2 is sometimes taken as evidence against individual bioequivalence; see Chow and Liu (2000, Chapter 15). A value of σ_D^2 bigger than 0.30 is sometimes taken to be large.

    (a) Explain how you will use the three sums of squares SS_D, SS_{WT} and SS_{WR}, having the distributions in (6.6.5), to compute an upper confidence limit for σ_D^2. Use the generalized confidence interval idea, as well as the MLS procedure.

    (b) For the data in Example 6.5, compute a 95% upper confidence limit for σ_D^2, applying both the generalized confidence interval idea, and the MLS procedure. How different are the results? Based on the confidence limits so obtained, can you conclude that the value of σ_D^2 can be significantly larger than 0.30?

6.7.8. In the bioequivalence problem, suppose a two-sided (0.90, 0.95) tolerance interval is required for the distribution of $\bar{D}$, given in (6.5.5). Explain how you will compute such an interval. Also explain how you will compute a two-sided (0.90, 0.95) tolerance interval for the distribution $N(\mu_T - \mu_R, c^2\sigma_D^2)$, where the various quantities are as defined in Section 6.6.

Chapter 7

Some Non-Normal Distributions

7.1 Introduction

All the methods discussed in earlier chapters for constructing tolerance regions and other related problems are based on the normality assumption. It has been now well recognized that there is nothing inherently normal about the normal distribution, and its common use in statistics is due to its simplicity, or due to the fact that it very often is a good approximation. Indeed, in many practical applications the normality assumption is not tenable. In particular, exposure data, lifetime data (time to event data), and other data such as personal income data that are typically skewed, do not satisfy the normality assumption. A strategy for handling non-normal data is to transform the sample so that the transformed sample fits a normal distribution at least approximately. For example, if X follows a lognormal distribution, then $\ln(X)$ follows a normal distribution exactly; if X has a gamma distribution, then $X^{\frac{1}{3}}$ has an approximate normal distribution (Wilson and Hilferty, 1931). Therefore, normal based approaches given in Chapter 2 can be used to construct tolerance intervals for a lognormal distribution after taking logarithmic transformation of the data, and for a gamma distribution after taking cube root transformation of the samples.

In the following sections, we describe methods for constructing tolerance limits and setting lower confidence limits for a survival probability for lognormal, gamma, two-parameter exponential, Weibull and other related distributions. For lognormal and gamma distributions, we describe the procedures for obtaining tolerance limits using the aforementioned transformations. For a two-parameter exponential distribution, an exact method and a generalized variable method are presented. For the Weibull distribution, tolerance limits and other related problems are addressed using the generalized variable method.

7.2   Lognormal Distribution

A random variable Y is said to be lognormally distributed, i.e., $Y \sim \text{lognormal}(\mu, \sigma^2)$ if $X = \ln(Y) \sim N(\mu, \sigma^2)$. The pdf of Y is given by

$$f(y|\mu, \sigma) = \frac{1}{\sqrt{2\pi} \, y\sigma} \exp\left[-\frac{(\ln y - \mu)^2}{2\sigma^2}\right], \quad y > 0, \ \sigma > 0, \ -\infty < \mu < \infty. \quad (7.2.1)$$

Let $Y_1, ..., Y_n$ be a sample from a lognormal(μ, σ^2) distribution. Then $X_1 = \ln(Y_1), ..., X_n = \ln(Y_n)$ is a sample from a normal distribution with mean μ and variance σ^2. Thus, normal based approaches described in Chapter 2 can be readily applied to construct one-sided tolerance limits, tolerance interval or equal-tailed tolerance interval based on the sample $X_1, ..., X_n$. In order to illustrate this, let

$$\bar{X} = \frac{1}{n} \sum_{i=1}^{n} X_i \ \text{ and } \ S^2 = \frac{1}{n-1} \sum_{i=1}^{n} (X_i - \bar{X})^2.$$

Recall that $\bar{X} + k_1 \frac{S}{\sqrt{n}}$, where the tolerance factor k_1 is defined in (2.2.3), is a one-sided upper tolerance limit for the normal distribution or for the distribution of the log-transformed samples. So, $\exp\left(\bar{X} + k_1 \frac{S}{\sqrt{n}}\right)$ is a one-sided upper tolerance limit for the sampled lognormal population. Tolerance intervals or equal-tailed tolerance intervals can be obtained similarly. Furthermore, in order to assess the survival probability at a time point t, we note that $P(Y > t) = P(\ln(Y) > \ln(t)) = P(X > \ln(t))$, and so the results of Section 1.1.3 with t replaced by $\ln(t)$ can be used. Specifically, we can obtain a lower confidence limit for $P(Y > t)$ (see Example 2.1).

Tolerance Limits for Y_1/Y_2

Let Y_1 and Y_2 be independent lognormal random variables, and it is desired to find tolerance limits for the distribution of $\frac{Y_1}{Y_2}$. We note that $\ln\left(\frac{Y_1}{Y_2}\right) = \ln(Y_1) - \ln(Y_2)$, and so $\ln\left(\frac{Y_1}{Y_2}\right)$ is distributed as $X_1 - X_2$, where X_1 and X_2

are independent normal random variables. Thus, we can simply apply the procedures in Sections 2.4.1 and 2.4.2 (after taking log-transformation of samples on Y_1 and Y_2) to get tolerance limits. Furthermore, one-sided lower tolerance limits for $\frac{Y_1}{Y_2}$ can be used to find lower confidence limits for the stress-strength reliability $P(Y_1 > Y_2)$. For constructing tolerance limit for the ratio $\frac{Y_1}{Y_2}$ when (Y_1, Y_2) is bivariate normally distributed, see Section 12.3.

7.3 Gamma Distribution

The pdf of a gamma distribution with shape parameter a and scale parameter b, say gamma(a, b), is given by

$$f(y|a, b) = \frac{1}{\Gamma(a)\ b^a} e^{-y/b} y^{a-1}, \ y > 0, \ a > 0, \ b > 0. \tag{7.3.1}$$

There is no exact method available for constructing tolerance intervals for a gamma distribution. Several approximate methods are proposed in the literature. Bain, Engelhardt and Shiue (1984) have obtained approximate tolerance limits by assuming first that the scale parameter b is known and the shape parameter a is unknown, and then replacing the scale parameter by its sample estimate. Ashkar and Ouarda (1998) developed an approximate method of setting confidence limits for the gamma quantile by transforming the tolerance limits for the normal distribution. The transformed distribution, however, is not independent of the parameters, and eventually the unknown parameters have to be replaced by their sample estimates to obtain approximate tolerance limits. Reiser and Rocke (1993) compared several approximate methods, and concluded that the delta method on logits and the bootstrap percentile method are the best. Aryal et. al. (2007) argued that the distribution of Y can be approximated by a lognormal distribution for large values of a. Their suggestion is to use normal based tolerance limits if the maximum likelihood estimate $\hat{a} > 7$. For $0 < \hat{a} \le 7$, they provided table values to construct tolerance factors. Recently, Krishnamoorthy, Mathew and Mukherjee (2008) have proposed approximate tolerance limits based on a suitable normal approximation.

Among all the approximate methods, the one proposed by Krishnamoorthy et al. (2008) seems to be not only simple to use, but also very satisfactory for practical applications. These authors considered two approximate methods, one based on the normal approximation to the cube root of a gamma random variable (Wilson and Hilferty, 1931) and the other based on the fourth root of a gamma random variable (Hawkins and Wixley, 1986). Krishnamoorthy, Mathew and Mukherjee's (2008) extensive numerical studies showed that the results based on the cube root transformation are in general better than those based on the fourth root transformation, and so we shall describe the procedure on the basis of the cube root transformation.

7.3.1 Normal Approximation to a Gamma Distribution

Wilson and Hilferty (1931) provided a normal approximation to the cube root of a chi-square random variable using moment matching method. Let Y_a be a gamma$(a, 1)$ random variable. As Y_a is distributed as $\frac{1}{2}\chi^2_{2a}$, we shall explain the moment matching approach for approximating the distribution of a chi-square variate raised to the λ power. Towards this, we note that the mean and variance of Y_a^λ are respectively given by

$$\mu_\lambda = \frac{\Gamma(a + \lambda)}{\Gamma(a)} \quad \text{and} \quad \sigma^2_\lambda = \frac{\Gamma(a + 2\lambda)}{\Gamma(a)} - \mu^2_\lambda. \qquad (7.3.2)$$

Wilson and Hilferty's (1931) choice for λ is $\frac{1}{3}$, and in this case $Y_a^{\frac{1}{3}} \sim N\left(\mu_{\frac{1}{3}}, \sigma^2_{\frac{1}{3}}\right)$ approximately. Hawkins and Wixley (1986) argued that the approximation can be improved for smaller values of a by using $\lambda = \frac{1}{4}$.

For the reasons noted earlier, we shall use the Wilson-Hilferty cube root approximation for constructing tolerance limits for a gamma(a, b) distribution. We first note that if $Y_{a,b}$ denotes such a gamma random variable, then $Y_{a,b}$ is distributed as bY_a. The Wilson-Hilferty approximation now states that $Y_{a,b}^{\frac{1}{3}}$ is approximately normal with mean and variance

$$\mu = \frac{b^{\frac{1}{3}}\Gamma(a + 1/3)}{\Gamma(a)} \quad \text{and} \quad \sigma^2 = \frac{b^{\frac{2}{3}}\Gamma(a + 2/3)}{\Gamma(a)} - \mu^2,$$

respectively. It turns out that the functional forms of μ and σ^2 (as functions of a and b) can be ignored for constructing tolerance limits, with negligible loss of accuracy. If $Y_1, ..., Y_n$ is a sample from a gamma(a, b) distribution, we simply consider the transformed sample $X_1 = Y_1^{\frac{1}{3}}, ..., X_n = Y_n^{\frac{1}{3}}$ as a sample from a normal distribution with an arbitrary mean μ and arbitrary variance σ^2, and then develop tolerance intervals, and procedures for other related problems, as though we have a sample from a normal distribution.

7.3.2 Tolerance Intervals and Survival Probability

Let $Y_1, \ldots, Y_n$ be a sample from a gamma(a, b) distribution. In order to apply the Wilson-Hilferty approximation, let $X_i = Y_i^{\frac{1}{3}}$, $i = 1, \ldots, n$, and define

$$\bar{X} = \frac{1}{n} \sum_{i=1}^{n} X_i \quad \text{and} \quad S_x^2 = \frac{1}{n-1} \sum_{i=1}^{n} (X_i - \bar{X})^2. \qquad (7.3.3)$$

If U is a $(p, 1 - \alpha)$ upper tolerance limit based on $\bar{X}$ and S_x^2, then U^3 is an approximate $(p, 1 - \alpha)$ upper tolerance limit for the gamma(a, b) distribution. Recall that for a normal distribution, a $(p, 1 - \alpha)$ upper tolerance limit U is given by (see Section 2.2)

$$U = \bar{X} + k_1 S_x, \quad \text{with} \quad k_1 = \frac{1}{\sqrt{n}} t_{n-1;1-\alpha}(z_p \sqrt{n}), \qquad (7.3.4)$$

where z_p is the p quantile of a standard normal distribution, and $t_{m;\alpha}(\delta)$ denotes the α quantile of a noncentral t distribution with df $= m$ and noncentrality parameter δ. Also note that U^3 is an approximate $1 - \alpha$ upper confidence limit for the p quantile of the gamma(a, b) distribution. Similarly, a $(p, 1 - \alpha)$ lower tolerance limit is also a $1 - \alpha$ lower confidence limit for the $(1 - p)$ quantile. Thus, in particular, the upper and lower tolerance limits derived above also provide approximate confidence limits for the appropriate percentiles of the gamma distribution. Here and elsewhere for the gamma distribution, if an approximate tolerance limit comes out to be negative, the limit is taken to be zero.

To obtain an approximate two-sided tolerance interval for a gamma(a, b) distribution, let $L = \bar{X} - k_2 S_x$ and $U = \bar{X} + k_2 S_x$, where k_2 is determined by (2.3.4), and the values of k_2 are given in Table B2, Appendix B. The interval (L^3, U^3) is a $(p, 1 - \alpha)$ two-sided tolerance interval for the gamma(a, b) distribution.

Assessing Survival Probability

Suppose we want to estimate the survival probability (reliability) at time t based on a sample of lifetime data $Y_1, \ldots, Y_n$ from a gamma distribution. As the survival probability $S_t = P(Y > t) = P(Y^{\frac{1}{3}} > t^{\frac{1}{3}}) = P(X > t^{\frac{1}{3}})$, approximately, where X is a normal random variable, the normal approximation method can be used to make inferences about S_t. Indeed, an approximate lower confidence limit for S_t can be obtained as the solution (with respect to p) of the equation

$$t_{n-1;1-\alpha}(z_p \sqrt{n}) = \frac{\bar{X} - t^{\frac{1}{3}}}{S_x / \sqrt{n}}, \qquad (7.3.5)$$

where $\bar{X}$ and S_x are as defined in (7.3.3). Once n, $1 - \alpha$ and the quantity on the right hand side of (7.3.5) are given, the above equation can be solved using the PC calculator *StatCalc* by Krishnamoorthy (2006). Notice that the above equation (7.3.5) is the same as the one for the normal survival probability in (2.2.6) with t replaced by $t^{\frac{1}{3}}$.

Using Monte Carlo simulation, Krishnamoorthy et al. (2008) evaluated the accuracy of the above approximate procedures for computing tolerance limits for the gamma distribution. Their simulation studies indicate that, for one-sided tolerance limits, the Wilson-Hilferty approximation provides satisfactory coverage probabilities except when a is very small. For two-sided tolerance intervals, the Wilson-Hilferty approximation is entirely satisfactory regardless of the value of a.

7.3.3   Applications with an Example

The gamma distribution is one of the waiting time distributions that may offer a good fit to time to failure data. However, this distribution is not widely used as a lifetime distribution model, but it is used in many other important practical problems. Gamma related distributions are used to model the amounts of daily rain fall in a region (Das, 1955 and Stephenson et al., 1999), and to fit hydrological data sets (Ashkar and Bobée, 1988, Ashkar and Ouarda, 1998 and Aksoy, 2000). In particular, Ashkar and Ouarda (1998) used a two-parameter gamma distribution to fit annual maximum flood series in order to construct confidence intervals for a quantile. Two-parameter gamma tolerance limits and predictions limits are used in monitoring and control problems. For example, in environmental monitoring, upper tolerance limits are often constructed based on background data (regional surface water, ground water or air monitoring data) and used to determine if a potential source of contamination (for example, landfill by a waste management facility, hazardous material storage facility, factory, etc.) has adversely impacted the environment (Bhaumik and Gibbons, 2006). The gamma distribution has also found a number of applications in occupational and industrial hygiene. In a recent article, Maxim et al. (2006) have observed that the gamma distribution is a possible distribution for concentrations of carbon/coke fibers in plants that produce green or calcined petroleum coke. In a study of tuberculosis risk and incidence, Ko et al. (2001) have noted that the gamma distribution is appropriate for modeling the length of time spent in the waiting room at primary care sites. We shall now illustrate the approximate procedures that we have developed using an environmental monitoring application.

Example 7.1 (Alkalinity concentrations in ground water)

The measurements in Table 7.1 represent alkalinity concentrations (mg/L) in ground water obtained from a "greenfield" site (the site of a waste disposal landfill prior to disposal of waste). The data are taken from Gibbons (1994, p. 261). Probability plots of original measurements and cube root transformed measurements are given in Figure 7.1. Notice that these two probability plots are almost identical, and they indicate that a gamma distribution fits the data very well.

Table 7.1: Alkalinity concentrations in ground water (mg/L)

Y:	28	32	39	40	40	42	42	42	49	51
	51	52	54	54	55	58	59	59	60	63
	66	70	79	82	89	96	118			

In order to apply the Wilson-Hilferty approximation, the mean and standard deviation of the cube root transformed samples are computed as $\bar{X} = 3.8274$ and $S_x = 0.4298$.

Tolerance Limits: In Table 7.2, we present 95% one-side tolerance limits and two-sided tolerance intervals along with the corresponding tolerance factors.

Table 7.2: Tolerance limits based on the Wilson-Hilferty approximation

$(p, 1 - \alpha)$	Factor for one-sided	Lower limit	Upper limit	Factor for two-sided	Two-sided tolerance interval
(.9,.95)	1.8114	28.341	97.7129	2.1841	(24.104, 108.27)
(.95,.95)	2.2601	23.296	110.507	2.6011	(19.890, 120.95)
(.99,.95)	3.1165	15.400	137.94	3.4146	(13.141, 148.46)

Probability of Exceeding a Threshold Value: Suppose we want to find a 95% lower limit for the probability that a sample alkalinity concentration exceeds 41 mg/L, that is, $P(Y > t) = P(X > 41^{\frac{1}{3}})$. Using (7.3.5), we get

$$t_{26;.95}(z_p\sqrt{27}) = \frac{3.8274 - 41^{\frac{1}{3}}}{.4298/\sqrt{27}} = 4.584.$$

Solving for the noncentrality parameter (using *StatCalc*), we get $z_p\sqrt{27} = 2.601$. This implies that $z_p = 0.5006$ or $p = \Phi(0.5006) = 0.692$. Thus, the probability that the alkalinity concentration exceeds 41 mg/L in a sample is at least 0.692 with confidence 95%.

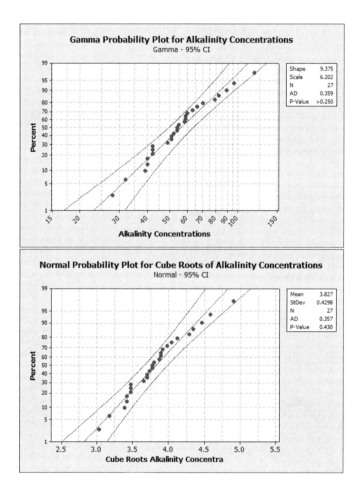

Figure 7.1: Probability plots of alkalinity concentrations

7.3.4   Stress-Strength Reliability

Let $Y_1 \sim \mathrm{gamma}(a_1, b_1)$ independently of $Y_2 \sim \mathrm{gamma}(a_2, b_2)$. If Y_1 is a strength variable and Y_2 is a stress variable, then the reliability parameter is given by

$$R = P(Y_1 > Y_2) = P\left(F_{2a_1, 2a_2} > \frac{a_2 b_2}{a_1 b_1}\right), \tag{7.3.6}$$

where $F_{m,n}$ denotes the F random variable with df $= (m, n)$. If a_1 and a_2 are known, then inferential procedures can be readily obtained (see Kotz et. al. 2003, p. 114). No exact procedure is available if a_1 and a_2 are unknown.

Suppose we are interested in testing

$$H_0 : R \le R_0 \quad \text{vs.} \quad H_a : R > R_0, \tag{7.3.7}$$

where R_0 is a specified probability. In order to use the Wilson-Hilferty approximation, we note that $R = P(X_1 - X_2 > 0)$, where $X_1 = Y_1^{\frac{1}{3}}$ and $X_2 = Y_2^{\frac{1}{3}}$ are independent normal random variables. Thus a level α test rejects the null hypothesis when a $(R_0, 1 - \alpha)$ lower tolerance limit for the distribution of $X_1 - X_2$ is positive; see Section 1.1.3. As X_1 and X_2 are approximately normally distributed, normal based tolerance limits for $X_1 - X_2$ can be used to test the above hypotheses.

Approximate methods for constructing one-sided tolerance limits (or estimating the stress-strength reliability involving two independent normal random variables) are given in Section 2.4.2. These methods can be readily applied to find a lower tolerance limit for the distribution of $X_1 - X_2$ based on the cube root transformed data. The accuracy study by Krishnamoorthy et al. (2008) shows that these approximate procedures are very satisfactory provided both sample sizes (samples on Y_1 and Y_2) are 10 or larger. Such a sample size condition is required in view of the approximations used in Section 2.4.2.

Example 7.2 (Simulated data)

We shall use the simulated data given in Basu (1981) to illustrate the computation of a lower confidence limit for the stress-strength reliability parameter R. The data are reproduced here in Table 7.3.

After taking cube root transformation, we computed the means and variances as $\bar{X}_1 = 1.02135$, $\bar{X}_2 = 0.35363$, $S_1^2 = 0.110025$, and $S_2^2 = 0.006823$. Now, we shall use the normal based methods for constructing lower tolerance limits for the distribution of $X_1 - X_2$ given in Section 2.4.2. The required quantities are

Table 7.3: Basu's (1981) simulated data

Y_1	1.7700	0.9457	1.8985	2.6121	1.0929	0.0362	1.0615	2.3895
	0.0982	0.7971	0.8316	3.2304	0.4373	2.5648	0.6377	
Y_2	0.0352	0.0397	0.0677	0.0233	0.0873	0.1156	0.0286	0.0200
	0.0793	0.0072	0.0245	0.0251	0.0469	0.0838	0.0796	

$$n_1 = 15, \ \widehat{q}_1 = 13.82174, \ \widehat{m}_1 = 15, \ \widehat{f}_1 = 16.01525,$$
$$n_2 = 15, \ \widehat{q}_2 = 0.053155, \ \widehat{m}_2 = 15, \ \text{and} \ \widehat{f}_2 = 15.4841.$$

A 95% lower confidence limit for R can be obtained by setting (2.4.9) equal to zero, and then solving the resulting equation for p. This yields $t_{16.0153,.95}(z_p\sqrt{15}) = 7.565$. Solving this equation for the noncentrality parameter, we get $z_p\sqrt{15} = 4.7611$. This implies that $z_p = 1.2293$ or $R_{1L} = p = 0.891$. Similarly, using (2.4.11), we get $R_{2L} = 0.889$. Therefore, $\min\{R_{1L}, R_{2l}\} = 0.889$ is our 95% lower confidence limit for R. Reiser and Rocke (1993) computed the lower limits using two recommended procedures; they are 0.898 (delta method on logits) and 0.904 (bootstrap percentile). Note the closeness of our lower limit with these two values.

7.4 Two-Parameter Exponential Distribution

A two-parameter exponential distribution has the probability density function (pdf) given by

$$f(x; \mu, \theta) = \frac{1}{\theta} e^{-\frac{(x-\mu)}{\theta}}, \quad x > \mu, \ \mu \geq 0, \ \theta > 0, \tag{7.4.1}$$

where μ is the location parameter and θ is the scale parameter. In lifetime data analysis, μ is referred to as the threshold or "guarantee time" parameter, and θ is the mean time to failure.

The results that we shall describe for the two-parameter exponential distribution are applicable to Pareto and power distributions because of the one-one relations among them. In particular, if X follows a Pareto distribution with pdf

$$\frac{\lambda \sigma^\lambda}{x^{\lambda+1}}, \quad x > \sigma, \ \lambda > 0, \tag{7.4.2}$$

then $Y = \ln(X)$ has the pdf in (7.4.1) with $\mu = \ln(\sigma)$ and $\theta = 1/\lambda$. If X follows a power distribution with pdf

$$\frac{\lambda x^{\lambda-1}}{\sigma^\lambda}, \quad 0 < x < \sigma, \ \lambda > 0, \tag{7.4.3}$$

then $Y = \ln(1/X)$ has the pdf in (7.4.1) with $\mu = \ln(1/\sigma)$ and $\theta = 1/\lambda$. Therefore, the methods of obtaining tolerance limits, and solutions to other related
problems involving two-parameter exponential distribution, can be readily extended to Pareto and power distributions.

7.4.1   Some Preliminary Results

Let $X_1, ..., X_n$ be a sample of observations from an exponential distribution with
the pdf in (7.4.1). The maximum likelihood estimators of μ and θ are given by

$$\widehat{\mu} = X_{(1)} \quad \text{and} \quad \widehat{\theta} = \frac{1}{n}\sum_{i=1}^{n}(X_i - X_{(1)}) = \bar{X} - X_{(1)}, \qquad (7.4.4)$$

where $X_{(1)}$ is the smallest of the X_i's. It is known that (see Lawless, 1982,
Section 3.5) $\widehat{\mu}$ and $\widehat{\theta}$ are independent with

$$\frac{(\widehat{\mu} - \mu)}{\theta} \sim \frac{\chi_2^2}{2n} \quad \text{and} \quad \frac{\widehat{\theta}}{\theta} \sim \frac{\chi_{2n-2}^2}{2n}. \qquad (7.4.5)$$

Generalized Pivotal Quantities for μ and θ

Let $\widehat{\mu}_0$ and $\widehat{\theta}_0$ be observed values of $\widehat{\mu}$ and $\widehat{\theta}$, respectively. A GPQ for μ is given
by

$$\begin{aligned} G_\mu &= \widehat{\mu}_0 - \frac{(\widehat{\mu} - \mu)}{\widehat{\theta}}\widehat{\theta}_0 \\ &= \widehat{\mu}_0 - \frac{\chi_2^2}{\chi_{2n-2}^2}\widehat{\theta}_0. \end{aligned} \qquad (7.4.6)$$

To get the last step, we used the distributional results in (7.4.5). To verify that
G_μ is a valid GPQ, we first note that the value of G_μ at $(\widehat{\mu}, \widehat{\theta}) = (\widehat{\mu}_0, \widehat{\theta}_0)$ is μ;
secondly, we see from the second equation of (7.4.6) that the distribution of G_μ,
when $\widehat{\mu}_0$ and $\widehat{\theta}_0$ are fixed, does not depend on any parameter. Thus, G_μ satisfies
the two conditions in (C.1) of Section 1.4.1.

A GPQ for θ is given by

$$G_\theta = \frac{\theta}{2n\widehat{\theta}}2n\widehat{\theta}_0 = \frac{2n\widehat{\theta}_0}{\chi_{2n-2}^2}. \qquad (7.4.7)$$

It is easy to check that G_θ satisfies the two conditions in (C.1) of Section 1.4.1.

7.4.2 One-Sided Tolerance Limits

Exact Methods for Computing Tolerance Factors

Guenther (1971) has investigated one-sided tolerance limits of the form $\widehat{\mu} + kS$, where $S = n\widehat{\theta}$ and k is the tolerance factor to be determined. Here we shall present the methods of finding lower and upper tolerance factors due to Guenther, Patil and Uppuluri (1976).

Let us consider an upper tolerance limit of the form $\widehat{\mu} + k_2 S$, where k_2 is to be determined so that

$$P_{\widehat{\mu},S}\left\{P\left(X \le \widehat{\mu} + k_2 S | \widehat{\mu}, S\right) \ge p\right\} = 1 - \alpha,$$

where X is an exponential(μ, θ) random variable independent of $\widehat{\mu}$ and S. It is easy to check that the above equation simplifies to

$$P\left(\frac{\widehat{\mu} - \mu}{\theta} + k_2 \frac{S}{\theta} \ge -\ln(1-p)\right) = 1 - \alpha. \tag{7.4.8}$$

Similarly, it can be easily checked that the $(p, 1-\alpha)$ lower tolerance factor k_1 is the solution of

$$P\left(\frac{\widehat{\mu} - \mu}{\theta} + k_1 \frac{S}{\theta} \le -\ln(p)\right) = 1 - \alpha. \tag{7.4.9}$$

As the tolerance factors could be negative, it follows from (7.4.8) and (7.4.9) that the distributions of the quantities

$$Z = \frac{\widehat{\mu} - \mu}{\theta} + \lambda \frac{S}{\theta}, \quad \lambda > 0, \tag{7.4.10}$$

and

$$Y = \frac{\widehat{\mu} - \mu}{\theta} - \lambda \frac{S}{\theta}, \quad \lambda > 0, \tag{7.4.11}$$

are required to determine the tolerance factors k_1 and k_2. Using the distributional results in (7.4.5), we see that

$$Z \sim \frac{U}{2n} + \lambda \frac{V}{2} \quad \text{and} \quad Y \sim \frac{U}{2n} - \lambda \frac{V}{2}, \tag{7.4.12}$$

where $U \sim \chi_2^2$ independently of $V \sim \chi_{2n-2}^2$.

To express the distributions of Z and Y, as given in Guenther et al. (1976), let us denote the cdf of $\chi_{2\nu}^2$ by

$$H_{2\nu}(x) = P(\chi_{2\nu}^2 \le x) = \frac{1}{2^\nu \Gamma(\nu)} \int_0^x e^{-t/2} t^{\nu-1} dt. \tag{7.4.13}$$

The distributions of Z and Y can be written in terms of H as given in the following theorem.

Theorem 7.4.1 Let Z and Y be as defined in (7.4.10) and (7.4.11). Then the cdf of Z is given by

$$F_Z(z) = \begin{cases} -\frac{e^{-nz}}{(1-n\lambda)^{n-1}} H_{2n-2}\left[\frac{2z}{\lambda}(1-n\lambda)\right] + H_{2n-2}\left(\frac{2z}{\lambda}\right), & n\lambda < 1, \\[2ex] -\frac{e^{-nz}(-1)^{n-1}}{(n\lambda-1)^{n-1}}\left[1 - \sum_{i=0}^{n-2}\frac{e^c(-c)^i}{i!}\right] + H_{2n-2}\left(\frac{2z}{\lambda}\right), & n\lambda > 1, \\[2ex] H_{2n}(2nz) & n\lambda = 1, \end{cases}$$

$$(7.4.14)$$

and

$$F_Y(y) = \begin{cases} 1 - \frac{e^{-ny}}{(1+n\lambda)^{n-1}}, & y \geq 0, \\[2ex] 1 - H_{2n-2}\left(-\frac{2y}{\lambda}\right) - \frac{e^{-ny}}{(1+n\lambda)^{n-1}} \\[1ex] \times\left\{1 - H_{2n-2}\left[-\frac{2y}{\lambda}(1+n\lambda)\right]\right\}, & y \leq 0. \end{cases}$$

$$(7.4.15)$$

The one-sided upper tolerance factor k_2 is the solution of the equation $1 - F_Z(-\ln(1-p)) = 1 - \alpha$ (with λ replaced by k_2) and lower tolerance factor k_1 is the solution of $F_Y(-\ln(p)) = 1 - \alpha$ (with λ replaced by k_1). Since H_m is the χ_m^2 cdf, the required expressions can be evaluated using software packages that compute the chi-square cdf. Later we shall comment further on the computation.

A Closed Form Lower Tolerance Limit

Guenther et al. (1976) and Engelhardt and Bain (1978) provided a closed form expression for k_1 under a special case where k_1 will come out as negative. This requires a condition on the sample size n, and this condition will be obtained shortly. Using (7.4.9) and (7.4.12), we see that k_1 is the solution of

$$P(Y \leq -\ln(p)) = P\left(\frac{U}{2n} + k_1\frac{V}{2} \leq -\ln(p)\right) = 1 - \alpha,$$

or equivalently, the solution of

$$P\left(\frac{U + 2n\ln(p)}{nV} \leq k_1\right) = 1 - \alpha. \tag{7.4.16}$$

In order to derive a closed form expression for k_1 in situations where k_1 will be negative, we note that $U \sim \chi_2^2$ with the cdf $1 - e^{-x/2}$; thus k_1 will be negative if and only if

$$1 - \alpha = P\left(\frac{U + 2n\ln(p)}{nV} \le k_1\right) < P\left(\frac{U + 2n\ln(p)}{nV} \le 0\right) = 1 - p^n,$$

or $n < \frac{\ln(\alpha)}{\ln(p)}$. Substituting $\lambda = -k_1$ and $y = -\ln(p)$ in the first equation of (7.4.15), we get

$$P\left(\frac{U + 2n\ln(p)}{nV} \le k_1\right) = 1 - \frac{p^n}{(1 - nk_1)^{n-1}} = 1 - \alpha,$$

which yields

$$k_1 = \frac{1}{n}\left[1 - \left(\frac{p^n}{\alpha}\right)^{1/(n-1)}\right]. \tag{7.4.17}$$

Thus, a $(p, 1 - \alpha)$ lower tolerance limit for an exponential(μ, θ) distribution is given by

$$\hat{\mu} + \left[1 - \left(\frac{p^n}{\alpha}\right)^{1/(n-1)}\right]\hat{\theta} \text{ for } n \le \frac{\ln(\alpha)}{\ln(p)}. \tag{7.4.18}$$

Notice that no table value is required to find a $(p, 1 - \alpha)$ lower tolerance limit using the above formula, but for a given p and α, n should satisfy the above condition. For example, when $p = 0.90$ and $1 - \alpha = 0.95$, $n \le 28$, and for $(p, 1 - \alpha) = (0.95, 0.99)$, $n \le 89$.

Remark 7.1 Suppose we seek a negative value of k_2 so that $\hat{\mu} + k_2 S$ is a $(p, 1 - \alpha)$ upper tolerance limit. Then, the solution of (7.4.8) is given by (see Exercise 7.6.2)

$$k_2 = \frac{1}{n}\left[1 - \left(\frac{(1 - p)^n}{1 - \alpha}\right)^{1/(n-1)}\right]. \tag{7.4.19}$$

The above k_2 is negative if and only if $n < \frac{\ln(1-\alpha)}{\ln(1-p)}$, which is not satisfied for conventional choices of p and $1 - \alpha$. Thus, for practical choices of p and $1 - \alpha$, k_2 is usually positive for $\hat{\mu} + k_2 S$ to be a $(p, 1 - \alpha)$ upper tolerance limit.

Generalized Variable Approach

The p quantile of a two-parameter exponential distribution is given by $q_p = \mu - \theta\ln(1 - p)$. Thus a GPQ for q_p can be obtained by replacing the parameters by their GPQs, and is given by

$$
\begin{aligned}
G_{q_p} &= G_\mu - G_\theta \ln(1-p) \\
&= \widehat{\mu}_0 - \left[\frac{\chi_2^2 + 2n\ln(1-p)}{\chi_{2n-2}^2}\right]\widehat{\theta}_0,
\end{aligned}
\tag{7.4.20}
$$

where $(\widehat{\mu}_0, \widehat{\theta}_0)$ is the observed value of $(\widehat{\mu}, \widehat{\theta})$. Let $E_{p;\alpha}$ denote the α quantile of $E_p = \frac{\chi_2^2 + 2n\ln(1-p)}{\chi_{2n-2}^2}$. Then

$$
\widehat{\mu}_0 - E_{p;\alpha}\widehat{\theta}_0
\tag{7.4.21}
$$

is a $1 - \alpha$ upper confidence limit for q_p, which in turn is a $(p, 1 - \alpha)$ upper tolerance limit for the exponential(μ, θ) distribution. Similarly, we see that

$$
\widehat{\mu}_0 - E_{1-p,1-\alpha}\widehat{\theta}_0
\tag{7.4.22}
$$

is a $1-\alpha$ lower confidence limit for $q_{1-p} = \mu - \theta\ln(p)$, or equivalently, a $(p, 1-\alpha)$ lower tolerance limit for the exponential(μ, θ) distribution.

Interestingly, it can be shown that the upper and lower tolerance limits obtained using the generalized variable approach are actually exact. In other words, $\widehat{\mu}_0 - E_{p;\alpha}\widehat{\theta}_0$ is an exact upper confidence limit for q_p, and $\widehat{\mu}_0 - E_{1-p,1-\alpha}\widehat{\theta}_0$ is an exact lower confidence limit for q_{1-p}. That is, their coverage probabilities are equal to $1-\alpha$. The proof of this observation is quite simple, and is taken from Roy and Mathew (2005). Here we shall give the proof to show that $\widehat{\mu}_0 - E_{p;\alpha}\widehat{\theta}_0$ is an exact upper confidence limit for q_p. Let $\chi_{2,0}^2$ and $\chi_{2n-2,0}^2$ denote observed values of a χ_2^2 random variable and a χ_{2n-2}^2 random variable, respectively. In view of the distributional results (7.4.5), we have the following representations for the observed values $\widehat{\mu}_0$ and $\widehat{\theta}_0$:

$$
\widehat{\mu}_0 = \mu + \frac{\chi_{2,0}^2}{2n}\theta, \quad \text{and} \quad \widehat{\theta}_0 = \frac{\chi_{2n-2,0}^2}{2n}\theta.
$$

Since $\widehat{\mu}_0 - E_{p;\alpha}\widehat{\theta}_0$ is the $1 - \alpha$ quantile of G_{q_p} given in (7.4.20), the coverage probability associated with the upper confidence limit $\widehat{\mu}_0 - E_{p;\alpha}\widehat{\theta}_0$ for q_p is given by

$$
P_{\widehat{\mu}_0,\widehat{\theta}_0}\left\{P_{\chi_2^2,\chi_{2n-2}^2}\left(G_{q_p} \leq q_p | \widehat{\mu}_0, \widehat{\theta}_0\right) \leq 1 - \alpha\right\}.
$$

Using the representation for $\widehat{\mu}_0$ and $\widehat{\theta}_0$ given above, and using the expression for G_{q_p} given in (7.4.20), the expression for the coverage probability simplifies to

$$
P_{\chi_{2,0}^2,\chi_{2n-2,0}^2}\left\{P_{\chi_2^2,\chi_{2n-2}^2}\left(\chi_{2,0}^2 - \frac{\chi_2^2 + 2n\ln(1-p)}{\chi_{2n-2}^2}\chi_{2n-2,0}^2\right.\right.
$$
$$
\left.\left. \leq -2n\ln(1-p)\Big|\chi_{2,0}^2, \chi_{2n-2,0}^2\right) \leq 1 - \alpha\right\},
$$

where we have also used the expression $q_p = \mu - \theta \ln(1-p)$. Rearranging terms, the coverage probability can be written as

$$P_{\chi^2_{2,0}, \chi^2_{2n-2,0}} \left\{ P_{\chi^2_2, \chi^2_{2n-2}} \left(- \frac{\chi^2_2 + 2n\ln(1-p)}{\chi^2_{2n-2}} \right. \right.$$
$$\left. \left. \leq - \frac{\chi^2_{2,0} + 2n\ln(1-p)}{\chi^2_{2n-2,0}} \right| \chi^2_{2,0}, \chi^2_{2n-2,0} \right) \leq 1 - \alpha \right\}.$$

Since $-\frac{\chi^2_{2,0} + 2n\ln(1-p)}{\chi^2_{2n-2,0}}$ is an observed value of $-\frac{\chi^2_2 + 2n\ln(1-p)}{\chi^2_{2n-2}}$, it is clear that as a function of the random variable $-\frac{\chi^2_{2,0} + 2n\ln(1-p)}{\chi^2_{2n-2,0}}$, the cdf

$$P_{\chi^2_2, \chi^2_{2n-2}} \left(- \frac{\chi^2_2 + 2n\ln(1-p)}{\chi^2_{2n-2}} \leq - \frac{\chi^2_{2,0} + 2n\ln(1-p)}{\chi^2_{2n-2,0}} \right| \chi^2_{2,0}, \chi^2_{2n-2,0} \right)$$

has a uniform distribution. Hence the coverage probability is equal to $1 - \alpha$.

Once we obtain $-E_{1-p,1-\alpha}$ and $-E_{p,\alpha}$, it can be easily checked that the lower tolerance factor k_1 that satisfies (7.4.16) is $\frac{-E_{1-p,1-\alpha}}{n}$, and the upper tolerance factor k_2 that satisfies (7.4.8) is $\frac{-E_{p;\alpha}}{n}$.

Computation of Tolerance Factors

For a given n, p and $1 - \alpha$, one-sided upper tolerance factor k_2 is the solution of the equation $1 - F_Z(-\ln(1-p)) = 1 - \alpha$, and lower tolerance factor k_1 is the solution of $F_Y(-\ln(p)) = 1 - \alpha$, where $F_Z(.)$ and $F_Y(.)$ are given in (7.4.14) and (7.4.15), respectively. Guenther et al. (1976) computed factors k_1 and k_2 for $n = 2(1)30, 40, 50$, $p = 0.80, 0.90, 0.95, 0.99, 0.999$ and $1 - \alpha = 0.90, 0.95$. These factors are given in Tables B10 and B11, Appendix B. Our own experience suggests that computational skills are necessary to find the roots of the aforementioned equations, and one may encounter some convergence problems if n is large. Note that $k_1 = -\frac{E_{1-p;1-\alpha}}{n}$ and $k_2 = -\frac{E_{p;\alpha}}{n}$, where $E_p = \frac{\chi^2_2 + 2n\ln(1-p)}{\chi^2_{2n2}}$, and the distribution of E_p does not depend on any unknown parameters. So Monte Carlo simulation can be used to approximate the values of k_1 and k_2. Our own Monte Carlo simulation estimates of $-\frac{E_{p,\alpha}}{n}$ based on 100,000 runs turned out to be almost identical to those exact factors listed in Tables B10 and B11, Appendix B.

7.4.3 Estimation of Survival Probability

Suppose we want to find a $1 - \alpha$ lower confidence limit for $S_t = P(X > t)$ based on a sample from an exponential(μ, θ) distribution. Setting the lower tolerance limit in (7.4.18) to t, and solving the resulting equation for p (see Section 1.1.3), we get

$$\alpha^{\frac{1}{n}} \left[1 - \frac{t - \widehat{\mu}}{\widehat{\theta}} \right]^{\frac{n-1}{n}} \tag{7.4.23}$$

as a $1 - \alpha$ lower confidence limit for S_t when $n \leq \frac{\ln(\alpha)}{\ln(p)}$. We recall that the condition $n \leq \frac{\ln(\alpha)}{\ln(p)}$ is needed here, since the exact closed form lower tolerance limit given above is derived under this condition; see (7.4.18).

To apply the generalized variable method, we first note that $S_t = \exp(-(t - \mu)/\theta)$. Replacing the parameters by their GPQs, and after some simplification, we get

$$G_{S_t} = \exp \left\{ -\frac{1}{2n} \left[\left(\frac{t - \widehat{\mu}_0}{\widehat{\theta}_0} \right) \chi^2_{2n-2} + \chi^2_2 \right] \right\} = \exp \left\{ -\frac{1}{2n} A \right\}, \tag{7.4.24}$$

where $A = \left(\frac{t - \widehat{\mu}_0}{\widehat{\theta}_0} \right) \chi^2_{2n-2} + \chi^2_2$. If A_α is the α quantile of A, then $\exp \left(-\frac{1}{2n} A_\alpha \right)$ is a $1 - \alpha$ upper confidence limit for S_t, and $\exp \left(-\frac{1}{2n} A_{1-\alpha} \right)$ is a $1 - \alpha$ lower confidence limit for S_t. This lower confidence limit is also exact, as proved in Roy and Mathew (2005). The proof is similar to the corresponding proof given earlier for the lower tolerance limit.

It is also possible to develop an accurate approximation for the lower confidence limit for S_t obtained using the generalized variable method. The approximation is obtained by noting that in the expression for G_{S_t} given in (7.4.24), the quantity $\left(\frac{t - \widehat{\mu}_0}{\widehat{\theta}_0} \right) \chi^2_{2n-2} + \chi^2_2$ is a linear combination of two independent chi-squares; thus the distribution of this quantity can be approximated by that of a multiple of a chi-square, whenever the coefficient $\frac{t - \widehat{\mu}_0}{\widehat{\theta}_0}$ is nonnegative. Here we shall simply give the approximation; a detailed derivation is given in Roy and Mathew (2005). Note that what is required is an approximation for $G_{S_t;\alpha}$, the α quantile of G_{S_t}. In order to describe the approximation, let

$$a_1 = \frac{1}{2n\widehat{\theta}_0} \times \frac{(t - \widehat{\mu}_0)^2(n - 1) + \widehat{\theta}_0^2}{(t - \widehat{\mu}_0)(n - 1) + \widehat{\theta}_0}$$

$$a_2 = \frac{2 \left[(t - \widehat{\mu}_0)(n - 1) + \widehat{\theta}_0 \right]^2}{(t - \widehat{\mu}_0)^2(n - 1) + \widehat{\theta}_0^2}. \tag{7.4.25}$$

Furthermore, let

$$
\tilde{A}_\alpha =
\begin{cases}
-\frac{1}{n}\left[\ln \alpha + (n-1)\ln\left\{1 + \frac{(\hat\mu_0 - t)}{\hat\theta_0}\right\}\right], & \text{if } t \le \hat\mu_0 \\[2mm]
a_1 \chi^2_{a_2;1-\alpha}, & \text{if } t > \hat\mu_0.
\end{cases}
\tag{7.4.26}
$$

Note that the second part of the expression for $\tilde{A}_\alpha$ involves a multiple of the $1 - \alpha$ quantile of a chi-square distribution with a_2 df, where the df a_2 and the multiple a_1 are given in (7.4.25). The approximation for $G_{S_t;\alpha}$, the α quantile of G_{S_t}, is given by

$$
G_{S_t;\alpha} = \exp\left\{-\tilde{A}_\alpha\right\},
\tag{7.4.27}
$$

where $\tilde{A}_\alpha$ is given in (7.4.26). The numerical results in Roy and Mathew (2005) show that the above approximation is very accurate.

Example 7.3 (Failure mileages of military careers)

We shall use the data given in Grubbs (1971) that represent the failure mileages of 19 military carriers. The failure mileages given in Table 7.4 fit a two-parameter exponential distribution (see Figure 7.2).

Table 7.4: Failure mileages of 19 military carriers

162	200	271	302	393	508	539	629	706	777
884	1008	1101	1182	1463	1603	1984	2355	2880	

Reproduced with permission from *Technometrics*. Copyright [1971] by the American Statistical Association.

For these data, the estimates are $\hat\mu = X_{(1)} = 162$, $\hat\theta = 835.21$, and $S = n\hat\theta = 15869$. To compute a $(0.90, 0.95)$ lower tolerance limit given in (7.4.22) based on the generalized variable method, we estimated $E_{1-p;1-\alpha} = E_{.10,.95}$ using Monte Carlo simulation with 100,000 runs as 0.0568. This yields $162 - 0.0568 \times 835.21 = 114.56$. To apply the exact method, we found $k_1 = -0.0030$ (Table B10, Appendix B) and the exact lower tolerance limit is $162 - 0.0030 \times 15869 = 114.39$. Even though the upper tolerance limit is not of interest here, for the sake of illustration we shall also compute a $(0.90, 0.95)$ upper tolerance limit. To use the generalized confidence limit (7.4.21), we computed $E_{0.90,0.05}$ using Monte Carlo simulation as -3.8620, and the upper tolerance limit is $162 + 3.8620 \times 835.21 = 3237.24$. Applying the exact method, we get $\hat\mu_0 + k_2 S = 162 + 0.1937 \times 15869 = 3235.83$. We see that the solutions are practically the same.

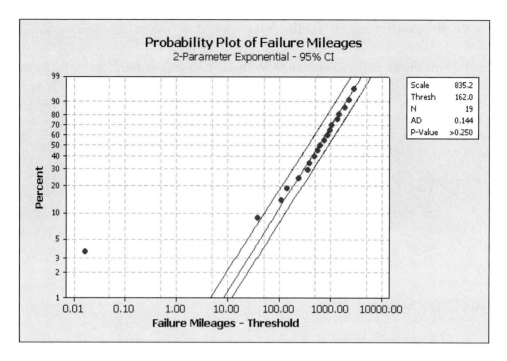

Figure 7.2: Exponential probability plot of failure mileages of military carriers

Suppose we want to find 95% lower limit for the survival probability at $t = 200$ miles. Using the exact formula in (7.4.23), we have

$$\alpha^{\frac{1}{n}} \left[1 - \frac{t - \widehat{\mu}}{\widehat{\theta}}\right]^{\frac{n-1}{n}} = (.05)^{\frac{1}{19}} \left[1 - \frac{200 - 162}{835.21}\right]^{\frac{18}{19}} = 0.817.$$

To apply the generalized variable approach, the quantity A in (7.4.24) simplifies to

$$A = \left[\left(\frac{200 - 162}{835.21}\right) \chi_{2n-2}^2 + \chi_2^2\right].$$

Using simulation with 100,000 runs, we estimated $A_{.95}$ as 7.6690, and so $\exp(-7.6690/38) = 0.817$ is a lower confidence limit for S_t at $t = 200$. In order to apply the approximation (7.4.27), we note that the observed values are $\widehat{\mu}_0 = X_{(1)} = 162$, $\widehat{\theta}_0 = 835.21$. Using (7.4.25), we get $a_1 = 0.015006$ and $a_2 = 6.3795$. Since $t > \widehat{\mu}_0$, we have $\tilde{A}_{.95} = a_1 \chi_{a_2;.95}^2 = 0.1974$. Hence the approximate lower confidence limit is given by $G_{S_t;.95} = \exp(-\tilde{A}_{.95}) = 0.821$. Notice that we get essentially the same result by all the approaches.

7.4.4 Stress-Strength Reliability

Let $X \sim \text{exponential}(\mu_1, \theta_1)$ independently of $Y \sim \text{exponential}(\mu_2, \theta_2)$. That is, the pdf of X is $f(x; \mu_1, \theta_1)$ and the pdf of Y is $f(y; \mu_2, \theta_2)$, where f is given in (7.4.1). Then the stress-strength reliability parameter can be expressed as follows.

If $\mu_1 > \mu_2$, the reliability parameter R is given by

$$
\begin{aligned}
P(X > Y) &= P_Y P_{X|Y}(X > Y | \mu_2 < Y < \mu_1) + P_Y P_{X|Y}(X > Y | Y > \mu_1) \\
&= \frac{1}{\theta_2} \int_{\mu_2}^{\mu_1} e^{-\frac{(y-\mu_2)}{\theta_2}} \, dy + \frac{1}{\theta_2} \int_{\mu_1}^{\infty} e^{-\frac{(y-\mu_1)}{\theta_1}} e^{-\frac{(y-\mu_2)}{\theta_2}} \, dy \\
&= 1 - \frac{\theta_2 e^{\frac{(\mu_2-\mu_1)}{\theta_2}}}{\theta_1 + \theta_2}.
\end{aligned}
$$

If $\mu_1 \leq \mu_2$, then R is given by

$$
\begin{aligned}
P(X > Y) &= E_Y P_{X|Y}(X > Y | Y) \\
&= \frac{1}{\theta_2} \int_{\mu_1}^{\infty} e^{-\frac{(y-\mu_1)}{\theta_1}} e^{-\frac{(y-\mu_2)}{\theta_2}} \, dy \\
&= \frac{\theta_1 e^{\frac{(\mu_1-\mu_2)}{\theta_1}}}{\theta_1 + \theta_2}.
\end{aligned}
$$

Thus, the reliability parameter R can be expressed as

$$
R = \left(1 - \frac{\theta_2 e^{(\mu_2-\mu_1)/\theta_2}}{\theta_1 + \theta_2}\right) I(\mu_1 > \mu_2) + \left(\frac{\theta_1 e^{(\mu_1-\mu_2)/\theta_1}}{\theta_1 + \theta_2}\right) I(\mu_1 \leq \mu_2), \quad (7.4.28)
$$

where $I(.)$ is the indicator function.

An Asymptotic Approach

An approximate confidence interval for R in (7.4.28) is based on an asymptotic distribution of the MLE of R. Specifically, Kotz et al. (2003) proposed this approach by deriving an asymptotic mean squared error (MSE) of $\widehat{R}$. Let $w_1 = n_1/(n_1 + n_2)$ and define

$$
C_j = \begin{cases}
\dfrac{\widehat{\theta}_i}{(\widehat{\theta}_i + \widehat{\theta}_j)^2} \exp\left[-\dfrac{(\widehat{\mu}_j - \widehat{\mu}_i)}{\widehat{\theta}_i}\right], & \text{if } \widehat{\mu}_j > \widehat{\mu}_i, \\[3mm]
\left[\dfrac{\widehat{\theta}_i}{(\widehat{\theta}_i + \widehat{\theta}_j)^2} + \dfrac{\widehat{\mu}_i - \widehat{\mu}_j}{\widehat{\theta}_j(\widehat{\theta}_i + \widehat{\theta}_j)}\right] \exp\left[-\dfrac{(\widehat{\mu}_i - \widehat{\mu}_j)}{\widehat{\theta}_j}\right], & \text{if } \widehat{\mu}_j \leq \widehat{\mu}_i,
\end{cases}
$$

where $i = 2$ if $j = 1$ and $i = 1$ if $j = 2$. Using these terms, an estimate of the asymptotic MSE of $\widehat{R}$ is given by

$$\widehat{\sigma}_{\widehat{R}}^2 = \frac{1}{n_1 + n_2}(w_1^{-1}C_1^2\widehat{\theta}_1^2 + (1 - w_1)^{-1}C_2^2\widehat{\theta}_2^2).$$

Using this estimate, we have

$$\frac{\sqrt{n_1 + n_2}(\widehat{R} - R)}{\widehat{\sigma}_{\widehat{R}}} \sim N(0, 1) \quad \text{approximately,}$$

for large $n_1 + n_2$. A $1 - \alpha$ lower limit for R based on the above asymptotic distribution is given by

$$\widehat{R} - z_{1-\alpha}\frac{\widehat{\sigma}_{\widehat{R}}}{\sqrt{n_1 + n_2}}, \tag{7.4.29}$$

where z_p denotes the p quantile of the standard normal distribution.

Generalized Confidence Limits for R

Let $(\widehat{\mu}_{i0}, \widehat{\theta}_{i0})$ be the observed value of the MLE $(\widehat{\mu}_i, \widehat{\theta}_i)$ based on a sample of size n_i from an exponential(μ_i, θ_i) distribution, $i = 1, 2$. Define

$$G_{\mu_i} = \widehat{\mu}_{i0} - \frac{U_i}{V_i}\widehat{\theta}_{i0}, \quad G_{\theta_i} = \frac{2n_i\widehat{\theta}_{i0}}{V_i}, \quad i = 1, 2, \tag{7.4.30}$$

where U_1, U_2, V_1 and V_2 are independent random variables with $U_i \sim \chi_2^2$ and $V_i \sim \chi_{2n_i - 2}^2$, $i = 1, 2$.

A GPQ for R can be obtained by replacing the parameters by their GPQs in (7.4.28). Suppose the location parameters are unknown, but equal, i.e., $\mu_1 = \mu_2$. Then the reliability parameter R simplifies to $\frac{\theta_1}{\theta_1 + \theta_2} = (1 + \frac{\theta_2}{\theta_1})^{-1}$. In this case, it is enough to find a confidence limit for θ_2/θ_1. A GPQ for $\frac{\theta_2}{\theta_1}$ is given by

$$\frac{G_{\theta_2}}{G_{\theta_1}} = \frac{\widehat{\theta}_{20}n_2(n_1 - 1)}{\widehat{\theta}_{10}n_1(n_2 - 1)}F_{2n_1-2,2n_2-2},$$

where $F_{m,n}$ denotes the F random variable with dfs m and n. Let $F_{m,n;\alpha}$ be the α quantile of $F_{m,n}$, and let $\xi_\alpha = \frac{\widehat{\theta}_{20}n_2(n_1-1)}{\widehat{\theta}_{10}n_1(n_2-1)}F_{2n1-2,2n_2-2;1-\alpha}$. Then, $\frac{1}{1+\xi_\alpha}$ is a $1 - \alpha$ lower confidence limit for R. This generalized confidence limit is equal to the exact one [see Bhattacharyya and Johnson (1974, Section 5)] for the reliability parameter.

If the location parameters are unknown and arbitrary, then a GPQ for R can be obtained by replacing the parameters in (7.4.28) by their GPQs. Denoting the resulting GPQ by G_R, we have

$$
G_R = \left(1 - \frac{G_{\theta_2} e^{\frac{(G_{\mu_2} - G_{\mu_1})}{G_{\theta_2}}}}{G_{\theta_1} + G_{\theta_2}} \right) I(G_{\mu_1} > G_{\mu_2}) + \left(\frac{G_{\theta_1} e^{\frac{(G_{\mu_1} - G_{\mu_2})}{G_{\theta_1}}}}{G_{\theta_1} + G_{\theta_2}} \right) I(G_{\mu_1} \leq G_{\mu_2}),
$$

$$(7.4.31)$$

In particular, for given $\widehat{\mu}_{10}, \widehat{\mu}_{20}, \widehat{\theta}_{10}$ and $\widehat{\theta}_{20}$, the distribution of G_R does not depend on any unknown parameters. So, the Monte Carlo method given in Algorithm 7.1 can be used to find confidence limits for R.

Algorithm 7.1

1. For a given data set, compute the MLEs $\widehat{\mu}_{10}, \widehat{\theta}_{10}, \widehat{\mu}_{20}, \widehat{\theta}_{20}$ using the formulas in (7.4.4).

2. Generate $U_1 \sim \chi_2^2$, $U_2 \sim \chi_2^2$, $V_1 \sim \chi_{2n_1-2}^2$, $V_2 \sim \chi_{2n_2-2}^2$.

3. Compute $G_{\mu_1}, G_{\mu_2}, G_{\theta_1}, G_{\theta_2}$ and G_R (see (7.4.30) and (7.4.31)).

4. Repeat the steps 2 and 3 a large number of times, say, 10,000

5. The 100α percentile of the generated G_R's is a $1 - \alpha$ lower limit for the reliability parameter R.

Krishnamoorthy, Mukherjee and Guo (2007) evaluated the coverage probabilities of the asymptotic lower confidence limits (7.4.29) and the generalized confidence limits for R, using Monte Carlo simulation. These authors observed that the asymptotic approach is very liberal even for samples as large as 100. The coverage probabilities of the asymptotic confidence limits go as low as 0.77 when the nominal level is 0.95. So the asymptotic approach is not recommended for applications. On the other hand, the generalized confidence limits for R are slightly conservative for small samples, and their coverage probabilities are very close to the nominal level for moderate to large samples. In general, the generalized limits can be recommended for practical applications.

Example 7.3 (Simulated data)

In order to illustrate the procedures for obtaining a lower bound on the stress-strength reliability parameter R, we shall use the simulated data given in Krishnamoorthy, Mukherjee and Guo (2007). The data on X were generated from exponential(4, 5) and on Y were generated from exponential(1, 2). The value of the reliability parameter R is 0.936. The ordered data are given in Table 7.5.

Table 7.5: Simulated data from two exponential distributions

X	4.21	4.88	5.17	5.64	6.31	7.42	7.89	8.14	8.27	9.92
	10.45	10.59	11.37	12.98	13.94	14.18	14.19	14.94	18.83	20.91
Y	1.07	1.09	1.16	1.17	1.65	1.98	2.12	2.13	2.54	3.18
	3.19	3.30	3.33	3.40	3.62	4.29	5.80	5.95	6.39	6.74

The MLEs are computed as

$$\widehat{\mu}_{10} = 4.21, \ \widehat{\theta}_{10} = 6.298, \ \widehat{\mu}_{20} = 1.07, \ \widehat{\theta}_{20} = 2.138 \text{ and } \widehat{R} = 0.942.$$

Using Algorithm 7.1 with 100,000 simulation runs, we computed the 95% lower confidence limit for R as 0.849. Thus, we conclude that the true stress-strength reliability is at least 0.849 with confidence 95%.

7.5 Weibull Distribution

Before we address the tolerance interval problem for the Weibull distribution, we shall provide some distributional results for the maximum likelihood estimators (MLEs) along with some pivotal quantities, and some details on the numerical computation of the MLEs.

7.5.1 Some Preliminaries

For a Weibull distribution with scale parameter b and the shape parameter c, denoted by Weibull(b, c), the pdf is given by

$$f(x|b, c) = \frac{c}{b} \left(\frac{x}{b}\right)^{c-1} \exp\left\{-\left[\frac{x}{b}\right]^c\right\}, \ x > 0, \ b > 0, \ c > 0. \tag{7.5.1}$$

The statistical problems involving Weibull distributions are usually not simple, since the MLEs of the parameters b and c do not have closed form, and they have to be obtained numerically. Therefore, exact analytical procedures were obtained only for a few problems (e.g., Lawless, 1973, 1978), and these are also not simple to implement. Some approximate results and Monte Carlo procedures were developed by Thoman, Bain and Antle (1969), based on the distributions of certain pivotal quantities involving the MLEs. These distributional results are

given in Lemma 7.1 in the following section. These results for the MLEs allow us
to find the distributions of some pivotal quantities empirically, based on which
inferential procedures for Weibull parameters, and other related problems can
be developed. Using this approach, Thoman et al. (1970) developed methods for
setting confidence limits for reliability, and for constructing one-sided tolerance
limits. These procedures are exact except for simulation errors.

Several approximate methods have been proposed until late 1980s, which do
not require simulation (e.g., Mann and Fertig, 1975, 1977; Engelhardt and Bain,
1977; Bain and Engelhardt, 1981). These procedures also require some table
values to compute confidence limits for a parameter, or to compute tolerance
limits. Nowadays, as computing technologies and software are widely available,
the Monte Carlo procedures certainly have an edge over the approximate methods. Thus here we will describe Monte Carlo procedures for computing one-sided
tolerance limits, for estimating a survival probability, and for constructing lower
limits for the stress-strength reliability involving Weibull distributions.

7.5.2   The Maximum Likelihood Estimators and Their Distributions

Let $X_1, \ldots, X_n$ be a sample from a Weibull(b, c) distribution. The MLEs for b
and c can be obtained from Cohen (1965) as follows. The MLE $\widehat{c}$ of c is the
solution to the equation

$$\frac{1}{\widehat{c}} - \frac{\sum\limits_{i=1}^{n} x_i^{\widehat{c}} \ln(x_i)}{\sum\limits_{i=1}^{n} x_i^{\widehat{c}}} + \frac{1}{n} \sum\limits_{i=1}^{n} \ln(x_i) = 0, \tag{7.5.2}$$

and the MLE of b is given by $\widehat{b} = \left(\frac{1}{n} \sum_{i=1}^{n} x_i^{\widehat{c}}\right)^{1/\widehat{c}}$ (see Exercise 7.6.7). In the
following lemma we exhibit some pivotal quantities.

Lemma 7.1 Let $X_1, \ldots, X_n$ be a sample from a Weibull(b, c) distribution, and
let $\widehat{b}$ and $\widehat{c}$ be the MLEs based on this sample. Then, $\widehat{c}/c$ and $\widehat{c}\ln\left(\frac{\widehat{b}}{b}\right)$ are pivotal
quantities.

Proof. Thoman et al. (1969) argued that $\widehat{c}/c$ can be viewed as the solution
of the likelihood equation (7.5.2) based on a sample from a Weibull(1, 1) distribution, and so its distribution does not depend on any parameters. Arguing

similarly, they showed the distribution of $\widehat{c}\ln\left(\widehat{b}/b\right)$ also does not depend on b and c.

This lemma can also be proved using Result 1.4.1 for the location-scale family. Toward this, we note that if $X \sim$ Weibull(b, c), then $Y = \ln(X) \sim$ extreme-value(μ, σ) distribution having the pdf

$$f(y; \mu, \sigma) = \frac{1}{\sigma}\exp\left(\frac{y - \mu}{\sigma}\right)\exp\left(-\exp\left(\frac{y - \mu}{\sigma}\right)\right), \qquad (7.5.3)$$

where $\mu = \ln(b)$ is the location parameter, and $\sigma = c^{-1}$ is the scale parameter. Thus, $Y_i = \ln(X_i)$, $i = 1, ..., n$, can be regarded as a sample from the extreme-value(μ, σ) distribution given in (7.5.3), and $\widehat{\mu} = \ln(\widehat{b})$ and $\widehat{\sigma} = \widehat{c}^{-1}$ are the MLEs of μ and σ, respectively. Also, it can be verified that $\widehat{\mu}$ and $\widehat{\sigma}$ are equivariant estimators, and so it follows from Result 1.4.1 that $\frac{\widehat{\mu}-\mu}{\sigma}$, $\frac{\widehat{\sigma}}{\sigma}$ and $\frac{\widehat{\mu}-\mu}{\widehat{\sigma}}$ are pivotal quantities. Replacing $(\mu, \sigma, \widehat{\mu}, \widehat{\sigma})$ by $\left(\ln(b), \frac{1}{c}, \ln(\widehat{b}), \frac{1}{\widehat{c}}\right)$, we see that $\frac{\widehat{c}}{c}$ and $\widehat{c}\ln\left(\frac{\widehat{b}}{b}\right)$ are pivotal quantities. □

The distributions of the aforementioned pivotal quantities are parameter free but still their distributions are difficult to find, and they can be obtained only empirically. For instance, $\frac{\widehat{c}}{c}$ is distributed as $\widehat{c}^*$ and $\widehat{c}\ln\left(\frac{\widehat{b}}{b}\right)$ is distributed as $\widehat{c}^*\ln\left(\widehat{b}^*\right)$, where $\widehat{c}^*$ and $\widehat{b}^*$ are the MLEs based on a sample from the Weibull$(1, 1)$ distribution. That is, $\widehat{c}^*$ is the solution of the equation (7.5.2) with $X_1, ..., X_n$ being a sample from the Weibull$(1, 1)$ distribution. Note that $b = 1$ and $c = 1$ is a choice, and in fact one can choose any positive values for b and c to study the distributions of the pivotal quantities empirically.

Computation of the MLEs: Let $X_1, ..., X_n$ be a sample from a Weibull(b, c) distribution. Let $Y_i = \ln(X_i)$, $i = 1, ..., n$. Menon (1963) showed that the estimator

$$\widehat{c}_u = \frac{\pi}{\sqrt{6}}\left(\frac{\sum\limits_{i=1}^{n}(Y_i - \bar{Y})^2}{n - 1}\right)^{-\frac{1}{2}} \qquad (7.5.4)$$

is asymptotically unbiased, having the asymptotic $N(c, 1.1c^2/n)$ distribution. Using $\widehat{c}_u$ as an initial value, the Newton-Raphson iterative method given in the following algorithm can be applied to find the root of the equation (7.5.2). As noted in the literature (Thoman et al., 1969), this iterative method is stable and converges rapidly.

Algorithm 7.2

1. For a given sample $x_1, ..., x_n$, let $y_i = \ln(x_i)$, $i = 1, ..., n$, and compute $\widehat{c}_u$ using (7.5.4). Let the computed value be c_0.
2. Compute $s_1 = \frac{1}{n}\sum_{i=1}^{n} y_i$.
3. Set $z_i = x_i^{c_0}$, $i = 1, ..., n$.
4. Compute $s_2 = \sum_{i=1}^{n} z_i$, $s_3 = \sum_{i=1}^{n} z_i y_i$ and $s_4 = \sum_{i=1}^{n} z_i y_i^2$.
5. Set $f = 1/c_0 + s_1 - s_3/s_2$.
6. Set $c_0 = c_0 + f/(1/c_0^2 + (s_2 s_4 - s_3^2)/s_2^2)$.
7. Repeat the steps 2 through 6, until f is zero.

The final value of c_0 is the MLE of c; the MLE of b is $\left(\frac{1}{n}s_2\right)^{1/c_0}$.

7.5.3 Generalized Pivotal Quantities for Weibull Parameters

Krishnamoorthy, Lin and Xia (2008) and Lin (2009) considered inferences based on the GPQs for the Weibull parameters, and showed that the results based on the GPQs and the Monte Carlo procedures given in Thoman and Bain (1969), and Thoman, Bain and Antle (1969) are the same for the one-sample as well as some two-sample problems. Furthermore, unlike the Monte Carlo approach, GPQs can be used to develop inferential procedures for any real valued function of b and c in a straightforward manner. We shall first describe GPQs on the basis of the distributional results in Lemma 7.1.

Let $\widehat{b}_0$ and $\widehat{c}_0$ be the observed values of the MLEs based on a sample of n observations from a Weibull(b, c) distribution. A GPQ for the scale parameter b is given by

$$G_b = \left(\frac{b}{\widehat{b}}\right)^{\frac{\widehat{c}}{\widehat{c}_0}} \widehat{b}_0 = \left(\frac{1}{\widehat{b}^*}\right)^{\frac{\widehat{c}^*}{\widehat{c}_0}} \widehat{b}_0, \qquad (7.5.5)$$

where $\widehat{c}^*$ and $\widehat{b}^*$ are the MLEs based on a sample from the Weibull($1, 1$) distribution. The second expression in (7.5.5) is obtained by recalling (from Lemma 7.1) that $\widehat{c}\ln\left(\frac{\widehat{b}}{b}\right)$ is a pivotal quantity. We shall now verify that G_b satisfies the two conditions in (C.1) of Section 1.4.1. First of all, the value of G_b at $(\widehat{b}, \widehat{c}) = (\widehat{b}_0, \widehat{c}_0)$ is b. Secondly, for a given $(\widehat{b}_0, \widehat{c}_0)$, we see from the second expression in (7.5.5) that the distribution of G_b does not depend on any unknown parameters.

A GPQ for c can be obtained as

$$G_c = \frac{c}{\widehat{c}}\widehat{c}_0 = \frac{\widehat{c}_0}{\widehat{c}^*}. \qquad (7.5.6)$$

It can be easily checked that the GPQ G_c satisfies the conditions in (C.1) of Section 1.4.1.

7.5.4 One-Sided Tolerance Limits

The cdf of a Weibull(b, c) distribution is given by

$$F_X(x|b, c) = 1 - \exp\left\{-\left(\frac{x}{b}\right)^c\right\}. \tag{7.5.7}$$

Using the above cdf, we see that the p quantile of a Weibull(b, c) distribution is given by $q_p = b\left(-\ln(1 - p)\right)^{\frac{1}{c}} = b\theta_p^{\frac{1}{c}}$, where $\theta_p = -\ln(1 - p)$. By substituting the GPQs for the parameters, a GPQ for q_p can be obtained as

$$G_{q_p} = G_b\theta^{\frac{1}{G_c}} = \left(\frac{1}{\widehat{b^*}}\right)^{\frac{\widehat{c}^*}{\widehat{c}_0}} \widehat{b}_0 \; \theta_p^{\frac{\widehat{c}^*}{\widehat{c}_0}}, \tag{7.5.8}$$

where $\widehat{b}^*$ and $\widehat{c}^*$ are the MLEs based on a sample from a Weibull$(1, 1)$ distribution. It is easy to see that G_{q_p} satisfies the two conditions in (C.1) of Section 1.4.1. Appropriate percentiles of G_{q_p} give confidence limits for q_p. To find a $1 - \alpha$ percentile of G_{q_p}, we write

$$\ln(G_{q_p}) = (\widehat{c}_0)^{-1}\left[\widehat{c}^*(-\ln(\widehat{b}^*) + \ln(\theta_p))\right] + \ln(\widehat{b}_0). \tag{7.5.9}$$

Notice that, for a given $(\widehat{b}_0, \widehat{c}_0)$, the joint distribution of the terms inside the square brackets does not depend on any unknown parameters, and so its percentiles can be estimated using Monte Carlo simulation. Let $w_{p;\gamma}$ be the γ percentile of $\widehat{c}^*(-\ln(\widehat{b}^*) + \ln(\theta_p))$. For a given p, a $1 - \alpha$ upper confidence limit for q_p is given by

$$\widehat{b}_0 \exp\left(w_{p;1-\alpha}/\widehat{c}_0\right) \tag{7.5.10}$$

which is a $(p, 1 - \alpha)$ upper tolerance limit for the Weibull(b, c) distribution. Recall that a $(p, 1 - \alpha)$ lower tolerance limit is a $1 - \alpha$ lower confidence limit for q_{1-p}. A $1 - \alpha$ lower confidence limit for q_{1-p} is given by

$$\widehat{b}_0 \exp\left(w_{1-p;\alpha}/\widehat{c}_0\right), \tag{7.5.11}$$

which is a $(p, 1 - \alpha)$ lower tolerance limit for the Weibull(b, c) distribution. This lower tolerance limit is the same as the one given in Equation (13) of Thoman et al. (1970). This procedure of obtaining confidence limits for q_p is exact except for simulation errors. For other approaches, see Lawless (1975).

The distribution of $w = \widehat{c}^*(-\ln(\widehat{b}^*) + \ln(\theta_p))$ in (7.5.9) depends only on the sample size n, p and $1 - \alpha$. We computed the percentiles of w_p and those of w_{1-p} for values of p and $1 - \alpha$ from the set $\{0.90, 0.95, 0.99\}$ using Monte Carlo simulation. Due to the increase in the computational complexity for large values of the sample size n, we used simulation consisting of 100,000 runs when $n \leq 25$, and 10,000 runs for $n > 25$. These values are presented in Table B12, Appendix B, for constructing upper tolerance limits, and in Table B13, Appendix B, for constructing lower tolerance limits.

7.5.5   A GPQ for a Survival Probability

We shall now develop a GPQ for the survival probability

$$S(t) = P(X > t) = \exp\left[-\left(\frac{t}{b}\right)^c\right], \quad t > 0. \qquad (7.5.12)$$

As $\ln[-\ln(S(t))] = c\ln\left(\frac{t}{b}\right) = S'(t)$, say, it is enough to find a GPQ for $S'(t)$. Replacing b and c by their GPQs, we get

$$
\begin{aligned}
G_{S'(t)} &= \frac{\widehat{c}_0}{\widehat{c}^*}\ln\left(\frac{t}{\widehat{b}_0}\right) + \ln(\widehat{b}^*) \\
&= \widehat{c}^{*-1}\ln(-\ln(\widehat{S}(t))) + \ln(\widehat{b}^*), \qquad (7.5.13)
\end{aligned}
$$

where $\widehat{c}^*$ and $\widehat{b}^*$ are the MLEs based on a sample from the Weibull$(1, 1)$ distribution, and $\widehat{S}(t)$ is the MLE of $S(t)$, which can be obtained by substituting the MLEs for the parameters in (7.5.12). Thus, for a given $\widehat{S}(t)$, the distribution of $G_{S'(t)}$ does not depend on any unknown parameters, and so Monte Carlo simulation can be used to find a confidence limit for $S'(t)$, from which a bound for $S(t)$ can be obtained. Specifically, if U is a $1 - \alpha$ upper limit for $S'(t)$, then $\exp(-\exp(S'(t)))$ is a $1 - \alpha$ lower limit for $S(t)$.

The above GPQ is similar to the one in Thoman et al. (1970), and so the generalized confidence limits based on (7.5.13) are exact. Indeed, Equation (7.5.13) is a version of Equation (4c) of Thoman et al. (1970). Furthermore, Thoman et al. provided lower confidence limits for $S(t)$ for values of $\widehat{S}(t)$ ranging from .5(.02).98, $\gamma = .75, .90, .95, .98$ and for some selected sample sizes ranging from 8 to 100.

Example 7.4 (Number of million revolutions before failure for ball bearings)

The data in Table 7.6 represent the number of million revolutions before failure for each of 23 ball bearings. The data were analyzed by Thoman et al. (1969) using a Weibull distribution.

Table 7.6: Number of million revolutions of 23 ball bearings before failure

17.88	28.92	33.00	41.52	42.12	45.60	48.40	51.84
51.96	54.12	55.56	67.80	68.64	68.64	68.88	84.12
93.12	98.64	105.12	105.84	127.92	128.04	173.40	

The MLEs are $\widehat{c}_0 = 2.103$ and $\widehat{b}_0 = 81.876$.

Tolerance Limits: To compute a (0.90, 0.95) lower tolerance limit using (7.5.11), we obtained $w_\alpha = w_{.05} = -3.350$; this value is obtained from Table B12, Appendix B, using $p = 0.90$, $1 - \alpha = 0.95$ and $n = 23$. Thus, the (0.90, 0.95) lower tolerance limit is given by $\widehat{b}_0 \exp\left(w_{1-p;\alpha}/\widehat{c}_0\right) = 81.876 \exp(-3.350/2.103) = 16.65$. Hence we can conclude with confidence 95% that at least 90% of ball bearings survive 16.65 million revolutions.

Estimating Survival Probability: We shall now find a 95% lower confidence limit for the probability that a bearing will last at least 50 million revolutions. Substituting the MLEs $\widehat{b}_0$ and $\widehat{c}_0$, we obtained $\widehat{S}(50) = 0.701$. Using this value in (7.5.13)), and using Monte Carlo simulation with 10,000 runs, we computed a 95% upper limit for $S'(50)$ as -0.499. Thus, a 95% lower limit for $S(50)$ as $\exp(-\exp(-.499)) = 0.545$.

7.5.6 Stress-Strength Reliability

Let $X_1 \sim \text{Weibull}(b_1, c_1)$ independently of $X_2 \sim \text{Weibull}(b_2, c_2)$. If X_1 is a strength variable and X_2 is a stress variable, then the stress-strength reliability parameter is given by

$$R = P(X_1 > X_2) = \frac{c_2}{b_2} \int_0^\infty e^{-\left(\frac{x_2}{b_1}\right)^{c_1}} \left(\frac{x_2}{b_2}\right)^{c_2-1} e^{-\left(\frac{x_2}{b_2}\right)^{c_2}} dx_2. \qquad (7.5.14)$$

The integral in (7.5.14) can be evaluated analytically as an infinite series expression, but it is quite complex to obtain a lower confidence limit. Here we consider only the case of $c_1 = c_2$, since a simple expression for R can be obtained if $c_1 = c_2$. In this case,

$$R = \frac{b_1^c}{b_1^c + b_2^c} = \frac{1}{1 + b_2^c/b_1^c} = \frac{1}{1 + R^*}, \qquad (7.5.15)$$

where $R^* = \frac{b_2^c}{b_1^c}$. To get a GPQ for the common shape parameter c, we note that the MLE of the common shape parameter is the solution $\widehat{c}$ of the equation

$$\widehat{c}^{-1} = \left[\frac{n_1}{n_1 + n_2} \frac{\sum_{i=1}^{n_1} x_{1i}^{\widehat{c}} \ln(x_{1i})}{\sum_{i=1}^{n_1} x_{1i}^{\widehat{c}}} + \frac{n_2}{n_1 + n_2} \frac{\sum_{i=1}^{n_1} x_{2i}^{\widehat{c}} \ln(x_{2i})}{\sum_{i=1}^{n_2} x_{2i}^{\widehat{c}}} \right]$$
$$- \frac{\sum_{i=1}^{n_1} \ln(x_{1i}) + \sum_{i=1}^{n_2} \ln(x_{2i})}{n_1 + n_2}, \tag{7.5.16}$$

where $x_{k1}, \ldots, x_{kn_k}$ is a sample from the Weibull(b_k, c_k) distribution, $k = 1, 2$. The MLEs of the scale parameters are given by

$$\widehat{b}_1 = \left(\frac{\sum_{i=1}^{n_1} x_{1i}^{\widehat{c}}}{n_1} \right)^{1/\widehat{c}} \quad \text{and} \quad \widehat{b}_2 = \left(\frac{\sum_{i=1}^{n_2} x_{2i}^{\widehat{c}}}{n_2} \right)^{1/\widehat{c}}. \tag{7.5.17}$$

The above equations are generalizations of the log-likelihood equations for the case $n_1 = n_2$ given in Schafer and Sheffield (1976). These authors also argued that the distributions of $\frac{\widehat{c}}{c}$, $\widehat{c} \ln \left(\widehat{b}_1/b_1 \right)$ and $\widehat{c} \ln \left(\widehat{b}_2/b_2 \right)$ do not depend on any parameters, i.e., they are pivots.

Let $\widehat{c}^*$, $\widehat{b}_1^*$ and $\widehat{b}_2^*$ be the MLEs based on independent samples from a Weibull(1, 1) distribution, and let $(\widehat{c}_0, \widehat{b}_{10}, \widehat{b}_{20})$ be the observed value of $(\widehat{c}, \widehat{b}_1, \widehat{b}_2)$. In terms of these MLEs and $(\widehat{c}^*, \widehat{b}_1^*, \widehat{b}_2^*)$, we get a GPQ for c as $\widehat{c}_0/\widehat{c}^*$. Replacing the parameters in (7.5.15) by their GPQs, we get a GPQ for R^* as

$$G_{R^*} = \frac{\widehat{b}_1^* \left(\widehat{b}_{20} \right)^{\widehat{c}_0/\widehat{c}^*}}{\widehat{b}_2^* \left(\widehat{b}_{10} \right)^{\widehat{c}_0/\widehat{c}^*}} \tag{7.5.18}$$

or equivalently,

$$\ln(G_{R^*}) = \ln \left(\frac{\widehat{b}_1^*}{\widehat{b}_2^*} \right) + \widehat{c}^{*-1} \left[\ln \left(\frac{1 - \widehat{R}_0}{\widehat{R}_0} \right) \right], \quad \text{with} \quad \widehat{R}_0 = \frac{\left(\widehat{b}_{10} \right)^{\widehat{c}_0}}{\left(\widehat{b}_{10} \right)^{\widehat{c}_0} + \left(\widehat{b}_{20} \right)^{\widehat{c}_0}}. \tag{7.5.19}$$

For a given $\widehat{R}_0$, the distribution of $\ln(G_{R^*})$ does not depend on any unknown parameters, and so Monte Carlo simulation can be used to estimate the percentiles of $\ln(G_{R^*})$. If d_p denotes the p quantile of $\ln(G_{R^*})$, then $(1 + \exp(d_{1-\alpha}))^{-1}$ is a $1 - \alpha$ lower confidence limit for the stress-strength reliability parameter in (7.5.15).

Notice that the expression (7.5.19) enables us to compute confidence limits for R for given values of the confidence level, n and $\widehat{R}_0$. Krishnamoorthy and

Lin (2009) computed 95% lower confidence limits of R for $\widehat{R}_0 = .50(.02).98$ and for some selected values of n ranging from 8 to 50. These lower limits are given in Table B14, Appendix B.

Remark 7.2 All the results for the Weibull distribution can be extended to an extreme-value distribution in a straightforward manner. We simply apply the methods for the Weibull distribution to log-transformed samples from an extreme-value distribution. For example, to find a $(p, 1 - \alpha)$ one-sided tolerance limit based on a sample $Y_1, ..., Y_n$ from an extreme-value distribution, we apply the method for the Weibull distribution to the log-transformed sample $\ln(Y_1), ..., \ln(Y_n)$. The antilog of the resulting tolerance limit is the tolerance limit for the sampled extreme-value distribution.

Example 7.5 (Comparison of electrical cable insulations)

The data are taken from Example 5.4.2 of Lawless (2003), and they represent failure voltage levels of two types of electrical cable insulation when specimens were subjected to an increasing voltage stress in a laboratory test. Twenty specimens of each type were tested and the failure voltages are given in Table 7.7. Krishnamoorthy and Lin (2009) used the data for illustrating the preceding generalized variable approach for constructing a confidence limit for the stress-strength reliability parameter, and their results are given below.

Table 7.7: Fatigue voltages (in kilovolts per millimeter) for two types of electric cable insulation

Type I (X_1)	39.4	45.3	49.2	49.4	51.3	52.0	53.2	53.2	54.9	55.5
	57.1	57.2	57.5	59.2	61.0	62.4	63.8	64.3	67.3	67.7
Type II (X_2)	32.0	35.4	36.2	39.8	41.2	43.3	45.5	46.0	46.2	46.4
	46.5	46.8	47.3	47.3	47.6	49.2	50.4	50.9	52.4	56.3

The MLEs are $\widehat{c}_1 = 9.141, \widehat{b}_1 = 59.125, \widehat{c}_2 = 9.383$, and $\widehat{b}_2 = 47.781$. Closeness of $\widehat{c}_1$ and $\widehat{c}_2$ indicates that the assumption of $c_1 = c_2$ is tenable, and so our approach can be used to set confidence bound for the stress-strength reliability parameter $R = P(X > Y)$. To estimate R, we computed the MLEs using (7.5.16) and (7.5.17), and the solutions are $\widehat{c}_0 = 9.261, \widehat{b}_{10} = 59.161$ and $\widehat{b}_{20} = 47.753$. Using these MLEs, we obtained $\widehat{R}_0 = 0.88$. For this value of $\widehat{R}_0$ and for $n = 20$, we get the 95% lower confidence limit for R from Table B14, Appendix B, as 0.778. That is, with 95% confidence, we can assert that the probability that a Type I electrical cable insulation lasts longer than a Type II electrical cable insulation is at least 0.778.

7.6   Exercises

7.6.1. Let $Y_{i1}, ..., Y_{in_i}$ be a sample from a lognormal(μ_i, σ_i^2) distribution, $i = 1, 2$. Assume that Y_{ij}'s are mutually independent. Give a procedure for finding a $(p, 1-\alpha)$ lower tolerance limits for the distribution of $\ln\left(\frac{Y_1}{Y_2}\right)$ using Hall's method in Section 2.4.2.

7.6.2. Let k_1 be a $(p, 1 - \alpha)$ lower tolerance factor for a sample of size n from an exponential(μ, θ) distribution. Write k_1 as $k_1(n, p, 1 - \alpha)$. Let k_2 be a $(p, 1 - \alpha)$ upper tolerance factor for the same sample size.

(a) Show that $k_2(n, p, 1 - \alpha) = k_1(n, 1 - p, \alpha)$.

(b) Consider a $(p, 1-\alpha)$ upper tolerance limit of the form $\hat{\mu} + k_2 S$. If k_2 is negative, then show that k_2 that satisfies (7.4.8) is given by (7.4.19). Aslo, show that for many conventional choices of $1 - \alpha$ and p (such as 0.90, 0.95, 0.99), no positive integer value of n yields a negative value for k_2.

7.6.3. A random vector $\mathbf{Y}$ is said to have bivariate lognormal distribution if the joint distribution of logarithm of components $\mathbf{Y}$ is bivariate normal. Let $\mathbf{Y}_1, ..., \mathbf{Y}_n$ be a sample on $\mathbf{Y}$. Describe a procedure for constructing a tolerance interval for the distribution of $\frac{Y_1}{Y_2}$, where Y_1 and Y_2 are the components of $\mathbf{Y}$ (see also Bebu and Mathew, 2008).

7.6.4. The data in the following table represent vinyl chloride concentrations collected from clean upgradient monitoring wells, and are taken from Bhaumik and Gibbons (2006). The Q-Q plot by these authors showed that a gamma distribution provides an excellent fit to this data.

5.1	2.4	.4	.5	2.5	.1	6.8	1.2	.5	.6
5.3	2.3	1.8	1.2	1.3	1.1	.9	3.2	1.0	.9
.4	.6	8.0	.4	2.7	.2	2.0	.2	.5	.8
2.0	2.9	.1	4.0						

Reprinted with permission from *Technometrics*. Copyright [2006] by the American Statistical Association.

Assuming a gamma distribution,

(a) find $(0.90, 0.95)$ and $(0.95, 0.95)$ lower tolerance limits for the vinyl chloride concentration.

(b) Find a 95% lower confidence limit for the probability that the vinyl chloride concentration of a sample exceeds $5\mu g/L$.

(c) Krishnamoorthy, Lin and Xia (2008) observed that a Weibull model also fits the data very well. Assuming a Weibull distribution, verify that the MLEs are $\widehat{c} = 1.010$ and $\widehat{b} = 1.884$. Answer parts (a) and (b), and compare the results with those based on a gamma distribution.

7.6.5. Let $X_1, ..., X_n$ be a sample from a gamma(a, b) distribution with known shape parameter a.

(a) Find the distribution of $\frac{\sum_{i=1}^{n} X_i}{b}$.

(b) Using the distributional result in part(a), find a $1 - \alpha$ confidence interval for b.

(c) Let $Y_1 \sim$ gamma(a_1, b_1) independently of $Y_2 \sim$ gamma(a_2, b_2), where a_1 and a_2 are known. Assume that a sample of n_1 observations on Y_1 and a sample of n_2 observations on Y_2 are available. Find a $1-\alpha$ lower confidence limit for the stress-strength reliability parameter $P(Y_1 > Y_2)$ given in (7.3.6).

7.6.6. Let $X_1, ..., X_n$ be a sample from a two-parameter exponential distribution with known threshold parameter μ. Give methods for constructing one-sided tolerance limits.

7.6.7. Let $X_1, ..., X_n$ be a sample from a Weibull(θ, c) distribution with the pdf

$$\frac{c}{\theta} x^{c-1} e^{-x^c/\theta}, \quad x \geq 0, \ c > 0, \ \theta > 0.$$

(a) Show that the MLEs are determined by the equations

$$\left[\frac{\sum_{i=1}^{n} X_i^{\widehat{c}} \ln X_i}{\sum_{i=1}^{n} X_i^{\widehat{c}}} - \frac{1}{\widehat{c}} \right] = \frac{1}{n} \sum_{i=1}^{n} \ln X_i$$

$$\widehat{\theta} = \frac{1}{n} \sum_{i=1}^{n} X_i^{\widehat{c}}.$$

(b) Using part (a), show that the MLEs for a Weibull(b, c) distribution are as given in Section 7.5.2.

7.6.8. Let $X_1, ..., X_n$ be a sample from a Weibull(b, c) distribution. Assume that c is known. Find $(p, 1-\alpha)$ one-sided tolerance limits for the Weibull(b, c) distribution. [Hint: Find the distribution of $Y = X^c$, where $X \sim$ Weibull(b, c)]

7.6.9. Give procedures for finding $(p, 1 - \alpha)$ one-sided tolerance limits, and for finding a $1 - \alpha$ lower confidence limit for a survival probability $P(X > t)$, for the power distribution with the pdf (7.4.3).

7.6.10. The following is a sample of 30 observations simulated from a Pareto distribution with $\sigma = 4$ and $\lambda = 1.2$ (see the pdf in (7.4.2)).

6.43	37.58	4.27	5.97	4.02	5.58	4.28	12.28	18.33	9.56
19.73	10.93	6.51	11.45	23.79	4.09	4.15	7.84	4.06	9.30
14.69	12.20	5.86	12.28	19.62	4.89	29.99	23.30	6.07	7.85

(a) Find a $(0.90, 0.95)$ lower tolerance limit for the Pareto distribution.

(b) Find the 0.10 quantile of the Pareto distribution, and check if the above tolerance limit is less than this quantile.

(c) Find a $(0.95, 0.99)$ upper tolerance limit for the Pareto distribution.

(d) Find the 0.95 quantile of the Pareto distribution, and check if the above tolerance limit is greater than this quantile.

Chapter 8

Nonparametric Tolerance Intervals

8.1   Notations and Preliminaries

If a sample is from a continuous population, and does not fit a parametric model, or fits a parametric model for which tolerance intervals are difficult to obtain, then one may seek nonparametric tolerance intervals for an intended application. The nonparametric procedures that we shall describe in this chapter are applicable to find tolerance limits for any continuous population. However, for a given sample size, a nonparametric tolerance interval that satisfies specified content and coverage requirements may not exist. Furthermore, nonparametric tolerance intervals are typically wider than their parametric counterparts. The nonparametric methods are based on a result due to Wilks (1941) which states that if a sample is from a continuous distribution, then the distribution of the proportion of the population between two order statistics is independent of the population sampled, and is a function of only the particular order statistics chosen. Thus, nonparametric tolerance intervals are based on order statistics, and the intervals are defined as follows. Let $\boldsymbol{X} = (X_1, ..., X_n)$ be a random sample from a continuous distribution $F_X(x)$, and let $X_{(1)} < ... < X_{(n)}$ be the order statistics for the sample. Recall that a $(p, 1 - \alpha)$ tolerance interval $(L(\boldsymbol{X}), U(\boldsymbol{X}))$ is such that

$$P_{\boldsymbol{X}}\left\{ P_X\left(L(\boldsymbol{X}) \leq X \leq U(\boldsymbol{X}) \Big| \boldsymbol{X} \right) \geq p \right\} = 1 - \alpha,$$

where X also follows the same continuous distribution F_X independent of the sample $\boldsymbol{X}$. Wilk's result allows us to choose $L(\boldsymbol{X}) = X_{(r)}$ and $U(\boldsymbol{X}) = X_{(s)}$,

$r < s$, and the problem is to determine the values of r and s so that

$$P_{X_{(r)}, X_{(s)}} \left\{ P_X \left(X_{(r)} \leq X \leq X_{(s)} \middle| X_{(r)}, X_{(s)} \right) \geq p \right\} = 1 - \alpha. \tag{8.1.1}$$

A one-sided tolerance limit is similarly defined based on a single order statistic. In the following section, we shall give some preliminary results on the distribution of order statistics; results that are useful to find the values of r and s that satisfy (8.1.1), and also useful to find one-sided tolerance limits.

8.2   Order Statistics and Their Distributions

Let $X_1, ..., X_n$ be a sample from a population with a continuous distribution function $F_X(x)$. Let $X_{(i)}$ denote the ith smallest of $X_1, ..., X_n$. Then

$$X_{(1)} < X_{(2)} < \dots < X_{(n)}$$

are collectively referred to as the order statistics for the sample. Notice that the above arrangement is unique, because F_X is continuous, and so the probability that any two random variables assume the same value is zero. The statistic $X_{(r)}$ is called the rth order statistic.

Given below are the distributional results for order statistics, required to construct tolerance limits. For more details and results, see the book by David and Nagaraja (2003).

Result 8.1 (*Probability Integral Transform*)

Let X be a random variable with a continuous distribution function $F_X(x)$. Let $Y = F_X(X)$. Then $Y \sim$ uniform$(0, 1)$ distribution.

Proof.   Note that the inverse of the distribution function is defined by

$$F_X^{-1}(y) = \inf\{x : P(X \leq x) \geq y\}, \quad 0 < y < 1.$$

Also, for a uniform$(0, 1)$ random variable U, $F_U(u) = u$. For any $0 < y < 1$,

$$P_Y(Y \leq y) = P_X(F_X(X) \leq y) = P_X(X \leq F_X^{-1}(y)) = F_X(F_X^{-1}(y)) = y,$$

and so $Y \sim$ uniform$(0, 1)$.

As an obvious consequence of Result 8.1, we have

Result 8.2   If $X_1, ..., X_n$ is a sample from a continuous distribution F_X, then $U_1 = F_X(X_1), \ \dots, \ U_n = F_X(X_n)$ is a sample from a uniform$(0, 1)$ distribution.

Furthermore, if $X_{(1)} < X_{(2)} < \ldots < X_{(n)}$ are the order statistics for the sample $X_1, \ldots, X_n$, then

$$U_{(1)} = F_X(X_{(1)}), \quad \ldots, \quad U_{(n)} = F_X(X_{(n)})$$

can be regarded as order statistics for the sample $U_1, \ldots, U_n$ from a uniform$(0, 1)$ distribution. That is, $F_X(X_{(r)})$ is distributed as $U_{(r)}$.

Result 8.3 (*Empirical Distribution Function*) Let $X_1, \ldots, X_n$ be a sample from a continuous distribution F_X. Consider the empirical distribution function given by

$$\widehat{F}_n(x) = \frac{\text{number of } X_i\text{'s} \leq x}{n}.$$

Then $n\widehat{F}_n(x) \sim \text{binomial}(n, F_X(x))$.

Proof. Let $Q_i = 1$ if $X_i \leq x$, 0 otherwise. As X_i's are independent and identically distributed, Q_i's are independent Bernoulli random variables with the "success probability" $F_X(x)$. Hence $n\widehat{F}_n(x) = \sum_{i=1}^{n} Q_i \sim \text{binomial}(n, F_X(x))$.

Result 8.4 Let $X_{(r)}$ be the rth order statistic for a sample of n observations from a continuous distribution F_X. The pdf of $X_{(r)}$ is given by

$$f_{X_{(r)}}(x) = \frac{n!}{(r-1)!(n-r)!}[F_X(x)]^{r-1}[1 - F_X(x)]^{n-r} f_X(x), \qquad (8.2.1)$$

where $f_X(x)$ is the pdf of X.

Proof. The cdf of $X_{(r)}$ is given by

$$
\begin{aligned}
F_{X_{(r)}}(x) &= P(X_{(r)} \leq x) \\
&= P(\text{number of } X_i\text{'s} \leq x \text{ is at least } r) \\
&= P(n\widehat{F}_n(x) \geq r) \\
&= \sum_{k=r}^{n} \binom{n}{k} F_X(x)^k (1 - F_X(x))^{n-k} \quad [\text{Result 8.3}] \\
&= n \binom{n-1}{r-1} \int_0^{F_X(x)} t^{r-1}(1-t)^{n-r} dt.
\end{aligned}
$$

Equality of the last two expressions can be established using integration by parts. Differentiating the last expression with respect to x, we get the pdf in (8.2.1).

Using the pdf in (8.2.1), we get

Corollary 8.1 If $U_{(1)} < ... < U_{(n)}$ are order statistics for a sample from a uniform$(0,1)$ distribution, then $U_{(r)}$ follows a beta$(r, n-r+1)$ distribution with the pdf

$$f_{U_{(r)}}(u) = \frac{1}{B(r, n-r+1)} u^{r-1}(1-u)^{n-r+1-1}, \quad 0 < u < 1, \quad (8.2.2)$$

where $B(x,y) = \frac{\Gamma(x)\Gamma(y)}{\Gamma(x+y)}$ is the usual beta function.

Result 8.5 The joint pdf of $(X_{(r)}, X_{(s)})$ is given by

$$f_{X_{(r)}, X_{(s)}}(x,y) = \begin{cases} \frac{n!}{(r-1)!(s-r-1)!(n-s)!}[F_X(x)]^{r-1}[F_X(y) - F_X(x)]^{s-r-1} \\ \times [1 - F_X(y)]^{n-s} f_X(x) f_X(y), \quad -\infty < x < y < \infty, \\ 0, \qquad\qquad\qquad\qquad\qquad\qquad\quad x \geq y. \end{cases}$$

$$(8.2.3)$$

Proof. Let $r < s$ so that $X_{(r)} < X_{(s)}$. Also, let $x \geq y$. Note that $X_{(s)} \leq y \Rightarrow X_{(r)} \leq x$, and so

$$P(X_{(r)} \leq x, X_{(s)} \leq y) = P(X_{(s)} \leq y) = F_{X_{(s)}}(y) \text{ for } x \geq y.$$

If $x < y$ and $r < s$, then

$$
\begin{aligned}
P(X_{(r)} \leq x, X_{(s)} \leq y) &= P(n\widehat{F}_n(x) \geq r, n\widehat{F}_n(y) \geq s) \\
&= \sum_{i=r}^{n} \sum_{j=s}^{n} P(n\widehat{F}_n(x) = i, n\widehat{F}_n(y) = j) \\
&= \sum_{i=r}^{n} \sum_{j=s}^{n} P(i \text{ of the } X\text{'s} \leq x, \ j \text{ of the } X\text{'s} \leq y) \\
&= \sum_{i=r}^{n} \sum_{j=s}^{n} P(i \text{ of the } X\text{'s} \leq x, \ x < (j-i) \text{ of the } X\text{'s} \leq y, \\
&\qquad n-j \text{ of the } X\text{'s} \geq y) \\
&= \sum_{i=r}^{n} \sum_{j=s}^{n} \frac{n!}{i!(j-i)!(n-j)!}[F(x)]^i [F(y) - F(x)]^{j-i} \\
&\quad \times \ [1 - F(y)]^{n-j} \text{ for } x < y.
\end{aligned}
$$
$$(8.2.4)$$

Notice that the probability expression in (8.2.4) is the probability mass function of a trinomial distribution. The above expression can also be written as (see Exercise 8.8.1)

$$
\begin{aligned}
F_{X_{(r)}, X_{(s)}}(x,y) &= \frac{n!}{(r-1)!(s-r-1)!(n-s)!} \\
&\quad \times \int_0^{F(x)} \int_0^{F(y)} v^{r-1}(t-v)^{s-r-1}(1-t)^{n-s} dt\, dv. \quad (8.2.5)
\end{aligned}
$$

By differentiating with respect to (x, y), we get (8.2.3). □

Using the above result, we see that if $U_{(r)}$ and $U_{(s)}$ are respectively the rth and the sth order statistics for a sample of size n from a uniform$(0, 1)$ distribution, then

$$
\begin{aligned}
f_{U_{(r)}, U_{(s)}}(x, y) &= \frac{n!}{(r-1)!(s-r-1)!(n-s)!} x^{r-1} [y - x]^{s-r-1} \\
&\times [1 - y]^{n-s}, \quad 0 < x < y < 1.
\end{aligned}
\tag{8.2.6}
$$

The following relation between the beta and binomial distributions is required in the sequel.

Result 8.6 Let X be a binomial(n, p) random variable, and U be a beta$(k, n-k+1)$ random variable. Then, for a given k,

$$
P(X \geq k | n, p) = P(U \leq p), \quad k = 1, ..., n.
$$

Furthermore,

$$
P(X \leq k - 1 | n, p) = P(U \geq p), \quad k = 1, ..., n.
$$

8.3 One-Sided Tolerance Limits and Exceedance Probabilities

Let $X_1, ..., X_n$ be a sample from a continuous distribution F_X. In order to construct a nonparametric $(p, 1 - \alpha)$ lower tolerance limit, we need to find the positive integer k so that

$$
P_{X_{(k)}}[P_X(X \geq X_{(k)} | X_{(k)}) \geq p] = 1 - \alpha.
$$

The probability on the left-hand side can be expressed as

$$
P_{X_{(k)}}[1 - F(X_{(k)}) \geq p] = P_{X_{(k)}}[F(X_{(k)}) \leq 1 - p] = P(U_{(k)} \leq 1 - p),
$$

where $U_{(k)} = F(X_{(k)}) \sim$ beta$(k, n - k + 1)$ distribution (see Result 8.2). Using Result 8.6, we see that

$$
\begin{aligned}
P(U_{(k)} \leq 1 - p) &= 1 - P(U_{(k)} \geq 1 - p) \\
&= 1 - P(Y \leq k - 1 | n, 1 - p) \\
&= P(Y \geq k | n, 1 - p) \\
&= P(n - Y \leq n - k | n, 1 - p) \\
&= P(W \leq n - k | n, p),
\end{aligned}
\tag{8.3.1}
$$

where Y is a binomial$(n, 1-p)$ random variable, and $W = n-Y$ is a binomial(n, p) random variable. Thus, if k is the largest integer for which

$$P(Y \geq k|n, 1-p) \geq 1-\alpha, \qquad (8.3.2)$$

then $X_{(k)}$ is the desired $(p, 1-\alpha)$ lower tolerance limit.

To construct a $(p, 1-\alpha)$ upper tolerance limit, we need to find the positive integer m so that

$$P_{X_{(m)}}[P_X(X \leq X_{(m)}|X_{(m)}) \geq p] = 1-\alpha.$$

Proceeding as in the case of the lower tolerance limit, it can be shown that $X_{(n-k+1)}$ is a $(p, 1-\alpha)$ upper tolerance limit, where k is the largest integer satisfying (8.3.2); see Exercise 8.8.2.

For any fixed sample size, there may not exist order statistics that satisfy the requirements of one-sided tolerance limits. The sample size issue will be addressed in Section 8.6.

Lower Limits for an Exceedance Probability: For a continuous random variable X, suppose it is desired to estimate $P(X > t)$, where t is a specified number. If $t \leq X_{(n)}$, then a conservative $1-\alpha$ lower limit for $P(X > t)$ can be obtained as follows (see Section 1.1.3). Let $X_{(r)}$ be the smallest order statistic which is larger than t. Let p be such that $X_{(r)}$ is a $(p, 1-\alpha)$ lower tolerance limit. Since $t < X_{(r)}$,

$$P(X > t) \geq p \quad \text{holds with probability at least } 1-\alpha.$$

Therefore, we need to find the value of p so that $X_{(r)}$ is $(p, 1-\alpha)$ lower tolerance limit for the distribution of X. In view of (8.3.1), we see that p is to be determined so that

$$P(W \leq n - r|n, p) \geq 1-\alpha \Leftrightarrow P(U > p \mid n-r+1, r) \geq 1-\alpha, \qquad (8.3.3)$$

where W is a binomial(n, p) random variable, and U is a beta$(n-r+1, r)$ random variable. The above probability relation was obtained using Result 8.6. Thus p is given by beta$(\alpha; n-r+1, r)$, the α quantile of a beta$(n-r+1, r)$ distribution. That is, a $1-\alpha$ lower confidence limit for $P(X > t)$ is beta$(\alpha; n-r+1, r)$.

8.4   Tolerance Intervals

As mentioned earlier, to construct a $(p, 1-\alpha)$ nonparametric tolerance interval for a continuous distribution, we have to determine a pair of order statistics

$X_{(r)}$ and $X_{(s)}$, $r < s$, so that the interval $(X_{(r)}, X_{(s)})$ would contain at least a proportion p of the population with confidence $1 - \alpha$. That is, we have to determine the values of $r < s$ so that

$$P_{X_{(r)}, X_{(s)}} \left\{ P_X[X_{(r)} \leq X \leq X_{(s)} | X_{(r)}, X_{(s)}] \geq p \right\} = 1 - \alpha. \tag{8.4.1}$$

Because of the discreteness of r and s, we need to determine them so that $s - r$ is minimum, and the coverage probability is at least $1 - \alpha$, and as close to $1 - \alpha$ as possible. Note that the inner probability is

$$P_X[X_{(r)} \leq X \leq X_{(s)})] = F(X_{(s)}) - F(X_{(r)}).$$

Thus we can write (8.4.1) as

$$P_{X_{(r)}, X_{(s)}} \left\{ F(X_{(s)}) - F(X_{(r)}) \geq p \right\} = 1 - \alpha, \tag{8.4.2}$$

or equivalently,

$$P_{U_{(r)}, U_{(s)}} \left\{ U_{(s)} - U_{(r)} \geq p \right\} = 1 - \alpha. \tag{8.4.3}$$

The pdf of $U_{(s)} - U_{(r)}$ can be obtained from the joint pdf of $U_{(r)}$ and $U_{(s)}$ given in (8.2.6). Let $u = y - x$ and $v = y$ so that the inverse transformation is $x = v - u$ and $y = v$, $0 < u < v < 1$. The Jacobian of the transformation is one. Using these observations in (8.2.6), we obtain

$$f_{U,V}(u, v) = c(v - u)^{r-1} u^{s-r-1} (1 - v)^{n-s}, \quad 0 < u < v < 1,$$

where $c = \frac{n!}{(r-1)!(s-r-1)!(n-s)!}$, which is the constant term in (8.2.6). Thus, the marginal pdf of $U = U_{(s)} - U_{(r)}$ is given by

$$f_U(u) = c u^{s-r-1} \int_u^1 (v - u)^{r-1} (1 - v)^{n-s} dv.$$

This is actually a beta distribution. In order to see this, let $v - u = t(1 - u)$, $0 < t < 1$. This implies $(1 - v) = (1 - u) - t(1 - u) = (1 - u)(1 - t)$. Also, $dv = (1 - u)dt$. In terms of the new variables, we can write the pdf as

$$\begin{aligned} f_U(u) &= c u^{s-r-1} \int_0^1 (1 - u)^{r-1} t^{r-1} (1 - u)^{n-s} (1 - t)^{n-s} (1 - u) dt \\ &= c u^{s-r-1} (1 - u)^{n-s+r} \int_0^1 t^{r-1} (1 - t)^{n-s} dt. \end{aligned}$$

Notice that $\int_0^1 t^{r-1}(1 - t)^{n-s} dt = B(r, n - s + 1) = \frac{\Gamma(r)\Gamma(n-s+1)}{\Gamma(n-s+r+1)}$. Substituting this expression in the above equation, we get

$$\begin{aligned} f_U(u) &= \frac{n!}{(r-1)!(s-r-1)!(n-s)!} u^{s-r-1}(1-u)^{n-s+r} \frac{(r-1)!(n-s)!}{(n-s+r)!} \\ &= \frac{1}{B(s-r, n-s+r+1)} u^{s-r-1}(1-u)^{n-s+r}, \quad 0 < u < 1. \tag{8.4.4} \end{aligned}$$

Thus, $U = U_{(s)} - U_{(r)}$ is distributed as a beta random variable with shape parameters $s - r$ and $n - s + r + 1$. Using Lemma 8.3.1, we now conclude that

$$P_{U_{(r)}, U_{(s)}} \{U_{(s)} - U_{(r)} \geq p\} = P(X \leq s - r - 1), \qquad (8.4.5)$$

where $X \sim \text{binomial}(n, p)$. Notice that the above probability depends only on the difference $s - r$, and not on the actual values of s and r. Let $k = s - r$ be the least value for which

$$P(X \leq s - r - 1) \geq 1 - \alpha. \qquad (8.4.6)$$

Then any interval $(X_{(r)}, X_{(s)})$ is a $(p, 1 - \alpha)$ tolerance interval, provided $1 \leq r < s \leq n$ and $s - r = k$.

It is customary to take $s = n - r + 1$, so that $(X_{(r)}, X_{(n-r+1)})$ is a $(p, 1 - \alpha)$ tolerance interval (see Wilks, 1941 and David, 1981, p. 19). Wilks referred to this interval as the *truncated sample range* as it is formed by the rth smallest and the rth largest observations from a sample of n observations. However, a shorter interval is possible if we ignore this convention. For example, when $n = 38$ and $(p, 1 - \alpha) = (.80, .90)$, the only shortest interval of the form $(X_{(r)}, X_{(n-r+1)})$ that satisfies the probability requirement in (8.4.6) is $(X_{(2)}, X_{(37)})$ with true coverage probability 0.9613. It can be checked that $(X_{(2)}, X_{(36)})$ is also a $(.80, .90)$ tolerance interval with actual coverage probability 0.9014. Similarly, when $n = 69$ and $(p, 1 - \alpha) = (.80, .99)$, the only shortest interval of the form $(X_{(r)}, X_{(n-r+1)})$ that satisfies (8.4.6) is $(X_{(3)}, X_{(67)})$ with actual coverage probability 0.9968 while $(X_{(3)}, X_{(66)})$ has coverage probability 0.9908. We shall describe a method of determining order statistics that form a shortest tolerance interval in Section 8.6.

8.5   Confidence Intervals for Population Quantiles

Let κ_p denote the p quantile of a continuous distribution F_X, and let $X_{(1)}, ..., X_{(n)}$ be a set of order statistics from F_X. For $0 < p < 1$, let $r = np$ if np is an integer, and $r = [np]$ otherwise, where $[x]$ is the largest integer not exceeding x. The order statistic $X_{(r)}$ is the sample pth quantile, which is a point estimator of κ_p.

To construct a $1 - \alpha$ confidence interval for κ_p based on order statistics, we need to determine the values of r and s, $r < s$, so that $P(X_{(r)} \leq \kappa_p \leq X_{(s)}) = 1 - \alpha$. Notice that the event $X_{(r)} \leq \kappa_p \leq X_{(s)}$ is equivalent to "the number of

X_i's $< \kappa_p$ is at least r and at most $s - 1$." Using this relation, we can see that

$$
\begin{aligned}
P(X_{(r)} < \kappa_p < X_{(s)}) &= P(r \leq n\hat{F}_n(\kappa_p) \leq s - 1) \\
&= \sum_{i=r}^{s-1} \binom{n}{i} [F(\kappa_p)]^i [1 - F(\kappa_p)]^{n-i} \qquad \text{[Result 8.3]} \\
&= \sum_{i=r}^{s-1} \binom{n}{i} p^i [1 - p]^{n-i} \quad (\text{since } F(\kappa_p) = p) \\
&= 1 - \alpha. \qquad (8.5.1)
\end{aligned}
$$

Thus, $(X_{(r)}, X_{(s)})$ is a $1 - \alpha$ confidence interval for κ_p provided r and s satisfy (8.5.1). Ideally, r and s have to be determined so that they satisfy (8.5.1) and $s - r$ is as small as possible. For a given n and $1 - \alpha$, there may not exist r and s so that $P(X_{(r)} < \kappa_p < X_{(s)}) \geq 1 - \alpha$.

Recall that a $1 - \alpha$ one-sided confidence limit for κ_p is also the $(p, 1 - \alpha)$ one-sided tolerance limit for F_X. In particular, the order statistic $X_{(k)}$, where k is determined by (8.3.2), is a $1 - \alpha$ lower confidence limit for κ_{1-p} and $X_{(n-k+1)}$ is a $1 - \alpha$ upper confidence limit for κ_p.

8.6 Sample Size Calculation

For any fixed sample size, there may not exist order statistics that provide the required $(p, 1 - \alpha)$ tolerance intervals. So, for a given p and $1 - \alpha$, it is important to determine the sample size so that the required order statistics exist for computing a tolerance interval. In the following, we shall determine the sample size in various scenarios.

8.6.1 Sample Size for Tolerance Intervals of the Form $(X_{(1)}, X_{(n)})$

We first determine the sample size n so that $(X_{(1)}, X_{(n)})$ would contain at least a proportion p of the population with confidence level $1 - \alpha$. That is,

$$
P_{X_{(1)}, X_{(n)}} \left\{ P_X[X_{(1)} \leq X \leq X_{(n)} | X_{(1)}, X_{(n)}] \geq p \right\} = 1 - \alpha. \qquad (8.6.1)
$$

Substituting $s = n$ and $r = 1$ in (8.4.6), we see that the above probability requirement simplifies to

$$
P(X \leq n - 2) \geq 1 - \alpha \Leftrightarrow (n - 1)p^n - np^{n-1} + 1 \geq 1 - \alpha, \qquad (8.6.2)
$$

where X is a binomial(n, p) random variable. Let n_0 be the least value of n that satisfies (8.6.2). Then, the smallest order statistic $X_{(1)}$ and the largest order statistic $X_{(n_0)}$, from a sample of size n_0, form a $(p, 1 - \alpha)$ tolerance interval. Furthermore, for any sample of size $n \geq n_0$, there is at least a pair of order statistics that form a $(p, 1 - \alpha)$ tolerance interval.

Now consider the case of a one-sided upper tolerance limit (that is, one-sided upper confidence limit for κ_p). Such a limit can be obtained by finding the least value of n for which

$$P(\kappa_p < X_{(n)}) \geq 1 - \alpha.$$

Using (8.5.1) with $r = 0$ and $s = n$, we see that

$$P(\kappa_p < X_{(n)}) = \sum_{i=0}^{n-1} \binom{n}{i} p^i [1 - p]^{n-i} = 1 - p^n. \tag{8.6.3}$$

Thus $X_{(n)}$ is a $(p, 1 - \alpha)$ upper tolerance limit when the sample size n is the least value for which

$$1 - p^n \geq 1 - \alpha \quad \text{or} \quad n \geq \frac{\ln(\alpha)}{\ln(p)}.$$

To ensure the coverage probability, we choose $n = \left[\frac{\ln(\alpha)}{\ln(p)}\right]_+$, where $[x]_+$ is the smallest integer greater than or equal to x. Also, for this choice of n, $X_{(1)}$, the smallest order statistic in a sample of size n is a $(p, 1 - \alpha)$ lower tolerance limit, or equivalently, $X_{(1)}$ is a $1 - \alpha$ lower confidence limit for κ_{1-p}.

It should be noted that, because of the discreteness of the sample size, the true coverage probability will be slightly more than the specified confidence level $1 - \alpha$. For example, when $p = 0.90$ and $1 - \alpha = 0.90$, the required sample size for the two-sided tolerance interval is 38. Substituting 38 for n, and 0.90 for p in (8.6.2), we get the actual coverage probability as 0.9047.

Values of n are given in Table 8.1 so that the extreme order statistics form $(p, 1 - \alpha)$ one-sided tolerance limits; also, values of n are provided so that $(X_{(1)}, X_{(n)})$ is a $(p, 1 - \alpha)$ two-sided tolerance interval. As an example, if a sample of 5 observations is available from F_X, then the interval $(X_{(1)}, X_{(5)})$ will include at least 50% of the population with confidence 0.80. If a sample of 459 observations is drawn from a population, then at least 99% of the population exceed $X_{(1)}$ with confidence 0.99.

8.6.2 Sample Size for Tolerance Intervals of the Form $(X_{(r)}, X_{(s)})$

As noted earlier, for a pair of order statistics to form a $(p, 1 - \alpha)$ tolerance interval, the sample size n should satisfy

$$(n - 1)p^n - np^{n-1} + 1 \geq 1 - \alpha. \tag{8.6.4}$$

If n_0 is the least value of n that satisfies the above inequality, then $(X_{(1)}, X_{(n_0)})$ is a $(p, 1 - \alpha)$ tolerance interval. Furthermore, for a given $(p, 1 - \alpha)$ and $n \geq n_0$, there is a pair of order statistics that form a $(p, 1 - \alpha)$ tolerance interval. Thus, our goal is to find the least sample size so that a pair of order statistics form a tolerance interval, and for any n larger than the least sample size, determine the order statistics that form a $(p, 1-\alpha)$ tolerance interval with coverage probability not much in excess of $1 - \alpha$.

Table 8.1: Values of n so that (a) $(X_{(1)}, X_{(n)})$ is a two-sided $(p, 1 - \alpha)$ tolerance interval, (b) $X_{(1)}$ is a $(p, 1 - \alpha)$ one-sided lower tolerance limit; equivalently, $X_{(n)}$ is a $(p, 1 - \alpha)$ one-sided upper tolerance limit

p	Interval type	$1 - \alpha$ 0.80	0.90	0.95	0.99
0.50	one-sided	3	4	5	7
	two-sided	5	7	8	11
0.75	one-sided	6	9	11	17
	two-sided	11	15	18	24
0.80	one-sided	8	11	14	21
	two-sided	14	18	22	31
0.90	one-sided	16	22	29	44
	two-sided	29	38	46	64
0.95	one-sided	32	45	59	90
	two-sided	59	77	93	130
0.99	one-sided	161	230	299	459
	two-sided	299	388	473	662

We computed the least sample size n so that for $m = 1(1)50$, $p = .50, .75, .80,$ $.90, .95, .99$ and $1 - \alpha = .90, .95, .99$, the interval $(X_{(m)}, X_{(n)})$ is a $(p, 1 - \alpha)$ tolerance interval. These sample sizes are presented in Table B15, Appendix B. For a given m, the sample size n is computed as the least value that satisfies

$$P(X \leq n - m - 1 | n, p) \geq 1 - \alpha. \tag{8.6.5}$$

Any pair of order statistics, say $X_{(l)}$ and $X_{(o)}$, then form a $(p, 1 - \alpha)$ tolerance interval provided $l - o = n - m$. Furthermore, $\left(X_{\left(\frac{m}{2}\right)}, X_{\left(n - \frac{m}{2}\right)}\right)$ if m is even, and $\left(X_{\left(\frac{m+1}{2}\right)}, X_{\left(n - \frac{m+1}{2} + 1\right)}\right)$ if m is odd, are $(p, 1 - \alpha)$ tolerance intervals. Obviously the sample size n obtained subject to (8.6.5) depends on m. Notice that for a given $(p, 1 - \alpha)$, if n_{m_1} is the sample size obtained when $m = m_1$ and n_{m_1+1} is the sample size for $m = m_1 + 1$, then $(X_{(m_1)}, X_{(n^*)})$ is a $(p, 1 - \alpha)$ tolerance interval for any n^* satisfying $n_{m_1} \leq n^* < n_{m_1+1}$. As an example, suppose one wants to compute a $(0.90, 0.95)$ tolerance interval based on a sample of 245 observations. The immediate value smaller than 245 under the column $(0.90, 0.95)$ in Table B15, Appendix B, is 239, and the corresponding value of m is 16. Therefore, $(X_{(16)}, X_{(245)})$ and $(X_{(8)}, X_{(237)})$ are $(0.90, 0.95)$ tolerance intervals when $n = 245$. Using (8.4.6), it can be readily verified that the coverage probability of these tolerance intervals is 0.9616, and for any value of m greater than 16, the coverage probability will be less than 0.95. The sample sizes listed under the column $(0.90, 0.95)$ are ranging from 46 to 627, and so order statistics for constructing $(0.90, 0.95)$ tolerance intervals can be obtained for any sample size between them. Thus, Table B15, Appendix B, is preferable to the one that provides order statistics for a given sample size. For example, Somerville (1958) provides table values for a given sample size so that $(X_{(r)}, X_{(s)})$ is a $(p, 1 - \alpha)$ tolerance interval. However, such a table is of limited use, as the order statistics for the sample sizes that are not listed cannot be obtained from the table.

Sample Size for One-Sided Tolerance Limits: Table B15 can also be used to find $(p, 1 - \alpha)$ one-sided tolerance limits. To construct a $(p, 1 - \alpha)$ one-sided lower tolerance limit, we need to find the positive integer k so that

$$P_{X_{(k)}}(P_X(X \geq X_{(k)}|X_{(k)}) \geq p) = 1 - \alpha.$$

As noted earlier (see 8.3.2), this probability requirement simplifies to

$$P(X \leq n - (k - 1) - 1|n, p) \geq 1 - \alpha, \tag{8.6.6}$$

where $X \sim \text{binomial}(n, p)$. It follows from (8.6.5) and (8.6.6) that if m satisfies (8.6.5) then $k - 1 = m$ or $k = m + 1$ satisfies (8.6.6). Thus, it is interesting to note that, if $(X_{(m)}, X_{(n)})$ is a $(p, 1 - \alpha)$ tolerance interval, then $X_{(m+1)}$ is a $(p, 1 - \alpha)$ one-sided lower tolerance limit and $X_{(n-m)}$ is a $(p, 1 - \alpha)$ one-sided upper tolerance limit. As an example, we see from Table B15, Appendix B, that $(X_{(12)}, X_{(86)})$ is a $(0.80, 0.90)$ tolerance interval when $n = 86$, $X_{(13)}$ is a $(0.80, 0.90)$ one-sided lower tolerance limit, and $X_{(74)}$ is a $(0.80, 0.90)$ one-sided upper tolerance limit. So, Table B15, Appendix B, can also be used to determine the order statistics that serve as one-sided tolerance limits.

Example 8.1 (Example 7.1 continued)

We shall construct some nonparametric tolerance intervals for the alkalinity concentration data given in Table 7.1 of Example 7.1. Recall that the measurements represent alkalinity concentrations in ground water obtained from a "greenfield" site. Here we have a sample of size $n = 27$. Let's first compute a 95% upper confidence limit for the population proportion of alkalinity concentrations contained in the interval $(X_{(1)}, X_{(n)})$. Since the left hand side of (8.6.4) is a decreasing function of p, this amounts to finding the maximum value of p that satisfies the inequality (8.6.4). That is, the maximum value of p so that

$$26p^{27} - 27p^{27-1} + 1 \geq 0.95.$$

By trying a few values of p, it can be found that the largest value of p that satisfies the above inequality is 0.836. Note that for the alkalinity data $X_{(1)} = 28$ and $X_{(n)} = 118$. So we can conclude that at most 83.6% of the population alkalinity concentrations are between 28 and 118, with confidence level 0.95.

For this example, there is no pair of order statistics that would form a $(p, 0.95)$ tolerance interval when $p > 0.836$. Let's now compute a (0.75, 0.95) tolerance interval. Examining the column under $(p, 1 - \alpha) = (0.75, 0.95)$ in Table B15, we see that the listed sample size immediately smaller than $n = 27$ is 23, and the value of $m = 2$. So the interval $(X_{(2)}, X_{(n)}) = (32, 118)$ or $(X_{(1)}, X_{(n-1)}) = (28, 96)$ is a (0.75, 0.95) tolerance interval. The one-sided lower tolerance limit is given by $X_{(m+1)} = X_{(3)} = 39$ and the one-sided upper tolerance limit is $X_{(n-m)} = X_{(25)} = 89$.

We shall now construct normal based (0.75, 0.95) tolerance limits as described in Example 7.1. For the cube root transformed data, the mean $\bar{Y} = 3.8274$ and the standard deviation $S_y = 0.4298$. The two-sided tolerance factor from Table B2, Appendix B, is 1.529. Thus, $(3.8274 \pm 1.529 \times 0.4298)^3 = (31.87, 90.18)$. As anticipated, the parametric tolerance interval is narrower than both nonparametric tolerance intervals given in the preceding paragraph. The factor for constructing (0.75, 0.95) one-sided tolerance limits can be found from Table B1, Appendix B, and is 1.083. Thus, $(3.8274 - 1.083 \times 0.4298)^3 = 38$ is the desired one-sided lower tolerance limit, and $(3.8274 + 1.083 \times 0.4298)^3 = 79.11$ is the one-sided upper tolerance limit for the alkalinity concentrations. Observe that the normal based lower tolerance limit of 38 is smaller than the nonparametric limit of 39 (for lower tolerance limit, larger is better), and the normal based upper tolerance limit of 79.11 is smaller than the nonparametric limit of 89 (for upper tolerance limit, the smaller is better).

Probability of Exceeding a Threshold Value: Suppose we want to find a 95% lower limit for the probability that a sample alkalinity concentration exceeds 41 mg/L, that is, $P(X > 41)$. Notice that the smallest order statistic that is larger than 41 is $X_{(6)}$. So it follows from Section 8.3 (see the result that follows (8.3.3)) that

$$P(X > 41) \geq \text{beta}(\alpha; n - k + 1, k) = \text{beta}(0.05; 22, 6) = 0.649.$$

Note that the lower confidence limit for $P(X > 41)$ based on a gamma distribution is 0.692 (see Example 7.1) which is larger than the nonparametric lower confidence limit 0.649.

Remark 8.1. So far we have considered nonparametric tolerance intervals that are computed based on order statistics. Another option is to use a tolerance interval having a given form, and then estimate its content nonparametrically. For example, consider the interval $\bar{X} \pm kS$, where k is the factor determined as if the sample is from a normal distribution. That is, for a given content p and coverage $1 - \alpha$, k is the factor determined by (2.3.4) of Chapter 2. Here $\bar{X}$ and S^2, respectively, denote the mean and variance based on a sample of size n from the relevant population. If the sampled population is not normal, then the true content of such a tolerance interval need not be p. Fernholz and Gillespie (2001) have suggested that the true content be estimated by the bootstrap. If p^* denotes the estimate so obtained, then the interval $\bar{X} \pm kS$ can be used as an approximate $(p^*, 1 - \alpha)$ tolerance interval for the sampled population. Fernholz and Gillespie (2001) refers to this as a content corrected tolerance interval. We refer to their paper for further details.

8.7   Nonparametric Multivariate Tolerance Regions

To describe multivariate nonparametric tolerance regions, we shall first explain the concept of *equivalent blocks*, due to Tukey (1947), for a univariate sample. Let $X_{(1)} < ... < X_{(n)}$ be order statistics for a sample of size n from a continuous univariate distribution F_X. Define

$$B_1 = (-\infty, X_{(1)}], \quad B_j = (X_{(j)}, X_{(j-1)}], \quad j = 2, ..., n \text{ and } B_{n+1} = (X_{(n)}, \infty).$$

Let C_j denote the content level of the jth block. That is,

$$C_1 = F_X(X_{(1)}), \quad C_j = F_X(X_{(j)}) - F_X(X_{(j-1)}), \quad j = 2, ..., n, \quad C_{n+1} = 1 - F_X(X_{(n)}).$$

Note that $F_X(X_{(j)})$ is distributed as $U_{(j)}$, the jth order statistic for a sample of size n from a uniform$(0, 1)$ distribution. Thus it follows from Corollary 8.1 that $E(C_1) = E(C_{n+1}) = \frac{1}{n+1}$. Also, note that C_j is distributed as $U_{(j)} - U_{(j-1)}$,

$j = 2, ..., n$, and its pdf is given by (8.4.4) with $s = j$ and $r = j - 1$. That is,

$$f_{C_j}(x) = \frac{1}{B(1,n)}(1 - x)^{n-1}, \quad 0 < x < 1.$$

Using the above pdf, it can readily verified that $E(C_j) = \frac{1}{n+1}$, $j = 2, ..., n$. Since the expected contents of the blocks are the same, Tukey referred to the non-overlapping intervals $B_1, ..., B_{n+1}$ as equivalent blocks. A tolerance interval is composed of the union of a few consecutive blocks. For example, if the interval is chosen to be the union of the first r blocks, then the content is $\sum_{j=1}^{r} C_j$. If r is chosen so that

$$P\left(\sum_{j=1}^{r} C_j > p\right) = \frac{1}{B(r, n - r + 1)} \int_p^1 x^{r-1}(1 - x)^{n-r}du = 1 - \alpha, \quad (8.6.7)$$

then the union of the first r blocks is a $(p, 1 - \alpha)$ tolerance interval.

The above construction principles can be extended to the multivariate case also. Tukey (1947) has pointed out that the sample space of a continuous q dimensional random vector can be partitioned into $n + 1$ blocks on the basis of n realizations of that random vector. However, there is no unique way of partitioning a multivariate sample space into $n + 1$ equivalent blocks. We shall describe a procedure due to Murphy (1948) for the bivariate case, which is based on the abstract formulation by Tukey (1947).

Let $X_1, ..., X_n$ be a sample from a continuous bivariate distribution. Let $f_1, ..., f_n$ be a set of real valued functions of a bivariate variable x. Let $X_1^{(1)}, ..., X_{(n)}^{(1)}$ be the arrangement of the X_i's such that $f_1(X_{i+1}^{(1)}) > f_1(X_i^{(1)})$, $i = 1, ..., n - 1$. The first block is defined by

$$B_1 = \left\{X : f_1(X) < f_1(X_1^{(1)})\right\}.$$

That is, on the (x, y) plane, the first block is bounded by the curve $f_1(X) - f_1(X_1^{(1)}) = 0$. The observation $X_1^{(1)}$ is discarded, and the second block is determined on the basis of f_2 and the remaining $n - 1$ of the X_i's. The remaining X_i's are arranged in a sequence $X_1^{(2)}, ..., X_{n-1}^{(2)}$ such that $f_2(X_{i+1}^{(2)}) > f_2(X_i^{(2)})$, and the second block is defined as

$$B_2 = \left\{X : f_1(X_1^{(1)}) \leq f_1(X) \text{ and } f_2(X) < f_2(X_1^{(2)})\right\}.$$

In other words, the second block is bounded by the curves $f_1(X) - f_1(X_1^{(1)}) = 0$ and $f_2(X) - f_2(X_1^{(2)}) = 0$. If the process of discarding and rearranging the remaining X_i's is continued until all the functions are used, we shall obtain a set of $n + 1$ non-overlapping blocks (the $(n + 1)$st block at the last step of the process). Furthermore, as in the univariate case, the proportion of the bivariate population that are covered by any r blocks has the beta$(r, n-r+1)$ distribution. Thus, if r is defined subject to (8.6.7), then the union of r consecutive blocks form a $(p, 1 - \alpha)$ tolerance region.

In general, there is no unique way of finding a nonparametric multivariate tolerance region. For example, the functions $f_1, ..., f_n$ in Murphy's approach are arbitrary, and no simple guideline is available to choose these functions. Certainly, different choices lead to different tolerance regions. We refer to Chatterjee and Patra (1976) for an attempt to derive optimum choice of the functions $f_1, ..., f_n$; see also Chatterjee and Patra (1980). Rode and Chinchilli (1988) suggested the Box-Cox transformations to develop tolerance regions for non-normal distributions. Many methods proposed in the literature are not only difficult to apply, but also, a noted by Abt (1982), lead to irregular multiplanar shaped tolerance regions. Abt has also noted that some of the earlier procedures are not scale invariant, and proposed a "parallelogram method" to construct a scale invariant tolerance region.

For more recent theoretical investigations on multivariate non-parametric tolerance regions, we refer to Bucchianico, Einmahl and Mushkudiani (2001) and Li and Liu (2008).

8.8 Exercises

8.8.1. Using integration by parts, show that (8.2.5) is equal to (8.2.4).

8.8.2. Let k be the largest integer so that $P(X \geq k|n, 1 - p) \geq 1 - \alpha$, where X is a binomial$(n, 1 - p)$ random variable. Let m be the smallest integer such that $P(W \leq m - 1|n, p) \geq 1 - \alpha$, where W is a binomial(n, p) random variable.

(a) Show that $m = n - k + 1$.

(b) Let $X_{(r)}$ denote the rth order statistic for a sample $X_1, ..., X_n$ from a continuous population. Show that $X_{(n-k+1)}$ is a $(p, 1 - \alpha)$ upper tolerance limit for the sampled population.

8.8.3. For alkalinity concentration data in Table 7.1 of Example 7.1, compute a 95% confidence interval for $\kappa_{.6}$, where κ_p denotes the p quantile of the alkalinity concentration distribution.

8.8.4. The following data represent sodium contents (in mg) of a sample of 47 5g butter cookies.

10	11	13	13	14	14	14	14	15	15	15	16
16	16	16	17	17	17	17	18	18	18	18	18
19	19	19	19	20	20	20	20	20	21	21	22
22	22	23	23	24	26	26	26	26	28	28	

(a) Compute a 95% upper confidence limit for the population proportion of cookies with sodium contents that falls in the interval $(X_{(1)}, X_{(47)})$.

(b) Find the values of r and s, $r < s$ so that $(X_{(r)}, X_{(s)})$ is a $(0.90, 0.90)$ tolerance interval.

(c) Find a $(0.90, 0.95)$ one-sided tolerance limit for the sodium content.

(d) Construct a 95% confidence interval for median sodium concentration.

(e) Compute a 95% lower confidence limit for the population proportion of cookies with sodium contents 24 mg or more.

8.8.5. The following is a sample of 40 shaft holes diameters (in centimeters):

3.20	3.08	2.77	2.86	2.68	2.93	3.10	2.92
2.60	2.98	3.05	2.80	3.11	3.11	2.87	3.04
3.30	2.93	2.86	2.80	3.13	2.76	3.04	3.19
3.02	3.14	3.39	2.82	3.03	2.92	2.97	2.72
3.07	3.05	2.86	3.22	2.97	3.16	3.16	2.72

(a) Find a 95% confidence interval for the 80th percentile of the population shaft hole diameters.

(b) Find a $(0.80, 0.90)$ tolerance interval for the shaft hole diameters.

(c) Compute a 95% upper confidence limit for the population proportion of measurements that falls below 3.1.

8.8.6. Let $X = (X_1, X_2)'$ be a continuous bivariate random vector, and let $X_1, ..., X_n$ be a sample of realizations of X. Find a $(p, 1 - \alpha)$ tolerance interval for the distribution of $D = X_1 - X_2$ based on the sample.

Chapter 9

The Multivariate Normal Distribution

9.1 Introduction

In this chapter we generalize some of the results for the univariate normal distribution given in Chapter 2, to arrive at tolerance regions for a multivariate normal population. As in the univariate case, a multivariate tolerance region can be used to assess whether a production process that can be described by a set of process variables is under control, or to assess whether the proportion of engineering products in a lot satisfies tolerance specifications. For example, characteristics (such as length, width and diameter) of engineering products are required to be within certain tolerance specifications so that they meet quality standards. Typically, the characteristics are measured on a sample of products drawn from a production lot (i.e., the population), and a multivariate tolerance region is constructed based on the sample. Decision regarding acceptance or rejection of the lot maybe made by comparing the tolerance region with the region determined by the tolerance specifications. If the latter region contains the tolerance region, we may conclude, with a high confidence, that the tolerance specifications are met by most of the products in the lot. To check if a production process is under control, manufacturing plants initially produce items under a controlled or supervised environment (for example, using a group of highly skilled workers). The resulting data can be used to obtain a multivariate tolerance region that can serve as a standard, or a reference region, for future products from other manufacturing facilities. In this context, see also the discussion on reference limits given in Chapter 12.

Some specific applications of tolerance regions for a multivariate normal population are given in Hall and Sheldon (1979) and Fuchs and Kenett (1987, 1988). Hall and Sheldon (1979) have discussed an application dealing with ballistic miss distances. Here, one is interested in estimating the probability that the impact point of a missile, or a shot, falls within a circle of given radius, or in estimating the radius of a circle that will contain a specified proportion of hits. Fuchs and Kenett (1987) have applied multivariate tolerance regions to two examples: one dealing with testing adulteration in citrus juice (where the data is six dimensional) and another dealing with the diagnosis of atopic diseases based on the levels of immunoglobulin in blood (where the data is three dimensional). In the citrus juice example, the six dimensional data consisted of measurements of six attributes of the juice: total soluble solids produced at 20^0C, acidity, total sugars, potassium, formol number and total pectin. The reference sample consisted of 80 specimens of pure juice. Once the tolerance region is computed, we can check if a new specimen to be tested falls in the tolerance region, in order to decide if it is adulterated or not. Fuchs and Kenett (1988) also used multivariate tolerance regions in a quality control application for deciding whether ceramic substrate plates used in the microelectronics industry conform to an accepted standard. The five dimensional data used for this purpose consist of measurements of physical dimensions of individual substrates. A tolerance region for this problem is a region that covers a specified proportion of the target standard population, with a certain confidence. Decision regarding conformity of the new substrates, to the accepted standard, is decided based on whether the corresponding five dimensional observations fall within the tolerance region.

9.2 Notations and Preliminaries

Let $X_1, ..., X_n$ be a sample from a q-variate normal distribution $N_q(\mu, \Sigma)$, having mean vector μ and covariance matrix Σ. The sample mean vector $\bar{X}$, the sample sums of squares and cross-product matrix A, and the sample covariance matrix S, are defined as

$$\bar{X} = \frac{1}{n}\sum_{i=1}^n X_i, \quad A = \sum_{i=1}^n (X_i - \bar{X})(X_i - \bar{X})' \text{ and } S = \frac{1}{n-1}A. \quad (9.2.1)$$

Then $\bar{X}$ and $\frac{1}{n}A$ are the maximum likelihood estimators of μ and Σ, and S is an unbiased estimator of Σ. Furthermore,

$$\bar{X} \sim N_q\left(\mu, \frac{1}{n}\Sigma\right), \quad A \sim W_q(\Sigma, n-1), \text{ and } S \sim W_q\left(\frac{1}{n-1}\Sigma, n-1\right),$$

where $W_q(\Sigma, m)$ denotes a q-dimensional Wishart distribution with scale matrix Σ and df $= m$. Note that $\bar{X}$ and A are independently distributed, and so are

$\bar{X}$ and S. Let X denote a future observation from $N_q(\mu, \Sigma)$, where X is independent of $X_1, X_2, ..., X_n$. A tolerance region for $N_q(\mu, \Sigma)$ is taken to be the region

$$\{X : (X - \bar{X})'S^{-1}(X - \bar{X}) \leq c\}, \qquad (9.2.2)$$

where c is the tolerance factor to be determined so that the region is a $(p, 1-\alpha)$ tolerance region. Specifically, c satisfies

$$P_{\bar{X},S}\left[P_X\left\{(X - \bar{X})'S^{-1}(X - \bar{X}) \leq c \middle| \bar{X}, S\right\} \geq p\right] = 1 - \alpha. \qquad (9.2.3)$$

To begin with, we shall show that the tolerance factor c satisfying (9.2.3) does not depend on any unknown parameters. Let

$$Y = \Sigma^{-\frac{1}{2}}(X - \mu), \quad U = \Sigma^{-\frac{1}{2}}(\bar{X} - \mu) \quad \text{and} \quad V = \Sigma^{-\frac{1}{2}}S\Sigma^{-\frac{1}{2}}. \qquad (9.2.4)$$

Note that Y, U and V are independent with

$$Y \sim N_q(0, I_q), \quad U \sim N_q\left(0, \frac{1}{n}I_q\right) \quad \text{and} \quad V \sim W_q\left(\frac{1}{n-1}I_q, n-1\right). \qquad (9.2.5)$$

In terms of these transformed variables, (9.2.3) can be expressed as

$$P_{U,V}\left[P_Y\left\{(Y - U)'V^{-1}(Y - U) \leq c \middle| U, V\right\} \geq p\right] = 1 - \alpha. \qquad (9.2.6)$$

It is clear from (9.2.5) that the distributions of U, Y and V do not depend on any parameters. Thus the tolerance factor c satisfying (9.2.6) is also free of unknown parameters. An explicit analytic form is not available for computing c satisfying (9.2.3). Thus c has to be determined using Monte Carlo simulation or by a satisfactory approximate method; the representation (9.2.6) will be used for this purpose.

The following expression is useful in finding an approximation to c that satisfies (9.2.6). Let Γ be a $q \times q$ orthogonal matrix such that $\Gamma'V\Gamma = \text{diag}(l_1, ..., l_q)$, where l_i's are the eigenvalues of V. We can assume that $l_1 > ... > l_q > 0$. Let $Z = \Gamma'Y$ and $Q = \Gamma'U$. Using this transformation, we can write (9.2.6) as

$$P_{Q,l}\left[P_Z\left\{\sum_{i=1}^{p} \frac{(Z_i - Q_i)^2}{l_i} \leq c \middle| Q, l\right\} \geq p\right] = 1 - \alpha, \qquad (9.2.7)$$

where $Z = (Z_1, ..., Z_q)'$, $Q = (Q_1, ..., Q_q)'$ and $l = (l_1, l_2, ..., l_q)'$. Since $Y \sim N_q(0, I_q)$ and since Γ is orthogonal, we have $Z = \Gamma'Y \sim N_q(0, I_q)$. That is the Z_i's are independent $N(0, 1)$ random variables. Hence, conditionally given Q, we have $(Z_i - Q_i)^2 \sim \chi_1^2(Q_i^2)$, and are independent for $i = 1, ..., q$, where $\chi_m^2(\delta)$ denotes the noncentral chi-square random variable with df $= m$ and noncentrality parameter δ. The representation (9.2.7) will be exploited for developing approximations for c.

9.3 Some Approximate Tolerance Factors

In the following theorem, we present a general method for finding an approximation to c satisfying (9.2.7). The method is due to John (1963).

Result 9.1 Let $\boldsymbol{l} = (l_1, ..., l_q)'$, $l_1 > \cdots > l_q$, where l_i's are the eigenvalues of the Wishart matrix $\boldsymbol{V}$ given in (9.2.5), and let $\xi(\boldsymbol{l})$ be a real valued function of $\boldsymbol{l}$ such that $l_q < \xi(\boldsymbol{l}) < l_1$. Furthermore, let $\chi_{m;p}^2(\delta)$ denote the p quantile of a noncentral chi-square distribution with df $= m$ and noncentrality parameter δ, and let $\xi_\alpha(\boldsymbol{l})$ denote the α quantile of $\xi(\boldsymbol{l})$. Then, an approximate expression for c that satisfies (9.2.3) or (9.2.7) is given by

$$c = \frac{\chi_{q;p}^2\left(\frac{q}{n}\right)}{\xi_\alpha(\boldsymbol{l})}. \tag{9.3.1}$$

Proof. Since $l_1 > \cdots > l_q > 0$, we see that the inner probability in (9.2.7) satisfies

$$P_{\boldsymbol{Z}}\left[\sum_{i=1}^q \frac{(Z_i - Q_i)^2}{l_q} \leq c \middle| \boldsymbol{Q}, \boldsymbol{l}\right] \leq P_{\boldsymbol{Z}}\left[\sum_{i=1}^q \frac{(Z_i - Q_i)^2}{l_i} \leq c \middle| \boldsymbol{Q}, \boldsymbol{l}\right]$$

$$\leq P_{\boldsymbol{Z}}\left[\sum_{i=1}^q \frac{(Z_i - Q_i)^2}{l_1} \leq c \middle| \boldsymbol{Q}, \boldsymbol{l}\right]. \tag{9.3.2}$$

Since $l_q < \xi(\boldsymbol{l}) < l_1$, we also have

$$P_{\boldsymbol{Z}}\left[\sum_{i=1}^q \frac{(Z_i - Q_i)^2}{l_q} \leq c \middle| \boldsymbol{Q}, \boldsymbol{l}\right] \leq P_{\boldsymbol{Z}}\left[\sum_{i=1}^q \frac{(Z_i - Q_i)^2}{\xi(\boldsymbol{l})} \leq c \middle| \boldsymbol{Q}, \boldsymbol{l}\right]$$

$$\leq P_{\boldsymbol{Z}}\left[\sum_{i=1}^q \frac{(Z_i - Q_i)^2}{l_1} \leq c \middle| \boldsymbol{Q}, \boldsymbol{l}\right]. \tag{9.3.3}$$

Note from (9.2.5) that the Wishart distribution of $\boldsymbol{V}$ has a scale matrix that is a multiple of the identity matrix. Thus for large n, the l_i's are expected to be nearly equal. Hence for large n,

$$P_{\boldsymbol{Z}}\left[\sum_{i=1}^q \frac{(Z_i - Q_i)^2}{l_1} \leq c \middle| \boldsymbol{Q}, \boldsymbol{l}\right] \simeq P_{\boldsymbol{Z}}\left[\sum_{i=1}^q \frac{(Z_i - Q_i)^2}{l_q} \leq c \middle| \boldsymbol{Q}, \boldsymbol{l}\right]. \tag{9.3.4}$$

Consequently, (9.3.3) implies

$$P_{\boldsymbol{Z}}\left[\sum_{i=1}^{q}\frac{(Z_i-Q_i)^2}{l_i}\leq c\bigg|\boldsymbol{Q},\boldsymbol{l}\right]\simeq P_{\boldsymbol{z}}\left[\sum_{i=1}^{q}\frac{(Z_i-Q_i)^2}{\xi(\boldsymbol{l})})\leq c\bigg|\boldsymbol{Q},\boldsymbol{l}\right]. \qquad (9.3.5)$$

Substituting the right side of (9.3.5) in (9.2.7), we conclude that c satisfies

$$P_{\boldsymbol{Q},\boldsymbol{l}}\left[P_{\boldsymbol{Z}}\left\{\sum_{i=1}^{q}(Z_i-Q_i)^2\leq c\xi(\boldsymbol{l})\bigg|\boldsymbol{Q},\boldsymbol{l}\right\}\geq p\right]=1-\alpha, \qquad (9.3.6)$$

approximately. Since $\sum_{i=1}^{q}(Z_i-Q_i)^2\sim\chi_q^2(\boldsymbol{Q}'\boldsymbol{Q})$ for a fixed $\boldsymbol{Q}$, (9.3.6) is equivalent to

$$P_{\boldsymbol{Q},\boldsymbol{l}}\left[c\xi(\boldsymbol{l})\geq\chi_{q;p}^2(\boldsymbol{Q}'\boldsymbol{Q})\right]=1-\alpha. \qquad (9.3.7)$$

Recall that $\boldsymbol{Q}=\boldsymbol{\Gamma}'\boldsymbol{U}$, where $\boldsymbol{\Gamma}$ is an orthogonal matrix, and so $\boldsymbol{Q}'\boldsymbol{Q}=\boldsymbol{U}'\boldsymbol{U}$. It now follows from the distribution of $\boldsymbol{U}$ in (9.2.5) that $\boldsymbol{Q}'\boldsymbol{Q}=\sum_{i=1}^{q}Q_i^2\sim\chi_q^2/n$. Hence $E(\boldsymbol{Q}'\boldsymbol{Q})=\frac{q}{n}$. As a further approximation, we replace $\chi_{q;p}^2(\boldsymbol{Q}'\boldsymbol{Q})$ in (9.3.7) by $\chi_{q;p}^2\left(\frac{q}{n}\right)$; John (1963) justifies this approximation for large n. With this approximation, (9.3.6) becomes

$$P_{\boldsymbol{l}}\left[c\xi(\boldsymbol{l})\geq\chi_{q;p}^2\left(\frac{q}{n}\right)\right]=1-\alpha. \qquad (9.3.8)$$

It is now clear that if $\xi_\alpha(\boldsymbol{l})$ is the α quantile of $\xi(\boldsymbol{l})$, then

$$c\xi_\alpha(\boldsymbol{l})=\chi_{q;p}^2\left(\frac{q}{n}\right)$$

or c is as given in (9.3.1).   □

Note that the value of c in (9.3.1) depends on the $\xi(\boldsymbol{l})$ that we choose. Several choices are discussed below.

As long as we choose the real valued function $\xi(\boldsymbol{l})$ such that $l_q<\xi(\boldsymbol{l})<l_1$, where l_1 and l_q are, respectively, the largest and the smallest eigenvalues of the Wishart matrix $\boldsymbol{V}$ in (9.2.5), Theorem 9.1 can be used to find an approximate tolerance factor. Several approximations are suggested in John (1963); see also Siotani (1964) and Chew (1966). Usual choices of $\xi(\boldsymbol{l})$ are the arithmetic mean, geometric mean and harmonic mean of the eigenvalues $l_1,...,l_q$. For these choices, the distribution of $\xi(\boldsymbol{l})$ is tractable. Krishnamoorthy and Mathew (1999) compared the approximate tolerance factors based on the arithmetic mean, geometric mean, and the harmonic mean. Their extensive simulation studies indicated that none of these approximations is satisfactory for all configurations

of $(n, q, p, 1 - \alpha)$. In particular, the one based on the arithmetic mean is the worst and the one based on the harmonic mean is the best. However, these approximate tolerance factors are crude, and cannot be recommended for practical applications.

We shall now describe a better approximation, due to Krishnamoorthy and Mathew (1999). We begin with an approximation based on the harmonic mean of the l_i's.

Harmonic Mean Approximation

Let $\xi(l)$ be the harmonic mean of the eigenvalues $l_1, ..., l_q$. Then $\xi(l) = \frac{q}{\sum_{i=1}^{q} l_i^{-1}} = \frac{q}{\text{tr}(V^{-1})}$. We shall consider approximating the distribution of $\text{tr}(V^{-1})$ by that of $\frac{a}{\chi_b^2}$, where the constants a and b will be determined by equating the first and second moments of $\text{tr}(V^{-1})$ to the corresponding moments of $\frac{a}{\chi_b^2}$. Towards this, we note that, for $n > q + 4$,

$$E[\text{tr}(V^{-1})] = \frac{q}{n - q - 2},$$

$$E[\{\text{tr}(V^{-1})\}^2] = \frac{q[(n - q - 3)q + 2]}{(n - q - 1)(n - q - 2)(n - q - 4)}. \tag{9.3.9}$$

The expression for $E[\text{tr}(V^{-1})]$ given above can be obtained using the result that $E[V^{-1}] = (n - q - 2)^{-1} I_q$ (e.g., see Anderson 1984, p. 270). The expression for $E[\{\text{tr}(V^{-1})\}^2]$ can be obtained from von Rosen (1988, Corollary 3.1 (v)). Also,

$$E\left(\frac{a}{\chi_b^2}\right) = \frac{a}{(b - 2)},$$

$$E\left[\left(\frac{a}{\chi_b^2}\right)^2\right] = \frac{a^2}{(b - 2)(b - 4)}. \tag{9.3.10}$$

Equating the first two moments in (9.3.9) to those in (9.3.10), and then solving the resulting equations for a and b, we get

$$b = \frac{q(n - q - 1)(n - q - 4) + 4(n - 2)}{n - 2} \quad \text{and} \quad a = \frac{q(b - 2)}{n - q - 2}. \tag{9.3.11}$$

Thus, $\xi_\alpha(l)$ can be approximated by the α quantile of $\frac{q}{a}\chi_b^2$, where a and b are given in (9.3.11). This gives the following approximate tolerance factor, denoted by c_{mhm}, and is given by

$$c_{\text{mhm}} = \frac{a\chi_{q;p}^2\left(\frac{q}{n}\right)}{q\chi_{b;\alpha}^2}. \tag{9.3.12}$$

The above approximation is referred to as a modified harmonic mean approximation in Krishnamoorthy and Mathew (1999), hence the notation "mhm".

An Approximation Based on $V_{11.2}$

Recall that, in order to use Theorem 9.1, $\xi(l)$ should satisfy $l_q \leq \xi(l) \leq l_1$. We shall now consider a function of V, say $\xi(V)$, that satisfies $l_q \leq \xi(V) \leq l_1$, where $\xi(V)$ itself is not a function of l. The choice that we shall make is

$$\xi(V) = V_{11.2}, \tag{9.3.13}$$

where $V_{11.2}^{-1}$ is the first element of V^{-1}. More explicitly, suppose V is partitioned as

$$V = \begin{pmatrix} V_{11} & V_{12} \\ V_{21} & V_{22} \end{pmatrix}, \tag{9.3.14}$$

where V_{11} is a scalar, V_{21} is a $(q-1) \times 1$ vector, $V_{12} = V_{21}'$ and V_{22} is a $(q-1) \times (q-1)$ matrix. Then $V_{11.2} = V_{11} - V_{12} V_{22}^{-1} V_{21}$. Since any diagonal element of a positive definite matrix is less than or equal to the largest eigenvalue, we get $V_{11.2} \leq V_{11} \leq l_1$. Furthermore, since $V_{11.2}^{-1}$ is the first diagonal element of V^{-1}, and $1/l_q$ is the largest eigenvalue of V^{-1}, we also have $V_{11.2}^{-1} \leq 1/l_q$, that is, $V_{11.2} \geq l_q$. In other words, $l_q \leq V_{11.2} \leq l_1$. Since $V_{11.2} \sim \chi^2_{n-q}$ (see Anderson, 1984, Theorem 7.3.6), the choice $\xi(V) = V_{11.2}$ gives the following approximate tolerance factor

$$c_v = \frac{\chi^2_{q;p}\left(\frac{q}{n}\right)}{\chi^2_{n-q;\alpha}}. \tag{9.3.15}$$

An Approximation Based on the Harmonic Mean and $V_{11.2}$

Monte Carlo studies by Krishnamoorthy and Mathew (1999) indicated that, in general, the tolerance factor c_{mhm} in (9.3.12) is somewhat smaller than the actual tolerance factor, whereas c_v in (9.3.15) is somewhat large. This suggests that a tolerance factor between c_{mhm} and c_v may be quite satisfactory. Recall that c_{mhm} is derived based on the choice $\xi(l) = \frac{q}{\text{tr}(V^{-1})}$ and c_v is derived based on $\xi(V) = V_{11.2}$. We shall now consider a choice of $\xi(V)$ between $\frac{q}{\text{tr}(V^{-1})}$ and $V_{11.2}$. The choice that we shall make is the harmonic mean of $\frac{q}{\text{tr}(V^{-1})}$ and $V_{11.2}$. Specifically, let

$$\xi(V) = \frac{2}{\left[\frac{\text{tr}(V^{-1})}{q} + V_{11.2}^{-1}\right]}. \tag{9.3.16}$$

The main reason for choosing the harmonic mean (as opposed to the arithmetic mean or geometric mean) is that we can easily approximate the distribution of $\frac{1}{\xi(V)}$ with that of $\frac{d}{\chi_e^2}$, where d and e are obtained by matching the first two moments. Krishnamoorthy and Mathew (1999) derived the first two moments of $\frac{1}{\xi(V)}$, and these are given by

$$E[(\xi(V))^{-1}] = \frac{1}{n-q-2}, \quad n > q+2,$$

$$E[(\xi(V))^{-2}] = \frac{2q(n-q-1)-3(q-1)}{2q(n-q-1)(n-q-2)(n-q-4)}, \quad n > q+4.$$

The first two moments of $\frac{d}{\chi_e^2}$ are similar to those in (9.3.10). Equating the two sets of moments we get

$$e = \frac{4q(n-q-1)(n-q)-12(q-1)(n-q-2)}{3(n-2)+q(n-q-1)}, \quad d = \frac{e-2}{n-q-2}. \qquad (9.3.17)$$

Thus, the α quantile of $\xi(V)$ in (9.3.16), can be approximated by the α quantile of $\frac{\chi_e^2}{d}$, where d and e are given in (9.3.17). The resulting approximate tolerance factor will be denoted by c_{vhm}, and is given by

$$c_{vhm} = \frac{d\chi_{q;p}^2\left(\frac{q}{n}\right)}{\chi_{e;\alpha}^2}. \qquad (9.3.18)$$

Remark 9.1 When $q = 1$, all the approximate tolerance factors, including the ones based on the usual averages of the eigenvalues of V, coincide with the approximate tolerance factor in (2.3.5). However, when $q \geq 2$, these approximations are quite different.

9.4   Methods Based on Monte Carlo Simulation

Even though the numerical results in Krishnamoorthy and Mathew (1999) show that the approximate factor in (9.3.18) is to be preferred over various other approximations available in the literature, it is not satisfactory for all sample size, content and confidence level configurations. An alternative approach is the Monte Carlo simulation on the basis of the distributional results in (9.2.5), and the representation (9.2.6). The following algorithm can be used to perform the required computations.

Algorithm 9.1

For a given n, q, p and $1 - \alpha$:

1. Generate a $\boldsymbol{U}_i \sim N_q\left(0, \frac{1}{n}\boldsymbol{I}_q\right)$ and a $\boldsymbol{V}_i \sim W_q\left(\frac{1}{n-1}\boldsymbol{I}_q, n-1\right)$, $i = 1, ..., m_1$.

2. For each i,
   generate a $\boldsymbol{Y}_j \sim N_q(0, \boldsymbol{I}_q)$, and compute $(\boldsymbol{Y}_j - \boldsymbol{U}_i)'\boldsymbol{V}_i^{-1}(\boldsymbol{Y}_j - \boldsymbol{U}_i)$, $j = 1, ..., m_2$.

3. Set $T_i = 100p$th percentile of the $(\boldsymbol{Y}_j - \boldsymbol{U}_i)'\boldsymbol{V}_i^{-1}(\boldsymbol{Y}_j - \boldsymbol{U}_i)$, $j = 1, ..., m_2$.

The $100(1 - \alpha)$ percentage point of the T_i's is a Monte Carlo estimate of the tolerance factor c. To generate Wishart matrices in step 1, algorithm by Smith and Hocking (1972) can be used.

The above Monte Carlo method is simple to implement, but it poses some problems while computing the tolerance factor. First, it needs a total of $m_1 \times m_2$ simulation runs. For example, Hall and Sheldon (1979) and Krishnamoorthy and Mathew (1999) used $m_1 = m_2 = 1200$, which amounts to a total of 1,440,000 runs. Even with these many runs, the final estimate of the tolerance factor depends on the initial seed used for the random number generators. In other words, for a given $(n, q, p, 1 - \alpha)$, two different seeds may produce tolerance factors that are appreciably different. This instability is severe when the sample sizes are small relative to the dimension. As a consequence, the coverage probabilities of the tolerance regions based on Monte Carlo estimates of the tolerance factors are sometimes quite different from the nominal level $1 - \alpha$. To overcome these shortcomings, Krishnamoorthy and Mondal (2006) approximated the inner probability in (9.2.7) using a chi-square approximation due to Imhof (1961), and then used simulation to compute the probability with respect to the joint distribution of $(\boldsymbol{U}, \boldsymbol{V})$. Because this approach use simulation with only one "do loop", its accuracy can be increased by increasing the number of runs.

To approximate the inner probability in (9.2.7), that is, the conditional probability that $\sum_{i=1}^q l_i^{-1}(Z_i - Q_i)^2 \le c$, conditionally given Q_i's and l_i's, we need the following lemma, due to Imhof (1961, Section 4).

Lemma 9.1 Let $\chi_{m_1}^2(\delta_1), ..., \chi_{m_k}^2(\delta_k)$ be independent noncentral chi-square random variables, and let $W = \sum_{i=1}^k \lambda_i \chi_{m_i}^2(\delta_i)$, where $\lambda_i > 0$ for $i = 1, ..., k$. Then

$$P(W > w) \simeq P(\chi_a^2 > y),$$

where

$$y = \frac{c_2}{c_3}(w - c_1) + a, \quad a = \frac{c_2^3}{c_3^2}, \quad \text{and } c_j = \sum_{i=1}^k \lambda_i^j(m_i + j\delta_i), \quad j = 1, 2, 3. \quad (9.4.1)$$

Furthermore, for $0 < \alpha < 1$, let $\chi^2_{a;\alpha}$ denote the α quantile of χ^2_a, then the α quantile of W is approximated by

$$w_\alpha = \frac{c_3}{c_2}(\chi^2_{a;\alpha} - a) + c_1. \tag{9.4.2}$$

The above approximation was obtained by matching the moments. In particular, letting $R = \frac{\chi^2_a - a}{\sqrt{2a}}\sqrt{\mathrm{Var}(W)} + E(W)$, we see that the first three central moments of W are equal to those of R when a is as defined in (9.4.1).

The approximation in Lemma 9.1 is remarkably accurate for computing the upper percentiles of W; however, it could be very crude for approximating the lower percentiles of W, as shown in Table 9.1. In particular, we observe from the table that there are discrepancies between the 5th percentiles (as well as the first percentiles) obtained by Monte Carlo simulation and by applying Lemma 9.1, while the 95th percentiles (as well as the 99th percentiles) based on these two methods are in good agreement.

Table 9.1: $100p$th percentiles of $\sum_{i=1}^{k} c_i \chi^2_1(\delta_i)$ based on simulation (denoted by simul.) and based on the approximation in Lemma 9.1 (denoted by approx.)

c_1	c_2	c_3	δ_1	δ_2	δ_3	$p = 0.05$		$p = 0.95$	
						simul.	approx.	simul.	approx.
19.58	24.68	22.00	0.02	2.20	0.15	16.52	14.72	304.6	304.0
13.24	15.73	14.61	0.33	3.12	0.01	15.09	13.72	227.9	227.1
8.32	19.44	37.29	0.19	0.63	3.18	24.42	20.47	488.8	486.6
28.57	11.09	14.31	0.22	0.04	0.55	7.67	10.02	183.6	185.7
24.97	10.23	14.64	3.45	0.03	0.25	18.38	16.04	334.0	339.5
27.46	24.57	7.13	0.17	2.28	1.05	17.81	15.69	313.8	311.5
22.43	16.53	22.77	0.74	0.21	1.82	17.45	15.53	297.3	296.5
						$p = 0.01$		$p = 0.99$	
13.66	9.26	16.85	1.99	0.81	3.51	10.18	5.81	387.1	387.4
15.88	10.03	13.12	1.23	0.66	4.99	12.11	7.59	364.6	364.5
5.46	19.91	15.77	0.60	5.06	2.07	15.08	8.53	502.8	503.0
14.78	16.49	9.28	1.85	1.95	1.53	8.00	4.04	340.2	342.8
11.82	16.28	16.18	0.43	2.52	0.57	5.17	0.59	318.3	320.4
14.33	11.19	27.28	1.64	0.95	4.14	15.71	9.20	593.6	596.8
11.23	13.60	23.55	0.76	0.08	3.59	7.50	1.04	463.1	463.7

To approximate the inner probability in (9.2.7), we recall that, for fixed Q and l, $W = \sum_{i=1}^{q} l_i^{-1}(Z_i - Q_i)^2$ is distributed as $\sum_{i=1}^{q} l_i^{-1}\chi^2_1(Q_i^2)$. Therefore, the results of Lemma 9.1 can be readily applied with $\lambda_i = l_i^{-1}$, $m_i = 1$ and $\delta_i = Q_i^2$,

$i = 1, ..., q$. In this case, we see that

$$c_j = \sum_{i=1}^{q} \frac{1 + jQ_i^2}{l_i^j}, \quad j = 1, 2, 3 \quad \text{and} \quad a = \frac{c_2^3}{c_3^2}. \tag{9.4.3}$$

Thus, it follows from (9.4.2) that

$$\sum_{i=1}^{q} \frac{(Z_i - Q_i)^2}{l_i} \sim \frac{c_3}{c_2}(\chi_a^2 - a) + c_1 \quad \text{approximately.} \tag{9.4.4}$$

Using this approximation, we see that the inner probability inequality in (9.2.7) is equivalent to

$$P\left[\left(\frac{c_3}{c_2}(\chi_a^2 - a) + c_1\right) \leq c \middle| \mathbf{q}, \mathbf{1}\right] \geq p, \tag{9.4.5}$$

where a, c_1, c_2 and c_3 are given in (9.4.3). For fixed $\mathbf{Q}$ and $\mathbf{l}$, the probability inequality in (9.4.5) holds if and only if $c \geq \left(\frac{c_3}{c_2}(\chi_{a,p}^2 - a) + c_1\right)$, where $\chi_{a;p}^2$ denotes the p quantile of the χ_a^2 distribution. Using this relation in (9.2.7), we see that an approximation to the tolerance factor c satisfies

$$P_{\mathbf{Q},\mathbf{l}}\left(\left(\frac{c_3}{c_2}(\chi_{a;p}^2 - a) + c_1\right) \leq c\right) = 1 - \alpha. \tag{9.4.6}$$

Noticing that Q_i's are independent with $Q_i^2 \sim \frac{\chi_1^2}{n}$ and l_i's are the eigenvalues of $\mathbf{V}$, the following algorithm can be used to compute c satisfying (9.4.6).

Algorithm 9.2

For given n, q, p and $1 - \alpha$:

1. Generate $Q_j \sim \frac{\chi_1^2}{n}$, $j = 1, ..., q$ and $\mathbf{V} \sim W_q\left(\frac{1}{n-1}\mathbf{I}_q, n - 1\right)$

2. Compute the eigenvalues $l_1, ..., l_q$ of $\mathbf{V}$

3. Compute c_1, c_2, c_3 and a using (9.4.3)

4. Set $T = \left(\frac{c_3}{c_2}(\chi_{a;p}^2 - a) + c_1\right)$

5. Repeat the steps 1–4 a large number of times, say, 10,000

The $100(1 - \alpha)$th percentile of the simulated T's is an estimate of the tolerance factor c.

Krishnamoorthy and Mondal (2006) used Algorithm 9.2 with $m_1 = 100,000$ runs to compute the tolerance factors. The tolerance factors are presented in Table B16, Appendix B, for $q = 2(1)10$, $p = 0.90, 0.95, 0.99$, $1 - \alpha = 0.90, 0.95, 0.99$

and various sample sizes ranging from $2q + 1$ to 1000. The tolerance factor for omitted sample sizes can be computed using the following interpolation: Suppose that the tolerance factor for the sample size n_2 is not listed in the table, and the tolerance factors f_1 for n_1 and f_3 for n_3 are listed, where $n_1 < n_2 < n_3$. Then, the factor f_2 for n_2 can be interpolated by $(n_3 f_1 + n_1 f_3)/(n_1 + n_3)$ if n_2 is closer to n_1 than it is to n_3; by $(n_1 f_1 + n_3 f_3)/(n_1 + n_3)$ otherwise. As an example, when $q = 2$, $p = 0.90$ and $1 - \alpha = 0.95$, the reported tolerance factor for $n = 37$ is 6.99 (see Table B16, Appendix B, $q = 2$). The interpolated value for $n = 37$ using the tolerance factors for $n = 35$ and $n = 40$ is given by $(40 \times 7.09 + 35 \times 6.84)/(40 + 35) = 6.97$ which is very close to the reported value 6.99.

Example 9.1 (Lumber data)

The data in Table 9.2 represent stiffness (X_1) and bending strength (X_2) for a sample of 30 pieces of a particular grade of lumber. The measurements are in pounds/inch^2. The data are taken from Johnson and Wichern (1992, Table 5.6) who used them for illustrating confidence estimation procedure for the means of the stiffness and the bending strength. To apply our methods of constructing tolerance regions, we should first check the assumption of normality. A simple approach to check the bivariate normality assumption is to check if

$$d_j = (\boldsymbol{X}_j - \bar{\boldsymbol{X}})' \boldsymbol{S}^{-1} (\boldsymbol{X}_j - \bar{\boldsymbol{X}}), \quad j = 1, ..., n.$$

fit a χ^2_{q-1} distribution (see Johnson and Wichern, 1992, Section 4.8). As $q = 2$ for the present example, a χ^2_1 distribution (gamma distribution with shape parameter 1 and scale parameter 2) is fitted for the d_j's, and the plot is given in Figure 9.1. The plot indicates that the assumption of normality is tenable.

The summary statistics are

$$\bar{\boldsymbol{X}} = \begin{pmatrix} \bar{X}_1 \\ \bar{X}_2 \end{pmatrix} = \begin{pmatrix} 1860.47 \\ 8354.13 \end{pmatrix}, \boldsymbol{S} = \begin{pmatrix} 124049.8 & 361673.4 \\ 361673.4 & 3486334.0 \end{pmatrix},$$

and

$$\boldsymbol{S}^{-1} = \begin{pmatrix} 1.155673 \times 10^{-5} & -1.198899 \times 10^{-6} \\ -1.198899 \times 10^{-6} & 4.1120845 \times 10^{-7} \end{pmatrix} = (S^{ij}), \text{say.}$$

The reported tolerance factor corresponding to $(n, q, p, 1-\alpha) = (30, 2, 0.90, 0.95)$ in Table 16, Appendix, is 7.49. Thus, a $(0.90, 0.95)$ tolerance region is the set

$$\{\boldsymbol{X} : (\boldsymbol{X} - \bar{\boldsymbol{X}})' \boldsymbol{S}^{-1} (\boldsymbol{X} - \bar{\boldsymbol{X}}) \le 7.49\} \tag{9.4.7}$$

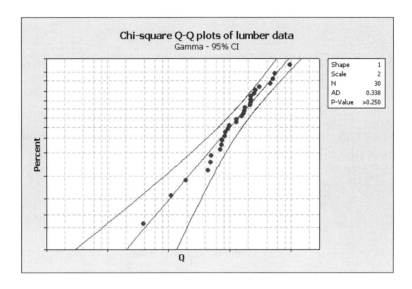

Figure 9.1: Chi-square Q-Q plot for the lumber data

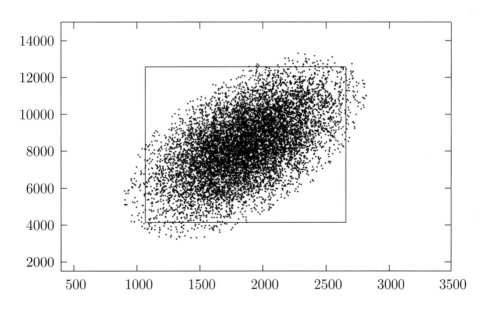

Figure 9.2: The (0.90, 0.95) ellipsoidal tolerance region (the scatter plot) and the rectangular region formed by (0.90, 0.95) Bonferroni simultaneous tolerance intervals, for the lumber data

or

$$\left\{ \boldsymbol{X} : (X_1 - \bar{X}_1)^2 S^{11} + 2S^{12}(X_1 - \bar{X}_1)(X_2 - \bar{X}_2) + S^{22}(X_2 - \bar{X}_2)^2 \leq 7.49 \right\},$$

where $\bar{X}_1$, $\bar{X}_2$ and S^{ij} are given above.

Table 9.2: Data on the stiffness (X_1) and the bending strength (X_2) for a sample of lumber

X_1	X_2	X_1	X_2
1232	4175	1712	7749
1115	6652	1931	6818
2205	7612	1820	9307
1897	10914	1900	6457
1932	10850	2426	10102
1612	7627	1558	7414
1598	6954	1470	7556
1804	8365	1858	7833
1752	9469	1587	8309
2067	6410	2208	9559
2365	10327	1487	6255
1646	7320	2206	10723
1579	8196	2332	5430
1880	9709	2540	12090
1773	10370	2322	10722

Reprinted with permission from Pearson Education, Inc.

The (0.90, 0.95) tolerance region is plotted in Figure 9.2; the region corresponds to the scatter plot in the figure. The region is constructed by generating $\boldsymbol{X}$'s from $N_2(\bar{\boldsymbol{X}}, \boldsymbol{S})$ and then plotting those $\boldsymbol{X}$'s that satisfy the inequality in (9.4.7). The region is an ellipsoidal region with center at $\bar{\boldsymbol{X}}$ and the length of the axes, and the directions, determined by the eigenvalues and the eigenvectors of $\boldsymbol{S}$, respectively. We observe from Figure 9.2 that majority of the $\boldsymbol{X}$'s are clustered around the mean (1860.47, 8354.13), which is the center of the ellipsoid. This means that if a future sample is really from the population from which the reference sample was drawn, then majority of the samples is expected to fall near the center of the tolerance region.

9.5   Simultaneous Tolerance Intervals

We shall now address the problem of computing simultaneous tolerance intervals for the individual variables in a multivariate normal population. A motivation for constructing such simultaneous tolerance intervals is the following. Suppose that a tolerance region is constructed based on a sample taken from a reference population. If a future observation falls outside the tolerance region, then we can conclude that the observation does not belong to the reference population. In this case it is of interest to identify the variable or variables that caused the observation to fall outside the reference population. This can be done by examining simultaneous tolerance intervals for the individual variables. We describe two methods for constructing simultaneous tolerance intervals for the components of a multivariate normal random vector so that each interval includes at least a proportion p of the measurements with overall coverage probability $1 - \alpha$. Specifically, $(p, 1 - \alpha)$ simultaneous tolerance intervals $\left(L_i(\bar{\boldsymbol{X}}, \boldsymbol{S}), U_i(\bar{\boldsymbol{X}}, \boldsymbol{S}) \right)$, $i = 1, ..., q$, are determined so that

$$P_{\bar{\boldsymbol{X}}, \boldsymbol{S}} \left\{ P_{\boldsymbol{X}} \left(L_i(\bar{\boldsymbol{X}}, \boldsymbol{S}) \leq X_i \leq U_i(\bar{\boldsymbol{X}}, \boldsymbol{S}) | \bar{\boldsymbol{X}}, \boldsymbol{S} \right) \geq p, \ i = 1, ..., q \right\} = 1 - \alpha,$$

where $\bar{\boldsymbol{X}}$ and $\boldsymbol{S} = \frac{1}{n-1} \boldsymbol{A}$ are the mean vector and variance-covariance matrix based on a sample of size n from a q-variate normal distribution, and $\boldsymbol{X} = (X_1, ..., X_q)'$ also follows the same normal distribution independently of the sample. A natural choice for (L_i, U_i) is $\bar{X}_i \pm c^* \sqrt{S_{ii}}$, where S_{ii} is the ith diagonal element of $\boldsymbol{S}$, and c^* is the factor to be determined to satisfy

$$P_{\bar{\boldsymbol{X}}, \boldsymbol{S}} \left\{ P_{X_i} \left(\bar{X}_i - c^* \sqrt{S_{ii}} \leq X_i \leq \bar{X}_i + c^* \sqrt{S_{ii}} | \bar{\boldsymbol{X}}, \boldsymbol{S} \right) \geq p, \ i = 1, ..., q \right\} = 1 - \alpha.$$
$$(9.5.1)$$

The factor c^* is referred to as a simultaneous tolerance factor. A choice of c^* on the basis of Scheffé's principle is $c_s^* = \sqrt{c}$, where c is the multivariate tolerance factor satisfying (9.2.6). In general, simultaneous intervals based on Scheffé's method are known to be very conservative. An alternative choice is based on the Bonferroni method (see Result 1.2.3 of Chapter 1), which produces results that are less conservative than those of the Scheffé method. The simultaneous tolerance factor c_b^* on the basis of Bonferroni's approach is the univariate tolerance factor corresponding to content p and the coverage $1 - \frac{\alpha}{q}$.

In order to assess the conservatism of both Scheffé's and Bonferroni's methods, we shall evaluate their coverage probabilities using Monte Carlo evaluation. Let's first compare the tolerance factors based on both approaches for a few values of $(n, q, p, 1 - \alpha)$. Recall that the Bonferroni factor c_b^* is the univariate tolerance factor corresponding to $\left(n, q, p, 1 - \frac{\alpha}{q} \right)$ and Scheffé's factor is the square root of the multivariate tolerance factor corresponding to $(n, q, p, 1 - \alpha)$ in Table B16. For $(n, q, p, 1 - \alpha) = (10, 2, 0.9, 0.9)$, we see that $c_s^* = 3.562$ and $c_b^* = 2.856$; when $(n, q, p, 1 - \alpha) = (20, 3, 0.9, 0.95)$, $c_s^* = 3.572$ and $c_b^* = 2.563$; when $(n, q, p, 1 - \alpha) = (300, 3, 0.9, 0.95)$, $c_s^* = 2.625$ and $c_b^* = 1.804$. In general, it is expected that c_s^* is larger than c_b^* (except when q is very large), and so Scheffé's simultaneous intervals are more conservative than the Bonferroni intervals.

As the Bonferroni procedure is invariant under the transformation $X \rightarrow AX + c$, where A is a diagonal matrix and c is a vector, without loss of generality, we can take $\mu = 0$ and Σ to be a correlation matrix while evaluating its coverage properties. For $n = 15$ and $q = 2$, the coverage probabilities are plotted in Figure 9.3 as a function of the correlation coefficient ρ. We observe from these plots that the Bonferroni procedure seems to be very satisfactory except for large ρ. In particular, the conservatism is less severe when the nominal coverage level is 0.95 or 0.99. The coverage probabilities of the Bonferroni tolerance intervals are also presented in Table 9.3 for several sample sizes no more than 30, $q = 2$ and $\rho = 0.6$. These table values and the plots in figure 9.3 suggest that the Bonferroni method is quite satisfactory even for small n as long as ρ is not too large.

Table 9.3: Bonferroni simultaneous tolerance factors and their coverage probabilities (in parentheses)

$q = 2; \rho = 0.60$

| | $1 - \alpha = 0.90$ | | $1 - \alpha = 0.95$ | | $1 - \alpha = 0.99$ | |
| | p | | p | | p | |
n	0.90	0.95	0.90	0.95	0.90	0.95
5	4.291(.91)	5.077(.91)	5.205(.95)	6.157(.95)	7.978(.99)	9.433(.99)
7	3.390(.91)	4.020(.91)	3.904(.95)	4.628(.95)	5.296(.99)	6.275(.99)
10	2.856(.91)	3.393(.91)	3.175(.95)	3.771(.95)	3.972(.99)	4.715(.99)
13	2.601(.91)	3.093(.91)	2.837(.95)	3.374(.95)	3.405(.99)	4.046(.99)
15	2.492(.91)	2.965(.91)	2.696(.95)	3.207(.95)	3.176(.99)	3.777(.99)
20	2.319(.91)	2.760(.91)	2.474(.95)	2.945(.95)	2.827(.99)	3.364(.99)
30	2.145(.90)	2.555(.90)	2.255(.95)	2.685(.95)	2.497(.99)	2.973(.99)

Example 9.1 (continued)

We compute (0.90, 0.95) simultaneous tolerance intervals for the lumber data in Table 9.2. The required univariate tolerance factor should be computed when

$$\left(n, q, p, 1 - \frac{\alpha}{2}\right) = (30, 2, 0.90, 0.975),$$

and its value is 2.255. The PC calculator *StatCalc* by Krishnamoorthy (2006) can be used to get this factor. Hence for stiffness, we have the tolerance interval $\bar{X}_1 \pm 2.255\sqrt{124049.8} = (1066.24, 2654.69)$, and for the strength it is $\bar{X}_2 \pm 2.255\sqrt{3486334} = (4143.7, 12564.6)$. Thus, at least 90% of stiffness measurements fall in the interval $(1066.24, 2654.69)$ and at least 90% of strength measurements fall in the interval $(4143.7, 12564.6)$ with confidence 95%. In other words,

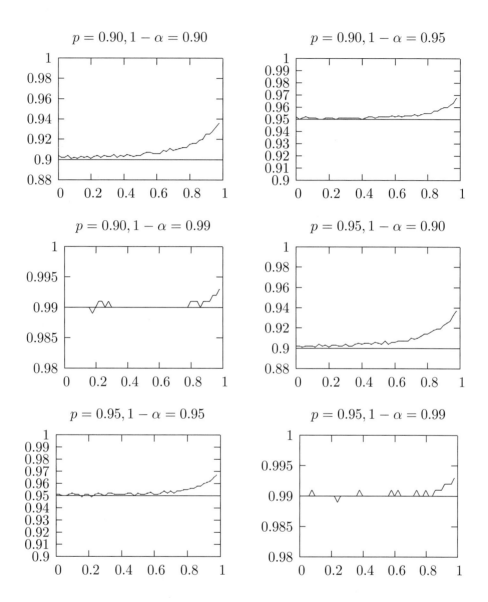

Figure 9.3: Coverage probabilities of $(p, 1 - \alpha)$ Bonferroni tolerance intervals as a function of correlation coefficient ρ; $n = 15$ and $q = 2$

the rectangle formed by $(x_1, y_1) = (1066.2, 4143.7)$, $(x_1, y_2) = (1066.2, 12564.6)$, $(x_2, y_1) = (2654.7, 4143.7)$ and $(x_2, y_2) = (2654.7, 12564.6)$ would include at least 90% of the stiffness and strength measurements; see Figure 9.2 for a visual comparison of the ellipsoidal tolerance region (discussed in the previous section) and the Bonferroni simultaneous tolerance region.

9.6   Tolerance Regions for Some Special Cases

There are applications where one needs to construct tolerance region for a bivariate normal population with known mean vector and unknown common variance. For example, ballistic experts are often interested in estimating the probability that a missile will fall within a circle of given radius (Hall and Sheldon, 1979). If we assume normality, then the known coordinates of the target are the means, and the square root of the common variance is the radius of the circle.

In this section, we shall address the problem of constructing a tolerance region for a $N_2(\mu, \Sigma)$, where $\mu = (\mu_1, \mu_2)'$ is a known mean vector and the covariance matrix $\Sigma = \text{diag}(\sigma_{11}, \sigma_{22})$. That is, Σ is a 2×2 diagonal matrix. In the following, we shall describe the procedures for finding tolerance factors under various assumptions about the parameters. We shall denote by $X_1, ..., X_n$ a sample from $N_2(\mu, \Sigma)$, and write $X_i = (X_{1i}, X_{2i})'$.

The Mean μ Is Known and $\sigma_{11} = \sigma_{22} = \sigma^2$ Is Unknown

Now $X_1, ..., X_n$ is a sample from a $N_2(\mu, \sigma^2 I_2)$ population. Define $S^2 = \frac{1}{2n}\left[\sum_{i=1}^n (X_{1i} - \mu_1)^2 + \sum_{i=1}^n (X_{2i} - \mu_2)^2\right]$, where $X_i = (X_{1i}, X_{2i})'$, $i = 1, ..., n$. As the means are known, S^2 is an unbiased estimator of the common variance σ^2, and $\frac{2nS^2}{\sigma^2} \sim \chi^2_{2n}$. The $(p, 1 - \alpha)$ tolerance factor k_1 is to be determined so that

$$P_{S^2}\left\{P_X\left(\frac{(X_1 - \mu_1)^2 + (X_2 - \mu_2)^2}{S^2} \le k_1 \middle| S^2\right) \ge p\right\} = 1 - \alpha, \qquad (9.6.1)$$

where $X = (X_1, X_2)' \sim N_2(\mu, \sigma^2 I_2)$ independently of the sample. Let $Z_1 = \frac{X_1 - \mu_1}{\sigma}$, $Z_2 = \frac{X_2 - \mu_2}{\sigma}$ and $V = \frac{S^2}{\sigma^2}$. In terms of these variables, (9.6.1) can be written as

$$P_V\left\{P_{Z_1, Z_2}\left(\frac{Z_1^2 + Z_2^2}{V} \le k_1 \middle| V\right) \ge p\right\} = 1 - \alpha. \qquad (9.6.2)$$

Using the fact that $Z_1^2 + Z_2^2 \sim \chi_2^2$ independently of $V \sim \frac{\chi^2_{2n}}{2n}$, it can be easily checked (see Exercise 9.7.2) that the factor k_1 satisfying (9.6.2) is given by

$$k_1 = \frac{-4n \ \ln(1 - p)}{\chi^2_{2n;\alpha}}. \qquad (9.6.3)$$

The Mean μ Is Known and σ_{11} and σ_{22} Are Unknown

Let $S_i^2 = \frac{1}{n} \sum_{i=1}^{n} (X_{ij} - \mu_i)^2$, $i = 1, 2$. The tolerance factor k_2 has to be determined so that

$$P_{S_1^2, S_2^2} \left\{ P_X \left(\frac{(X_1 - \mu_1)^2}{S_1^2} + \frac{(X_2 - \mu_2)^2}{S_2^2} \le k_2 \middle| S_1^2, S_2^2 \right) \ge p \right\} = 1 - \alpha, \quad (9.6.4)$$

where $X \sim N_2(\mu, \operatorname{diag}(\sigma_{11}, \sigma_{22}))$ independently of S_1^2 and S_2^2. Let $Z_i = \frac{X_i - \mu_i}{\sqrt{\sigma_{ii}}}$ and $V_i = \frac{S_i^2}{\sigma_{ii}}$, , $i = 1, 2$. In terms of these transformed variables, we can write (9.6.4) as

$$P_{V_1, V_2} \left\{ P_{Z_1, Z_2} \left(\frac{Z_1^2}{V_1} + \frac{Z_2^2}{V_2} \le k_2 \middle| V_1, V_2 \right) \ge p \right\} = 1 - \alpha. \quad (9.6.5)$$

Notice that Z_1, Z_2, V_1 and V_2 are mutually independent with $Z_i \sim N(0, 1)$ and $V_i \sim \frac{\chi_n^2}{n}$, $i = 1, 2$. As the distributions of these random variables do not depend on any unknown parameters, the factor k_2 satisfying (9.6.5) can be estimated using Monte Carlo simulation similar to the one in Algorithm 9.1. Using this approach, Hall and Sheldon (1979) provided values of k_2 for some values of $(n, p, 1 - \alpha)$. As mentioned earlier, this algorithm involves two nested "do loops" and may not be stable, and so we shall provide an approximate method similar to the one in Algorithm 9.2.

To find an approximation to k_2 satisfying (9.6.5), we first approximate the conditional distribution of $V_1^{-1} Z_1^2 + V_2^{-1} Z_2^2$ given (V_1, V_2). A natural choice is $a\chi_b^2$ distribution, where a and b are to be determined by matching the moments. Specifically, by matching the mean and variance of $V_1^{-1} Z_1^2 + V_2^{-1} Z_2^2$ with those of $a\chi_b^2$, we determine

$$a = \frac{V_1^{-2} + V_2^{-2}}{V_1^{-1} + V_2^{-1}} \quad \text{and} \quad b = \frac{(V_1^{-1} + V_2^{-1})^2}{V_1^{-2} + V_2^{-2}}. \quad (9.6.6)$$

Using this approximate distribution for $\frac{Z_1^2}{V_1} + \frac{Z_2^2}{V_2}$ in (9.6.5), an approximation to k_2 can be obtained as the solution of

$$P_{V_1, V_2} \left\{ P \left(a\chi_b^2 \le k_2 \middle| V_1, V_2 \right) \ge p \right\} = 1 - \alpha,$$

or equivalently,

$$P_{V_1, V_2} \left\{ a\chi_{b;p}^2 \le k_2 \right\} = 1 - \alpha. \quad (9.6.7)$$

Thus k_2 is the $1 - \alpha$ quantile of $a\chi_{b;p}^2$. Notice that $a\chi_{b;p}^2$, where a and b are given in (9.6.6), is a function of (V_1, V_2) whose joint distribution does not depend on any unknown parameters, and so the $1 - \alpha$ quantile of $a\chi^2_{b;p}$ can be estimated using Monte Carlo simulation

The coverage probabilities of $(p, 1-\alpha)$ tolerance regions based on k_2 satisfying (9.6.7), along with the values of k_2, are given in Table 9.4, when $p = 0.90, 0.95$, $1 - \alpha = 0.90, 0.95$ for a few values of n. It is clear from the table values that the estimated coverage probabilities are very close to the nominal level $1 - \alpha$ for all the cases considered. Thus, the approximate tolerance factors are very satisfactory even for sample size as small as 5.

Table 9.4: Approximate tolerance factors satisfying (9.6.7) and their coverage probabilities (in parentheses)

| | $1 - \alpha = 0.90$ | | $1 - \alpha = 0.95$ | | $1 - \alpha = 0.99$ | |
| | p | | p | | p | |
n	0.90	0.95	0.90	0.95	0.90	0.95
5	14.06(.90)	18.83(.90)	18.59(.95)	25.39(.95)	34.39(.99)	49.58(.99)
7	10.77(.91)	14.32(.90)	13.39(.95)	17.76(.95)	21.54(.99)	29.37(.99)
10	8.89 (.91)	11.62(.91)	10.44(.95)	13.84(.95)	14.75(.99)	19.52(.99)
13	7.91 (.90)	10.44(.90)	9.08 (.95)	11.94(.95)	11.96(.99)	16.07(.99)
15	7.50 (.90)	9.92 (.90)	8.49 (.95)	11.17(.95)	10.93(.99)	14.56(.99)
20	6.91 (.91)	9.05 (.90)	7.63 (.95)	10.00(.95)	9.38 (.99)	12.33(.99)

The Mean and Variances Are Unknown

In this case the $(p, 1 - \alpha)$ tolerance factor k_3 should be determined so that

$$P_{S_1^2, S_2^2}\left\{P_{\mathbf{X}}\left(\frac{(X_1 - \bar{X}_1)^2}{S_1^2} + \frac{(X_2 - \bar{X}_2)^2}{S_2^2} \le k_3 \Big| S_1^2, S_2^2\right) \ge p\right\} = 1 - \alpha, \quad (9.6.8)$$

where $\bar{X}_i = \frac{1}{n}\sum_{j=1}^n X_{ij}$, $i = 1, 2$. Setting $Z_i = \frac{X_i - \mu_i}{\sqrt{\sigma_{ii}}}$, $V_i = \frac{(n-1)S_i^2}{\sigma_{ii}}$, and $Q_i = \frac{\bar{X}_i - \mu_i}{\sqrt{\sigma_{ii}}}$, $i = 1, 2$, we see that k_3 is the solution of the equation

$$P_{V_1, V_2}\left\{P_{\mathbf{Z}}\left(\frac{(Z_1 - Q_1)^2}{V_1/(n - 1)} + \frac{(Z_2 - Q_2)^2}{V_2/(n - 1)} \le k_3 \Big| V_1, V_2\right) \ge p\right\} = 1 - \alpha. \quad (9.6.9)$$

We also note that Z_1, Z_2, Q_1, Q_2, V_1 and V_2 are mutually independent with $Z_i \sim N(0, 1)$, $Q_i \sim N(0, 1/n)$ and $V_i \sim \chi^2_{n-1}$, $i = 1, 2$. For a given Q_1 and Q_2, $(Z_1 - Q_1)^2 \sim \chi^2_1(Q_1^2)$ independently of $(Z_2 - Q_2)^2 \sim \chi^2_1(Q_2^2)$. Thus, the approximate approach due to Krishnamoorthy and Mondal (2006) can be readily used to find the factor k_3. Indeed, Algorithm 9.2 with $q = 2$ and l_i replaced by V_i can be used to estimate the tolerance factor k_3.

   Approximate tolerance factors satisfying (9.6.9) are computed using Algorithm 9.2 for some values of $(n, p, 1-\alpha)$, and they are reported in Table 9.5 along with their actual coverage probabilities. As the reported coverage probabilities are very close to the nominal level $1 - \alpha$, the approximate method based on Algorithm 9.2 is very satisfactory for computing the tolerance factor k_3 satisfying (9.6.8).

Table 9.5: Approximate tolerance factors satisfying (9.6.8) and their coverage probabilities (in parentheses)

| | $1 - \alpha = 0.90$ | | $1 - \alpha = 0.95$ | | $1 - \alpha = 0.99$ | |
| | p | | p | | p | |
n	0.90	0.95	0.90	0.95	0.90	0.95
5	20.89(.91)	28.10(.91)	29.72(.95)	40.50(.95)	65.04(.99)	91.12(.99)
7	13.67(.90)	18.31(.90)	17.55(.95)	23.88(.95)	29.81(.99)	41.63(.99)
10	10.25(.90)	13.68(.90)	12.25(.95)	16.33(.95)	17.88(.99)	24.55(.99)
13	8.74 (.90)	11.62(.90)	10.12(.95)	13.53(.95)	13.75(.99)	18.50(.99)
15	8.18 (.90)	10.85(.90)	9.39 (.95)	12.41(.95)	12.23(.99)	16.27(.99)
20	7.32 (.91)	9.66 (.90)	8.16 (.95)	10.75(.95)	10.03(.99)	13.40(.99)

Example 9.2 (Miss distance data)

   Let us consider the miss distance data (in meters) given in Example 1 of Hall and Sheldon (1979). The data are reproduced here in Table 9.6, where X_1 values represent the horizontal miss distances and X_2 values represent the vertical miss distances. Suppose that the target is $\boldsymbol{\mu} = (0,0)'$, $\sigma_{11} = \sigma_{22} = \sigma^2$

Table 9.6: Miss distance data

X_1	3.38	1.62	-5.83	-0.20	1.41	2.30	3.69	0.96	-0.58	-0.63
X_2	2.15	-2.17	0.21	1.94	0.92	-2.60	1.04	-0.79	-2.03	0.20

and the correlation between X_1 and X_2 is zero. Since Hall and Sheldon (1979) assume a common variance, this variance is estimated by

$$S^2 = \frac{\sum_{i=1}^{n} X_{1i}^2 + \sum_{i=1}^{n} X_{2i}^2}{2n} = \frac{70.6268 + 26.6121}{20} = 4.8619.$$

Let us construct a $(0.50, 0.95)$ tolerance region for the miss distances. The tolerance factor $k_1 = \frac{-4n \ln(1-p)}{\chi^2_{2n;\alpha}} = \frac{27.7259}{10.8508} = 2.5552$. The $(0.50, 0.95)$ tolerance region is enclosed by the circle centered at $(0,0)$ with radius $k_1 s = 2.5552 \times \sqrt{4.8619} = 5.6341$. Thus the tolerance region is $\{(X_1, X_2) : X_1^2 + X_2^2 \leq 5.6341\}$ (see (9.6.1)). We conclude that at least 50% of the hits fall in this region with 95% confidence.

For the sake of illustration, let us compute a $(0.50, 0.90)$ tolerance region for the distribution of miss distances when the variances are unknown and arbitrary. Noting that $n = 10$, we used Algorithm 9.2 to compute the tolerance factor k_2 an its value is 2.82 (see Section 9.4). Furthermore, $S_1^2 = \frac{\sum_{i=1}^{n} X_{1i}^2}{n} = 7.0627$ and $S_2^2 = \frac{\sum_{i=1}^{n} X_{2i}^2}{n} = 2.6612$. Thus, the $(0.50, 0.95)$ tolerance region for miss distances is the region enclosed by the ellipse $\frac{X_1^2}{7.0627} + \frac{X_2^2}{2.6612} = 2.82$.

9.7 Exercises

9.7.1. Let $A = (a_{ij})$ be an $m \times m$ positive definite matrix, and let $l_1 > \ldots > l_q$ be the ordered eigenvalues of A. Partition A as $\begin{pmatrix} a_{11} & A_{12} \\ A'_{12} & A_{22} \end{pmatrix}$, so that a is a scalar. Define $a_{11.2} = a_{11} - A_{12} A_{22}^{-1} A'_{12}$. Show that $a_{ii} \leq l_1$ for $i = 1, \ldots, q$, and $a_{11.2} \geq l_q$.

9.7.2. Let X be a continuous random variable, and let A denote the event that $P(X \leq t) \geq p$. The event A occurs if and only if $t \geq x_p$, where x_p is the $100p$ percentage point of X. Using this fact, show that the tolerance factor k_1 that satisfies (9.6.2) is given by (9.6.3).

9.7.3. Let $X = (X_1, X_2, X_3)' \sim N_3(\mu, \Sigma)$. Give a procedure for constructing a tolerance region for the distribution of $D = (X_1 - X_2, X_1 - X_3)'$.

9.7.4. Let $X_{i1}, \ldots, X_{in_i}$ be a sample from $N_p(\mu_i, \Sigma)$, $i = 1, 2$. Assume that both samples are independent. Construct a tolerance region for the distribution of $X_1 - X_2$, where $X_1 \sim N_p(\mu_1, \Sigma)$ independently of $X_2 \sim N_p(\mu_2, \Sigma)$.

9.7.5. Let X be $2q \times 1$ multivariate normal random vector. Partition X as $X = \begin{pmatrix} X_1 \\ X_2 \end{pmatrix}$ so that X_1 and X_2 are of order $q \times 1$. Let $X_1, \ldots, X_n$ be a sample on X. Construct a $(p, 1 - \alpha)$ tolerance region for the distribution of $X_1 - X_2$.

9.7.6. Ballistic experts are often interested in estimating the radius of a circle centered at the origin that will contain 50% of a bivariate normal distribution with equal variance σ^2 and correlation coefficient zero (Hall and Sheldon, 1979). Note that the radius is determined by $P(X^2 + Y^2 \leq r^2) = 0.5$, where (X, Y) has the above bivariate normal distribution.

(a) Show that $r = 1.1774\sigma$.

(b) Find a $1 - \alpha$ confidence interval for r based on a sample $(X_1, Y_1), ..., (X_n, Y_n)$.

Chapter 10

The Multivariate Linear Regression Model

10.1 Preliminaries

A multivariate linear regression model is often used to model the relationship between a vector response variable and one or more explanatory variables or covariates. This chapter is on the computation of tolerance regions in the context of such models. The results in this chapter generalize those in Chapter 3 and Chapter 9. In particular, we shall give multivariate generalizations of all the results in Chapter 3, including the calibration problem. Following the approach in Chapter 9, we shall also discuss several approximations for the tolerance factor. We shall first introduce the model and define the criterion based on which tolerance regions will be computed.

10.1.1 The Model

Let $\boldsymbol{Y}_1, \boldsymbol{Y}_2, \cdots, \boldsymbol{Y}_n$ be n independent $q \times 1$ random vectors having the distribution

$$\boldsymbol{Y}_i \sim N(\boldsymbol{\beta}_0 + \boldsymbol{B}\mathbf{x}_i, \boldsymbol{\Sigma}), \tag{10.1.1}$$

where the $\mathbf{x}_i$'s are known $m \times 1$ vectors, representing values of a vector explanatory variable, $\boldsymbol{\beta}_0$ is a $q \times 1$ intercept vector, $\boldsymbol{B}$ is a $q \times m$ matrix and $\boldsymbol{\Sigma}$ is a $q \times q$ positive definite matrix. Furthermore, $\boldsymbol{\beta}_0$, $\boldsymbol{B}$ and $\boldsymbol{\Sigma}$ are unknown parameters. Write

$$\boldsymbol{Y} = (\boldsymbol{Y}_1, \boldsymbol{Y}_2, \cdots, \boldsymbol{Y}_n), \quad \mathbf{X} = (\mathbf{x}_1, \mathbf{x}_2, \cdots, \mathbf{x}_n). \tag{10.1.2}$$

Then $\boldsymbol{Y}$ and $\boldsymbol{X}$ are $q \times n$ and $m \times n$ matrices, respectively, and

$$\boldsymbol{Y} \sim N(\boldsymbol{\beta}_0 \boldsymbol{1}_n' + \boldsymbol{B}\boldsymbol{X}, \boldsymbol{I}_n \otimes \boldsymbol{\Sigma}), \tag{10.1.3}$$

where $\boldsymbol{I}_n \otimes \boldsymbol{\Sigma}$ is actually the variance covariance matrix of $\text{vec}(\boldsymbol{Y})$, i.e., the vector obtained by writing the columns of $\boldsymbol{Y}$ as an $nq \times 1$ vector. Now let $\boldsymbol{Y}(\mathbf{x})$ denote another observation, independent of $\boldsymbol{Y}$, and having the distribution

$$\boldsymbol{Y}(\mathbf{x}) \sim N(\boldsymbol{\beta}_0 + \boldsymbol{B}\mathbf{x}, \boldsymbol{\Sigma}), \tag{10.1.4}$$

where $\mathbf{x}$ is a known value of the explanatory variable. Our problem is the construction of a tolerance region for $\boldsymbol{Y}(\mathbf{x})$, using the data matrix $\boldsymbol{Y}$ having the distribution in (10.1.3). Throughout, we shall assume that the $(m+1) \times n$ matrix $(\boldsymbol{1}_n, \boldsymbol{X}')'$ has rank $m+1$.

Let $\widehat{\boldsymbol{\beta}}_0$ and $\widehat{\boldsymbol{B}}$ denote the least squares estimators of $\boldsymbol{\beta}_0$ and $\boldsymbol{B}$, under the model (10.1.3). Then

$$\widehat{\boldsymbol{B}} = \boldsymbol{Y}\boldsymbol{P}_n\boldsymbol{X}'[\boldsymbol{X}\boldsymbol{P}_n\boldsymbol{X}']^{-1}, \quad \text{with} \quad \boldsymbol{P}_n = \left(\boldsymbol{I}_n - \frac{1}{n}\boldsymbol{1}_n\boldsymbol{1}_n'\right)$$

$$\widehat{\boldsymbol{\beta}}_0 = \bar{\boldsymbol{Y}} - \widehat{\boldsymbol{B}}\bar{\mathbf{x}}, \tag{10.1.5}$$

where $\bar{\boldsymbol{Y}} = \frac{1}{n}\sum_{i=1}^{n}\boldsymbol{Y}_i$ and $\bar{\mathbf{x}} = \frac{1}{n}\sum_{i=1}^{n}\mathbf{x}_i$. Note that the matrix $\boldsymbol{X}\boldsymbol{P}_n\boldsymbol{X}'$ is nonsingular (in fact, positive definite) in view of our assumption that the matrix $(\boldsymbol{1}_n, \boldsymbol{X}')$ has rank $(m+1)$. Also, $\widehat{\boldsymbol{B}} \sim N[\boldsymbol{B}, \{\boldsymbol{X}\boldsymbol{P}_n\boldsymbol{X}'\}^{-1} \otimes \boldsymbol{\Sigma}]$. Let $\boldsymbol{A}$ denote the residual sum of squares and sum of products matrix computed based on the model (10.1.3). Then

$$\begin{aligned}
\boldsymbol{A} &= [\boldsymbol{Y} - \widehat{\boldsymbol{\beta}}_0\boldsymbol{1}_n' - \widehat{\boldsymbol{B}}\boldsymbol{X}][\boldsymbol{Y} - \widehat{\boldsymbol{\beta}}_0\boldsymbol{1}_n' - \widehat{\boldsymbol{B}}\boldsymbol{X}]' \\
&= \boldsymbol{Y}\left[\boldsymbol{I}_n - \boldsymbol{P}_n\boldsymbol{X}'\{\boldsymbol{X}\boldsymbol{P}_n\boldsymbol{X}'\}^{-1}\boldsymbol{X}\boldsymbol{P}_n\right]\boldsymbol{Y}' \\
&\sim W_q(\boldsymbol{\Sigma}, f_m), \tag{10.1.6}
\end{aligned}$$

where $f_m = n - m - 1$ and $W_q(\boldsymbol{\Sigma}, r)$ denotes the q-dimensional Wishart distribution with df $= r$ and scale matrix $\boldsymbol{\Sigma}$. We also assume that $f_m \geq q$, so that $\boldsymbol{A}$ is positive definite with probability one. Note that $\widehat{\boldsymbol{\Sigma}} = \frac{1}{f_m}\boldsymbol{A}$ is an unbiased estimator of $\boldsymbol{\Sigma}$. For $\boldsymbol{Y}(\mathbf{x})$ having the distribution in (10.1.4), the tolerance region that we shall construct is of the form

$$\begin{aligned}
&\left\{\boldsymbol{Y}(\mathbf{x}) : [\boldsymbol{Y}(\mathbf{x}) - \widehat{\boldsymbol{\beta}}_0 - \widehat{\boldsymbol{B}}\mathbf{x}]'\widehat{\boldsymbol{\Sigma}}^{-1}[\boldsymbol{Y}(\mathbf{x}) - \widehat{\boldsymbol{\beta}}_0 - \widehat{\boldsymbol{B}}\mathbf{x}] \leq k(\mathbf{x})\right\} \\
&= \left\{\boldsymbol{Y}(\mathbf{x}) : (f_m)[\boldsymbol{Y}(\mathbf{x}) - \bar{\boldsymbol{Y}} - \widehat{\boldsymbol{B}}(\mathbf{x} - \bar{\mathbf{x}})]' \right. \\
&\qquad \left. \boldsymbol{A}^{-1}[\boldsymbol{Y}(\mathbf{x}) - \bar{\boldsymbol{Y}} - \widehat{\boldsymbol{B}}(\mathbf{x} - \bar{\mathbf{x}})] \leq k(\mathbf{x})\right\}, \tag{10.1.7}
\end{aligned}$$

where $k(\mathbf{x})$ is the tolerance factor to be determined. In order to define the criterion to be satisfied by tolerance regions, let

$$C_{\bar{Y},\widehat{B},A}(\mathbf{x}) = P_{Y(\mathbf{x})}\Big\{(f_m)[Y(\mathbf{x}) - \bar{Y} - \widehat{B}(\mathbf{x} - \bar{\mathbf{x}})]'$$

$$A^{-1}[Y(\mathbf{x}) - \bar{Y} - \widehat{B}(\mathbf{x} - \bar{\mathbf{x}})] \le k(\mathbf{x})|\bar{Y},\widehat{B}, A\Big\}(10.1.8)$$

Note that $C_{\bar{Y},\widehat{B},A}(\mathbf{x})$ is the conditional content of the region (10.1.7), conditionally given $\bar{Y}$, $\widehat{B}$ and A. The region (10.1.7) is a $(p, 1 - \alpha)$ tolerance region if $k(\mathbf{x})$ satisfies

$$P_{\bar{Y},\widehat{B},A}[C_{\bar{Y},\widehat{B},A}(\mathbf{x}) \ge p] = 1 - \alpha. \tag{10.1.9}$$

In this chapter, we shall develop some approximations for the tolerance factor, i.e., approximations for $k(\mathbf{x})$ satisfying (10.1.9), similar to the approximations developed in Chapter 9. Conclusions regarding the adequacy of the approximations are also reported, based on numerical results. Instead of using approximations, the tolerance factor can also be estimated by Monte Carlo simulation. The necessary framework for this Monte Carlo estimation is also developed in this chapter. We would like to emphasize here that we consider the derivation of tolerance factors only, i.e., the derivation of $k(\mathbf{x})$ for a fixed $\mathbf{x}$. The computation of simultaneous tolerance factors is not addressed here. Some limited results in this direction are available in Lee (1999).

10.1.2   Some Examples

In order to illustrate the concepts of tolerance regions in the context of a multivariate linear regression model, we shall use an example dealing with blood glucose measurements given in Andrews and Herzberg (1985, p. 211). The problem and the data first appeared in O'Sullivan and Mahan (1966). Here the vector $\mathbf{x} = (x_1, x_2, x_3)'$ represents the fasting glucose measurements of an individual on three occasions and $(Y_1, Y_2, Y_3)'$ are glucose measurements one hour after sugar intake, on the three occasions. The data given in Table 35.2 of Andrews and Herzberg (1985) consist of measurements for 52 women. We shall denote these by $(\mathbf{x}_i, Y_i)$, $i=1,2,...,$ 52. We shall assume that a multivariate linear regression model explains the relationship between Y_i and $\mathbf{x}_i$, and Y_i follows a multivariate normal distribution. That is

$$Y_i \sim N(\boldsymbol{\beta}_0 + B\mathbf{x}_i, \Sigma), \tag{10.1.10}$$

where $\boldsymbol{\beta}_0$, B and Σ are unknown, and the Y_i's are independent. Analysis of this data based on a multivariate linear regression model is also discussed in Rencher

(1995, Problem 10.15). Now suppose $\boldsymbol{Y}(\mathbf{x})$ is another observation, independent of the $\boldsymbol{Y}_i$'s in (10.1.10), corresponding to a known value $\mathbf{x}$ of the fasting blood glucose measurement. Then

$$\boldsymbol{Y}(\mathbf{x}) \sim N(\boldsymbol{\beta}_0 + \boldsymbol{B}\mathbf{x}, \boldsymbol{\Sigma}). \qquad (10.1.11)$$

A tolerance region in this context is a region that is expected to contain most of the $\boldsymbol{Y}(\mathbf{x})$-distribution, i.e., most of the glucose measurements one hour after sugar intake, for those individuals having a fixed vector $\mathbf{x}$ of fasting blood glucose measurements. Later in this chapter, we will consider this example once again.

Tolerance regions in the context of a multivariate growth curve model have been investigated by Bowden and Steinhorst (1973). The growth curve was assumed to be a polynomial in time, and the problem was to construct tolerance bands for the population mean growth curves. The model, though similar to (10.1.10), also contained random effects, and Bowden and Steinhorst (1973) also assumed $\boldsymbol{\Sigma}$ to be a multiple of the identity matrix. Their article also contains an example dealing with tolerance bands for the growth curves of fish, where the growth turned out to be linear in time.

10.2   Approximations for the Tolerance Factor

We shall now derive some approximations for $k(\mathbf{x})$ satisfying (10.1.9), using approaches similar to those described in Section 9.3. Along the way, we shall obtain a representation for the tolerance interval condition (10.1.9), which will facilitate the Monte Carlo estimation of the tolerance factor. An interesting and practically useful observation is that $k(\mathbf{x})$ depends on $\mathbf{x}$ only through the scalar d^2 given in (10.2.7) below. Recall a similar observation in the context of the univariate linear regression model; see Section 3.2 of Chapter 3.

We shall first obtain a simplified expression for $P_{\bar{Y},\widehat{B},A}[C_{\bar{Y},\widehat{B},A}(\mathbf{x}) \geq p]$, where $C_{\bar{Y},\widehat{B},A}(\mathbf{x})$ is defined in (10.1.8). Using (10.1.4) and the distributions $\widehat{B} \sim N[B, \{\boldsymbol{X}(\boldsymbol{I}_n - \frac{1}{n}\mathbf{1}_n\mathbf{1}'_n)\boldsymbol{X}'\}^{-1} \otimes \boldsymbol{\Sigma}]$, $\bar{Y} \sim N(\boldsymbol{\beta}_0 + \boldsymbol{B}\bar{\mathbf{x}}, \frac{1}{n}\boldsymbol{\Sigma})$ and $\boldsymbol{A} \sim W_q(\boldsymbol{\Sigma}, f_m)$, we get

$$
\begin{aligned}
\boldsymbol{Z}_1 &= \boldsymbol{\Sigma}^{-\frac{1}{2}}[\boldsymbol{Y}(\mathbf{x}) - \boldsymbol{\beta}_0 - \boldsymbol{B}\mathbf{x}] \sim N(\mathbf{0}, \boldsymbol{I}_q), \\
\boldsymbol{G}_1 &= \boldsymbol{\Sigma}^{-\frac{1}{2}}(\widehat{B} - B) \sim N(0, \{\boldsymbol{X}\boldsymbol{P}_n\boldsymbol{X}'\}^{-1} \otimes \boldsymbol{I}_q), \\
\boldsymbol{U}_1 &= \boldsymbol{\Sigma}^{-\frac{1}{2}}(\bar{Y} - \boldsymbol{\beta}_0 - \boldsymbol{B}\bar{\mathbf{x}}) \sim N\left(\mathbf{0}, \frac{1}{n}\boldsymbol{I}_q\right), \\
\boldsymbol{V} &= \boldsymbol{\Sigma}^{-\frac{1}{2}}\boldsymbol{A}\boldsymbol{\Sigma}^{-\frac{1}{2}} \sim W_q(\boldsymbol{I}_q, f_m), \qquad (10.2.1)
\end{aligned}
$$

where Z_1, G_1, U_1 and V are also independently distributed, and we also recall that $P_n = (I_n - \frac{1}{n}\mathbf{1}_n\mathbf{1}_n')$; see (10.1.5). Hence

$$H_1 = U_1 + G_1(\mathbf{x} - \bar{\mathbf{x}}) \sim N\left(\mathbf{0}, \left(\frac{1}{n} + c^2\right)I_q\right), \tag{10.2.2}$$

where,

$$c^2 = (\mathbf{x} - \bar{\mathbf{x}})'\{XP_nX'\}^{-1}(\mathbf{x} - \bar{\mathbf{x}}). \tag{10.2.3}$$

Straightforward algebra gives

$$[Y(\mathbf{x}) - \bar{Y} - \widehat{B}(\mathbf{x}-\bar{\mathbf{x}})]'A^{-1}[Y(\mathbf{x}) - \bar{Y} - \widehat{B}(\mathbf{x}-\bar{\mathbf{x}})] = (Z_1 - H_1)'V^{-1}(Z_1 - H_1), \tag{10.2.4}$$

where Z_1 and V are defined in (10.2.1) and H_1 is given in (10.2.2). Thus the condition (10.1.9) can be expressed as

$$P_{H_1, V}\left[P_{Z_1}\left\{(f_m)(Z_1 - H_1)'V^{-1}(Z_1 - H_1) \le k(\mathbf{x})\Big| H_1, V\right\} \ge p\right] = 1 - \alpha. \tag{10.2.5}$$

The equation (10.2.5) can be used for the Monte Carlo estimation of $k(\mathbf{x})$; this will be discussed in the next section.

In order to facilitate approximate calculation of the tolerance factor $k(\mathbf{x})$, we shall now provide yet another simplification of the above condition. Let Γ be an orthogonal matrix such that

$$\Gamma'V\Gamma = L_0 = \mathrm{diag}(l_1, l_2, \cdots, l_p), \tag{10.2.6}$$

where $l_1 > l_2 > \cdots > l_p$ denote the ordered eigenvalues of V. Clearly,

$$\begin{aligned} Z &= \Gamma'Z_1 \sim N(\mathbf{0}, I_q), \\ H &= \Gamma'H_1 \sim N(\mathbf{0}, d^2 I_q), \quad \text{where} \quad d^2 = \frac{1}{n} + c^2, \end{aligned} \tag{10.2.7}$$

and, from (10.2.4),

$$\begin{aligned} [Y(\mathbf{x}) - \bar{Y} - \widehat{B}(\mathbf{x} - \bar{\mathbf{x}})]' \quad A^{-1} \quad &[Y(\mathbf{x}) - \bar{Y} - \widehat{B}(\mathbf{x} - \bar{\mathbf{x}})] \\ &= (Z - H)'L_0^{-1}(Z - H) \\ &= \sum_{i=1}^{p} \frac{(Z_i - H_i)^2}{l_i}, \end{aligned} \tag{10.2.8}$$

where $Z = (Z_1, Z_2, \cdots, Z_p)'$ and $H = (H_1, H_2, \cdots, H_p)'$. Note that in the distribution of the statistic in (10.2.8), the vector $\mathbf{x}$ comes into the picture through the scalar quantity d^2 in (10.2.7), since the distribution of H involves

d^2; see (10.2.7). Let $L = (l_1, l_2, \cdots, l_p)'$. Then the tolerance region condition (10.1.9) can be written as

$$P_{H,L}\left[P_Z\left\{(f_m)\sum_{i=1}^{q}\frac{(Z_i - H_i)^2}{l_i} \leq k_1(d)|H,L\right\} \geq p\right] = 1 - \alpha, \qquad (10.2.9)$$

where we have used the notation $k_1(d)$ instead of $k(x)$, and d^2 is defined in (10.2.7). If we were to prepare tables of $k_1(d)$, estimated by Monte Carlo, fairly extensive tables will have to be prepared since $k_1(d)$ depends on several variables: q, f_m, p, $1 - \alpha$ and d. Hence it is highly desirable to have good approximations for $k_1(d)$. Obviously, the purpose of the approximation is to come up with an analytic expression for $k_1(d)$, so that $k_1(d)$ can be easily computed. Let $\xi(V)$ be a scalar valued function of V satisfying

$$l_p < \xi(V) < l_1. \qquad (10.2.10)$$

Some choices for $\xi(V)$ will be discussed later. As an approximation, we shall replace all the l_i's in (10.2.9) by $\xi(V)$. In the sequel, we shall use the notation $\chi_r^2(\eta)$ to denote a noncentral chi-square random variable with df $= r$ and noncentrality parameter η, and χ_r^2 to denote a central chi-square random variable with df $= r$. Using the fact that conditionally given H, $\sum_{i=1}^{q}(Z_i - H_i)^2 \sim \chi_q^2(H'H)$, an approximate version of the condition (10.2.9) is

$$P_{H,\xi(V)}\left[P\left\{(f_m)\frac{1}{\xi(V)}\chi_q^2(H'H) \leq k_1(d)|H,\xi(V)\right\} \geq p\right] = 1 - \alpha. \qquad (10.2.11)$$

Note from (10.2.7) that

$$H'H \sim d^2\chi_q^2, \quad \text{where} \quad d^2 = \frac{1}{n} + c^2, \qquad (10.2.12)$$

as defined in (10.2.7). We shall now approximate the distribution of $\chi_q^2(H'H)$ with a scalar multiple of the product of a central chi-square random variable and a noncentral chi-square random variable. This is accomplished by equating the conditional first moments (conditionally given H) and the unconditional second moments. The result is stated below.

Lemma 10.1. Suppose $H'H$ has the distribution in (10.2.12). Then the distribution of $\chi_q^2(H'H)$ can be approximated by that of $\frac{f}{q+\delta}\chi_e^2\chi_q^2(\delta)$, where

$$
\begin{aligned}
f &= \frac{d^4}{1+d^2}, \quad e = \frac{q(1+d^2)^2}{d^4}, \\
\delta &= d^2\left[\frac{d^2(q+2) + \sqrt{d^4(q+2)^2 + (2d^2+1)q(q+2)}}{2d^2+1}\right]. \qquad (10.2.13)
\end{aligned}
$$

Furthermore, χ_e^2 depends only on H, $\chi_q^2(\delta)$ does not depend on H and χ_e^2 and $\chi_q^2(\delta)$ are independently distributed.

Proof: Approximating the distribution of $\chi_q^2(H'H)$ by that of $a\chi_q^2(\delta)$, we get $a = \frac{q+H'H}{q+\delta}$, by equating the first moments, conditionally given H. Here δ is a non-random quantity to be determined. We thus have the approximation

$$\chi_q^2(H'H) \approx \frac{q + H'H}{q + \delta} \chi_q^2(\delta). \tag{10.2.14}$$

Using the distribution of $H'H$ in (10.2.12) and equating the unconditional second moments of the random variables on both sides of (10.2.14), we get

$$E\left[2(q + 2H'H) + (q + H'H)^2\right] = \frac{E[(q + H'H)^2]}{(q + \delta)^2}[2(q + 2\delta) + (q + \delta)^2]. \tag{10.2.15}$$

Since $E(H'H) = d^2q$ and $E[(H'H)^2] = 2d^4q + (d^2q)^2$, (10.2.15) gives

$$\frac{2q + 4qd^2 + q^2 + 2d^2q^2 + 2d^4q + d^4q^2}{q^2 + 2d^2q^2 + 2d^4q + d^4q^2} = \frac{(q + \delta)^2 + 4(q + \delta) - 2q}{(q + \delta)^2}.$$

Solving, we get

$$\delta = d^2\left[\frac{d^2(q + 2) + \sqrt{d^4(q + 2)^2 + (2d^2 + 1)q(q + 2)}}{2d^2 + 1}\right].$$

Finally, we approximate the distribution of $q + H'H$ by that of $f\chi_e^2$, where f and e are determined by equating the first and second moments. This gives

$$f = \frac{d^4}{1 + d^2} \quad \text{and} \quad e = \frac{q(1 + d^2)^2}{d^4}. \tag{10.2.16}$$

Thus we finally have the approximation $\chi_q^2(H'H) \approx \frac{f}{q+\delta}\chi_e^2\chi_q^2(\delta)$, where f, e and δ are as given above. This completes the proof of the lemma. $\square$

It should be noted that it is not our intention to study the accuracy of the approximation in the above lemma. Our purpose is to use the above approximation towards deriving an approximate tolerance factor. The accuracy of the approximate tolerance factor will then be investigated.

Let

$$W_1 \sim \chi_e^2 \quad \text{and} \quad W_2 \sim \chi_q^2(\delta), \tag{10.2.17}$$

where W_1 and W_2 are also independent. In view of Lemma 10.1, along with the results in (10.2.11) and (10.2.17), we conclude that $k_1(d)$ satisfies

$$P_{W_1,\xi(V)}\left[P_{W_2}\left\{\frac{(f_m)}{\xi(V)}\frac{f}{q+\delta}W_1W_2 \le k_1(d)|W_1,\xi(V)\right\} \ge p\right] = 1-\alpha. \quad (10.2.18)$$

The choices of $\xi(V)$ that we shall make will be such that

$$\xi(V) \sim e_1\chi^2_{e_2}, \quad (10.2.19)$$

at least as an approximation, for some positive constants e_1 and e_2. The constants e_1 and e_2 can be determined by equating the first and second moments. Before we look at specific choices for $\xi(V)$, and the values of e_1 and e_2, we shall show that when (10.2.19) holds, $k_1(d)$ satisfying (10.2.18) can be obtained analytically. Note that the distributions in (10.2.17) and (10.2.19), along with the independence of W_1 and $\xi(V)$ give

$$W_0 = \frac{W_1/e}{\xi(V)/(e_1e_2)} \sim F_{e,e_2}, \quad (10.2.20)$$

the central F-distribution with df $=(e, e_2)$. Using W_0 in (10.2.20), (10.2.18) can be expressed as

$$P_{W_0}\left[P_{W_2}\left\{\frac{(f_m)ef}{e_1e_2(q+\delta)}W_0W_2 \le k_1(d)|W_0\right\} \ge p\right] = 1-\alpha. \quad (10.2.21)$$

$k_1(d)$ satisfying (10.2.21) can be derived explicitly and is given by

$$k_1(d) = \frac{(f_m)ef}{e_1e_2(q+\delta)}\chi^2_{q;p}(\delta)F_{e,e_2;1-\alpha}, \quad (10.2.22)$$

where $F_{r_1,r_2;\alpha}$ denotes the α quantile of a central F-distribution with (η_1, η_2) df, and $\chi^2_{r,\alpha;\eta}$ denotes the α quantile of a noncentral chi-square distribution with r df and noncentrality parameter η. Finally, $k_1(d)$ in (10.2.22) is our approximate tolerance factor.

In order to compute (10.2.22), the quantities required are f, e and δ in (10.2.13), the constants e_1 and e_2 in (10.2.19) and the percentiles $\chi^2_{q;p}(\delta)$ and $F_{e,e_2;1-\alpha}$. Thus we need to make an appropriate choice of $\xi(V)$ so that (10.2.19) will hold at least approximately, and the constants e_1 and e_2 can be obtained. The following choices of $\xi(V)$, to be denoted by $\xi_1(V)$ and $\xi_2(V)$, are recommended in Chapter 9:

$$\xi_1(V) = V_{11.2}$$
$$\xi_2(V) = 2\left[\frac{\text{tr}(V^{-1})}{q} + V_{11.2}^{-1}\right]^{-1}, \quad (10.2.23)$$

where $V_{11.2}^{-1}$ is the first element of V^{-1}. When $\xi_1(V) = V_{11.2}$, we have $e_1 = 1$ and $e_2 = n - m - q$ in (10.2.19), since $V_{11.2} \sim \chi_{n-m-q}^2$; see Anderson (1984, Theorem 7.3.6). The value of $k_1(d)$, say $k_{11}(d)$, resulting from the choice $\xi(V) = V_{11.2}$ is

$$k_{11}(d) = \frac{(f_m)ef}{(n - m - q)(q + \delta)}\chi_{q;p}^2(\delta)F_{e,n-m-q;1-\alpha}. \qquad (10.2.24)$$

Regarding the choice $\xi_2(V)$ in (10.2.23), we shall compute e_1 and e_2 so that (10.2.19) will hold approximately. Following the arguments in Chapter 9, we shall compute e_1 and e_2 in (10.2.19) by equating the first two moments of $\frac{1}{\xi_2(V)}$ to those of $\frac{1}{e_1\chi_{e_2}^2}$. The first two moments of $\frac{1}{\xi_2(V)} = \frac{1}{2}\left[\frac{tr(V^{-1})}{q} + V_{11.2}^{-1}\right]$ as given in Chapter 9 are:

$$E\left(\frac{1}{\xi_2(V)}\right) = \frac{1}{n - m - q - 2}$$
$$E\left(\frac{1}{[\xi_2(V)]^2}\right) = \frac{2q(n - m - q - 1) - 3(q - 1)}{2q(n - m - q - 1)(n - m - q - 2)(n - m - q - 4)}$$
$$(10.2.25)$$

The first two moments of $\frac{1}{e_1\chi_{e_2}^2}$ are $\frac{1}{e_1(e_2-2)}$ and $\frac{1}{e_1^2(e_2-2)(e_2-4)}$. Equating these two quantities to the respective moments in (10.2.25), we get the values of e_1 and e_2, to be denoted by e_{12} and e_{22}:

$$e_{22} = \frac{4q(n - m - q - 1)(n - m - q) - 12(q - 1)(n - m - q - 2)}{3(n - m - 2) + q(n - m - q - 1)},$$
$$e_{12} = \frac{n - m - q - 2}{e_{22} - 2}. \qquad (10.2.26)$$

Thus the value of $k_1(d)$, say $k_{12}(d)$, resulting from the choice $\xi(V) = \xi_2(V)$ in (10.2.23) is given by

$$k_{12}(d) = \frac{(f_m)ef}{e_{12}e_{22}(q + \delta)}\chi_{q;p}^2(\delta)F_{e,e_{22};1-\alpha}, \qquad (10.2.27)$$

where e_{12} and e_{22} are given in (10.2.26).

10.3   Accuracy of the Approximate Tolerance Factors

We compared the approximate tolerance factors $k_{11}(d)$ and $k_{12}(d)$ in (10.2.24) and (10.2.27), respectively, with a Monte Carlo estimate of the actual tolerance factor $k_1(d)$. Recall that $k_1(d)$ satisfies (10.2.9). Equivalently, $k_1(d)$ satisfies

$$P_{\boldsymbol{H}_1,\boldsymbol{V}}\left[P_{\boldsymbol{Z}_1}\left\{(f_m)(\boldsymbol{Z}_1-\boldsymbol{H}_1)'\boldsymbol{V}^{-1}(\boldsymbol{Z}_1-\boldsymbol{H}_1)\le k_1(d)|\boldsymbol{H}_1,\boldsymbol{V}\right\}\ge p\right]=1-\alpha,$$
$$(10.3.1)$$

where

$$\boldsymbol{Z}_1\sim N(\boldsymbol{0},\boldsymbol{I}_q),\quad \boldsymbol{H}_1\sim N(\boldsymbol{0},d^2\boldsymbol{I}_q),\quad\text{and}\quad \boldsymbol{V}\sim W_q(\boldsymbol{I}_q,f_m).\qquad(10.3.2)$$

We also obtained Monte Carlo estimates of the coverage probabilities of the tolerance regions obtained using $k_{11}(d)$ and $k_{12}(d)$. Note that this is simply an estimate of the left hand side of (10.3.1). When these estimates are close to $1-\alpha$, we conclude that the corresponding tolerance factor is accurate. The specific numerical results are not reported here; see Lee (1999) and Lee and Mathew (2004). The conclusions that emerge from the numerical results are as follows.

For a fixed value of f_m, the approximations $k_{11}(d)$ and $k_{12}(d)$ become less satisfactory as the dimension q increases. In this regard, $k_{12}(d)$ appears to be worse. If f_m is not large, $k_{11}(d)$ is a satisfactory approximation, provided that q is not large. In situations where the tolerance regions based on $k_{11}(d)$ is liberal (i.e., the estimated coverage probability is less than the nominal level $1-\alpha$), it is more so when p is large and $1-\alpha$ is small. In fact if p is not too big, $k_{11}(d)$ mostly provides a conservative tolerance region. The approximation $k_{12}(d)$ is unsatisfactory unless f_m is large; it becomes a particularly poor approximation especially when p and the dimension q get large. However, $k_{11}(d)$ is somewhat conservative in this case. When f_m is rather big, both $k_{11}(d)$ and $k_{12}(d)$ turned out to be very satisfactory.

In situations where the approximate tolerance factors are unsatisfactory, the tolerance factor can be estimated by Monte Carlo simulation. The representation (10.3.1) can be conveniently used for this purpose.

10.4 Methods Based on Monte Carlo Simulation

The Monte Carlo simulation explained in Section 9.4 of the previous chapter can be easily adopted for computing $k_1(d)$ satisfying (10.2.5). We shall briefly explain this in this section, as described in Krishnamoorthy and Mondal (2008).

To begin with, an algorithm similar to Algorithm 9.1 can be used for the estimation of $k_1(d)$, without using any further approximations. The random variables to be generated are specified in (10.2.1)–(10.2.3). The algorithm is given below:

Algorithm 10.1

1. For a given n, m, q, p, $1 - \alpha$, $\mathbf{X}$, and for a specified value of the vector $\mathbf{x}$, compute c^2 given in (10.2.3), and $d^2 = \frac{1}{n} + c^2$.

2. Generate $\mathbf{H}_{1i} \sim N_q\left(0, d^2 \mathbf{I}_q\right)$ and $\mathbf{V}_i \sim W_q(\mathbf{I}_q, f_m)$, $i = 1, ..., m_1$.

3. For each i, generate $\mathbf{Z}_{1j} \sim N_q(0, \mathbf{I}_q)$
   compute $Q_j = (f_m)(\mathbf{Z}_{1j} - \mathbf{H}_1)'\mathbf{V}^{-1}(\mathbf{Z}_{1j} - \mathbf{H}_1)$, $j = 1, ..., m_2$.

4. Set $T_i = 100p$th percentile of the Q_j's, $i = 1, ..., m_1$.

The $100(1 - \alpha)$ percentage point of the T_i's is a Monte Carlo estimate of the tolerance factor $k_1(d)$.

Algorithm 9.2 of Chapter 9 can be easily adopted to get an accurate approximation for $k_1(d)$. For this, define

$$c_j = \sum_{i=1}^{q} \frac{1 + jH_i^2}{l_i^j}, \quad j = 1, 2, 3 \quad \text{and} \quad a = \frac{c_2^3}{c_3^2}, \tag{10.4.1}$$

where $d^2 = \frac{1}{n} + c^2$, $H_i^2 \sim d^2 \chi_1^2$ ($i = 1, 2, ..., q$) are independent chi-square random variables with one df each, and the l_i's denote the ordered eigenvalues of $\mathbf{V}$. We now have the following algorithm similar to Algorithm 9.2.

Algorithm 10.2

1. For a given n, m, q, p, $1 - \alpha$, $\mathbf{X}$, and for a specified value of the vector $\mathbf{x}$, compute c^2 given in (10.2.3), and $d^2 = \frac{1}{n} + c^2$.

2. Generate $H_1^2 \sim d^2 \chi_1^2, ..., H_q^2 \sim d^2 \chi_1^2$ and $\mathbf{V} \sim W_q(\mathbf{I}_q, f_m)$.

3. Compute the eigenvalues $l_1, ..., l_q$ of $\mathbf{V}$.

4. Compute c_1, c_2, c_3 and a using (10.4.1).

5. Set $T = (f_m)\left(\sqrt{\frac{c_2}{a}}(\chi_{a,\alpha}^2 - a) + c_1\right)$.

6. Repeat Steps 2–5 m_1 times.

The 100γth percentile of the m_1 simulated T's is a Monte Carlo estimate of the tolerance factor $k_1(d)$.

The numerical results reported in Krishnamoorthy and Mondal (2008) show that the approximate tolerance factor obtained using Algorithm 10.2 practically coincides with the exact ones for the case of simple linear regression model. For $q \geq 2$, their simulation studies showed that the factors based on Algorithm 10.2 are very satisfactory. In the previous section we did identify the situations where the tolerance factors $k_{11}(d)$ and $k_{12}(d)$, having closed form expressions, are accurate enough for practical use. When this is not the case, the computation of the tolerance factor using Algorithm 10.2 appears to be the only practical solution that is easy to compute, and is also accurate.

10.5 Application to the Example

We shall now illustrate the construction of the tolerance region for the example in Section 10.1, dealing with blood glucose measurements. The data on the measurements of blood glucose levels for 52 women are given in Andrews and Herzberg (1985, Table 35.2). Here $\mathbf{x} = (x_1, x_2, x_3)'$ represents fasting glucose measurements on three occasions and $(Y_1, Y_2, Y_3)'$ represents glucose measurements one hour after sugar intake on the three occasions. We take $p = 0.95$, and $1 - \alpha = 0.99$. The quantities needed for constructing the tolerance region are

$$
\widehat{\boldsymbol{B}} = \begin{bmatrix} 0.5933 & -0.1378 & 0.5499 \\ 0.1046 & 0.7688 & 0.3043 \\ 0.5245 & -0.3815 & 0.8257 \end{bmatrix}, \ \bar{\boldsymbol{Y}} = \begin{bmatrix} 109.1346 \\ 104.5769 \\ 109.8462 \end{bmatrix},
$$

$$
\bar{\mathbf{x}} = \begin{bmatrix} 70.1154 \\ 73.4231 \\ 75.1154 \end{bmatrix}, \ \boldsymbol{A} = \begin{bmatrix} 13676.3931 & 12664.9718 & 13139.9057 \\ 12664.9718 & 12317.8247 & 12572.0367 \\ 13139.9057 & 12572.0367 & 13534.1261 \end{bmatrix}.(10.5.1)
$$

Suppose we want to find a tolerance region when $\mathbf{x} = (103, 103, 100)'$. From (10.2.3) and (10.2.7), we get $d^2 = 0.4752$. From (10.2.24) and (10.2.27), we have

$$
k_{11}(d) = \frac{(f_m)ef}{(n - m - q)(q + \delta)} \chi^2_{q;p}(\delta) F_{e,n-m-q;1-\alpha} = 24.3434, \qquad (10.5.2)
$$

and

$$
k_{12}(d) = \frac{(f_m)ef}{e_{12}e_{22}(q + \delta)} \chi^2_{q;p}(\delta) F_{e,e_{22};1-\alpha} = 22.4505. \qquad (10.5.3)
$$

Therefore, a tolerance region for $\boldsymbol{Y}(\mathbf{x})$ is

$$
\{\boldsymbol{Y}(\mathbf{x}) : 48[\boldsymbol{Y}(\mathbf{x}) - \bar{\boldsymbol{Y}} - \widehat{\boldsymbol{B}}(\mathbf{x} - \bar{\mathbf{x}})]' \boldsymbol{A}^{-1}[\boldsymbol{Y}(\mathbf{x}) - \bar{\boldsymbol{Y}} - \widehat{\boldsymbol{B}}(\mathbf{x} - \bar{\mathbf{x}})] \leq k_1(d)\}, \quad (10.5.4)
$$

where $\widehat{B}$, $\bar{Y}$, $\bar{\mathbf{x}}$ and A are given in (10.5.1). In the place of $k_1(d)$, we can use $k_{11}(d)$ or $k_{12}(d)$ given in (10.5.2) and (10.5.3), respectively, when $\mathbf{x} = (103, 103, 100)'$. Our numerical results showed that $k_{12}(d)$ is a more satisfactory tolerance factor in this situation; thus we recommend the tolerance region based on $k_{12}(d)$. Note also that $k_{12}(d)$ is smaller than $k_{11}(d)$ in this example.

10.6 Multivariate Calibration

In Chapter 3 we discussed the computation of *multiple use confidence intervals* for the univariate calibration problem in a linear regression context. We now take up the same problem in the multivariate case. More specifically, we shall address the problem of computing multiple use confidence regions for the multivariate calibration problem. For reviews and discussions on the multivariate calibration problem, see the papers by Brown (1982), Brown and Sundberg (1987) and Sundberg (1994, 1999), and the book by Brown (1993).

10.6.1 Problem Formulation and the Pivot Statistic

In order to formulate the problem, consider the model (10.1.3), and recall that we derived tolerance regions for the distribution of the random variable $Y(\mathbf{x})$ as specified in (10.1.4), where the vector $\mathbf{x}$ consists of known values of the explanatory variables. In the calibration problem, we have an observation $Y(\mathbf{x})$ corresponding to an unknown value $\mathbf{x}$ of the explanatory variable. The problem of interest is the computation of a confidence region for the unknown vector $\mathbf{x}$ using the data matrix Y having the normal distribution specified in (10.1.3), along with the observation $Y(\mathbf{x})$ having the distribution in (10.1.4).

While addressing the univariate calibration problem in Section 3.4 of Chapter 3, a multiple use confidence region was derived by inverting a simultaneous tolerance interval for the observation $Y(x)$. In the same section, it was also noted that a multiple use confidence region can be derived by inverting a tolerance interval, rather than a simultaneous tolerance interval. The justification for this observation came from numerical results only. It is this latter observation that we shall use for the derivation of multiple use confidence regions in the multivariate calibration problem. Once again, this approach will be justified based on numerical results. In this context, we note that satisfactory simultaneous tolerance regions are not available in the multivariate regression set up.

Recall that our models are given by

$$Y \sim N(\beta_0 \mathbf{1}'_n + BX, I_n \otimes \Sigma), \quad Y(\mathbf{x}) \sim N(\beta_0 + B\mathbf{x}, \Sigma), \tag{10.6.1}$$

where the above quantities are as defined in Section 10.1.1, except that $Y(\mathbf{x})$ has been observed corresponding to an unknown $\mathbf{x}$. In the calibration problem, the data matrix Y is referred to as the calibration data. The dimensions of Y, $Y(\mathbf{x})$ and X are $q \times n$, $q \times 1$ and $m \times n$, respectively, and those of the parameters β_0, B, Σ and $\mathbf{x}$ are $q \times 1$, $q \times m$, $q \times q$ and $m \times 1$, respectively. In order that $\mathbf{x}$ be identifiable, we shall also assume that $q \geq m$. In the context of multiple use confidence regions, it is understood that there will be a sequence of independent $Y(\mathbf{x})$ values corresponding to a sequence of possibly different $\mathbf{x}$ values having the distribution specified in (10.6.1); see also Section 3.4.

Consider the least squares estimators $\widehat{B}$ and $\widehat{\beta}_0$ given in (10.1.5) and the Wishart matrix A given in (10.1.6). The pivot statistic that we shall use for constructing a confidence region for $\mathbf{x}$ is motivated by the following observation. From the model for $Y(\mathbf{x})$ given in (10.6.1), it follows that if β_0, B and Σ are known, the weighted least squares estimator of $\mathbf{x}$ is $\tilde{\mathbf{x}} = (B'\Sigma^{-1}B)^{-1}B'\Sigma^{-1}(Y(\mathbf{x}) - \beta_0)$ with covariance matrix $(B'\Sigma^{-1}B)^{-1}$. Hence the quantity $(\tilde{\mathbf{x}} - \mathbf{x})'B'\Sigma^{-1}B(\tilde{\mathbf{x}} - \mathbf{x})$ is a natural pivot for constructing a confidence region for $\mathbf{x}$. Since β_0, B and Σ are unknown, we shall replace them by the corresponding estimators, namely $\widehat{\beta}_0$, $\widehat{B}$ and $\frac{1}{f_m}A$. Thus, let $\widehat{\mathbf{x}} = (\widehat{B}'A^{-1}\widehat{B})^{-1}\widehat{B}'A^{-1}(Y(\mathbf{x}) - \widehat{\beta}_0)$, and a pivot for constructing a confidence region for the m−dimensional vector $\mathbf{x}$ is

$$
\begin{aligned}
T(\mathbf{x}) &= \frac{n-m-q}{m}(\widehat{\mathbf{x}} - \mathbf{x})'\widehat{B}'A^{-1}\widehat{B}(\widehat{\mathbf{x}} - \mathbf{x}) \\
&= \frac{n-m-q}{m}[Y(\mathbf{x}) - \bar{Y} - \widehat{B}(\mathbf{x} - \bar{\mathbf{x}})]'A^{-1}\widehat{B}(\widehat{B}'A^{-1}\widehat{B})^{-1}\widehat{B}'A^{-1} \times \\
&\quad [Y(\mathbf{x}) - \bar{Y} - \widehat{B}(\mathbf{x} - \bar{\mathbf{x}})]. \tag{10.6.2}
\end{aligned}
$$

We note that if $q = m$, then the above pivot simplifies to

$$T(\mathbf{x}) = \frac{n-2m}{m}[Y(\mathbf{x}) - \bar{Y} - \widehat{B}(\mathbf{x} - \bar{\mathbf{x}})]'A^{-1}[Y(\mathbf{x}) - \bar{Y} - \widehat{B}(\mathbf{x} - \bar{\mathbf{x}})]. \tag{10.6.3}$$

The confidence region for $\mathbf{x}$ that we shall construct is given by

$$\{\mathbf{x} : T(\mathbf{x}) \leq k_2(\mathbf{x})\}, \tag{10.6.4}$$

where $k_2(\mathbf{x})$, a function of $\mathbf{x}$, is to be determined subject to appropriate coverage probability requirements.

10.6.2   The Confidence Region

We shall now explain the coverage probability condition to be satisfied by multiple use confidence regions. The condition is similar to the corresponding condition in Chapter 3, Section 3.4; thus we shall not elaborate upon the interpretations of the condition. Let

$$C_2(\mathbf{x}; \bar{\boldsymbol{Y}}, \widehat{\boldsymbol{B}}, \boldsymbol{A}) = P_{\boldsymbol{Y}(\mathbf{x})} \left\{ T(\mathbf{x}) \le k_2(\mathbf{x}) \middle| \bar{\boldsymbol{Y}}, \widehat{\boldsymbol{B}}, \boldsymbol{A} \right\}. \tag{10.6.5}$$

Let $\{\mathbf{x}_j\}$, $j = 1, 2, ..., M$, denote a sequence of M values of $\mathbf{x}$, and let $\{Y(\mathbf{x}_j)\}$, $j = 1, 2, ..., M$, denote the corresponding sequence of M independent $\boldsymbol{Y}(\mathbf{x})$ values. The function $k_2(\mathbf{x})$ is to be determined subject to the following condition:

$$P_{\bar{\boldsymbol{Y}}, \widehat{\boldsymbol{B}}, \boldsymbol{A}} \left[\frac{1}{M} \sum_{j=1}^{M} C_2(\mathbf{x}_j; \bar{\boldsymbol{Y}}, \widehat{\boldsymbol{B}}, \boldsymbol{A}) \ge p \right] \ge 1 - \alpha, \tag{10.6.6}$$

for every sequence $\{\mathbf{x}_j\}$, $j = 1, 2, ..., M$, and for every positive integer M. Obviously, a sufficient condition for (10.6.6) to hold is that

$$P_{\bar{\boldsymbol{Y}}, \widehat{\boldsymbol{B}}, \boldsymbol{A}} \left[\min_{\mathbf{x}} C_2(\mathbf{x}; \bar{\boldsymbol{Y}}, \widehat{\boldsymbol{B}}, \boldsymbol{A}) \ge p \right] \ge 1 - \alpha, \tag{10.6.7}$$

where the minimum in (10.6.7) is to be computed subject to available bounds on $\mathbf{x}$.

The conditions given in (10.6.6) and (10.6.7) are both difficult to work with, and solutions based on them are currently not available. Thus we shall now exhibit a solution based on a tolerance region condition. This approach is pursued in Mathew, Sharma and Nordström (1998), who also report numerical results to verify if the solution so obtained satisfies the condition (10.6.6). The tolerance region condition that we shall work with is given by

$$P_{\bar{\boldsymbol{Y}}, \widehat{\boldsymbol{B}}, \boldsymbol{A}} \left[C_2(\mathbf{x}; \bar{\boldsymbol{Y}}, \widehat{\boldsymbol{B}}, \boldsymbol{A}) \ge p \right] \ge 1 - \alpha. \tag{10.6.8}$$

It turns out that the left hand side of (10.6.8) depends on $\boldsymbol{B}$. Consequently, we shall obtain a conservative solution. That is, we shall derive $k_2(\mathbf{x})$ so that for the corresponding $C_2(\mathbf{x}; \bar{\boldsymbol{Y}}, \widehat{\boldsymbol{B}}, \boldsymbol{A})$ defined in (10.6.5), the left hand side of (10.6.8) will be greater than or equal to $1 - \alpha$. We shall now give the theoretical result that shows the existence of a $k_2(\mathbf{x})$ that provides such a conservative solution. The computation of $k_2(\mathbf{x})$ will then be explained. Actually the $k_2(\mathbf{x})$ that we compute turns out to be a function of c^2 defined in (10.2.3). Thus from now on, we shall use the notation $k_2(c^2)$ instead of $k_2(\mathbf{x})$. We also recall the identifiability assumption $q \ge m$.

Lemma 10.2.   Let $W_1 \sim W_m(I_m, n - q - 1)$, $W_2 \sim W_{q-m}(I_{q-m}, f_m)$, $W \sim N(0, I_m \otimes I_{q-m})$, and $W_0 \sim \chi_q^2$, where W_1, W_2, W and W_0 are also independently distributed. For $q > m$, let $\boldsymbol{\lambda} = (\lambda_1, \lambda_2, ..., \lambda_m)'$ denote the eigenvalues of $(I + WW_2^{-1}W')W_1^{-1}$, and for $q = m$, let $\boldsymbol{\lambda} = (\lambda_1, \lambda_2, ..., \lambda_m)'$ denote the eigenvalues of W_1^{-1}, where it is assumed that $\lambda_1 \geq \lambda_2 \geq .... \geq \lambda_m$. Let $k_2(c^2)$ satisfy

$$
P_{\boldsymbol{\lambda}, W_0} \left[P \left(\frac{n - m - q}{m} \left\{ \lambda_1 \chi_{1,1}^2 \left((\frac{1}{n} + c^2)W_0 \right) + \sum_{i=2}^{m} \lambda_i \chi_{1,i}^2 \right\} \right. \right.
$$

$$
\leq \left. \left. k_2(c^2) \middle| \boldsymbol{\lambda}, W_0 \right) \geq p \right] = 1 - \alpha, \qquad (10.6.9)
$$

where $\chi_{1,1}^2 \left((\frac{1}{n} + c^2)W_0 \right)$ denotes a non-central chi-square random variable with 1 df and non-centrality parameter $((\frac{1}{n} + c^2)W_0)$, $\chi_{1,i}^2$, $i = 2, 3, ..., m$, denote central chi-square random variables each having 1 df, and $\chi_{1,1}^2 \left((\frac{1}{n} + c^2)W_0 \right)$ and $\chi_{1,i}^2$ are all independently distributed. Then

$$
P_{\bar{Y}, \widehat{B}, A} \left[P_Y \left\{ T(\mathbf{x}) \leq k_2(c^2) \middle| \bar{Y}, \widehat{B}, A \right\} \geq p \right] \geq 1 - \alpha. \qquad (10.6.10)
$$

In the statement of Lemma 10.2, we note that the non-central chi-square random variable $\chi_{1,1}^2 \left((n^{-1} + c^2)W_0 \right)$ has a non-centrality parameter that is also random, since $W_0 \sim \chi_q^2$. Also, recall that c^2 is a function of $\mathbf{x}$, the quantity for which a confidence region is required; see (10.2.3). The proof of Lemma 10.2 is given in Mathew, Sharma and Nordström (1998). The proof is technically involved, and is not included here.

10.6.3   Computation of the Confidence Region

In order to implement the above confidence region for $\mathbf{x}$ the major practical problem is the computation of $k_2(c^2)$ satisfying (10.6.9). Note that the functional form of $k_2(c^2)$ is required. We shall now explain the numerical computation of $k_2(c^2)$.

Note that in practical applications of calibration where the models (10.6.1) are applicable, the parameter of interest, namely $\mathbf{x}$, represents a physical quantity and a parameter space for $\mathbf{x}$ will be known. In particular, an upper bound, say δ, will be available on c^2. Thus we shall assume

$$
0 \leq c^2 \leq \delta. \qquad (10.6.11)
$$

Our numerical computation of $k(c^2)$ will be for c^2 satisfying (10.6.11). It can be shown that $k_2(c^2)$ is an increasing function of c^2 (this follows from (10.6.9), along with the property that a noncentral chi-square random variable is stochastically increasing in the noncentrality parameter). Our approach for numerically obtaining the functional form of $k_2(c^2)$ is as follows. Numerically evaluate the value of $k_2(c^2)$ satisfying (10.6.9) for a few specified values of c^2 satisfying (10.6.11) and try to fit a suitable function to the values of $k_2(c^2)$ so obtained. The fitted function will approximately give the functional form of $k_2(c^2)$, which can be used to compute the region (10.6.10) for $\mathbf{x}$.

Thus, our problem is the numerical evaluation of $k_2(c^2)$ satisfying (10.6.9) for various values of c^2 subject to (10.6.11), for specified values of p and $1 - \alpha$. For any given value of c^2, say c_1^2, we will need a starting value, say $k_{2*}(c_1^2)$, for evaluating $k_2(c_1^2)$. The starting value $k_{2*}(c_1^2)$ that we shall use will be such that $k_{2*}(c_1^2) < k_2(c_1^2)$ and its choice for any $c_1^2 \geq 0$ will be exhibited shortly. Once $k_{2*}(c_1^2)$ is chosen for a specified value of c_1^2, such that $k_{2*}(c_1^2) < k_2(c_1^2)$, we shall increase the value of $k_{2*}(c_1^2)$ in steps, each time numerically evaluating the left hand side (lhs) of (10.6.9), until the lhs of (10.6.9) is equal to $1 - \alpha$ up to a desired level of accuracy. The lhs of (10.6.9) can be evaluated for a given value of c_1^2 and $k_{2*}(c_1^2)$ in the following manner. Generate one set of values of the Wishart matrices $\mathbf{W}_1$ and $\mathbf{W}_2$, the normal matrix $\mathbf{W}$ and the chi-square random variable W_0 specified in Lemma 10.2, and compute the ordered eigenvalues $\lambda_1 > \lambda_2 > \ldots > \lambda_m$ of the matrix $(\mathbf{I}_m + \mathbf{W}\mathbf{W}_2^{-1}\mathbf{W}')\mathbf{W}_1^{-1}$. (If $q = m$, we need to generate only the Wishart matrix $\mathbf{W}_1$ and the λ_i's are the ordered eigenvalues of $\mathbf{W}_1^{-1}$). For the value of $\boldsymbol{\lambda} = (\lambda_1, \ldots, \lambda_m)'$ and W_0 so obtained, let $i(\boldsymbol{\lambda}, W_0)$ be an indicator function that takes the value one if

$$P\left[\frac{n-m-q}{m}\left\{\lambda_1\chi_{11}^2\left((n^{-1}+c^2)W_0\right) + \sum_{i=2}^{m}\lambda_i\chi_{i1}^2\right\} \leq k_{2*}(c^2)\right] \geq p.$$

(10.6.12)

If (10.6.12) does not hold, assign the value zero to $i(\boldsymbol{\lambda}, W_0)$. Note that the computation of $i(\boldsymbol{\lambda}, W_0)$ requires the computation of the lhs of (10.6.12), which is a probability involving linear combinations of independent chi-square random variables. For any specified value of $\boldsymbol{\lambda}$ and W_0, the lhs of (10.6.12) can be evaluated using simulation. The values of $\boldsymbol{\lambda}$ and W_0 can be generated a large number of times, and the value of $i(\boldsymbol{\lambda}, W_0)$ can be evaluated each time. The lhs of (10.6.9) is the proportion of times $i(\boldsymbol{\lambda}, W_0)$ takes the value one. If this proportion is less than $1 - \alpha$, then $k_{2*}(c_1^2) < k_2(c_1^2)$ and the value of $k_{2*}(c_1^2)$ has to be increased in order to get a better approximation of $k_2(c_1^2)$. As already pointed out, we increase the value of $k_{2*}(c_1^2)$ in steps, each time computing the proportion of times $i(\boldsymbol{\lambda}, W_0)$ takes the value one, until this proportion is approximately equal to $1 - \alpha$. Once $k_2(c^2)$ is thus computed for a few values of c^2, an approximate functional form of $k_2(c^2)$ can be obtained. Later in this section, we shall illustrate this procedure using an example.

For $c^2 = 0$, a starting value $k_{2*}(0)$ satisfying $k_{2*}(0) < k_2(0)$ is exhibited in Mathew, Sharma and Nordström (1998), and is given by

$$k_{2*}(0) = \max \left[\frac{n-m-q}{m} \frac{\chi^2_{m;p}}{\chi^2_{m(n-q-1);\alpha}}, \quad F_{m,n-m-q;p(1-\alpha)} \right], \qquad (10.6.13)$$

where $F_{m_1,m_2;\epsilon}$ denotes the 100ϵ percentile of an F-distribution with (m_1, m_2) df. The quantity $k_{2*}(0)$ given above can be taken as a starting value for computing $k_2(0)$. The value of $k_{2*}(0)$ can be increased in steps in order to arrive at $k_2(0)$, as already explained. Now consider a finite sequence of values

$$0 < c_1^2 < c_2^2 < \ldots < c_s^2 = \delta. \qquad (10.6.14)$$

Since $k_2(c^2)$ is an increasing function of c^2, we have $k_2(0) < k_2(c_1^2)$. In other words, once $k_2(0)$ is numerically obtained, it can serve as a starting value for the evaluation of $k_2(c_1^2)$. In general $k_2(c_i^2)$ is a starting value for the evaluation of $k_2(c_{i+1}^2)$ when $c_i^2 < c_{i+1}^2$. The value of s and the choice of the c_i^2's in the interval $[0, \delta]$ satisfying (10.6.14) are clearly subjective. If δ is small, the interval $[0, \delta]$ will be narrow and perhaps the numerical evaluation of $k_2(c_i^2)$ for a small number of c_i^2's may be enough to approximately determine the functional form of $k_2(c^2)$. Hopefully, the example given later in this section will provide further insight into the various aspects of the above numerical procedure.

Here is a summary of the numerical procedure for the evaluation of $k_2(c^2)$.

(i) Start with $k_{2*}(0)$ in (10.6.13) for the evaluation of $k_2(0)$. Since $k_{2*}(0) \leq k_2(0)$, the value of $k_{2*}(0)$ can be increased in steps, each time evaluating the lhs of (10.6.12), until its value is approximately equal to $1 - \alpha$. The numerical evaluation of the lhs of (10.6.13) is already explained above.

(ii) Fix s values of c^2 satisfying (10.6.14). For $c_{i+1}^2 > c_i^2$, $k_2(c_i^2)$ can be taken as a starting value for the computation of $k_2(c_{i+1}^2)$. The value of $k_2(c_i^2)$ can be increased in steps, as mentioned before, in order to arrive at $k_2(c_{i+1}^2)$, $i = 0, 1, 2, \ldots, s - 1$.

(iii) The pairs $(c_i^2, k_2(c_i^2))$, $i = 0, 1, 2, \ldots, s$ (where $c_0^2 = 0$) can be plotted and a suitable function can be fitted. Since we have a finite interval $[0, \delta]$, a polynomial of appropriate degree should provide a good fit.

10.6.4 A Generalization

A model that is more general than (10.6.1) is one where $\mathbf{x}$ in (10.6.1) is a nonlinear function of fewer unknown parameters, denoted by an $r \times 1$ vector $\boldsymbol{\xi}$ $(r \leq m)$. Let $\mathbf{h}(\boldsymbol{\xi})$ denote this $m \times 1$ vector valued function. Here the functional form of $\mathbf{h}(\boldsymbol{\xi})$ is assumed to be known. We then have the model

$$\mathbf{Y} \sim N(\boldsymbol{\beta}_0 \mathbf{1}_n' + \mathbf{BX}, \mathbf{I}_n \otimes \boldsymbol{\Sigma}), \quad \mathbf{Y}(\boldsymbol{\xi}) \sim N(\boldsymbol{\beta}_0 + \mathbf{Bh}(\boldsymbol{\xi}), \boldsymbol{\Sigma}), \quad (10.6.15)$$

The columns of $\mathbf{X}$ are now the values of $\mathbf{h}(.)$ evaluated at known design points. The problem now is the construction of a confidence region for $\boldsymbol{\xi}$. Note that the model (10.6.15) will arise when we have polynomial regression. We shall assume that the components of $\mathbf{h}(\boldsymbol{\xi})$ are differentiable functions of $\boldsymbol{\xi}$. Under this assumption, let $\boldsymbol{H}(\boldsymbol{\xi})$ be the $q \times r$ matrix defined as

$$\boldsymbol{H}(\boldsymbol{\xi}) = \widehat{\boldsymbol{B}} \frac{\partial \mathbf{h}(\boldsymbol{\xi})}{\partial \boldsymbol{\xi}}, \quad (10.6.16)$$

where $\widehat{\boldsymbol{B}}$ is the least squares estimator defined in (10.1.5). We also assume that $q \geq r$, and that the $\boldsymbol{H}(\boldsymbol{\xi})$ have rank r (with probability one) for all $\boldsymbol{\xi}$ belonging to the appropriate parameter space. Now define

$$\begin{aligned} T(\boldsymbol{\xi}) &= \frac{n-m-q}{r}[\boldsymbol{Y}(\boldsymbol{\xi}) - \bar{\boldsymbol{Y}} - \widehat{\boldsymbol{B}}(\mathbf{h}(\boldsymbol{\xi}) - \bar{\mathbf{x}})]' \boldsymbol{A}^{-1} \boldsymbol{H}(\boldsymbol{\xi}) \\ &\quad \times (\boldsymbol{H}(\boldsymbol{\xi})' \boldsymbol{A}^{-1} \boldsymbol{H}(\boldsymbol{\xi}))^{-1} \boldsymbol{H}(\boldsymbol{\xi})' \boldsymbol{A}^{-1}[\boldsymbol{Y}(\boldsymbol{\xi}) - \bar{\boldsymbol{Y}} - \widehat{\boldsymbol{B}}(\mathbf{h}(\boldsymbol{\xi}) - \bar{\mathbf{x}})], \end{aligned}$$
$$(10.6.17)$$

where $\boldsymbol{A}$ is the Wishart matrix defined in (10.1.6). We note that if $q = r$, then the above pivot simplifies to

$$T(\boldsymbol{\xi}) = \frac{n-m-r}{r}[\boldsymbol{Y}(\boldsymbol{\xi}) - \bar{\boldsymbol{Y}} - \widehat{\boldsymbol{B}}(\mathbf{h}(\boldsymbol{\xi}) - \bar{\mathbf{x}})]' \boldsymbol{A}^{-1}[\boldsymbol{Y}(\boldsymbol{\xi}) - \bar{\boldsymbol{Y}} - \widehat{\boldsymbol{B}}(\mathbf{h}(\boldsymbol{\xi}) - \bar{\mathbf{x}})].$$
$$(10.6.18)$$

The confidence region for $\boldsymbol{\xi}$ that we shall construct is given by

$$\{\boldsymbol{\xi} : T(\boldsymbol{\xi}) \leq k_2(\boldsymbol{\xi})\}, \quad (10.6.19)$$

where the function $k_2(\boldsymbol{\xi})$ is to be determined subject to the appropriate coverage probability requirement, similar to (10.6.8).

Towards the computation of $k_2(\boldsymbol{\xi})$, define

$$c^2 = (\mathbf{h}(\boldsymbol{\xi}) - \bar{\mathbf{x}})' \{\mathbf{X}(\mathbf{I}_n - \frac{1}{n}\mathbf{1}_n\mathbf{1}_n')\mathbf{X}'\}^{-1}(\mathbf{h}(\boldsymbol{\xi}) - \bar{\mathbf{x}}). \quad (10.6.20)$$

As before, $k_2(\boldsymbol{\xi})$ turns out to be a function of c^2, to be denoted by $k_2(c^2)$. Similar to Lemma 10.2, the following lemma gives the condition to be satisfied by $k_2(c^2)$.

Lemma 10.3. Let $\boldsymbol{W}_1 \sim W_r(\boldsymbol{I}_r, n - q - m + r - 1)$, $\boldsymbol{W}_2 \sim W_{q-r}(\boldsymbol{I}_{q-r}, f_m)$, $\boldsymbol{W} \sim N(0, \boldsymbol{I}_r \otimes \boldsymbol{I}_{q-r})$, and $W_0 \sim \chi_q^2$, where $\boldsymbol{W}_1$, $\boldsymbol{W}_2$, $\boldsymbol{W}$ and W_0 are also independently distributed. For $q > r$, let $\boldsymbol{\lambda} = (\lambda_1, \lambda_2, ..., \lambda_r)'$ denote the eigenvalues of $(\boldsymbol{I} + \boldsymbol{W}\boldsymbol{W}_2^{-1}\boldsymbol{W}')\boldsymbol{W}_1^{-1}$, and for $q = r$, let $\boldsymbol{\lambda} = (\lambda_1, \lambda_2, ..., \lambda_r)'$ denote the eigenvalues of $\boldsymbol{W}_1^{-1}$, where it is assumed that $\lambda_1 \geq \lambda_2 \geq .... \geq \lambda_r$. Let $k_2(c^2)$ satisfy

$$P_{\boldsymbol{\lambda}, W_0}\left[P\left(\frac{n - m - q}{r}\left\{\lambda_1\chi_{1,1}^2\left((\frac{1}{n} + c^2)W_0\right) + \sum_{i=2}^{r}\lambda_i\chi_{1,i}^2\right\}\right.\right.$$
$$\left.\left.\leq\ k_2(c^2)\Big|\boldsymbol{\lambda}, W_0\right) \geq p\right] = 1 - \alpha. \tag{10.6.21}$$

The computation of $k_2(c^2)$ satisfying (10.6.21) is similar to the computation of $k_2(c^2)$ satisfying (10.6.9). However, the starting value $k_{2*}(0)$ given in (10.6.13) now becomes

$$k_{2*}(0) = \max\left[\frac{n - m - q}{r}\frac{\chi_{r;p}^2}{\chi_{r(n-q-m+r-1);\alpha}^2},\ F_{r,n-m-q;p(1-\alpha)}\right]. \tag{10.6.22}$$

10.6.5   An Example and Some Numerical Results

The example is taken from Oman and Wax (1984) and deals with the estimation of gestational age (i.e., week of pregnancy) based on two fetal bone lengths. The model relating the bone lengths to the gestational age is given in Oman and Wax (1984) and the gestational age enters the model non-linearly so that the model (10.6.15) is applicable. The data that are available for this example consist of fetal bone length measurements for several women whose gestational ages are precisely known. This data can be used repeatedly in order to construct confidence regions for the unknown gestational age of women, after observing the corresponding fetal bone lengths. In other words, it is required to construct multiple use confidence regions. Here $r = 1$, i.e., the parameter $\boldsymbol{\xi}$ is a scalar (the gestational age), to be denoted by ξ. Furthermore, $q = 2$, and the bivariate observations consist of ultrasound measurements on two fetal bone lengths: the femur length (F) and the biparietal diameter (BPD). The model that relates the gestational age ξ to the observation vector $(F, BPD)'$ is

$$(F, BPD)' \sim N(\boldsymbol{\beta}_0 + \boldsymbol{B}\mathbf{h}(\xi), \boldsymbol{\Sigma}), \text{ where } \mathbf{h}(\xi) = \begin{pmatrix} \xi \\ \xi^2 \end{pmatrix},$$

see Oman and Wax (1984). The parameter space for ξ (in weeks) is the interval $[14, 41]$, and the data analyzed in Oman and Wax (1984) consists of (F, BPD) measurements for $n = 1114$ women for whom the value of ξ was precisely known. Let Y be the 2×1114 matrix whose columns are the $(F, BPD)'$ measurements for these women. As in (10.6.15), we shall use $Y(\xi)$ to denote the $(F, BPD)'$ measurement for a woman whose gestational age ξ is unknown. Then Y and $Y(\xi)$ are independent following the models

$$Y \sim N(\beta_0 1'_{1114} + BX, I_{1114} \otimes \Sigma), \quad Y(\xi) \sim N(\beta_0 + Bh(\xi), \Sigma). \quad (10.6.23)$$

The ith column of X in (10.6.23) is $h(\xi_i)$, ξ_i being the known gestational age for the ith woman $(i = 1, 2, \ldots, 1114)$.

Based on the data in Oman and Wax (1984), we have

$$X(I_n - n^{-1}1_n 1'_n)X' = \begin{pmatrix} 52877.52 & 2878329 \\ 2878329 & 159145978 \end{pmatrix}$$

$$\text{and } \bar{x} = \begin{pmatrix} 28.410233 \\ 854.607720 \end{pmatrix}.$$

In our computations, we have chosen $p = 0.95$ and $1 - \alpha = 0.95$. The pivotal quantity $T(\xi)$ is given by (10.6.17) with $r = 1, n - m - q = 1110$, and

$$H(\xi) = \hat{B} \begin{pmatrix} 1 \\ 2\xi \end{pmatrix}.$$

Since $H(\xi)$ is a column vector, $T(\xi)$ becomes

$$T(\xi) = 1110 \times \frac{\left[(Y(\xi) - \bar{Y} - \hat{B}(h(\xi) - \bar{x})'A^{-1}H(\xi)\right]^2}{[H(\xi)'A^{-1}H(\xi)]}. \quad (10.6.24)$$

For the gestational age data, the matrices $\hat{\beta}_0, \hat{B}$ and A as given in Oman (1988, p.182) are

$$\hat{\beta}_0 = \begin{pmatrix} -42.917 \\ -39.187 \end{pmatrix}, \quad \hat{B} = \begin{pmatrix} 4.514 & -0.0402 \\ 5.292 & -0.0492 \end{pmatrix}, \quad \text{and } A = \begin{pmatrix} 8281.00 & 4900.48 \\ 4900.48 & 14484.23 \end{pmatrix}.$$

Since $q = 2$ and $r = 1$, the Wishart matrices W_1 and W_2 in Lemma 10.3 are now scalars, to be denoted by W_1 and W_2, having the chi-square distributions $W_1 \sim \chi^2_{1110}$, $W_2 \sim \chi^2_{1111}$. Furthermore, $W_0 \sim \chi^2_2$, and W is a scalar, to be denoted by W, having a standard normal distribution. Thus $\lambda = \lambda_1$, a scalar given by

$$\lambda_1 = \left(1 + \frac{W^2}{W_2}\right) \Big/ W_1.$$

For c^2 given in (10.6.20) and $p = 1 - \alpha = 0.95$, we need to evaluate $k_2(c^2)$ satisfying (see (10.6.21)

$$P_{\lambda_1,W_0}\left[P\{1110 \times \lambda_1\chi_{11}^2\left((\frac{1}{1114} + c^2)W_0\right) \leq k_2(c^2)\big|\lambda_1, W_0\} \geq 0.95\right] = 0.95.$$

$$(10.6.25)$$

Here is a summary of the numerical procedure reported in Mathew, Sharma and Nordstrom (1998), for the computation of the function $k_2(c^2)$. For several values of $\xi \in [14, 41]$, c^2 was calculated and $k_2(c^2)$ was numerically obtained, as explained in Section 10.6.3. The starting value for the computation of $k_2(0)$ was $k_{2*}(0)$ given in (10.6.22). Since $\frac{n-m-q}{r}\frac{\chi_{r;p}^2}{\chi_{n-m-q+r-1);\alpha}^2} = 1110 \times \frac{\chi_{1;.95}^2}{\chi_{1110;.05}^2} = 1110 \times \frac{3.8415}{1032.493} = 4.1299$ and $F_{r,n-m-p;p(1-\alpha)} = F_{1,1110;.9025} = 2.7506$, $k_{2*}(0) = 4.1299$. This value was used as a starting value for the computation of $k_2(c^2)$ for $c^2 = 0.00059$, the smallest value of c^2 that we considered (see Table 10.1 below). The computation was carried out as follows. To begin with, 100,000 pairs of values of (λ_1, W_0) were generated. For a given value of c^2 and $k_2(c^2)$ and for a given pair of values of (λ_1, W_0), the quantity

$$P\left\{1110 \times \lambda_1\chi_{1,1}^2\left((\frac{1}{1114} + c^2)W_0\right) \leq k_2(c^2)\big|\lambda_1, W_0\right\}$$

was evaluated based on 100,000 simulations. The indicator variable $i(\lambda_1, W_0)$ was defined to be one, if this probability was at least 0.95, following the notation in Section 10.6.3. Otherwise $i(\lambda_1, W_0) = 0$. The value of $i(\lambda_1, W_0)$ was computed for each of the 100,000 pairs of values of (λ_1, W_0). The lhs of (10.6.25) was then the proportion of times $i(\lambda_1, W_0)$ took the value one. The quantity $k_2(c^2)$ was determined so as to make this proportion equal to 0.95, approximately. In other words, for computing $k_2(0.00059)$, the lhs of (10.6.25) was computed, starting with the value $k_{2*}(0) = 4.1299$. The value of $k_{2*}(0)$ was adjusted suitably and the lhs of (10.6.25) was evaluated repeatedly, until a value of $k_2(c^2)$ was obtained for which the lhs of (10.6.25) was 0.95, approximately. The value of $k_2(c^2)$ that was accepted was such that the lhs of (10.6.25) was between 0.9490 and 0.9510. The values so obtained are given in Table 10.1 below (the quantity $\tilde{k}_2(c^2)$ given in Table 10.1 is explained shortly). These values are reproduced from Mathew, Sharma and Nordström (1998). The value $k_{2*}(0) = 4.1299$ turned out to be a very satisfactory starting value for the computation of $k_2(0.00059)$, since, from Table 10.1, $k_2(0.00059) = 4.1418$, which is very close to $k_{2*}(0)$. A plot of the $(c^2, k_2(c^2))$ values in Table 10.1 is given in Figure 10.1.

The following function gave a good fit to the plot.

$$\tilde{k}_2(c^2) = 4.136977 + 8.62437c^2 + 295.4134c^4, \qquad (10.6.26)$$

Table 10.1: Values of ξ, c^2, $k_2(c^2)$ satisfying (10.6.25), and $\tilde{k}_2(c^2)$ satisfying (10.6.26)

ξ	c^2	$k_2(c^2)$	$\tilde{k}_2(c^2)$
14	0.01033	4.2572	4.2576
15	0.00747	4.2175	4.2180
16	0.00530	4.1899	4.1911
18	0.00257	4.1610	4.1611
22	0.00096	4.1458	4.1456
26	0.00102	4.1461	4.1461
30	0.00073	4.1435	4.1435
34	0.00059	4.1418	4.1422
38	0.00357	4.1719	4.1716
39	0.00539	4.1939	4.1921
40	0.00784	4.2225	4.2229
41	0.01105	4.2690	4.2684

where we use the notation $\tilde{k}_2(c^2)$ to denote the fitted function. Figure 10.1 also gives a plot of $\tilde{k}_2(c^2)$. The plot in Figure 10.1 and the values of $\tilde{k}_2(c^2)$ given in Table 10.1 show that $\tilde{k}_2(c^2)$ is a very good approximation to $k_2(c^2)$. Thus the region (10.6.19) is given by

$$\{\xi : T(\xi) \leq \tilde{k}_2(c^2)\}, \tag{10.6.27}$$

where $T(\xi)$ and $\tilde{k}_2(c^2)$ are given by (10.6.24) and (10.6.26), respectively. Recall that c^2 (and hence $\tilde{k}_2(c^2)$) is a function of ξ, the parameter of interest.

For a few $(F, BPD)'$−values, the region (10.6.27) is given in Table 10.2.

Table 10.2: The region (10.6.27) for ξ using the Oman and Wax (1984) data for a few values of $(F, BPD)'$

(F, BPD)	(14,27)	(32,47)	(45,62)
Region (10.6.27)	(12.968, 16.008)	(18.462, 21.955)	(23.052, 27.107)
(F, BPD)	(56,75)	(65,85)	
Region (10.6.27)	(27.563, 32.400)	(31.788, 37.763)	

Note that our first interval in Table 10.2 extends beyond the interval (14, 41), which is the parameter space for ξ. Strictly speaking the region

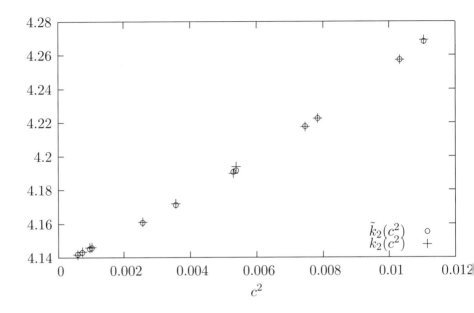

Figure 10.1: Plots of $k_2(c^2)$ and $\tilde{k}_2(c^2)$ in Table 10.1

should be (10.6.27) intersected with the parameter space, i.e., $\{\xi : T(\xi) \le \tilde{k}_2(c^2)\} \cap (14, 41)$. This will obviously bring all the intervals within the parameter space. We note that intersecting with the parameter space will not affect the coverage probability.

As in the univariate case, a natural question is whether the region derived using the tolerance region condition (10.6.8) will satisfy the condition (10.6.6) that is required for a multiple use confidence region. The numerical results reported in Mathew, Sharma and Nordström (1998) support this. Such numerical results are not reported here.

The results in this section (and also in this chapter) are for the case of an unknown positive definite covariance matrix $\mathbf{\Sigma}$. If $\mathbf{\Sigma}$ is known to be a multiple of the identity matrix, the calibration problem, and the construction of multiple use confidence regions are addressed in Mathew and Zha (1997). We also refer to Mathew and Sharma (2002) where the problem of obtaining multiple use confidence regions by combining information from different sources is addressed, in the context of univariate calibration. It should also be noted that several authors have investigated the problem of constructing *single use confidence regions* for the multivariate calibration problem. If the pivot $T(\mathbf{x})$ given in (10.6.2) or (10.6.3) is used for this purpose, then a $100(1 - \alpha)\%$ single use confidence region is of the form $\{\mathbf{x} : T(\mathbf{x}) \le k_3(\mathbf{x})\}$, where the function $k_3(\mathbf{x})$

satisfies the condition

$$P_{Y(\mathbf{x}), \bar{Y}, \widehat{B}, A}[T(\mathbf{x}) \le k_3(\mathbf{x})] = 1 - \alpha.$$

We refer to the articles by Brown (1982), Fujikoshi and Nishii (1984), Brown and Sundberg (1987), Davis and Hayakawa (1987), Oman (1988), Mathew and Kasala (1994) and Mathew and Zha (1996), for results concerning single use confidence regions.

10.7 Exercises

10.7.1. Consider the multivariate regression model (10.1.3) and let $Y(\mathbf{x}_1)$ and $Y(\mathbf{x}_2)$ be two independent observations corresponding to the values $\mathbf{x}_1$ and $\mathbf{x}_2$ of the explanatory variable, so that

$$Y(\mathbf{x}_1) \sim N(\boldsymbol{\beta}_0 + B\mathbf{x}_1, \boldsymbol{\Sigma}) \quad \text{and} \quad Y(\mathbf{x}_2) \sim N(\boldsymbol{\beta}_0 + B\mathbf{x}_2, \boldsymbol{\Sigma}).$$

Explain how you will compute a tolerance region for $Y(\mathbf{x}_1) - Y(\mathbf{x}_2)$. Develop approximations similar to those in Section 10.2 and Section 10.4.

10.7.2. For the blood glucose measurement example discussed in Section 10.1.2 and Section 10.5, let $\mathbf{x}_1 = (103, 103, 100)'$ and $\mathbf{x}_2 = (106, 105, 101)'$ denote the fasting glucose measurements on three occasions for two women. Use the data in Section 10.5 and compute a (0.95, 0.99) tolerance region for $Y(\mathbf{x}_1) - Y(\mathbf{x}_2)$, the difference between their blood glucose measurements on the three occasions, one hour after sugar intake. Use the methodologies developed in the previous problem.

10.7.3. Consider the approximation $\frac{f}{q+\delta} \chi_e^2 \chi_q^2(\delta)$ for the distribution of $\chi_q^2(H'H)$, as stated in Lemma 10.1, where the notations are as in the lemma. For various values of q and d, estimate the 5th, 10th, 50th, 90th and 95th percentiles of $\frac{f}{q+\delta} \chi_e^2 \chi_q^2(\delta)$ and that of $\chi_q^2(H'H)$, based 10,000 simulated values of these random variables. To what extent do the percentiles of $\frac{f}{q+\delta} \chi_e^2 \chi_q^2(\delta)$ and $\chi_q^2(H'H)$ agree? Based on the simulated values of the percentiles, what can you conclude regarding the accuracy of the approximation mentioned in Lemma 10.1?

10.7.4. Consider the multivariate regression model (10.1.3) and the model (10.1.4) for the observation vector $Y(\mathbf{x})$. Suppose it is known that $\boldsymbol{\Sigma} = \sigma^2 I_q$, where σ^2 is an unknown scalar. Develop procedures and approximations for computing a tolerance region for $Y(\mathbf{x})$.

10.7.5. Let $\boldsymbol{Y}_1, \boldsymbol{Y}_2, \cdots, \boldsymbol{Y}_n$ be n independent $q \times 1$ random vectors having the distribution $\boldsymbol{Y}_i \sim N(\boldsymbol{\beta}_0 + \boldsymbol{B}\mathbf{x}_i, \sigma^2 v(\mathbf{x}_i)I_q)$, where σ^2 is an unknown scalar, and $v(\mathbf{x}_i)$ is a known function of $\mathbf{x}_i$. Now let $\boldsymbol{Y}(\mathbf{x})$ be another observation having the distribution $\boldsymbol{Y}(\mathbf{x}) \sim N(\boldsymbol{\beta}_0 + \boldsymbol{B}\mathbf{x}, \sigma^2 v(\mathbf{x})I_q)$, where $\mathbf{x}$ is a known value of the explanatory variable. Develop procedures and approximations for computing a tolerance region for $\boldsymbol{Y}(\mathbf{x})$.

10.7.6. For the calibration problem (see Section 10.6.1), explain how confidence regions can be constructed for the two multivariate models in Exercises 10.7.4 and 10.7.5. Establish analogues of Lemma 10.2, and also explain the computational procedure similar to what is outlined in Section 10.6.3.

10.7.7. An example of the calibration problem where the models in Problem 10.7.5 is applicable is described in Smith and Corbett (1987), dealing with the accurate measurement of marathon running courses. The method of measurement is known as the *bicycle method*. In this method, bicycles are fitted with revolution counters that accurately record the number of revolutions of the front wheel when the bicycles are ridden on the running course. The distance of the running course is arithmetically computed based on the number of revolutions of the front wheel as recorded on the revolution counter. The bicycles are first ridden over several standard distances (that are precisely known) and the counter readings are recorded. This information is used to set up a model that relates the counter readings to the true distances. In the application considered in Smith and Corbett (1987), there are 8 cyclists and 13 standard distances. If Y_{ij} denotes the counter reading by the jth cyclist on the ith standard distance, and if x_i denotes the true distance (known), then the model considered by Smith and Corbett (1987) is $Y_{ij} \sim N(b_j x_i, \sigma^2 x_i)$, where the b_j's are unknown regression coefficients and $\sigma^2 > 0$ is also unknown; $i = 1, 2, ..., 13, \ j = 1, 2, ..., 8$. Express the model using the vector-matrix notation in Exercise 10.7.5. Let x denote the unknown distance of a marathon running course, and let $Y_j(x)$ denote the corresponding counter reading by the jth cyclist. Explain the construction of a multiple use confidence interval for x. Carry out the required computations using the data in Smith and Corbett (1987).

                                                                    (Mathew and Zha, 1997)

Chapter 11

Bayesian Tolerance Intervals

11.1   Notations and Preliminaries

This chapter is on the derivation of tolerance intervals in a Bayesian framework. We shall first give a brief sketch of the Bayesian approach, and then describe the procedure to derive tolerance intervals for a univariate normal distribution, and for a one-way random model with balanced data. Even though we take up only these models to illustrate the Bayesian methodology, the procedure explained in this section can be applied to obtain tolerance intervals in any parametric set up.

In order to introduce the basic idea, let X denote a random variable whose distribution depends on a parameter $\boldsymbol{\theta}$, and the realizations of X give the observed data. Here $\boldsymbol{\theta}$ could be a vector. Let $\boldsymbol{x}$ be a vector representing realizations of X (i.e., $\boldsymbol{x}$ is the observed sample), and let $L(\boldsymbol{x}|\boldsymbol{\theta})$ be the likelihood function. If $\pi(\boldsymbol{\theta})$ denotes a prior distribution for $\boldsymbol{\theta}$, then the posterior distribution of $\boldsymbol{\theta}$, say $p(\boldsymbol{\theta}|\boldsymbol{x})$, is given by

$$p(\boldsymbol{\theta}|\boldsymbol{x}) = \frac{L(\boldsymbol{x}|\boldsymbol{\theta})\pi(\boldsymbol{\theta})}{\int_{\boldsymbol{\Theta}} L(\boldsymbol{x}|\boldsymbol{\theta})\pi(\boldsymbol{\theta})d\boldsymbol{\theta}}, \tag{11.1.1}$$

where $\boldsymbol{\Theta}$ is the parameter space for $\boldsymbol{\theta}$. The posterior distribution of any function $g(\boldsymbol{\theta})$ of $\boldsymbol{\theta}$ can obviously be derived from $p(\boldsymbol{\theta}|\boldsymbol{x})$. We shall also use the notation $p(g(\boldsymbol{\theta})|\boldsymbol{x})$ to denote the posterior distribution of $g(\boldsymbol{\theta})$. Bayesian inference concerning $g(\boldsymbol{\theta})$ is accomplished using the posterior distribution $p(g(\boldsymbol{\theta})|\boldsymbol{x})$. For a more detailed discussion, along with examples, see, for example, the book by Gelman, Carlin, Stern and Rubin (2004). Here we shall concentrate only on the tolerance interval problem.

To begin with, suppose we want to derive a $(p, 1 - \alpha)$ upper tolerance limit for the distribution of the random variable X. If $g_p(\boldsymbol{\theta})$ denotes the p quantile of X, then what is required is a $1 - \alpha$ upper confidence limit for $g_p(\boldsymbol{\theta})$, based on the posterior distribution $p(g_p(\boldsymbol{\theta})|\boldsymbol{x})$. Once this posterior distribution is obtained, the problem reduces to the computation of $\hat{g}_{p;1-\alpha}$ that satisfies

$$P\left(g_p(\boldsymbol{\theta}) \leq \hat{g}_{p;1-\alpha}\right) = 1 - \alpha, \tag{11.1.2}$$

where the probability is computed with respect to the posterior distribution $p(g_p(\boldsymbol{\theta})|\boldsymbol{x})$. In situations where it is difficult to derive a convenient analytic form for $p(g_p(\boldsymbol{\theta})|\boldsymbol{x})$, a Bayesian simulation method can be used to obtain $\hat{g}_{p;1-\alpha}$, provided one can generate data from the posterior distribution of $\boldsymbol{\theta}$, i.e., the distribution $p(\boldsymbol{\theta}|\boldsymbol{x})$. The simulation method consists of first generating data from the posterior distribution $p(\boldsymbol{\theta}|\boldsymbol{x})$. Let the simulated data be denoted by $\boldsymbol{\theta}_1$, $\boldsymbol{\theta}_2$, ..., $\boldsymbol{\theta}_M$, where M denotes the size of the simulation sample. Now compute $g_p(\boldsymbol{\theta}_1)$, $g_p(\boldsymbol{\theta}_2)$, ..., $g_p(\boldsymbol{\theta}_M)$. The $100(1 - \alpha)$th percentile of the $g_p(\boldsymbol{\theta}_i)$–values provides an approximation to $\hat{g}_{p;1-\alpha}$.

The Bayesian approach for the computation of a $(p, 1-\alpha)$ two-sided tolerance interval is not that straightforward, except in some simple models. However, the interval can be numerically obtained by performing a Bayesian simulation. The idea, as explained in Wolfinger (1998), is as follows. As before, let $\boldsymbol{\theta}_1$, $\boldsymbol{\theta}_2$, ..., $\boldsymbol{\theta}_M$ be a simulation sample of size M, generated from the posterior distribution $p(\boldsymbol{\theta}|\boldsymbol{x})$. Similar to $g_p(\boldsymbol{\theta})$, let $g_{\frac{1+p}{2}}(\boldsymbol{\theta})$ and $g_{\frac{1-p}{2}}(\boldsymbol{\theta})$, respectively, denote the $\frac{1+p}{2}$ and $\frac{1-p}{2}$ quantiles. Now consider the pairs $\left(g_{\frac{1+p}{2}}(\boldsymbol{\theta}_i), g_{\frac{1-p}{2}}(\boldsymbol{\theta}_i)\right)$, $i = 1, 2, ..., M$, which form a simulation sample of size M from the bivariate posterior distribution of $\left(g_{\frac{1+p}{2}}(\boldsymbol{\theta}), g_{\frac{1-p}{2}}(\boldsymbol{\theta})\right)$. Suppose we decide to construct a $(p, 1-\alpha)$ two-sided tolerance interval that is symmetric about the mean $\bar{g}_p$ given by

$$\bar{g}_p = \frac{1}{M} \sum_{i=1}^{M} \left(\frac{g_{\frac{1+p}{2}}(\boldsymbol{\theta}_i) + g_{\frac{1-p}{2}}(\boldsymbol{\theta}_i)}{2} \right). \tag{11.1.3}$$

Now consider a scatter plot of $\left(g_{\frac{1+p}{2}}(\boldsymbol{\theta}_i), g_{\frac{1-p}{2}}(\boldsymbol{\theta}_i)\right)$, $i = 1, 2, ..., M$, with $g_{\frac{1-p}{2}}(\boldsymbol{\theta}_i)$ on the vertical axis and $g_{\frac{1+p}{2}}(\boldsymbol{\theta}_i)$ on the horizontal axis, and draw the line

$$g_{\frac{1-p}{2}}(\boldsymbol{\theta}_i) = -g_{\frac{1+p}{2}}(\boldsymbol{\theta}_i) + 2\bar{g}_p \tag{11.1.4}$$

on the scatter plot. Let $(\hat{g}_1, \hat{g}_2)$ be a point on the line with the property that $100(1 - \alpha)\%$ of the points in the scatter plot satisfies $g_{\frac{1-p}{2}}(\boldsymbol{\theta}_i) \geq \hat{g}_2$ and $g_{\frac{1+p}{2}}(\boldsymbol{\theta}_i) \leq \hat{g}_1$. In other words, the two lines drawn through $(\hat{g}_1, \hat{g}_2)$, parallel to the axes, are such that $100(1-\alpha)\%$ of the points in the scatter plot are included in the half rectangle open towards the upper left portion of the graph. The interval $(\hat{g}_1, \hat{g}_2)$ is the required $(p, 1 - \alpha)$ two-sided Bayesian tolerance interval. Note that the symmetry point of the tolerance interval can be different from the mean $\bar{g}_p$ given in (11.1.3). This approach will be graphically illustrated later in this chapter. We would like to point out that when Wolfinger (1998) proposed this methodology for the one-way random model, he determined $(\hat{g}_1, \hat{g}_2)$ subject to the condition that $100\alpha\%$ of the points in the scatter plot satisfy $g_{\frac{1-p}{2}}(\boldsymbol{\theta}_i) \le \hat{g}_2$ and $g_{\frac{1+p}{2}}(\boldsymbol{\theta}_i) \ge \hat{g}_1$. That is, $100\alpha\%$ of the points in the scatter plot are included in the half rectangle open towards the bottom right portion of the graph. However, the determination of $(\hat{g}_1, \hat{g}_2)$ should be done subject to the requirement we have indicated above.

The above descriptions provide general recipes for obtaining one-sided and two-sided Bayesian tolerance intervals. We shall now follow the recipes to derive Bayesian tolerance intervals for some specific models.

11.2 The Univariate Normal Distribution

Let $\bar{X}$ and S^2 denote the sample mean and sample variance based on a sample of size n from $N(\mu, \sigma^2)$. Bayesian tolerance intervals for the univariate normal distribution $N(\mu, \sigma^2)$ was originally derived by Aitchison (1964, 1966), and the results are also given in Guttman (1970, Chapters 7−9). The relevant results are summarized in this section. Two prior distributions will be considered for (μ, σ^2), where we use the notation $\pi(\mu, \sigma^2)$ for the prior:

(i) the non-informative prior distribution

$$\pi(\mu, \sigma^2) \propto \frac{1}{\sigma^2}, \tag{11.2.1}$$

(ii) a family of conjugate prior distributions with

$$\mu|\sigma^2 \sim N\left(\mu_0, \frac{\sigma^2}{n_0}\right), \quad \text{and} \quad \sigma^2 \sim \text{Inv-}\chi^2(m_0, \sigma_0^2), \tag{11.2.2}$$

where Inv-$\chi^2(m_0, \sigma_0^2)$ denotes a scaled inverted chi-square distribution with parameters m_0 and σ_0^2, having the density

$$\frac{\left(\frac{m_0}{2}\right)^{\frac{m_0}{2}}}{\Gamma\left(\frac{m_0}{2}\right)}(\sigma_0^2)^{\frac{m_0}{2}}(\sigma^2)^{-\left(\frac{m_0}{2}+1\right)} \exp\left(-\frac{m_0\sigma_0^2}{2\sigma^2}\right). \tag{11.2.3}$$

It should be clear that a scaled inverted chi-square distribution is a special case of an inverted gamma distribution. Such a prior is a very common choice for estimating variances; see Searle, Casella and McCulloch (1992, Section 3.9) or Gelman et al. (2004, Section 3.3). It can be easily verified that if $\sigma^2 \sim$ Inv-$\chi^2(m_0, \sigma_0^2)$, then $m_0 \sigma_0^2 / \sigma^2$ follows a chi-square distribution with m_0 df. Note that under (11.2.2), the prior parameters are the quantities μ_0, n_0, m_0 and σ_0^2. Also note that a conjugate prior is such that the posterior distribution has the same parametric form as the prior distribution. Under the prior given in (11.2.2), the joint prior density of (μ, σ^2) is thus given by

$$
\pi(\mu, \sigma^2) = \frac{\sqrt{n_0}}{\sqrt{2\pi}} \frac{\left(\frac{m_0}{2}\right)^{\frac{m_0}{2}}}{\Gamma\left(\frac{m_0}{2}\right)} (\sigma_0^2)^{\frac{m_0}{2}} (\sigma^2)^{-1/2} (\sigma^2)^{-\left(\frac{m_0}{2}+1\right)}
$$

$$
\times \exp\left(-\frac{1}{2\sigma^2}\left\{m_0\sigma_0^2 + n_0(\mu - \mu_0)^2\right\}\right). \tag{11.2.4}
$$

11.2.1   Tolerance Intervals Under the Non-Informative Prior

Let's first consider the derivation of tolerance limits under the non-informative prior distribution (11.2.1). Let $\bar{x}$ and s^2 denote the observed values of $\bar{X}$ and S^2, respectively. The posterior distribution of (μ, σ^2) is now given by

$$
p(\mu, \sigma^2 | \bar{x}, s^2) = \frac{\sqrt{n}}{2^{\frac{n}{2}}\sqrt{\pi}\,\Gamma(\frac{n-1}{2})} \times \frac{1}{\{(n-1)s^2\}^{-\frac{n-1}{2}}(\sigma^2)^{-\frac{n}{2}-1}}
$$

$$
\times \exp\left\{-\frac{1}{2\sigma^2}\left[(n-1)s^2 + n(\bar{x} - \mu)^2\right]\right\}. \tag{11.2.5}
$$

Under (11.2.1), we thus have the posterior distributions

$$
\mu | \bar{x}, \sigma^2 \sim N\left(\bar{x}, \frac{\sigma^2}{n}\right) \quad \text{and} \quad \left.\frac{(n-1)s^2}{\sigma^2}\right| s^2 \sim \chi^2_{n-1}. \tag{11.2.6}
$$

Thus the posterior distribution of $\frac{\sqrt{n}(\mu - \bar{x})}{\sigma}$ is $N(0,1)$, and $\frac{\sqrt{n}(\mu - \bar{x})}{\sigma}$ and $(n-1)s^2/\sigma^2$ are independently distributed. Suppose we want an upper tolerance limit for $N(\mu, \sigma^2)$, and let $a_1 = a_1(\bar{x}, s^2)$ denote the limit. The content of the corresponding one-sided tolerance interval is clearly $\Phi\left(\frac{a_1 - \mu}{\sigma}\right)$, where Φ denotes the standard normal cdf. We want to choose a_1 so that

$$
P_{\mu, \sigma^2}\left\{\Phi\left(\frac{a_1 - \mu}{\sigma}\right) \geq p\right\} = 1 - \alpha,
$$

where we note that the probability is calculated with respect to the posterior distribution of (μ, σ^2), given in (11.2.5). Note that the statement $\Phi\left(\frac{a_1 - \mu}{\sigma}\right) \geq p$

is equivalent to $\mu + z_p\sigma \leq a_1$, which can equivalently be expressed as

$$\frac{[\sqrt{n}(\mu - \bar{x})/\sigma] + z_p\sqrt{n}}{s/\sigma} \leq \frac{\sqrt{n}(a_1 - \bar{x})}{s}. \tag{11.2.7}$$

Under the posterior distributions in (11.2.5), the quantity on the left hand side of the inequality (11.2.7) has a noncentral t-distribution with $n - 1$ df and noncentrality parameter $z_p\sqrt{n}$. Hence, from the inequality (11.2.7), we get the solution

$$a_1 = \bar{x} + \frac{1}{\sqrt{n}}t_{n-1;1-\alpha}(z_p\sqrt{n}) \times s. \tag{11.2.8}$$

We note that this is the same solution that was obtained by the frequentist approach in Chapter 2; see Section 2.2. It can similarly be shown that under the non-informative prior distribution given in (11.2.1), the Bayesian $(p, 1 - \alpha)$ two-sided tolerance interval also coincides with the tolerance interval in Section 2.2.2 of Chapter 2. Note however that the two-sided interval obtained using the graphical procedure described towards the end of Section 11.1 will result in an equal-tail tolerance interval, as explained in Section 2.3.2 of Chapter 2; see also Problem 11.5.1.

11.2.2   Tolerance Intervals Under the Conjugate Prior

Let's now consider the derivation of tolerance limits under the prior distribution (11.2.2). In order to obtain the expression for the posterior distribution of (μ, σ^2), let

$$\bar{\bar{x}} = \frac{n_0\mu_0 + n\bar{x}}{n_0 + n}$$

$$q^2 = \frac{1}{m_0 + n - 1}\left[m_0\sigma_0^2 + (n - 1)s^2 + \frac{n_0 n}{n_0 + n}(\bar{x} - \mu_0)^2\right]. \tag{11.2.9}$$

It can be shown that

$$\mu|\sigma^2, \bar{x}, s^2 \sim N\left(\bar{\bar{x}}, \frac{\sigma^2}{n_0 + n}\right), \text{ and } \sigma^2|\bar{x}, s^2 \sim \text{Inv-}\chi^2(m_0 + n - 1, q^2), \tag{11.2.10}$$

see Gelman et al. (2004, pp. 79-80). Thus the posterior distribution of (μ, σ^2) is

$$p(\mu, \sigma^2|\bar{x}, s^2) = p(\mu|\sigma^2, \bar{x}, s^2) \times p(\sigma^2|\bar{x}, s^2), \tag{11.2.11}$$

where $p(\mu|\sigma^2, \bar{x}, s^2)$ and $p(\sigma^2|\bar{x}, s^2)$ are, respectively, the normal distribution and the scaled inverted chi-square distribution specified in (11.2.10). Note that the posterior distribution of $(m_0 + n - 1)q^2/\sigma^2$ is a chi-square distribution with

$m_0 + n - 1$ df, and that of $\sqrt{n_0 + n}(\mu - \bar{x})/\sigma$ is standard normal, and these two quantities are independently distributed.

Suppose we want to compute a $(p, 1 - \alpha)$ upper tolerance limit for $N(\mu, \sigma^2)$, and let $a_2 = a_2(\bar{x}, s^2)$ denote the limit. The content of the corresponding one-sided tolerance interval is $\Phi\left(\frac{a_2 - \mu}{\sigma}\right)$. We want to choose a_2 so that

$$P_{\mu,\sigma^2}\left\{\Phi\left(\frac{a_2 - \mu}{\sigma}\right) \geq p\right\} = 1 - \alpha,$$

where the probability is calculated with respect to the posterior distribution of (μ, σ^2), given in (11.2.11). The statement $\Phi\left(\frac{a_2-\mu}{\sigma}\right) \geq p$ is equivalent to $\mu + z_p\sigma \leq a_2$, which can equivalently be expressed as

$$\frac{[\sqrt{n_0 + n}(\mu - \bar{x})/\sigma] + z_p\sqrt{n_0 + n}}{q/\sigma} \leq \frac{\sqrt{n_0 + n}(a_2 - \bar{x})}{q}. \qquad (11.2.12)$$

Hence, similar to the derivation of (11.2.8), we get

$$a_2 = \bar{\bar{x}} + \frac{1}{\sqrt{n_0 + n}} t_{m_0+n-1;1-\alpha}(z_p\sqrt{n_0 + n}) \times q, \qquad (11.2.13)$$

where q^2 is defined in (11.2.9). We note the similarity between (11.2.8) and (11.2.13); instead of $n - 1$, the df associated with the non-central t-distribution is m_0+n-1, the factor $\sqrt{n}$ gets replaced with $\sqrt{n_0 + n}$ and q^2 defined in (11.2.9) takes the place of s^2. A two-sided Bayesian tolerance interval can be similarly computed by making the same changes in the frequentist solution.

Example 2.1 (continued)

For the purpose of illustrating the calculation of Bayesian tolerance limits, let us consider the air lead level data given in Table 2.1. Here the sample size is $n = 15$, and $\bar{x} = 4.333$ and $s = 1.739$, for the log-transformed data. For the distribution of the log-transformed air lead levels, the (0.95, 0.90) frequentist upper tolerance limit is 8.383. This is also the Bayesian (0.95, 0.90) upper tolerance limit, if we assume the non-informative prior in (11.2.1). Now suppose we assume the conjugate prior distribution in (11.2.2) with $\mu_0 = 2$, $n_0 = m_0 = 10$ and $\sigma_0^2 = 3$. From (11.2.9) we get $\bar{\bar{x}} = 3.399$, and $q^2 = 4.375$. Furthermore, $t_{m_0+n-1;1-\alpha}(z_p\sqrt{n_0 + n}) = t_{24;.90}(1.645 \times \sqrt{5}) = 10.662$. From (11.2.13), we get the Bayesian (0.95, 0.90) upper tolerance limit as 7.860.

We note that the Bayesian (0.95, 0.90) upper tolerance limit under the conjugate prior is smaller than that under the non-informative prior. The reason for this should be clear; the prior mean for μ is $\mu_0 = 2$, significantly smaller than $\bar{x}$. Obviously the use of such a prior can be recommended only if such prior information is available, either based on expert opinion, or based on the analysis from previous data. If no such prior information is available, one has to resort to using the non-informative prior, in which case the tolerance intervals for $N(\mu, \sigma^2)$ coincide with the frequentist solutions. An alternative is to use an empirical Bayes approach, where the prior parameters (or the prior itself) is estimated using past data, or the current data. Here we shall not pursue this further; see Miller (1989) for the derivation of empirical Bayes tolerance intervals for a normal distribution.

11.3 The One-Way Random Model with Balanced Data

The model and the notations are given in Chapter 4; see Section 4.1. Let $\bar{Y}_{..}$ be the mean, and SS_τ and SS_e be the sum of squares defined in (4.1.2). Then $\bar{Y}_{..}$, SS_τ and SS_e are independently distributed with

$$\bar{Y}_{..} \sim N\left(\mu, \frac{n\sigma_\tau^2 + \sigma_e^2}{an}\right), \quad \frac{SS_\tau}{n\sigma_\tau^2 + \sigma_e^2} \sim \chi_{a-1}^2, \text{ and } \frac{SS_e}{\sigma_e^2} \sim \chi_{a(n-1)}^2, \quad (11.3.1)$$

as specified in (4.1.3). The likelihood is taken to be the joint density of the above three random variables, to be denoted by $L(\bar{y}_{..}, ss_\tau, ss_e | \mu, \sigma_\tau^2, \sigma_e^2)$, where $\bar{y}_{..}$, ss_τ and ss_e denote the observed values of $\bar{Y}_{..}$, SS_τ and SS_e, respectively.

We shall illustrate the derivation of one-sided and two-sided Bayesian tolerance intervals assuming the following non-informative prior for $(\mu, \sigma_\tau^2, \sigma_e^2)$:

$$\pi(\mu, \sigma_\tau^2, \sigma_e^2) \propto \frac{1}{\sigma_e^2(n\sigma_\tau^2 + \sigma_e^2)}. \quad (11.3.2)$$

If we define

$$\eta_\tau = n\sigma_\tau^2 + \sigma_e^2 \text{ and } \eta_e = \sigma_e^2, \quad (11.3.3)$$

then the marginal posterior distributions of η_τ and η_e can be easily worked out if we ignore the natural restriction $\eta_\tau \geq \eta_e$. These posterior distributions, and the conditional posterior distribution of μ, are given by

$$\eta_\tau | \bar{y}_{..}, ss_\tau, ss_e \sim \text{Inv-}\chi^2(a-1, ms_\tau),$$
$$\eta_e | \bar{y}_{..}, ss_\tau, ss_e \sim \text{Inv-}\chi^2(a(n-1), ms_e),$$
$$\text{and } \mu | \bar{y}_{..}, ss_\tau, ss_e, \eta_\tau, \eta_e \sim N\left(\bar{y}_{..}, \frac{\eta_\tau}{an}\right), \quad (11.3.4)$$

where $ms_\tau = \frac{ss_\tau}{a-1}$ and $ms_e = \frac{ss_e}{a(n-1)}$ denote the mean squares. Furthermore, if we ignore the restriction $\eta_\tau \geq \eta_e$, then η_τ and η_e are independently distributed, and the joint posterior distribution of (μ, η_τ, η_e) is the product of the three distributions in (11.3.4). Subject to the restriction $\eta_\tau \geq \eta_e$, samples from the posterior distribution of (η_τ, η_e) can be generated using a rejection sampling procedure; see Guttman and Menzefricke (2003) and Van der Merewe, Pretorius and Mayer (2006). The procedure consists of generating (η_τ, η_e) from the independent scaled inverted chi-square distributions in (11.3.4), and retaining only those pairs (η_τ, η_e) that satisfy $\eta_\tau \geq \eta_e$. Once such a pair of values of (η_τ, η_e) is available, an observation from the conditional posterior distribution of μ can be generated using the conditional normal distribution given in (11.3.4).

A $(p, 1 - \alpha)$ Bayesian upper tolerance limit for $N(\mu, \sigma_\tau^2 + \sigma_e^2)$ can be constructed if we can generate samples from the marginal posterior density of $\mu + z_p \sqrt{\sigma_\tau^2 + \sigma_e^2}$. The $1 - \alpha$ quantile of the sample so generated will give an estimate of the $(p, 1 - \alpha)$ Bayesian upper tolerance limit. The following algorithm can be used to perform the required computation. Note that $\sigma_\tau^2 = (\eta_\tau - \eta_e)/n$.

Algorithm 11.1

1. Generate observations $(\eta_{\tau,i}, \eta_{e,i})$, $i = 1, 2, ..., M_0$, from the distribution of (η_τ, η_e) specified in (11.3.4), where M_0 is the simulation size. Retain the pairs that satisfy $\eta_\tau \geq \eta_e$. Suppose there are M such pairs, say $i = 1, 2, ..., M$.

2. Given each pair $(\eta_{\tau,i}, \eta_{e,i})$ satisfying $\eta_{\tau,i} \geq \eta_{e,i}$, generate μ_i from the conditional normal distribution given in (11.3.4).

3. Compute $\mu_i + z_p \sqrt{\sigma_{\tau,i}^2 + \sigma_{e,i}^2}$, $i = 1, 2, ..., M$, where $\sigma_{\tau,i}^2 = (\eta_{\tau,i} - \eta_{e,i})/n$ and $\sigma_{e,i}^2 = \eta_{e,i}$.

4. The $100(1 - \alpha)$th percentile of the M values $\mu_i + z_p \sqrt{\sigma_{\tau,i}^2 + \sigma_{e,i}^2}$ gives an estimate of the Bayesian upper tolerance limit.

We note that a plot of the M sample values gives an estimate of the marginal posterior distribution of $\mu + z_p \sqrt{\sigma_\tau^2 + \sigma_e^2}$.

A $(p, 1 - \alpha)$ Bayesian upper tolerance limit for the distribution of the true values, i.e., for the distribution $N(\mu, \sigma_\tau^2)$, can be similarly obtained by generating sample values from the posterior distribution of $\mu + z_p \sigma_\tau$, where $\sigma_\tau = \sqrt{(\eta_\tau - \eta_e)/n}$.

Now suppose we want to compute a $(p, 1 - \alpha)$ two-sided Bayesian tolerance interval for $N(\mu, \sigma_\tau^2 + \sigma_e^2)$. For this, we shall follow the numerical procedure explained towards end of the first section of this chapter. Here is an algorithm that will perform the necessary computations:

Algorithm 11.2

1. Generate a sample from the posterior distribution of (μ, η_τ, η_e) as described in Algorithm 11.1. Let the sample be $(\mu_i, \eta_{\tau,i}, \eta_{e,i})$, $i = 1, 2, ..., M$.

2. Obtain a scatter plot of the points

$$\left(\mu_i + z_{\frac{1+p}{2}} \sqrt{\sigma^2_{\tau,i} + \sigma^2_{e,i}}, \ \mu_i - z_{\frac{1+p}{2}} \sqrt{\sigma^2_{\tau,i} + \sigma^2_{e,i}} \right),$$

$i = 1, 2, ..., M$.

3. Compute $\bar{\mu} = \frac{1}{M} \sum_{i=1}^{M} \mu_i$, and plot the line

$$\mu_i - z_{\frac{1+p}{2}} \sqrt{\sigma^2_{\tau,i} + \sigma^2_{e,i}} = - \left(\mu_i + z_{\frac{1+p}{2}} \sqrt{\sigma^2_{\tau,i} + \sigma^2_{e,i}} \right) + 2\bar{\mu}$$

on the scatter plot.

4. Determine the point $(\hat{\mu}_1, \hat{\mu}_2)$ on the line so that $100(1 - \alpha)\%$ of points in the scatter plot satisfy

$$\mu_i - z_{\frac{1+p}{2}} \sqrt{\sigma^2_{\tau,i} + \sigma^2_{e,i}} \geq \hat{\mu}_1 \text{ and } \mu_i + z_{\frac{1+p}{2}} \sqrt{\sigma^2_{\tau,i} + \sigma^2_{e,i}}) \leq \hat{\mu}_2.$$

Then $(\hat{\mu}_1, \hat{\mu}_2)$ is the required $(p, 1 - \alpha)$ two-sided tolerance interval for $N(\mu, \sigma^2_\tau + \sigma^2_e)$. A $(p, 1 - \alpha)$ two-sided Bayesian tolerance interval for $N(\mu, \sigma^2_\tau)$ can be similarly obtained. Note that the symmetry point of the tolerance interval can be different from the mean $\bar{\mu}$, for example, the symmetry point can be taken to be $\bar{y}_{..}$.

The above procedure is described and illustrated in Wolfinger (1998). Apart from the non-informative prior (11.3.2), Wolfinger has also considered an informative prior consisting of a normal distribution for μ and scaled inverted chi-square distributions for σ^2_τ and σ^2_e.

It is quite straightforward to adapt the above procedure for the derivation of Bayesian tolerance intervals for any mixed effects model with balanced data. We refer to Van der Merwe and Hugo (2007), where tolerance intervals are derived for a two-factor nested random effects model with balanced data. One of the examples discussed in the next section is taken from this paper.

The procedures described in this section can also be adopted to the one-way random model with unbalanced data. The approaches are the same; however, the required computations are more involved, as expected. We refer to van der Merwe, Pretorius and Meyer (2006) for the details on such computation.

11.4   Two Examples

Example 11.1 (Continuation of Example 4.2)

We shall compute Bayesian lower tolerance limits and two-sided tolerance intervals for the distributions $N\left(\mu, \sigma_\tau^2 + \sigma_e^2\right)$ and $N\left(\mu, \sigma_\tau^2\right)$ in the context of the study of breaking strengths of cement briquettes, reported in Example 4.2 of Chapter 4, using the non-informative prior distributions mentioned in the previous section; see (11.3.2). In this example, we have the one-way random model and balanced data, with $a = 9$ and $n = 5$. The observed values of $\bar{Y}_{..}$, SS_τ and SS_e are, respectively,

$$\bar{y}_{..} = 543.2, \ \ ss_\tau = 5037 \ \ \text{and} \ \ ss_e = 18918.$$

Using Algorithm 11.1, we computed a $(0.90, 0.95)$ Bayesian lower tolerance limit for $N\left(\mu, \sigma_\tau^2 + \sigma_e^2\right)$ using 5000 simulated samples from the posterior distribution of (μ, η_τ, η_e), where η_τ and η_e are defined in (11.3.3). Rejection sampling was employed to generate the samples, as pointed out earlier. The $(0.90, 0.95)$ Bayesian lower tolerance limit came out to be 499.01. Algorithm 11.2 was followed to obtain a $(0.90, 0.95)$ two-sided tolerance interval; the relevant plot is given in Figure 11.1. The interval was obtained as $(485.14, 601.61)$.

For the distribution $N\left(\mu, \sigma_\tau^2\right)$ of the true values, the $(0.90, 0.95)$ Bayesian lower tolerance limit came out to be 517.42, and the $(0.90, 0.95)$ two-sided tolerance interval came out to be $(508.16, 578.24)$. We note that all of the Bayesian solutions obtained here are quite close to the corresponding frequentist solutions obtained in Chapter 4.

Example 11.2 (Monitoring the quality of synthetic yarn)

This example is taken from Van der Merwe and Hugo (2007), and deals with monitoring the quality of synthetic yarn. For this, data were obtained on the extension property of yarn, i.e., the percentage increase in the length of the yarn before breaking. The data were obtained at the SANS Fibres (Pty) Ltd., South Africa, a company that manufactures continuous filament polyester and nylon yarns. Samples of $b = 8$ packages were obtained per day at the manufacturing plant, and data on the percentage increase in length before breaking were obtained on $n = 5$ samples per package. The data were thus collected for $a = 15$ days during January 1995. The problem of interest is the computation of a tolerance interval for the average percentage increase in length for a specified number of packages, when a specified number of observations are made per package. For example, it could be of interest to compute a lower tolerance

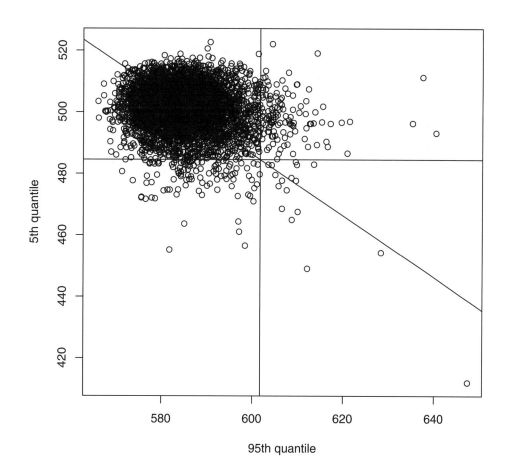

Figure 11.1: Computation of a two-sided (0.90, 0.95) tolerance interval for $N\left(\mu, \sigma_\tau^2 + \sigma_e^2\right)$ in Example 11.1; the intersection of the vertical and horizontal lines gives the lower and upper limits of the tolerance interval.

limit for the average percentage increase in length for $b^* = 8$ packages obtained on a single day, when $n^* = 5$ observations are made per package. One can of course compute a lower tolerance limit for a single observation on the percentage increase in length.

Since the packages are nested within the days, the data can be analyzed using a two-fold nested model that accounts for day-to-day variation and package-to-package variation. Let Y_{ijk} denote the percentage increase in length, before breaking, for the kth sample from the jth package during the ith day. Following Van der Merwe and Hugo (2007), we have the model

$$Y_{ijk} = \mu + \tau_i + \beta_{j(i)} + e_{ijk},$$

$i = 1, 2, ..., a$; $j = 1, 2, ..., b$; and $k = 1, 2, ..., n$. Here τ_i's represent the effects due to the days and $\beta_{j(i)}$'s represent the nested effects due to the packages. We make the usual assumptions: $\tau_i \sim N(0, \sigma_\tau^2)$, $\beta_{j(i)} \sim N(0, \sigma_\beta^2)$, $e_{ijk} \sim N(0, \sigma_e^2)$, and all the random variables are independent. Let

$$\bar{Y}_{ij.} = \frac{1}{n} \sum_{k=1}^{n} Y_{ijk}, \ \bar{Y}_{i..} = \frac{1}{bn} \sum_{j=1}^{b} \sum_{k=1}^{n} Y_{ijk}, \ \text{and} \ \bar{Y}_{...} = \frac{1}{abn} \sum_{i=1}^{a} \sum_{j=1}^{b} \sum_{k=1}^{n} Y_{ijk},$$

and define

$$SS_\tau = bn \sum_{i=1}^{a} (\bar{Y}_{i..} - \bar{Y}_{...})^2,$$

$$SS_\beta = n \sum_{j=1}^{b} \sum_{k=1}^{n} (\bar{Y}_{ij.} - \bar{Y}_{i..})^2,$$

$$\text{and} \ SS_e = \sum_{i=1}^{a} \sum_{j=1}^{b} \sum_{k=1}^{n} (Y_{ijk} - \bar{Y}_{ij.})^2. \tag{11.4.1}$$

Also define

$$\eta_\tau = bn\sigma_\tau^2 + n\sigma_\beta^2 + \sigma_e^2, \ \eta_\beta = n\sigma_\beta^2 + \sigma_e^2, \ \text{and} \ \eta_e = \sigma_e^2. \tag{11.4.2}$$

Note that we have the restriction $\eta_e \leq \eta_\beta \leq \eta_\tau$. Then we have the distributions

$$Z = \sqrt{abn} \left(\bar{Y}_{...} - \mu \right) / \eta_\tau \sim N(0, 1),$$

$$U_\tau^2 = \frac{SS_\tau}{\eta_\tau} \sim \chi_{a-1}^2,$$

$$U_\beta^2 = \frac{SS_\beta}{\eta_\beta} \sim \chi_{a(b-1)}^2,$$

$$\text{and} \ U_e^2 = \frac{SS_e}{\eta_e} \sim \chi_{ab(n-1)}^2, \tag{11.4.3}$$

where Z, U_τ^2, U_β^2 and U_e^2 are also independently distributed. Let $\bar{y}_{...}$, ss_τ, ss_β and ss_e, respectively, denote the observed values of $\bar{Y}_{...}$, SS_τ, SS_β and SS_e. The likelihood function is given by

$$L(\bar{y}_{...}, ss_\tau, ss_\beta, ss_e | \mu, \eta_\tau, \eta_\beta, \eta_e) \propto (\eta_\tau)^{-\frac{a}{2}} (\eta_\beta)^{-\frac{a(b-1)}{2}} (\eta_e)^{-\frac{ab(n-1)}{2}}$$

$$\times \exp\left\{-\frac{1}{2}\left[\frac{abn(\bar{y}_{...} - \mu)^2}{\eta_\tau} + \frac{ss_\tau}{\eta_\tau} + \frac{ss_\beta}{\eta_\beta} + \frac{ss_e}{\eta_e}\right]\right\}.$$

We shall assume the non-informative prior distribution

$$\pi(\mu, \eta_\tau, \eta_\beta, \eta_e) \propto \frac{1}{\eta_\tau \times \eta_\beta \times \eta_e}. \tag{11.4.4}$$

If we ignore the restriction $\eta_e \le \eta_\beta \le \eta_\tau$, the posterior distributions of η_τ, η_β, and η_e, and the conditional posterior distribution of μ are given by

$$\eta_\tau | \bar{y}_{...}, ss_\tau, ss_\beta, ss_e \sim \text{Inv-}\chi^2(a-1, ms_\tau),$$
$$\eta_\beta | \bar{y}_{...}, ss_\tau, ss_\beta, ss_e \sim \text{Inv-}\chi^2(a(b-1), ms_\beta),$$
$$\eta_e | \bar{y}_{...}, ss_\tau, ss_\beta, ss_e \sim \text{Inv-}\chi^2(ab(n-1), ms_e),$$
$$\text{and } \mu | \bar{y}_{...}, ss_\tau, ss_\beta, ss_e, \eta_\tau, \eta_\beta, \eta_e \sim N\left(\bar{y}_{...}, \frac{\eta_\tau}{abn}\right), \tag{11.4.5}$$

where ms_τ, ms_β and ms_e denote the respective mean squares. Subject to the restriction $\eta_e \le \eta_\beta \le \eta_\tau$, a rejection sampling approach can be used to generate observations from the joint posterior distribution of $(\eta_\tau, \eta_\beta, \eta_e)$.

Now suppose we want to compute tolerance limits for the average percentage increase in length for b^* packages, when n^* observations are obtained on each package on a specific day. If $\bar{Y}^*$ denotes such an average, then

$$\bar{Y}^* \sim N\left(\mu, \frac{b^* n^* \sigma_\tau^2 + n^* \sigma_\beta^2 + \sigma_e^2}{b^* n^*}\right).$$

Since $\sigma_\tau^2 = (\eta_\tau - \eta_\beta)/bn$, $\sigma_\beta^2 = (\eta_\beta - \eta_e)/n$, and $\sigma_e^2 = \eta_e$, the above distribution can be expressed as

$$\bar{Y}^* \sim N\left(\mu, \frac{1}{b^* n^*}\left\{\frac{b^* n^*}{bn}\eta_\tau + \frac{n^*}{n}\left(1 - \frac{b^*}{b}\right)\eta_\beta + \left(1 - \frac{n^*}{n}\right)\eta_e\right\}\right). \tag{11.4.6}$$

In order to compute a $(p, 1 - \alpha)$ Bayesian lower tolerance limit for the above distribution, we need to generate samples from the posterior distribution of

$$\mu - z_p\left[\frac{1}{b^* n^*}\left\{\frac{b^* n^*}{bn}\eta_\tau + \frac{n^*}{n}\left(1 - \frac{b^*}{b}\right)\eta_\beta + \left(1 - \frac{n^*}{n}\right)\eta_e\right\}\right]^{1/2}.$$

We can generate samples from the posterior distribution of $(\eta_\tau, \eta_\beta, \eta_e)$ using the posterior distributions of η_τ, η_β, and η_e given in (11.4.5), along with a rejection sampling approach. For each value of $(\eta_\tau, \eta_\beta, \eta_e)$ so generated, the conditional posterior distribution of μ, given in (11.4.5), can be used to generate a sample value from the posterior distribution of μ. Let $(\mu_i, \eta_{\tau,i}, \eta_{\beta,i}, \eta_{e,i})$, $i = 1, 2, ...,$ M, denote M sample values thus generated from the posterior distribution of $(\mu, \eta_\tau, \eta_\beta, \eta_e)$. Now compute

$$\mu_i - z_p \left[\frac{1}{b^*n^*} \left\{ \frac{b^*n^*}{bn}\eta_{\tau,i} + \frac{n^*}{n}\left(1 - \frac{b^*}{b}\right)\eta_{\beta,i} + \left(1 - \frac{n^*}{n}\right)\eta_{e,i} \right\} \right]^{1/2},$$

$i = 1, 2, ..., M$. The 100αth percentile of these values gives an estimate of the required lower tolerance limit.

In order to compute a two-sided $(p, a - \alpha)$ Bayesian tolerance interval for the distribution in (11.4.6), we can use an algorithm similar to Algorithm 11.2. Thus, obtain a scatter plot of

$$q_{1i} = \mu_i - z_{\frac{1+p}{2}} \left[\frac{1}{b^*n^*} \left\{ \frac{b^*n^*}{bn}\eta_{\tau,i} + \frac{n^*}{n}\left(1 - \frac{b^*}{b}\right)\eta_{\beta,i} + \left(1 - \frac{n^*}{n}\right)\eta_{e,i} \right\} \right]^{1/2} \tag{11.4.7}$$

against

$$q_{2i} = \mu_i + z_{\frac{1+p}{2}} \left[\frac{1}{b^*n^*} \left\{ \frac{b^*n^*}{bn}\eta_{\tau,i} + \frac{n^*}{n}\left(1 - \frac{b^*}{b}\right)\eta_{\beta,i} + \left(1 - \frac{n^*}{n}\right)\eta_{e,i} \right\} \right]^{1/2}, \tag{11.4.8}$$

$i = 1, 2, ..., M$. On the scatter plot, plot the line

$$q_1 = -q_2 + 2\bar\mu,$$

where $\bar\mu = \frac{1}{M}\sum_{i=1}^M \mu_i$. Now determine the point $(\hat\mu_2, \hat\mu_1)$ on the line so that $100(1 - \alpha)\%$ of points in the scatter plot satisfy $q_{1i} \geq \hat\mu_1$ and $q_{2i} \leq \hat\mu_2$. Then $(\hat\mu_1, \hat\mu_2)$ is the required $(p, 1 - \alpha)$ two-sided tolerance interval. A $(p, 1 - \alpha)$ two-sided Bayesian tolerance interval for the distribution $N\left(\mu, \frac{b^*n^*\sigma_\tau^2+n^*\sigma_\beta^2}{b^*n^*}\right)$ can be similarly obtained. Note that this is the distribution of the true average percentage increase in length for b^* packages, when n^* observations are obtained on each package on a specific day.

Now let's perform the computations based on the data given in Van der Merwe and Hugo (2007). The observed values of $\bar{Y}_{...}$, SS_τ, SS_β and SS_e are, respectively,

$$\bar{y}_{...} = 20.96, \ ss_\tau = 390.6720, \ ss_\beta = 132.6570, \ \text{and} \ ss_e = 395.4342. \tag{11.4.9}$$

Let's choose $b^* = b = 8$ and $n^* = n = 5$, so that the distribution in (11.4.6) becomes $\bar{Y}^* \sim N\left(\mu, \frac{\eta_\tau}{40}\right)$. Note that in order to compute a $(0.90, 0.95)$ lower tolerance limit for the above distribution, we have to compute a 95% lower confidence limit for $\mu - 1.28\sqrt{\eta_\tau}$. For this we generated 5000 samples from the posterior distribution of $(\mu, \eta_\tau, \eta_\beta, \eta_e)$, and used the rejection sampling approach. Following the numerical procedure indicated above, we obtained the $(0.90, 0.95)$ lower tolerance limit as 14.80. In order to obtain a two-sided $(0.90, 0.95)$ tolerance interval, we obtained a scatter-plot of the samples (q_{1i}, q_{2i}) obtained from the joint posterior distribution of q_1 and q_2, the 5th percentile and the 95th percentile, respectively, of the distribution $N\left(\mu, \frac{\eta_\tau}{40}\right)$. The plot is given in Figure 11.2. Once again, adopting the numerical approach outlined above, the two-sided $(0.90, 0.95)$ tolerance interval came out to be the interval $(12.74, 21.12)$. The interval was computed to be symmetric around $\bar{\mu}$. We also computed a $(0.90, 0.95)$ lower tolerance limit, and a $(0.90, 0.95)$ two-sided tolerance interval for the distribution of the true value $N\left(\mu, \frac{\eta_\tau - \eta_e}{40}\right)$. The lower tolerance limit came out to be 14.788 and the two-sided tolerance interval was obtained as $(12.75, 29.19)$.

All the computations in this chapter were done using the R programming language (2008).

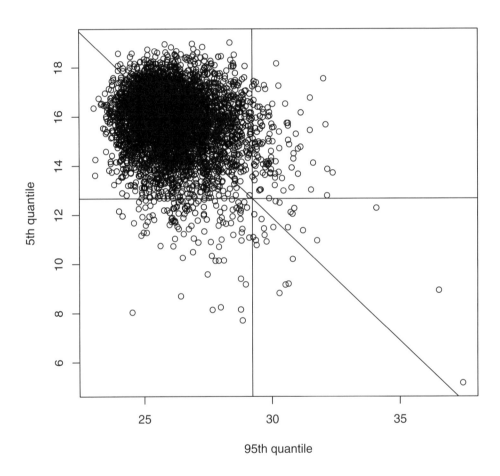

Figure 11.2: Computation of a two-sided $(0.90,\ 0.95)$ tolerance interval for $N\left(\mu,\frac{\eta_\tau}{40}\right)$ in Example 11.2; the intersection of the vertical and horizontal lines gives the lower and upper limits of the tolerance interval.

11.5 Exercises

11.5.1. For the univariate normal distribution $N(\mu, \sigma^2)$, suppose a tolerance interval is required for the distribution of the mean of b future observations.

(a) Explain how you will derive one-sided and two-sided $(p, 1-\alpha)$ Bayesian tolerance intervals under the prior distributions given in (11.2.1) and (11.2.2).

(b) In the set up of Example 2.1, as considered in Section 11.2.2, compute a (0.95, 0.90) Bayesian upper tolerance limit for the mean of $b = 10$ future observations under the prior distribution given in (11.2.2), where the prior parameters have the values specified in Section 11.2.2.

(c) Repeat the above for the mean of $b = 15$ and $b = 20$ future observations.

(d) When the future sample size is increased from $b = 10$ to $b = 15$ and 20, does the (0.95, 0.90) Bayesian upper tolerance limit for the mean increase or decrease? Should you expect this?

11.5.2. Consider the example on the breaking strengths of cement briquettes, given in Example 4.2.

(a) Suppose we want to construct a (0.90, 0.95) Bayesian lower tolerance limit for the distribution of the average breaking strength of 10 observations from a new batch. Compute such a lower tolerance limit under the non-informative prior distribution 11.3.2.

(b) Suppose we want to construct a (0.90, 0.95) Bayesian lower tolerance limit for the distribution of the "true average breaking strength" of 10 observations from a new batch. Compute such a lower tolerance limit under the non-informative prior distribution (11.3.2).

11.5.3. In the set up of the previous problem, suppose we want to compare the breaking strengths of observations from two new batches. The comparisons will be done based on a (0.90, 0.95) two-sided Bayesian tolerance interval for the distribution of $Y_1 - Y_2$, where Y_1 and Y_2 represent observations from two new batches. Compute the required tolerance interval under the non-informative prior distribution (11.3.2).

11.5.4. Let $X_1, X_2, \ldots, X_n$ be a random sample from the one parameter exponential distribution

$$\frac{1}{\theta} \exp\left(-\frac{x}{\theta}\right).$$

Consider the prior distribution $\pi(\theta) \propto \frac{1}{\theta}$.

(a) Derive the posterior distribution of θ.

(b) Show that the posterior distribution of $2n\bar{x}/\theta$ is a chi-square with $2n$ df.

(c) Explain how you will compute a $(p, 1 - \alpha)$ Bayesian lower tolerance limit for the exponential distribution, under the above prior.

[Guttman, 1970]

11.5.5. Consider the one parameter exponential distribution given in the previous problem, and suppose we have the prior distribution

$$\pi(\theta) \propto \theta^{-(n_0+1)} \exp\left(-\frac{\eta}{\theta}\right),$$

where n_0 and η are the prior parameters.

(a) Show that the above prior is a conjugate prior.

(b) Derive the posterior distribution of θ.

(c) Show that the posterior distribution of $2(n\bar{x} + \eta)/\theta$ is a chi-square with $2(n + n_0)$ df.

(d) Explain how you will compute a $(p, 1 - \alpha)$ Bayesian lower tolerance limit for the exponential distribution, under the above prior.

[Guttman, 1970]

11.5.6. The following data, taken from Proschan (1963), give the times between successive failures of the air conditioning equipment in a Boeing 720 airplane: 74, 57, 48, 29, 502, 12, 70, 21, 29, 386, 59, 27, 153, 26, and 326. (This is only part of the data; the full data is given in Proschan (1963, Table 1)). The times between successive failures follows a one-parameter exponential distribution. Assuming the conjugate prior given in the previous problem with prior parameters $n_0 = 100$ and $\eta = 1500$, compute a (0.95, 0.95) upper tolerance limit for the distribution of the times between successive failures.

11.5.7. Consider a two-way crossed classification model with interaction, and with random effects. Follow the arguments in Example 11.2 and outline procedures for computing one-sided and two-sided Bayesian tolerance intervals for the distribution of the observable random variable and the unobservable true values, under a non-informative prior distribution.

Chapter 12

Miscellaneous Topics

12.1 Introduction

This chapter deals with several topics related to tolerance intervals and tolerance regions, not covered in the previous chapters. Some of the topics discussed in this chapter have been fairly well investigated in the literature (for example, β-expectation tolerance intervals, and sample size determination), whereas some of the topics are based on very recent work (tolerance intervals for the ratio of normal random variables). This chapter also includes some important applications of tolerance intervals and regions. For example, a tolerance interval for the ratio of two normal random variables is motivated by a bioassay application related to an influenza vaccine. A number of applications involving censored data are included; environmental and exposure data are frequently of this nature, and so are lifetime data. Yet another important application of a tolerance interval is motivated by the possibility of using it as a reference interval, in order to characterize observations that belong to a reference population. We also note that all of the previous chapters dealt with the computation of tolerance intervals and regions for continuous populations. In this chapter, we shall also address the tolerance interval problem for two discrete distributions: binomial and Poisson.

12.2 β-Expectation Tolerance Regions

The tolerance intervals and regions that we have derived so far are p-content and $(1 - \alpha)$-confidence tolerance intervals and regions. Another type of tolerance intervals (regions) investigated in the literature is referred to as a β−expectation tolerance intervals (regions). A β−expectation tolerance interval or region is such that the average content is β. Such an interval or region is also a prediction interval or prediction region for a future observation. Wilks (1941) derived such an interval for the univariate normal distribution $N(\mu, \sigma^2)$; see also Paulson (1943) and Guttman (1970). Later investigators have derived such intervals in the context of mixed and random effects models; see Mee (1984b) and Lin and Liao (2006). In this section, we shall derive such intervals (one-sided as well as two-sided) for the univariate normal distribution, followed by the derivation for the one-way random model with balanced data, as well as a general mixed model with balanced data. The one-way random model with unbalanced data will also be discussed. In terms of notation, in all the previous chapters we have used p to denote the content of a $(p, 1 - \alpha)$ tolerance interval. However, in the present section, we shall use the terminology β−expectation tolerance interval, instead of a p−expectation tolerance interval, since the terminology β−expectation tolerance interval is now commonly used in the literature.

12.2.1 β−Expectation Tolerance Intervals for the Normal Distribution

Let $X_1, ..., X_n$ be a sample from a $N(\mu, \sigma^2)$ population with unknown mean μ and unknown variance σ^2. Let

$$\bar{X} = \frac{1}{n} \sum_{i=1}^{n} X_i \text{ and } S^2 = \frac{1}{n-1} \sum_{i=1}^{n} (X_i - \bar{X})^2.$$

If $X \sim N(\mu, \sigma^2)$, the interval $\bar{X} \pm kS$ is a β−expectation tolerance interval if the factor k satisfies the condition

$$E_{\bar{X}, S} \left\{ P \left(\bar{X} - kS \leq X \leq \bar{X} + kS | \bar{X}, S \right) \right\} = \beta, \qquad (12.2.1)$$

where $E_{\bar{X}, S}$ denotes expectation with respect to the distribution of $(\bar{X}, S)$. We note that there is only one probability, namely β, associated with a β−expectation tolerance interval.

It is quite easy to derive the factor k satisfying the condition (12.2.1). The derivation becomes particularly simple once it is noted that a β−expectation tolerance interval is simply a prediction interval that satisfies the condition

$$P_{X, \bar{X}, S} \left(\bar{X} - kS \leq X \leq \bar{X} + kS \right) = \beta. \qquad (12.2.2)$$

The proof of the above assertion is quite straightforward. A formal proof appears in Paulson (1943). In fact, Paulson (1943) proves a general result which states that if $(L(\mathcal{S}), U(\mathcal{S}))$ is a $100\beta\%$ prediction interval for a future observation on a random variable X, based on a random sample $\mathcal{S}$, then

$$E_\mathcal{S}\left\{P\left(L(\mathcal{S}) \le X \le U(\mathcal{S})|\mathcal{S}\right)\right\} = \beta,$$

showing that the usual $100\beta\%$ prediction interval is also a β–expectation tolerance interval.

Since $(X - \bar{X})/\left(S\sqrt{1 + \frac{1}{n}}\right)$ follows a t distribution with df $= n - 1$, the factor k satisfying (12.2.2), or equivalently (12.2.1), is given by $k = t_{n-1;\frac{(1+\beta)}{2}}\sqrt{1 + \frac{1}{n}}$. Thus a two-sided β–expectation tolerance interval for $N(\mu, \sigma^2)$ is given by

$$\bar{X} \pm t_{n-1;\frac{(1+\beta)}{2}}\sqrt{1 + \frac{1}{n}} \times S. \tag{12.2.3}$$

A β–expectation upper tolerance limit is easily seen to be $\bar{X} + t_{n-1;\beta}\sqrt{1 + \frac{1}{n}} \times S$ and a β–expectation lower tolerance limit is $\bar{X} - t_{n-1;\beta}\sqrt{1 + \frac{1}{n}} \times S$.

Example 2.1 (continued)

Let us consider Example 2.1 of Chapter 2 dealing with the air lead levels at a laboratory. The sample consists of 15 observations, and it was observed that the log-transformed data followed a normal distribution. The sample mean and standard deviation of the log-transformed data are computed as 4.333 and 1.739, respectively. Let us compute a .95–expectation tolerance interval for the lead levels. With df $= 14$, we have $t_{14;\frac{1+\beta}{2}} = t_{14;0.975} = 2.145$. Simplifying the formula (12.2.3), we get the required β–expectation tolerance interval for the logged data as $(0.4805, \ 8.1855)$. Hence the corresponding interval for the lead levels is $(1.617, \ 3588.537)$. Also, for $\beta = 0.95$, since $t_{14,0.95} = 1.761$, a β–expectation upper tolerance limit for the lead levels simplifies to 1800.485, and a β–expectation lower tolerance limit is 3.223.

12.2.2 β–Expectation Tolerance Intervals for the One-Way Random Model with Balanced Data

For a one-way random model with balanced data, β–expectation tolerance intervals have been derived by Mee (1984b) using the Satterthwaite approximation. The approximation was used in a manner similar to the derivation of a $(p, 1-\alpha)$ tolerance interval due to Mee and Owen (1983); see Section 4.3.1 of Chapter

4. In a recent paper, Lin and Liao (2006) have derived $\beta-$expectation tolerance intervals for a general mixed model with balanced data, which is the model considered in Section 6.3. In a later paper, Lin and Liao (2008) also derived simultaneous prediction intervals in the context of a general random effects model with balanced data. Here we shall consider the solutions due to Mee (1984b) and due to Lin and Liao (2006) for the one-way random model with balanced data.

We shall use the notations in Chapter 4. Thus we have a levels of a factor, randomly selected, with n observations per level. The model we shall consider is given in (4.1.1), involving the two variance components σ_τ^2 and σ_e^2. Let $\bar{Y}_{..}$, SS_τ and SS_e be as defined in (4.1.2), with the distributions specified in (4.1.3). Furthermore, let $\widehat{\sigma}_\tau^2$ and $\widehat{\sigma}_e^2$ denote the ANOVA estimators of σ_τ^2 and σ_e^2, respectively; see (4.1.4). When a random variable Y follows the one-way random model with the standard assumptions, we have $Y \sim N(\mu, \sigma_\tau^2 + \sigma_e^2)$. Since $\bar{Y}_{..}$ is an estimator of μ, we shall assume that the two sided $\beta-$expectation tolerance interval is of the form $\bar{Y}_{..} \pm k\sqrt{\widehat{\sigma}_\tau^2 + \widehat{\sigma}_e^2}$, where the factor k is to be determined.

Mee's Approach

In order to describe the approach due to Mee (1984b), let R be the variance ratio $\frac{\sigma_\tau^2}{\sigma_e^2}$ and $R_0 = \frac{R+1}{nR+1}$. We shall now use the Satterthwaite approximation to conclude that

$$\frac{\widehat{\sigma}_\tau^2 + \widehat{\sigma}_e^2}{\sigma_\tau^2 + \sigma_e^2} \sim \frac{\chi_f^2}{f},$$

where

$$f = \frac{(R+1)^2}{\{(R + \frac{1}{n})^2 /(a-1)\} + \{(1 - \frac{1}{n})/(an)\}}, \qquad (12.2.4)$$

as noted in Section 4.3.1. The condition to be satisfied by k is

$$P_{\bar{Y}_{..}, SS_\tau, SS_e}\left(\bar{Y}_{..} - k\sqrt{\widehat{\sigma}_\tau^2 + \widehat{\sigma}_e^2} \leq Y \leq \bar{Y}_{..} + k\sqrt{\widehat{\sigma}_\tau^2 + \widehat{\sigma}_e^2}\right) = \beta.$$

Since $\bar{Y}_{..} \sim N\left(\mu, \frac{n\sigma_\tau^2 + \sigma_e^2}{an}\right)$ independently of $Y \sim N(\mu, \sigma_\tau^2 + \sigma_e^2)$, we have

$$\frac{Y - \bar{Y}_{..}}{\sqrt{(\sigma_\tau^2 + \sigma_e^2) + \frac{(n\sigma_\tau^2 + \sigma_e^2)}{an}}} \sim N(0,1).$$

The condition to be satisfied by k can be expressed as

$$P\left(-k \leq \frac{Y - \bar{Y}_{..}}{\sqrt{(\sigma_\tau^2 + \sigma_e^2) + \frac{(n\sigma_\tau^2 + \sigma_e^2)}{an}}} \middle/ \left(\frac{\widehat{\sigma}_\tau^2 + \widehat{\sigma}_e^2}{\sigma_\tau^2 + \sigma_e^2}\right)^{\frac{1}{2}} \leq k\right) = \beta.$$

In view of the Satterthwaite approximation, the random quantity in the above equation has an approximate t distribution with df $= f$. Noting that

$$\frac{Y - \bar{Y}_{..}}{\sqrt{(\sigma_\tau^2 + \sigma_e^2) + \frac{(n\sigma_\tau^2 + \sigma_e^2)}{an}}} \Bigg/ \sqrt{\frac{\widehat{\sigma}_\tau^2 + \widehat{\sigma}_e^2}{\sigma_\tau^2 + \sigma_e^2}} = \frac{Y - \bar{Y}_{..}}{\sqrt{1 + \frac{1}{anR_0}}} \Bigg/ \sqrt{\widehat{\sigma}_\tau^2 + \widehat{\sigma}_e^2},$$

where $R_0 = \frac{R+1}{nR+1} = \frac{\sigma_\tau^2 + \sigma_e^2}{n\sigma_\tau^2 + \sigma_e^2}$, the factor k can be obtained by following the corresponding derivation for $N(\mu, \sigma^2)$. This gives

$$k = t_{f; \frac{1+\beta}{2}} \sqrt{1 + \frac{1}{anR_0}}.$$

However, a β–expectation tolerance interval cannot be computed using the above factor k since it depends on R_0, which is a function of the unknown variance ratio $\frac{\sigma_\tau^2}{\sigma_e^2}$. Mee (1984b) recommends replacing R by the estimate

$$\widehat{R} = \max\left\{0, \frac{1}{n}\left(\frac{MS_\tau}{MS_e} - 1\right)\right\}, \tag{12.2.5}$$

where MS_τ and MS_e are the mean squares in the ANOVA table; see Table 4.1 in Chapter 4. Thus an estimator of f is given by

$$\widehat{f} = \frac{(\widehat{R} + 1)^2}{\left\{\left(\widehat{R} + \frac{1}{n}\right)^2 / (a-1)\right\} + \left\{\left(1 - \frac{1}{n}\right) / (an)\right\}} \tag{12.2.6}$$

and the estimated factor $\widehat{k}$ is

$$\widehat{k} = t_{\widehat{f}; \frac{1+\beta}{2}} \sqrt{1 + \frac{1}{an\widehat{R}_0}}, \tag{12.2.7}$$

where $\widehat{R}_0 = \frac{\widehat{R}+1}{n\widehat{R}+1}$. The β–expectation tolerance interval is finally given by

$$\bar{Y}_{..} \pm t_{\widehat{f}; \frac{1+\beta}{2}} \sqrt{1 + \frac{1}{an\widehat{R}_0}} \times \sqrt{\widehat{\sigma}_\tau^2 + \widehat{\sigma}_e^2}. \tag{12.2.8}$$

Similarly, the β–expectation upper tolerance limit is

$$\bar{Y}_{..} + t_{\widehat{f}; \beta} \sqrt{1 + \frac{1}{an\widehat{R}_0}} \times \sqrt{\widehat{\sigma}_\tau^2 + \widehat{\sigma}_e^2},$$

and the β–expectation lower tolerance limit is

$$\bar{Y}_{..} - t_{\widehat{f}; \beta} \sqrt{1 + \frac{1}{an\widehat{R}_0}} \times \sqrt{\widehat{\sigma}_\tau^2 + \widehat{\sigma}_e^2}.$$

The Lin-Liao Approach

The idea proposed by Lin and Liao (2006) is for a general mixed linear model with balanced data. Here we shall adopt the procedure for the one-way random model with balanced data; the more general model will be taken up later. Assume that the β-expectation tolerance interval is of the form $\bar{Y}_{..} \pm D_1$, where D_1 is a margin of error statistic to be determined. The condition to be satisfied by D_1 is

$$
\begin{aligned}
\beta &= P\left(-D_1 \leq Y - \bar{Y}_{..} \leq D_1\right) \\
&= P\left(-D_1 \leq Z\sqrt{(\sigma_\tau^2 + \sigma_e^2) + \frac{(n\sigma_\tau^2 + \sigma_e^2)}{an}} \leq D_1\right), \qquad (12.2.9)
\end{aligned}
$$

where

$$
Z = \frac{Y - \bar{Y}_{..}}{\sqrt{(\sigma_\tau^2 + \sigma_e^2) + \frac{(n\sigma_\tau^2 + \sigma_e^2)}{an}}} \sim N(0, 1).
$$

Since $Z\sqrt{(\sigma_\tau^2 + \sigma_e^2) + \frac{(n\sigma_\tau^2 + \sigma_e^2)}{an}}$ has a distribution that is symmetric around zero, D_1 can be taken to be the $\left(\frac{1+\beta}{2}\right)$ quantile of this random variable. For such a choice of D_1, $-D_1$ will be the $\left(\frac{1-\beta}{2}\right)$ quantile of the same random variable, and consequently (12.2.9) will hold. However, since σ_τ^2 and σ_e^2 are unknown, Lin and Liao (2006) recommend that they be replaced by generalized pivotal quantities (GPQs) before computing the $\left(\frac{1+\beta}{2}\right)$ quantile. Noting that

$$
(\sigma_\tau^2 + \sigma_e^2) + \frac{(n\sigma_\tau^2 + \sigma_e^2)}{an} = \frac{1}{n}\left(1 + \frac{1}{a}\right)(n\sigma_\tau^2 + \sigma_e^2) + \left(1 - \frac{1}{n}\right)\sigma_e^2,
$$

a GPQ is given by

$$
\frac{1}{n}\left(1 + \frac{1}{a}\right)\frac{ss_\tau}{U_\tau^2} + \left(1 - \frac{1}{n}\right)\frac{ss_e}{U_e^2},
$$

where ss_τ and ss_e denote the observed values of SS_τ and SS_e, respectively, and U_τ^2 and U_e^2 denote independent random variables having chi-square distributions with $a - 1$ and $a(n-1)$ df, respectively; see (4.1.3) in Chapter 4. The derivation of this GPQ should be clear from the derivations in Section 4.3.3. The margin of error statistic D_1 is taken to be the $\left(\frac{1+\beta}{2}\right)$ quantile of the random variable

$$
Z\sqrt{\frac{1}{n}\left(1 + \frac{1}{a}\right)\frac{ss_\tau}{U_\tau^2} + \left(1 - \frac{1}{n}\right)\frac{ss_e}{U_e^2}}. \qquad (12.2.10)
$$

The percentile can be easily estimated by Monte Carlo simulation, and an algorithm (similar to Algorithm 4.1, for example) can be easily specified for doing this computation. The computation of one-sided β−expectation tolerance limits should be clear. For example, in order to obtain a β−expectation upper tolerance limit, we have to use the β quantile of the above random variable.

The numerical results reported in Lin and Liao (2006) show that the above procedure provides coverages close to β and is more accurate compared to the solution due to Mee (1984b). The solution obtained by Mee (1984b) appears to be somewhat conservative; see Table 2 in Lin and Liao (2006). However, from a practical point of view, the amount of conservatism appears to be somewhat insignificant. Note also that the approximate β−expectation tolerance interval due to Mee (1984b) has an explicit analytic expression, and hence is easily computed. A major advantage of the Lin and Liao (2006) methodology is that it can be easily applied to obtain β−expectation tolerance intervals in the context of general models with balanced data; this will be discussed later.

Let us also apply the Lin and Liao (2006) approach for computing a β-expectation tolerance interval for the true value $\mu + \tau_i$ in the model (4.1.1). Following a derivation similar to what is given above, we conclude that such an interval is given by $\bar{Y}_{..} \pm D_2$, where the margin of error statistic D_2 is taken to be the $100\left(\frac{1+\beta}{2}\right)$th percentile of the random variable

$$Z\sqrt{\left[\frac{1}{n}\left(1+\frac{1}{a}\right)\frac{ss_\tau}{U_\tau^2} - \frac{1}{n}\frac{ss_e}{U_e^2}\right]_+}, \qquad (12.2.11)$$

where for any scalar c, $c_+ = \max\{c, 0\}$. Similarly, a β−expectation upper tolerance limit, and a β−expectation lower tolerance limit can also be obtained.

Example 4.1 (continued)

As noted earlier, there are $a = 5$ batches, and each batch consists of $n = 5$ specimens. From the ANOVA Table 4.3, we have $ss_\tau = 4163.4$ and $ss_e = 1578.4$. We also have the observed values

$$\bar{y}_{..} = 388.36, \ ms_\tau = 1040.8, \ ms_e = 78.9, \ \hat{\sigma}_\tau^2 = 192.36, \text{ and } \hat{\sigma}_e^2 = 78.90,$$

where ms_τ and ms_e are the observed values of MS_τ and MS_e, respectively.

β-*Expectation Tolerance Intervals for* $N(\mu, \sigma_\tau^2 + \sigma_e^2)$: Let us first compute 0.95-expectation tolerance intervals using Mee's (1984b) approach. Towards this, we note that $\hat{R} = 2.4383$, $\hat{R}_0 = 0.2606$, $\hat{f} = 6.6710$ and $t_{\hat{f};.975} = 2.3885$. Using these quantities in (12.2.8), we get the .95-expectation tolerance interval as $388.36 \pm 42.25 = (346.11, 430.61)$.

We shall now compute .95-expectation tolerance intervals using the Lin and Liao (2006) approach. The 95th and 97.5th percentiles of the quantity in (12.2.10) are, respectively, 36.50 and 46.52; these percentile are obtained using a simulation consisting 10,000 runs. Thus, .95-expectation lower limit is $\bar{y}_{..} - 36.50 = 351.86$ and the upper limit is $\bar{y}_{..} + 36.50 = 424.86$. The 0.95-expectation tolerance interval is $\bar{y}_{..} \pm 46.52 = (341.84, 434.88)$.

β-Expectation Tolerance Intervals for $N(\mu, \sigma_\tau^2)$: To apply the Lin-Liao (2006) approach, the 95th and 97.5th percentiles of the quantity in (12.2.11) are 32.93 and 43.31 respectively. Thus, the 95% lower prediction limit for the distribution $N(\mu, \sigma_\tau^2)$ is $\bar{y}_{..} - 32.93 = 355.43$. Furthermore, the upper prediction limit is 431.67, and the 95% prediction interval is $\bar{y}_{..} \pm 43.31 = (345.05, 431.67)$.

12.2.3 β–Expectation Tolerance Intervals for the One-Way Random Model with Unbalanced Data

It should be clear that the derivations presented above in the context of the one-way random model with balanced data use many of the ideas and results in Chapter 4. Much the same way, some of the results obtained in Chapter 5 can be used to derive β–expectation tolerance intervals for the one-way random model with unbalanced data. In fact it is quite convenient to use the approximations described for the Krishnamoorthy and Mathew (2004) approach in Section 5.3.1. The details are given below for the derivation of two-sided β–expectation tolerance intervals. A β–expectation upper tolerance limit, or a β–expectation lower tolerance limit can be similarly obtained.

Consider the one-way random model with unbalanced data, given in (5.1.1) and let SS_e denote the error sum of squares. Furthermore, let $\tilde{n}$, $\bar{\bar{Y}}$ and $SS_{\bar{y}}$ be as defined in (5.3.1), and let $U_{\bar{y}}^2$ be the approximate chi-square random variable defined in (5.3.2). Suppose we take a β–expectation tolerance interval for a random variable Y, following the one-way random model, to be $\bar{\bar{Y}} \pm D_3$, where D_3 is the margin of error statistic, to be determined. Using the results in Section 5.3.1, and following the derivations for the case of balanced data, we conclude that D_3 can be taken to be the $\left(\frac{1+\beta}{2}\right)$ quantile of the random variable

$$Z\sqrt{\left(1 + \frac{1}{a}\right)\frac{ss_{\bar{y}}}{U_{\bar{y}}^2} + (1 - \tilde{n})\frac{ss_e}{U_e^2}}, \qquad (12.2.12)$$

where $ss_{\bar{y}}$ and ss_e denote the observed values of $SS_{\bar{y}}$ and SS_e, respectively, $U_e^2 \sim \chi_{N-a}^2$, $U_{\bar{y}}^2 \sim \chi_{a-1}^2$ and $Z \sim N(0, 1)$. Here $N = \sum_{i=1}^{a} n_i$. Note from (5.3.1) that $\tilde{n} < 1$ and hence $1 - \tilde{n} > 0$. Similarly, when we have unbalanced data, a

β–expectation tolerance interval for the true value $\mu + \tau_i$ in the model (5.1.1) can be obtained as $\bar{\bar{Y}} \pm D_4$, where the margin of error statistic D_4 can be taken to be the $\left(\frac{1+\beta}{2}\right)$ quantile of the random variable

$$Z\sqrt{\left[\left(1+\frac{1}{a}\right)\frac{ss_{\bar{y}}}{U_{\bar{y}}^2} - \tilde{n}\frac{ss_e}{U_e^2}\right]_+}. \qquad (12.2.13)$$

Example 5.1 (continued)

Based on the moisture content data in Table 5.1, let us compute a .95-expectation upper tolerance limit for the moisture content. Here $a = 5$ and $N = \sum_{i=1}^{a} n_i = 14$. The observed values are:

$\bar{y}_{1.} = 7.98,\ \ \bar{y}_{2.} = 6.63,\ \ \bar{y}_{3.} = 7.25,\ \ \bar{y}_{4.} = 9.13,\ \ \bar{y}_{5.} = 7.10,\ \ \bar{\bar{y}} = 7.62,\ \ ss_{\bar{y}} = 3.80$, and $ss_e = 7.17$.

To find .95-expectation tolerance interval for the distribution $N(\mu, \sigma_\tau^2 + \sigma_e^2)$, the 95th and 97.5th percentiles of the quantity in (12.2.12) are computed as 2.5797 and 3.2838, respectively. Using these, the .95-expectation lower tolerance limit is computed as $\bar{\bar{y}} - 2.5797 = 5.04$, and the upper tolerance limit is computed as $\bar{\bar{y}} + 2.5797 = 10.20$. The .95-expectation tolerance interval is computed as $\bar{\bar{y}} \pm 3.2838 = (4.34, 10.90)$.

To find .95-expectation tolerance interval for the distribution $N(\mu, \sigma_\tau^2)$, the 95th and 97.5th percentiles of the quantity in (12.2.13) are computed as 2.0136 and 2.7150, respectively. Using these percentiles, the .95-expectation lower tolerance limit is computed as $\bar{\bar{y}} - 2.0136 = 5.60$, and the upper tolerance limit is computed as $\bar{\bar{y}} + 2.0136 = 9.63$. The .95-expectation tolerance interval is computed as $\bar{\bar{y}} \pm 2.7150 = (4.90, 10.33)$.

12.2.4 β–Expectation Tolerance Intervals for a General Mixed Effects Model with Balanced Data

As noted in Chapter 6, a general formulation of the problem consists of deriving a β–expectation tolerance interval for the distribution $N(0, \sum_{i=1}^{q} c_i \sigma_i^2)$, based on the independent statistics $\hat{\theta}$ and SS_i $(i = 1, 2, ..., q)$, with

$$Z = \frac{\hat{\theta} - \theta}{\sqrt{\sum_{i=1}^{q} d_i \sigma_i^2}} \sim N(0, 1)$$

$$\text{and } U_i^2 = \frac{SS_i}{\sigma_i^2} \sim \chi_{f_i}^2, \tag{12.2.14}$$

where the c_i's and d_i's are known constants. The derivation given here, due to Lin and Liao (2006), is similar to the derivation given earlier for the case of the one-way random model with balanced data. Thus a two-sided β−expectation tolerance interval for the distribution $N(\theta, \sum_{i=1}^{q} c_i \sigma_i^2)$ is given by $\hat{\theta} \pm D_5$, where the margin of error statistic D_5 can be taken to be the $\left(\frac{1+\beta}{2}\right)$ quantile of the random variable

$$Z \sqrt{\left[\sum_{i=1}^{q}(c_i + d_i)\frac{ss_i}{U_i^2}\right]_+}, \tag{12.2.15}$$

with $Z \sim N(0, 1)$ and ss_i being the observed value of SS_i $(i = 1, 2, ..., q)$.

Example 6.1 (continued)

Let's now compute a β−expectation tolerance interval in the context of the glucose monitoring meter experiment described in Example 6.1, Section 6.2. The model and the notations are given in Section 6.2, and a sample of data are given in Table A1, Appendix A. We shall obtain a β−expectation tolerance interval for the distribution $N(\mu_T - \mu_R, \sigma_T^2)$, based on the independent random variables $\bar{W} - \bar{Y}$, SS_T, SS_R and SS_e, specified in the solution to Example 6.1, as given in Section 6.3. We shall choose $\beta = 0.95$. The observed values of $\bar{W} - \bar{Y}$, SS_T, SS_R and SS_e are -1.3791, 718.9790, 280.3167, and 8376.4525, respectively. In the notations of the present section, $\hat{\theta} = \bar{W} - \bar{Y}$. In terms of the notations for Example 6.1 as given in Section 6.2 and Section 6.3, and using the notations in the present section, we have $c_1 + d_1 = \frac{5}{132}$, $c_2 + d_2 = \frac{1}{540}$, $c_3 + d_3 = -\frac{1}{27}$. Furthermore, $ss_1 = $ observed value of $SS_T = 718.98$, $ss_2 = $ observed value of $SS_R = 280.32$, and $ss_3 = $ observed value of $SS_e = 8376.45$. Also, $U_1^2 \sim \chi_{43}^2$, $U_2^2 \sim \chi_{19}^2$, and $U_3^2 \sim \chi_{1656}^2$. Based on $\beta = 0.95$ and using the above quantities in (12.2.15), simulation consisting of 100,000 runs gave the 0.975 quantile of the random variable in (12.2.15) as 1.374. Since the observed value of $\hat{\theta}$ is -1.3791, the β−expectation tolerance interval for the distribution $N(\mu_T - \mu_R, \sigma_T^2)$ simplifies to the interval $(-2.7531, -0.0051)$.

A natural question that comes up is on the computation of β−expectation tolerance intervals for models with unbalanced data, other than the one-way random model. For example, in Section 6.4 of Chapter 6, $(p, 1 - \alpha)$ tolerance intervals were derived for a general linear mixed model involving only one random effect, and hence only two variance components. There are two main difficulties in deriving β–expectation tolerance intervals for such models. The first difficulty is on the choice of the estimator for estimating the fixed effects parameter vector. For the one-way random model with unbalanced data, we used the quantity $\bar{\bar{Y}}$ defined in (5.3.1). The second issue is on defining an appropriate sum of squares (similar to $SS_{\bar{y}}$ given in (5.3.1)) so that this sum of squares has an approximate distribution that is a scalar multiple of a chi-square, and furthermore, the sum of squares is also independent of the error sum of squares and also independent of the estimator of the fixed effects parameter. Since it is not clear how all of these can be achieved, we do not have a solution to the β–expectation tolerance interval problem for unbalanced data situations more general than the one-way random model with unbalanced data.

12.2.5   Multivariate β–Expectation Tolerance Regions

Let us consider the derivation of a multivariate β-expectation tolerance region in the context of the multivariate linear regression model discussed in Chapter 10. Thus we have the model (10.1.3) for the $q \times n$ data matrix $\boldsymbol{Y}$, and we shall derive the β-expectation tolerance region using the independent quantities $\bar{\boldsymbol{Y}}$, $\widehat{\boldsymbol{B}}$, and $\boldsymbol{A}$, having the distributions

$$\bar{\boldsymbol{Y}} \sim N\left(\boldsymbol{\beta}_0 + \boldsymbol{B}\mathbf{x}, \frac{1}{n}\boldsymbol{\Sigma}\right), \quad \widehat{\boldsymbol{B}} \sim N\left[\boldsymbol{B}, \left\{\mathbf{X}\left(\boldsymbol{I}_n - \frac{1}{n}\mathbf{1}_n\mathbf{1}_n'\right)\mathbf{X}'\right\}^{-1} \otimes \boldsymbol{\Sigma}\right],$$

and $\boldsymbol{A} \sim W_q(\boldsymbol{\Sigma}, n - m - 1);$

we refer to Section 10.1 of Chapter 10 for details. A tolerance region is required for the $q \times 1$ vector $\boldsymbol{Y}(\mathbf{x})$ having the distribution in (10.1.4):

$$\boldsymbol{Y}(\mathbf{x}) \sim N(\boldsymbol{\beta}_0 + \boldsymbol{B}\mathbf{x}, \boldsymbol{\Sigma}).$$

The statistic we shall use to construct the β-expectation tolerance region is the quantity used in (10.1.7):

$$[\boldsymbol{Y}(\mathbf{x}) - \bar{\boldsymbol{Y}} - \widehat{\boldsymbol{B}}(\mathbf{x} - \bar{\mathbf{x}})]'\boldsymbol{A}^{-1}[\boldsymbol{Y}(\mathbf{x}) - \bar{\boldsymbol{Y}} - \widehat{\boldsymbol{B}}(\mathbf{x} - \bar{\mathbf{x}})]. \qquad (12.2.16)$$

Note that

$$\boldsymbol{Y}(\mathbf{x}) - \bar{\boldsymbol{Y}} - \widehat{\boldsymbol{B}}(\mathbf{x} - \bar{\mathbf{x}}) \sim N\left(0, c(\mathbf{x})\boldsymbol{\Sigma}\right),$$

where $\quad c(\mathbf{x}) = 1 + \dfrac{1}{n} + (\mathbf{x} - \bar{\mathbf{x}})'\{\mathbf{X}(\boldsymbol{I}_n - n^{-1}\mathbf{1}_n\mathbf{1}_n')\mathbf{X}'\}^{-1}(\mathbf{x} - \bar{\mathbf{x}}).$ (12.2.17)

Hence a Hotelling's T^2 statistic can be constructed using (12.2.16), and can be used to obtain a β-expectation tolerance region for $\mathbf{Y}(\mathbf{x})$. The statistic, say $T^2(\mathbf{x})$, is given by

$$T^2(\mathbf{x}) = \frac{n-m-1}{c(\mathbf{x})}[\mathbf{Y}(\mathbf{x}) - \bar{\mathbf{Y}} - \widehat{\mathbf{B}}(\mathbf{x}-\bar{\mathbf{x}})]'\mathbf{A}^{-1}[\mathbf{Y}(\mathbf{x}) - \bar{\mathbf{Y}} - \widehat{\mathbf{B}}(\mathbf{x}-\bar{\mathbf{x}})]. \quad (12.2.18)$$

Let $F_{q,n-m-q;\beta}$ denote the β quantile of an F distribution with $(q, n-m-q)$ df. Then the β-expectation tolerance region for $\mathbf{Y}(\mathbf{x})$ is given by

$$\left\{ \mathbf{Y}(\mathbf{x}) : T^2(\mathbf{x}) \leq \frac{q(n-m-1)}{n-m-q}F_{q,n-m-q;\beta} \right\}.$$

An optimum property of the above tolerance region is noted in Evans and Fraser (1980). A β-expectation tolerance region can be similarly defined for a q-variate multivariate normal distribution $N_q(\boldsymbol{\mu}, \boldsymbol{\Sigma})$; see Fraser and Guttman (1956). Let $\mathbf{Y}_1, ..., \mathbf{Y}_n$ be a sample from $N_q(\boldsymbol{\mu}, \boldsymbol{\Sigma})$ and define

$$\bar{\mathbf{Y}} = \frac{1}{n}\sum_{i=1}^{n}\mathbf{Y}_i \text{ and } \mathbf{A} = \sum_{i=1}^{n}(\mathbf{Y}_i - \bar{\mathbf{Y}})(\mathbf{Y}_i - \bar{\mathbf{Y}})',$$

as defined in Chapter 9. If $\mathbf{Y} \sim N_q(\boldsymbol{\mu}, \boldsymbol{\Sigma})$, then a β-expectation tolerance region for $\mathbf{Y}$ is given by

$$\left\{ \mathbf{Y} : \left(1 + \frac{1}{n}\right)^{-1}(\mathbf{Y} - \bar{\mathbf{Y}})'\mathbf{A}^{-1}(\mathbf{Y} - \bar{\mathbf{Y}}) \leq \frac{q}{n-q}F_{q,n-q;\beta} \right\}.$$

12.2.6 Bayesian β-Expectation Tolerance Intervals

So far, we have derived β-expectation tolerance intervals from a frequentist perspective. Such intervals (and regions) can be obtained in the Bayesian framework as well. Here we shall give a very brief discussion of this. Using the notations in Section 11.1, let X denote a random variable whose distribution depends on a parameter $\boldsymbol{\theta}$, so that the realizations of X give the observed data. Let $\boldsymbol{x}$ represent the observed sample, and $\pi(\boldsymbol{\theta})$ denote a prior distribution for $\boldsymbol{\theta}$. If X_0 denote a future observation from the distribution of X, then the *posterior predictive distribution* of X_0, denoted by $p(x_0|\boldsymbol{x})$, is given by

$$p(x_0|\boldsymbol{x}) = \int p(x_0|\boldsymbol{\theta})p(\boldsymbol{\theta}|\boldsymbol{x})d\boldsymbol{\theta},$$

where $p(x_0|\boldsymbol{\theta})$ is the distribution of X_0, and $p(\boldsymbol{\theta}|\boldsymbol{x})$ is the posterior distribution of $\boldsymbol{\theta}$; see Gelman et al. (2003, p. 8). Bayesian β-expectation tolerance intervals are obtained when the expectation is with respect to the posterior predictive distribution $p(x_0|\boldsymbol{x})$.

For normal distributions, such tolerance intervals (one-sided as well as two sided) are given in Guttman (1970, Chapter 8). Bayesian simulation can be used to obtain such tolerance intervals in random effects models; see Wolfinger (1998), Van der Merwe and Hugo (2007) and Van der Merwe, Pretorius and Meyer (2006). Bayesian prediction intervals are constructed in Hamada et al. (2004) so that a specified proportion of observations in a finite sample will be contained in the interval, with a specified probability. The authors have also investigated the relationship of such intervals and tolerance intervals.

12.3 Tolerance Limits for a Ratio of Normal Random Variables

The procedure for computing a confidence interval for the ratio of two normal means is well known, and has important applications in bioassay. Here the means could be based on normal distributions that may or may not be independent. The computation of a tolerance interval for the ratio of two normally distributed random variables (independent or dependent) is a problem that has not received much attention in the literature, even though the problem does come up in some pharmaceutical applications and in the context of bioassays. The first attempt at deriving a tolerance interval for the ratio of two independent normal random variables is due to Hall and Sampson (1973), who gave an approximate solution assuming that the coefficient of variations are small. The authors also gave an application related to drug development. The same tolerance interval problem also comes up in the context of testing parallelism in immunoassays; see Yang, Zhang and Cho (2006). In a very recent article, Zhang et al. (2009) have obtained upper and lower tolerance limits for the ratio of two independent or dependent normal random variables, using the generalized confidence interval idea. In this section, we shall describe the solution due to Zhang et al. (2009). We shall also discuss a bioassay application related to an influenza vaccine.

Suppose X_1 and X_2 are independent normally distributed random variables, or $(X_1, X_2)'$ follows a bivariate normal distribution. We shall address the problem of computing an upper tolerance limit for X_1/X_2. It appears that a simple or direct approach is not possible for solving such a tolerance interval problem. We already noted that a $(p, 1-\alpha)$ upper tolerance limit for any distribution is a $1-\alpha$ upper confidence limit for the p quantile of the distribution. A further property we shall use in this section concerns the relation between one-sided tolerance limits for a random variable, and one-sided confidence limits for the corresponding survival probability; see, for example, Section 1.1.3. That is, if we can compute a $1 - \alpha$ lower confidence limit for the cumulative distribution function (cdf) of a continuous random variable, evaluated at a point t, then by equating the lower confidence limit so obtained to p, and solving for t, the solution provides the required $(p, 1 - \alpha)$ upper tolerance limit for the random variable. For our problem of computing an upper tolerance limit for X_1/X_2, the use of this property is necessary because we do have a convenient representation for the cdf of X_1/X_2, but not for the quantiles. Thus we shall compute a $1 - \alpha$ lower confidence limit for the cdf $P(X_1/X_2 \le t)$, equate the lower confidence limit to p, and numerically obtain the solution to t. This will provide the required $(p, 1 - \alpha)$ upper tolerance limit for X_1/X_2.

We shall first give a representation for the cdf $P(X_1/X_2 \le t)$. Suppose $(X_1, X_2)' \sim N_2(\boldsymbol{\mu}, \boldsymbol{\Sigma})$, a bivariate normal distribution with mean $\boldsymbol{\mu} = (\mu_1, \mu_2)'$ and covariance matrix $\boldsymbol{\Sigma} = \begin{pmatrix} \sigma_{11} & \sigma_{12} \\ \sigma_{12} & \sigma_{22} \end{pmatrix}$. It is well known that the conditional distribution of X_1 given X_2 is $N\left(\mu_1 + \sigma_{12}\sigma_{22}^{-1}(X_2 - \mu_2), \sigma_{11.2}\right)$, where $\sigma_{11.2} = \sigma_{11} - \frac{\sigma_{12}^2}{\sigma_{22}}$. Hence

$$
\begin{aligned}
P\left(\frac{X_1}{X_2} \le t\right) &= E_{X_2}\left\{P\left(\frac{X_1}{X_2} \le t | X_2\right)\right\} \\
&= E_{X_2}\left\{P(X_1 \le tX_2)|X_2 > 0) + P(X_1 \ge tX_2)|X_2 < 0)\right\}.
\end{aligned}
$$

Using this result, and using the standard normal cdf Φ, we get the following representation:

$$
\begin{aligned}
P\left(\frac{X_1}{X_2} \le t\right) &= \int_0^\infty \Phi\left(\frac{x_2 t - \mu_1 - \sigma_{12}\sigma_{22}^{-1}(x_2 - \mu_2)}{\sqrt{\sigma_{11.2}}}\right) f(x_2; \mu_2, \sigma_{22})dx_2 \\
&\quad + \int_{-\infty}^0 \left[1 - \Phi\left(\frac{x_2 t - \mu_1 - \sigma_{12}\sigma_{22}^{-1}(x_2 - \mu_2)}{\sqrt{\sigma_{11.2}}}\right)\right] f(x_2; \mu_2, \sigma_{22})dx_2 \\
&= \Phi\left(-\frac{\mu_2}{\sqrt{\sigma_{22}}}\right) \\
&\quad + \int_{-\frac{\mu_2}{\sqrt{\sigma_{22}}}}^\infty \Phi\left(\frac{(\mu_2 + z\sqrt{\sigma_{22}})t - (\mu_1 + z\sigma_{12}/\sqrt{\sigma_{22}})}{\sqrt{\sigma_{11.2}}}\right)\phi(z)dz \\
&\quad - \int_{-\infty}^{-\frac{\mu_2}{\sqrt{\sigma_{22}}}} \Phi\left(\frac{(\mu_2 + z\sqrt{\sigma_{22}})t - (\mu_1 + z\sigma_{12}/\sqrt{\sigma_{22}})}{\sqrt{\sigma_{11.2}}}\right)\phi(z)dz \\
&= g(\boldsymbol{\mu}, \boldsymbol{\Sigma}) \text{ say}, \quad\quad\quad\quad\quad\quad\quad\quad\quad\quad (12.3.1)
\end{aligned}
$$

where $f(x_2; \mu_2, \sigma_{22})$ is the normal density with mean μ_2 and variance σ_{22}, and $\phi(z)$ is the standard normal density. To get the second representation above, we have used the transformation $z = (x_2 - \mu_2)/\sqrt{\sigma_{22}}$. Note that in the independent case, the representation for $P(X_1/X_2 \leq t)$ can be obtained by putting $\sigma_{12} = 0$ in (12.3.1). The cdf $P(X_1/X_2 \leq t) = g(\boldsymbol{\mu}, \boldsymbol{\Sigma})$ is clearly a function of the parameters $\boldsymbol{\mu}$ and $\boldsymbol{\Sigma}$. In order to compute a $1 - \alpha$ lower confidence limit for the cdf, we shall first obtain a GPQ for the cdf; the α quantile of the GPQ will give the required lower confidence limit. In order to obtain the required GPQ, we shall actually exhibit a GPQ for the entire set of parameters $(\boldsymbol{\mu}, \boldsymbol{\Sigma})$. If $(G_{\boldsymbol{\mu}}, G_{\boldsymbol{\Sigma}})$ is a GPQ for $(\boldsymbol{\mu}, \boldsymbol{\Sigma})$, then $g(G_{\boldsymbol{\mu}}, G_{\boldsymbol{\Sigma}})$ is a GPQ for the function $g(\boldsymbol{\mu}, \boldsymbol{\Sigma})$. We shall now derive a GPQ for $(\boldsymbol{\mu}, \boldsymbol{\Sigma})$.

Suppose $(X_{1j}, X_{2j})', j = 1, \ldots, n$, is a random sample from the bivariate normal distribution $N_2(\boldsymbol{\mu}, \boldsymbol{\Sigma})$. Let $(\bar{X}_1, \bar{X}_2)'$ denote the sample mean vector, and let

$$\boldsymbol{A} = \sum_{j=1}^{n} \left(\begin{array}{c} X_{1j} - \bar{X}_1 \\ X_{2j} - \bar{X}_2 \end{array} \right) \left(\begin{array}{c} X_{1j} - \bar{X}_1 \\ X_{2j} - \bar{X}_2 \end{array} \right)' = \left(\begin{array}{cc} A_{11} & A_{12} \\ A_{21} & A_{22} \end{array} \right).$$

Clearly, $(\bar{X}_1, \bar{X}_2)' \sim N_2\left(\boldsymbol{\mu}, \frac{1}{n}\boldsymbol{\Sigma}\right)$ and $\boldsymbol{A} \sim W_2(\boldsymbol{\Sigma}, n - 1)$, the bivariate Wishart distribution with scale matrix $\boldsymbol{\Sigma}$ and degrees of freedom $n - 1$. Let $\bar{x}_i$ denote the observed value of $\bar{X}_i$, $i = 1, 2$, and let $\boldsymbol{a}$ denote the observed value of $\boldsymbol{A}$. Since $\boldsymbol{A} \sim W_2(\boldsymbol{\Sigma}, n - 1)$, it follows that when the observed value $\boldsymbol{a}$ of $\boldsymbol{A}$ is fixed,

$$\boldsymbol{H} = \boldsymbol{a}^{-1/2}(\boldsymbol{a}\boldsymbol{\Sigma}\boldsymbol{a})^{-1/2}(\boldsymbol{a}\boldsymbol{A}\boldsymbol{a})(\boldsymbol{a}\boldsymbol{\Sigma}\boldsymbol{a})^{-1/2}\boldsymbol{a}^{-1/2} \sim W_2(\boldsymbol{a}^{-1}, n - 1).$$

The value of $\boldsymbol{H}$ at $\boldsymbol{A} = \boldsymbol{a}$ is easily seen to be $\boldsymbol{\Sigma}^{-1}$. Thus

$$\boldsymbol{H}^{-1} = \boldsymbol{K} = \left(\begin{array}{cc} K_{11} & K_{12} \\ K_{12} & K_{22} \end{array} \right) \tag{12.3.2}$$

is a GPQ for $\boldsymbol{\Sigma}$. In order to derive a GPQ for $\boldsymbol{\mu}$, let

$$\boldsymbol{Y} = \left(\begin{array}{c} Y_1 \\ Y_2 \end{array} \right) = \left(\begin{array}{c} \bar{x}_1 \\ \bar{x}_2 \end{array} \right) - \left(\frac{\boldsymbol{K}}{n} \right)^{1/2} \left(\frac{\boldsymbol{\Sigma}}{n} \right)^{-1/2} \left(\begin{array}{c} \bar{X}_1 - \mu_1 \\ \bar{X}_2 - \mu_2 \end{array} \right)$$

$$= \left(\begin{array}{c} x_1 \\ x_2 \end{array} \right) - \left(\frac{\boldsymbol{K}}{n} \right)^{1/2} \boldsymbol{Z}, \tag{12.3.3}$$

where

$$\boldsymbol{Z} = \left(\frac{\boldsymbol{\Sigma}}{n} \right)^{-1/2} \left(\begin{array}{c} \bar{X}_1 - \mu_1 \\ \bar{X}_2 - \mu_2 \end{array} \right) \sim N_2(\boldsymbol{0}, \boldsymbol{I}_2),$$

and the matrix $\boldsymbol{K}$ is defined in (12.3.2). It is now easy to verify that $\boldsymbol{Y} = (Y_1, Y_2)'$ given in (12.3.3) is a GPQ for $\boldsymbol{\mu}$. Thus a GPQ for $(\boldsymbol{\mu}, \boldsymbol{\Sigma})$ is $(G_{\boldsymbol{\mu}}, G_{\boldsymbol{\Sigma}}) = (\boldsymbol{Y}, \boldsymbol{K})$. We are now ready to construct an upper tolerance limit for X_1/X_2 based on a confidence limit for the cdf.

12.3.1   An Upper Tolerance Limit Based on an Approximation to the cdf

A simple approximation can be obtained for the cdf $P(X_1/X_2 \le t)$ when it is known that $\mu_2 > 0$ and the coefficient of variation of X_2, namely $\sqrt{\sigma_{22}}/\mu_2$ is small; later we shall comment on how small $\sqrt{\sigma_{22}}/\mu_2$ should be. Under the above conditions, we have

$$
\begin{aligned}
P(X_1/X_2 \le t) &= P(X_1/X_2 \le t, X_2 > 0) + P(X_1/X_2 \le t, X_2 < 0) \\
&\approx P(X_1 \le tX_2) \\
&= \Phi\left(\frac{-(\mu_1 - t\mu_2)}{\sqrt{\sigma_{11} - 2t\sigma_{12} + t^2\sigma_{22}}}\right) \\
&= \Phi(-u(\boldsymbol{\mu}, \boldsymbol{\Sigma}, t)) \\
\text{where } u(\boldsymbol{\mu}, \boldsymbol{\Sigma}, t) &= \frac{(\mu_1 - t\mu_2)}{\sqrt{\sigma_{11} - 2t\sigma_{12} + t^2\sigma_{22}}},
\end{aligned}
$$

and $\Phi(.)$ is the cdf of the standard normal distribution. This approximation is also discussed in Hinkley (1969). A similar approximation can also be developed when $\mu_2 < 0$ and the CV of X_2 is small. Here we shall discuss the case of $\mu_2 > 0$, along with the CV being small; the other case is similar.

Now we need to find a $1 - \alpha$ upper confidence limit for $u(\boldsymbol{\mu}, \boldsymbol{\Sigma}, t)$, say $h(\bar{x}_1, \bar{x}_2, \boldsymbol{a}, t)$. Then $\Phi(-h(\bar{x}_1, \bar{x}_2, \boldsymbol{a}, t))$ is a $1 - \alpha$ lower confidence limit for the cdf $\Phi(-u(\boldsymbol{\mu}, \boldsymbol{\Sigma}, t))$. We can then solve $\Phi(-h(\bar{x}_1, \bar{x}_2, \boldsymbol{a}, t)) = p$, or equivalently

$$
h(\bar{x}_1, \bar{x}_2, \boldsymbol{a}, t) = -z_p, \tag{12.3.4}
$$

to obtain $t = g(\bar{x}_1, \bar{x}_2, \boldsymbol{a})$, a $(p, 1 - \alpha)$ upper tolerance limit for X_1/X_2.

We shall now construct a $1 - \alpha$ upper confidence limit for $u(\boldsymbol{\mu}, \boldsymbol{\Sigma}, t)$ using the generalized confidence interval idea. We shall use the GPQs $(G_{\boldsymbol{\mu}}, G_{\boldsymbol{\Sigma}}) = (\boldsymbol{Y}, \boldsymbol{K})$ defined earlier; see (12.3.2) and (12.3.3). In addition, define

$$
Z_1 = \frac{(\bar{X}_1 - t\bar{X}_2) - (\mu_1 - t\mu_2)}{\sqrt{(\sigma_{11} - 2t\sigma_{12} + t^2\sigma_{22})/n}} \sim N(0, 1), \tag{12.3.5}
$$

and let

$$
\begin{aligned}
G_{u(\boldsymbol{\mu}, \boldsymbol{\Sigma}, t)} &= \frac{\bar{x}_1 - t\bar{x}_2}{\sqrt{K_{11} - 2tK_{12} + t^2K_{22}}} - \frac{(\bar{X}_1 - t\bar{X}_2) - (\mu_1 - t\mu_2)}{\sqrt{(\sigma_{11} - 2t\sigma_{12} + t^2\sigma_{22})/n}} \frac{1}{\sqrt{n}} \\
&= \frac{\bar{x}_1 - t\bar{x}_2}{\sqrt{K_{11} - 2tK_{12} + t^2K_{22}}} - \frac{Z_1}{\sqrt{n}}, \tag{12.3.6}
\end{aligned}
$$

where $\bar{x}_1$ and $\bar{x}_2$ denote the observed values of $\bar{X}_1$ and $\bar{X}_2$, respectively, Z_1 is defined in (12.3.5), and the K_{ij}'s are defined in (12.3.2). It is readily verified that $G_{u(\boldsymbol{\mu},\boldsymbol{\Sigma},t)}$ is a GPQ for $u(\boldsymbol{\mu},\boldsymbol{\Sigma},t)$. Keeping the observed values $(\bar{x}_1, \bar{x}_2$ and $\boldsymbol{a})$ fixed, we can easily estimate the percentiles of $G_{u(\boldsymbol{\mu},\boldsymbol{\Sigma},t)}$ by Monte Carlo simulation. The $1-\alpha$ quantile of $G_{u(\boldsymbol{\mu},\boldsymbol{\Sigma},t)}$ so obtained gives a $1-\alpha$ generalized upper confidence limit for $u(\boldsymbol{\mu},\boldsymbol{\Sigma},t)$, which in turn can be used to obtain an approximate $(p, 1-\alpha)$ upper tolerance limit for X_1/X_2. Note that in the independent case (i.e., when $\sigma_{12} = 0$) we have $u(\boldsymbol{\mu},\boldsymbol{\Sigma},t) = \frac{\mu_1 - t\mu_2}{\sqrt{\sigma_{11} + t^2\sigma_{22}}}$, and a GPQ for $u(\boldsymbol{\mu},\boldsymbol{\Sigma},t)$ is now given by

$$G^*_{u(\boldsymbol{\mu},\boldsymbol{\Sigma},t)} = \frac{\bar{x}_1 - t\bar{x}_2}{\sqrt{V_{11} + t^2 V_{22}}} - \frac{Z_2}{\sqrt{n}}, \tag{12.3.7}$$

where

$$
\begin{aligned}
Z_2 &= \frac{(\bar{X}_1 - t\bar{X}_2) - (\mu_1 - t\mu_2)}{\sqrt{(\sigma_{11} + t^2\sigma_{22})/n}} \sim N(0,1), \\
V_{11} &= \frac{\sigma_{11}}{A_{11}}a_{11} = \frac{a_{11}}{U_{11}} \\
V_{22} &= \frac{\sigma_{22}}{A_{22}}a_{22} = \frac{a_{22}}{U_{22}},
\end{aligned}
\tag{12.3.8}
$$

U_{11} and U_{22} being independent chi-square random variables, each having df $n-1$. Here A_{11} and A_{22} are the diagonal elements of the sample sum of squares and sum of products matrix $\boldsymbol{A}$ defined earlier. Note that V_{11} and V_{22} are GPQs for σ_{11} and σ_{22}, respectively.

Recall that after obtaining a $1-\alpha$ generalized upper confidence limit (say, $h(\bar{x}_1, \bar{x}_2, \boldsymbol{a}, t)$) for $u(\boldsymbol{\mu},\boldsymbol{\Sigma},t)$, we need to solve for t from the equation

$$h(\bar{x}_1, \bar{x}_2, \boldsymbol{a}, t) + z_p = 0$$

to obtain the required upper tolerance limit; see (12.3.4). The half-interval method (also known as the bisection method) can be conveniently used for obtaining the solution to t. In order to implement this method, we obtain the lower confidence limit $h(\bar{x}_1, \bar{x}_2, \boldsymbol{a}, t)$ for a sequence of $t-$values until we find two values, say t_1 and t_2, such that

$$h(\bar{x}_1, \bar{x}_2, \boldsymbol{a}, t_2) + z_p < 0 < h(\bar{x}_1, \bar{x}_2, \boldsymbol{a}, t_1) + z_p.$$

Then $h(\bar{x}_1, \bar{x}_2, \boldsymbol{a}, t)+z_p$ must be zero for a value of t between t_1 and t_2. To locate such a value of t, let t_3 be the average of t_1 and t_2, and compute $h(\bar{x}_1, \bar{x}_2, \boldsymbol{a}, t_3)+ z_p$. If $h(\bar{x}_1, \bar{x}_2, \boldsymbol{a}, t_3)+z_p > 0$, then $h(\bar{x}_1, \bar{x}_2, \boldsymbol{a}, t)+z_p$ must be zero for a t between t_3 and t_2. If $h(\bar{x}_1, \bar{x}_2, \boldsymbol{a}, t_3) + z_p < 0$, then $h(\bar{x}_1, \bar{x}_2, \boldsymbol{a}, t) + z_p$ must be zero for a

t between t_1 and t_3. Continuing in this way, we can identify smaller and smaller intervals on which $h(\bar{x}_1, \bar{x}_2, \boldsymbol{a}, t) + z_p$ must assume the value zero. When the interval is small enough, the process stops.

12.3.2   Tolerance Limits Based on the Exact cdf

The results in the previous section have been derived under the assumptions that $\mu_2 > 0$ and the coefficient of variation of X_2 is small. We shall now do away with this assumption. For this we shall use the representation for the actual cdf of X_1/X_2 given in (12.3.1). A GPQ for the cdf is now easily obtained by replacing μ_1 and μ_2 with their respective GPQs, namely the components of $\boldsymbol{Y}$ in (12.3.3), and by replacing the elements of $\boldsymbol{\Sigma}$ with their respective GPQs, namely the components of $\boldsymbol{K}$ in (12.3.2). We note that in order to implement this method, it is necessary to evaluate the integral in (12.3.1). This integral can be evaluated by using numerical methods such as Simpson's rule, or by directly using the integration subroutines in standard software.

When X_1 and X_2 are independent, $\sigma_{12} = 0$. GPQs for σ_{11} and σ_{22} are obviously given by V_{11} and V_{22} given in (12.3.8). GPQs for μ_i, $i = 1,\ 2$, are given by

$$\bar{x}_i - \frac{\bar{X}_i - \mu_i}{\sqrt{\sigma_{ii}/n}} \sqrt{V_{ii}/n} = \bar{x}_i - Z_{0i}\sqrt{V_{ii}/n},$$

where $Z_{0i} = \frac{\bar{X}_i - \mu_i}{\sqrt{\sigma_{ii}/n}} \sim N(0,1)$ and are independent for $i = 1,\ 2$. Using these GPQs for μ_1, μ_2, σ_{11} and σ_{22}, we can easily obtain a GPQ for the cdf of X_1/X_2 in the independent case.

In order to study the performance of the approximate upper tolerance limits, it is enough to study the performance of the generalized lower confidence limit for $P(X_1/X_2 \leq t)$. Zhang et al. (2009) have reported simulated coverage probabilities of the generalized lower confidence limit for $P(X_1/X_2 \leq t)$. Their numerical results show that the approximation to the cdf (derived under the assumptions $\mu_2 > 0$ and the coefficient of variation of X_2 is small) is quite accurate if the coefficient of variation of X_2 is no more than 0.30. In general, the coverage probabilities turned out to be satisfactory except when the sample size was somewhat small. Zhang et al. (2009) recommend that a bootstrap calibration be carried out to improve the coverage; we refer to their paper for details.

12.3.3 An Application

We shall now illustrate the computation of an upper tolerance limit for X_1/X_2 using an application from bioassay.

The reverse transcriptase (RT) assay is used as a screening tool to detect potential retroviral contamination in the raw materials used in the manufacture of certain influenza vaccines. Prior to the blending of the final vaccine product, the materials are subject to sterility testing, to determine if retroviruses are present. The presence of the Reverse Transcriptase or RT enzyme can be used as a reliable indicator for the detection of retroviruses. In the RT assay, the larger amount of this enzyme in the sample will induce a larger radioactivity count. In order to calibrate the radioactivity count resulting from the cell culture itself, a negative control is usually included in the assay. One necessary condition for a sample to be classified as negative is that the ratio of radioactivity count induced from the sample to radioactivity count induced from the negative control be less than some pre-specified limit. An upper tolerance limit is used to ensure that a given proportion (say, 95%) of the future assay results will be less than the limit with a given confidence level, if the process is in control. Historical in-control data show that the radioactivity count from the sample and that from the negative control follow a bivariate normal distribution.

The historical data available to us consisted of forty five pairs of radioactivity counts of negative controls and samples, accumulated from RT assays. (Due to confidentiality issues, the actual data cannot be published). Based on the data, the observed values $\bar{x}_1$, $\bar{x}_2$ and $\boldsymbol{a}$ are given by

$$\bar{x}_1 = 38.1, \quad \bar{x}_2 = 38.9, \quad \text{and} \quad 44 \times \boldsymbol{a} = \begin{pmatrix} 56.3 & 36.0 \\ 36.0 & 35.1 \end{pmatrix}$$

We note that the estimated coefficient of variation for X_2 is only 0.15. Thus we decided to use the approximation method in Section 12.3.1 to compute the upper tolerance limit using $1 - \alpha = 0.95$. For this, we first computed a 95% generalized upper confidence limit for $u(\boldsymbol{\mu}, \boldsymbol{\Sigma}, t)$ given in Section 12.3.1; we used 5000 simulated values of the GPQ in order to obtain this upper confidence limit. Once the upper confidence limit was obtained, we solved the equation (12.3.4) by the half-interval method. For $p = 0.95$, the upper tolerance limit came out to be 1.230. Also, for $p = 0.99$, the limit came out as 1.343.

12.4   Sample Size Determination

Sample size determination in the context of tolerance intervals has been addressed by Wilks (1941), who considers both a non-parametric set up, as well as the parametric set up of a normal distribution. Wilks (1941) addressed the problem in the context of $\beta-$expectation tolerance intervals. Typically, the criterion used to determine the sample size for a $\beta-$expectation tolerance interval is as follows. Determine the sample size so that the actual proportion of the population (say, $N(\mu, \sigma^2)$) that is contained in the $\beta-$ expectation tolerance interval is in some sense close to β, with a high probability. In the case of a $p-$content tolerance interval, one can compute the sample size so that the probability is small that the interval covers too large a proportion of the population. This will guarantee that the tolerance interval is not too wide. Such a criterion was proposed by Faulkenberry and Weeks (1968). In this section, we shall briefly address the sample size determination for a $p-$content and $(1 - \alpha)-$confidence two-sided tolerance interval, and for a $\beta-$expectation two-sided tolerance interval, both in the context of a univariate normal distribution.

12.4.1   Sample Size Determination for a $(p, 1 - \alpha)$ Two-Sided Tolerance Interval for a Normal Population

Let $\bar{X}$ and S^2 denote the sample mean and sample variance, respectively, based on a sample of size n from $N(\mu, \sigma^2)$. A $(p, 1 - \alpha)$ two-sided tolerance interval for $N(\mu, \sigma^2)$ is derived in Section 2.3.1 of Chapter 2, and is of the form $\bar{X} \pm kS$, where the tolerance factor k has the approximate expression (see (2.3.5))

$$k \simeq \left(\frac{(n-1)\chi^2_{1;p}(1/n)}{\chi^2_{n-1;\alpha}} \right)^{\frac{1}{2}}.$$

Let $C(\bar{X}, S)$ denote the content of the tolerance interval, i.e., the proportion of the normal population that is contained in the tolerance interval, conditionally given $\bar{X}$ and S. That is

$$C(\bar{X}, S) = P_X \left(\bar{X} - kS \leq X \leq \bar{X} + kS | \bar{X}, S \right),$$

where $X \sim N(\mu, \sigma^2)$. Obviously, the $(p, 1 - \alpha)$ two-sided tolerance interval $\bar{X} \pm kS$ is such that

$$P_{\bar{X}, S} \left(C(\bar{X}, S) \geq p \right) = 1 - \alpha. \tag{12.4.1}$$

A formulation of the sample size problem, due to Faulkenberry and Weeks (1968) is as follows. For a given p, $1 - \alpha$, δ and α', determine the sample size n so that

$$P_{\bar{X}, S} \left(C(\bar{X}, S) \geq p + \delta \right) = \alpha'.$$

Equivalently,

$$P_{\bar{X},S}\left(C(\bar{X}, S) \leq p + \delta\right) = 1 - \alpha'. \tag{12.4.2}$$

Here α' is taken to be small, say 0.10, 0.05 or 0.01. Note that the tolerance interval is to be derived subject to (12.4.1) and the condition (12.4.2) is to be used to determine the sample size n. As already noted, the condition (12.4.2) implies that there is only a small probability α' that the $(p, 1 - \alpha)$ tolerance interval will have a content that exceeds p by a margin of δ.

The above sample size problem is addressed in the articles by Faulkenberry and Weeks (1968), Faulkenberry and Daly (1970) and Guenther (1972). However, they used inefficient approximations for the tolerance factor k. Later, Odeh, Chou and Owen (1987) addressed the same sample size problem. For some selected values of p, $1 - \alpha$, δ and α', they have given a table listing the sample sizes, and the tables are reproduced in Table B17, Appendix B.

For a given sample size n, it may also be of interest to determine the margin δ so that the $(p, 1 - \alpha)$ two-sided tolerance interval $\bar{X} \pm kS$ will satisfy (12.4.2), for specified values of p, $1 - \alpha$ and α'. Tables of such $\delta-$values are given in Odeh, Chou and Owen (1987); these are not reproduced here.

Example 2.2 (continued)

This example from Chapter 2 deals with the filling operation monitoring of a machine that is set to fill a liter of milk in plastic containers. Suppose a (0.90, 0.95) two-sided tolerance interval is to be used to assess the accuracy of the filling machine. We shall determine the required sample size for $\delta = 0.05$, and $\alpha' = 0.10$. From Table B17, Appendix B, the required sample size is $n = 145$. In other words, if we compute a (0.90, 0.95) two-sided tolerance interval based on a sample of size 145, we can claim that the content of the tolerance interval will exceed $p + \delta = 0.95$ with a probability of only 0.10.

The methodology described in this section can obviously be adopted to determine the required sample size in the context of one-sided tolerance limits. The sample size problem has also been addressed in the context of an equal-tailed tolerance interval for the normal distribution discussed in Section 2.3.2; see Chou and Mee (1984) for details. The article by Fountain and Chou (1991) address the sample size problem for finite populations.

12.4.2   Sample Size Determination for a β–Expectation Two-Sided Tolerance Interval for a Normal Population

In order to introduce the sample size problem for a β–expectation two sided tolerance interval for a normal distribution, let $C(\bar{X}, S)$ denote the content of the tolerance interval, defined earlier. Recall that the tolerance interval is once again of the form $\bar{X} \pm kS$, where the factor k is to be determined subject to the condition

$$E\left[C(\bar{X}, S)\right] = \beta.$$

We then have $k = t_{n-1; \frac{(1+\beta)}{2}} \sqrt{1 + \frac{1}{n}}$; see (12.2.3). The sample size problem formulated by Wilks (1941) is that of determining the minimum sample size n so that

$$P\left(\beta - \delta_1 \leq C(\bar{X}, S) \leq \beta + \delta_2\right) \geq 1 - \alpha', \tag{12.4.3}$$

for specified values of $\delta_1 > 0$, $\delta_2 > 0$ and $1 - \alpha'$. The above condition states that with probability at least $1 - \alpha'$, the content of the β–expectation tolerance interval belongs to the interval $(\beta - \delta_1,\ \beta - \delta_2)$.

For $\delta_1 = \delta_2 = \delta$, tables of the minimum sample size satisfying (12.4.3) are given in Odeh, Chou and Owen (1989). These are reproduced in Table B18, Appendix B. It is also possible to compute the value of δ so that (12.4.3) will hold for $\delta_1 = \delta_2 = \delta$, for a given sample size and for given values of β and $1 - \alpha'$. Such tables are also given in Odeh, Chou and Owen (1989). These authors also address the sample size problem when, instead of (12.4.3), we require

$$P\left(C(\bar{X}, S) \leq \beta + \delta\right) \geq 1 - \alpha'. \tag{12.4.4}$$

In other words, the requirement is that with probability at least $1 - \alpha'$, the coverage of the β–expectation tolerance interval is no more than $\beta + \delta$. Here we have given the tables of sample sizes corresponding to the criterion (12.4.3) only (with $\delta_1 = \delta_2 = \delta$). Other tables, for example corresponding to the criterion (12.4.4), can be found in Odeh, Chou and Owen (1989).

In an earlier article, Chou (1984) addressed the sample size problem for a β–expectation two sided tolerance interval for a normal distribution under the requirement that the proportion of the normal distribution below the lower tolerance limit, and the proportion above the upper tolerance limit are both suitably controlled. Clearly, the sample size problem can also be addressed in the context of one-sided β–expectation tolerance limits. We also note that in this section we have discussed sample size determination only in the context of a univariate normal distribution; see Wang and Iyer (1996) for a discussion of sampling plans in the context of a one-way random model with balanced data.

Example 2.2 (continued)

Suppose a β-expectation two-sided tolerance interval is to be used to assess the accuracy of the filling machine with $\beta = 0.90$. We shall determine the required sample size for $\delta = 0.05$, and $1 - \alpha' = 0.90$. From Table B18, Appendix B, the required sample size is $n = 161$. Thus if we compute a β-expectation two-sided tolerance interval with $\beta = 0.90$ based on a sample of size 161, we can claim that the content of the tolerance interval will be between 0.85 and 0.95, with a probability of 0.90.

12.5   Reference Limits and Coverage Intervals

Reference limits are widely used by clinicians to identify the measurement range expected from a reference population. For a reference population, the reference interval typically contains the central 95% of the values of the population. The reference population may represent a selected healthy population, and the reference limits thus provide a standard range that can be used to interpret measurements obtained from new patients. We refer to the book by Harris and Boyd (1995) for a detailed treatment of the topic, along with a discussion of the basic issues. An excellent review is also given in the recent paper by Trost (2006). Here we shall give a brief description of the topic, pointing out the relationship to tolerance intervals.

Point estimates of the reference limits can be easily obtained by simply estimating the 2.5th and 97.5th percentiles of the reference population. However, we cannot guarantee the intended coverage of 95% when we use a reference interval based on such estimated reference limits. If estimates with a certain desired coverage are desired, one possibility is to compute the estimated reference limits subject to the condition that the estimated reference limits contain at least a proportion p (say, $p = 0.90$) of the central values of the reference population, with confidence level $1 - \alpha$ (say, $1 - \alpha = 0.95$). This requirement results in a p-content equal-tailed tolerance interval, with confidence level $1 - \alpha$.

If the measurements from the reference population follow a univariate normal distribution $N(\mu, \sigma^2)$, the reference limits that contain the central $100p\%$ of the population are $\mu - z_{\frac{1+p}{2}}\sigma$ and $\mu + z_{\frac{1+p}{2}}\sigma$. Simply replacing μ and σ with estimates will obviously result in a reference interval whose coverage could be much less than $100p\%$. Thus we have to use an equal-tailed tolerance interval, namely, a tolerance interval that will contain the interval $\left(\mu - z_{\frac{1+p}{2}}\sigma, \ \mu + z_{\frac{1+p}{2}}\sigma\right)$ with confidence level, say $1 - \alpha$. A solution to this problem is given in Section 2.3.2 of

Chapter 2, and the interval is of the form $(\bar{X} - k_e S, \bar{X} + k_e S)$, where $\bar{X}$ and S are the sample mean and standard deviation based on a sample of size n, and the computation of the tolerance factor k_e is explained in Section 2.3.2. Depending on the purpose for which the reference limits will be used, one can also compute a β−expectation tolerance interval, to be used as a reference interval; see Harris and Boyd (1995, p. 110–111). Sometimes it is also required to have a coverage interval that is a β−expectation tolerance interval whose content belongs to the interval $(\beta - \delta, \ \beta + \delta)$, with a high probability, say $1 - \alpha'$. Here δ is a specified margin; see Poulson, Holst and Christensen (1997, p. 1604). We have already addressed the problem of determining the sample size necessary to meet this requirement.

The importance of having statistically valid reference limits has been widely recognized in Clinical Chemistry and Metrology, where the terms coverage interval and statistical coverage interval are used; see Poulson, Holst and Christensen (1997), Chen and Hung (2006), Chen, Huang and Chen (2007) and Willink (2004, 2006). The literature on reference limits and coverage intervals address the problems non-parametrically, as well as parametrically in the set up of a normal distribution.

Reference regions in the multivariate context can be similarly defined; see Harris and Boyd (1995, Chapter 4) and Trost (2006). Here we will have $(p, 1 - \alpha)$ tolerance regions as well as β−expectation tolerance regions. The role of tolerance regions as reference regions for the assessment of clinical observations is also described in the recent article by Eaton, Muirhead and Pickering (2006).

12.6 Tolerance Intervals for Binomial and Poisson Distributions

The problem of constructing tolerance intervals for a discrete distribution has not received much attention in the literature. Tolerance intervals on the basis of a discrete model is useful to assess the magnitude of discrete quality characteristics of a product, for example, the number of defective components in a system (see Example 12.1). To define a $(p, 1−\alpha)$ tolerance interval for a discrete distribution, let $\boldsymbol{X}$ be a sample from a discrete distribution, and let X follow the same distribution independently of $\boldsymbol{X}$. A $(p, 1 - \alpha)$ one-sided upper tolerance limit $U_1(\boldsymbol{X})$ is defined so that

$$P_{\boldsymbol{X}} \left\{ P_X \left[X \leq U_1(\boldsymbol{X}) | \boldsymbol{X} \right] \geq p \right\} \geq 1 - \alpha. \qquad (12.6.1)$$

A $(p, 1 - \alpha)$ one-sided lower tolerance limit $L_1(\boldsymbol{X})$ is defined similarly. Recall that these one-sided tolerance limits are also $1 - \alpha$ confidence limits on appropriate quantiles. Let k_p is the p quantile of the discrete distribution under consideration. A $(p, 1 - \alpha)$ equal-tailed tolerance interval $(L_e(\boldsymbol{X}), U_e(\boldsymbol{X}))$ is defined so that

$$P_{\boldsymbol{X}} \left(L_e(\boldsymbol{X}) \leq k_{\frac{1-p}{2}} \text{ and } k_{\frac{1+p}{2}} \leq U_e(\boldsymbol{X}) \right) \geq 1 - \alpha. \tag{12.6.2}$$

We shall now describe a method for constructing one-sided as well as equal-tailed tolerance intervals in a general set up. Consider a distribution $F_X(x|\theta)$ that depends only on a single parameter θ, and is stochastically nondecreasing in θ. That is, for every t, $P(X > t|\theta_1) \geq P(X > t|\theta_2)$ for all $\theta_1 > \theta_2$. Then the p quantile $k_p(\theta)$, defined as the smallest value for which $P(X \leq k_p(\theta)|\theta) \geq p$, is a nondecreasing function of θ. As a consequence, if θ_u is a $1 - \alpha$ upper confidence limit for θ, then $k_p(\theta_u)$, defined as the smallest value so that

$$P(X \leq k_p(\theta_u)|\theta_u) \geq p, \tag{12.6.3}$$

is a $1 - \alpha$ upper confidence limit for $k_p(\theta)$, and so it is a $(p, 1 - \alpha)$ upper tolerance limit for the distribution. Similarly, if θ_l is a $1 - \alpha$ lower confidence limit for θ, then $k_{1-p}(\theta_l)$, defined as the largest number so that

$$P(X \geq k_{1-p}(\theta_l)|\theta_l) \geq p, \tag{12.6.4}$$

is a $(p, 1 - \alpha)$ lower tolerance limit for the distribution. If (θ_l, θ_u) is a $1 - \alpha$ confidence interval for θ, then simultaneously the inequalities $k_{\frac{1-p}{2}}(\theta_l) \leq k_{\frac{1-p}{2}}(\theta)$ and $k_{\frac{1+p}{2}}(\theta) \leq k_{\frac{1+p}{2}}(\theta_u)$ hold with probability $1 - \alpha$. Therefore,

$$\left(k_{\frac{1-p}{2}}(\theta_l), \ k_{\frac{1+p}{2}}(\theta_u) \right) \tag{12.6.5}$$

is a $(p, 1 - \alpha)$ equal-tailed tolerance interval. Note that $k_{\frac{1-p}{2}}(\theta_l)$ is defined as the largest number for which

$$P\left(X \geq k_{\frac{1-p}{2}}(\theta_l) \Big| \theta_l \right) \geq \frac{1+p}{2}$$

and $k_{\frac{1+p}{2}}(\theta_u)$ is defined as the smallest number for which

$$P\left(X \leq k_{\frac{1+p}{2}}(\theta_u) \Big| \theta_u \right) \geq \frac{1+p}{2}.$$

As the binomial and Poisson distributions are stochastically increasing in their parameters, the above procedures can be used to obtain tolerance intervals.

Also, the above procedures with the exact confidence intervals for θ produce exact tolerance intervals. To construct tolerance intervals for the binomial and Poisson distributions, Hahn and Chandra (1981) used the above procedures with the exact confidence intervals for the parameters due to Clopper and Pearson (1934) and Garwood (1936).

12.6.1 Binomial Distribution

It is clear from the discussion in the preceding section that a good confidence interval for a binomial parameter π could produce a satisfactory tolerance interval. It is now well understood that the exact confidence intervals for a binomial π or for Poisson mean are too conservative, and many authors (see Agresti and Coull, 1998; Brown, Cai and Das Gupta, 2001) recommend approximate intervals for applications. There are several approximate confidence intervals available for a binomial π (see Brown, Cai and Das Gupta, 2001, and Cai, 2005, and the references therein). Among the approximate confidence intervals, the score confidence intervals seem to be satisfactory in terms of coverage probability and accuracy (see Agresti and Coull, 1998). Therefore, we shall consider tolerance intervals based on the exact confidence intervals and those based on the score confidence intervals.

Exact Confidence Intervals for π

The Clopper-Pearson (1934) approach for obtaining an exact confidence interval for a binomial proportion π is as follows. For a given sample size n and an observed number of successes m, the lower limit π_l for π is the solution of the equation

$$\sum_{i=m}^{n} \binom{n}{i} \pi_l^i (1 - \pi_l)^{n-i} = \frac{\alpha}{2},$$

and the upper limit π_u is the solution of the equation

$$\sum_{i=0}^{m} \binom{n}{i} \pi_u^i (1 - \pi_u)^{n-i} = \frac{\alpha}{2}.$$

Using a relation between the binomial and beta distributions it can be shown that

$$\pi_l = B_{m,n-m+1;\frac{\alpha}{2}} \quad \text{and} \quad \pi_u = B_{m+1,n-m;1-\frac{\alpha}{2}}, \tag{12.6.6}$$

where $B_{a,b;q}$ denotes the q quantile of a beta distribution with the shape parameters a and b. The interval (π_l, π_u) is an exact $1 - \alpha$ confidence interval for π, in the sense that the coverage probability is always greater than or equal the specified confidence level $1 - \alpha$. One-sided confidence limits for π_l and π_u can be obtained by replacing $\frac{\alpha}{2}$ by α. For the extreme case of $m = n$, the upper limit is 1 and the lower limit is $\alpha^{\frac{1}{n}}$; when $m = 0$, the lower limit is 0 and the upper limit is $1 - \alpha^{\frac{1}{n}}$.

Score Confidence Interval for π

Let $\widehat{\pi} = \frac{X}{n}$, where $X \sim \text{binomial}(n, \pi)$. The score interval is on the basis of the asymptotic result that

$$T(\widehat{\pi}, \pi) = \frac{\widehat{\pi} - \pi}{\sqrt{\pi(1 - \pi)/n}} \sim N(0, 1).$$

Let $c = z_{1-\frac{\alpha}{2}}$ be the $1 - \frac{\alpha}{2}$ quantile of the standard normal distribution. The endpoints of the score confidence interval are the roots of the quadratic equation $T^2(\widehat{\pi}, \pi) = c^2$, which are given by

$$(\pi_l, \pi_u) = \left(\frac{\widehat{\pi} + \frac{c^2}{2n}}{1 + \frac{c^2}{n}} \right) \pm \frac{\frac{c}{\sqrt{n}} \sqrt{\widehat{\pi}(1 - \widehat{\pi}) + c^2/(4n)}}{1 + \frac{c^2}{n}}. \qquad (12.6.7)$$

One-sided limits for π can be obtained by replacing $z_{1-\frac{\alpha}{2}}$ by $z_{1-\alpha}$.

Binomial Tolerance Intervals

Let m be an observed value of the binomial(n, π) random variable X. Let $k_p(\pi)$ denote the p quantile; that is, $k_p(\pi)$ is the smallest integer so that

$$P(X \le k_p(\pi)|n, \pi) = \sum_{i=0}^{k_p(\pi)} \binom{n}{i} \pi^i (1 - \pi)^{n-i} \ge p. \qquad (12.6.8)$$

If π_u is a $1 - \alpha$ upper confidence limit for π based on m, then $k_p(\pi_u)$, defined as the smallest integer so that $P(X \le k_p(\pi_u)|n, \pi_u) \ge p$, is a $(p, 1 - \alpha)$ upper tolerance limit for the binomial(n, π) distribution. Similarly, the largest number $k_{1-p}(\pi_l)$ for which $P(X \ge k_{1-p}(\pi_l)|n, \pi_l) \ge p$, where π_l is a $1-\alpha$ lower confidence limit for π, is a $(p, 1-\alpha)$ one-sided lower tolerance limit. A $(p, 1-\alpha)$ equal-tailed tolerance interval is given by

$$\left(k_{\frac{1-p}{2}}(\pi_l), k_{\frac{1+p}{2}}(\pi_u) \right), \qquad (12.6.9)$$

where (π_l, π_u) is a $1 - \alpha$ two-sided confidence interval for π.

To find $k_p(\pi_u)$, we compute the right-tail probabilities $P(X = n|n, \pi_u)$, $P(X = n - 1|n, \pi_u)$, ... until the sum of these probabilities is greater than or equal to $1 - p$. If this happens at $X = n - j$, that is, $\sum_{i=n-j}^{n} P(X = i|n, \pi_u) \geq 1 - p$ and $\sum_{i=n-j+1}^{n} P(X = i|n, \pi_u) < 1 - p$, then $n - j$ is the $(p, 1 - \alpha)$ upper tolerance limit. Furthermore, it can be checked that $P(X \leq n - j|n, \pi_u) \geq p$ (see Exercise 12.8.12). Similarly, the $(p, 1 - \alpha)$ lower tolerance limit $k_{1-p}(\pi_l)$ based on the $1 - \alpha$ lower confidence limit π_l can be computed by computing left-tail probabilities.

Tolerance Intervals based on Approximate Quantiles

Krishnamoorthy and Xia (2009) noted that tolerance intervals can be obtained by estimating normal based approximate quantiles of binomial(n, π). On the basis of the normal approximation to the quantity $\frac{X - n\pi}{n\pi(1-\pi)}$, the p quantile of a binomial(n, π) distribution is given by $k_p(\pi) \simeq n\pi + z_p\sqrt{n\pi(1 - \pi)}$, where z_p is the p quantile of the standard normal distribution. Noting that the above quantile is an increasing function of π, an approximate $(p, 1-\alpha)$ upper tolerance limit for the binomial(n, π) distribution obtained by replacing the π by a $1 - \alpha$ upper confidence limit π_u. More specifically,

$$k_p(\pi_u) \simeq \min\left\{ n, [n\pi_u + z_p\sqrt{n\pi_u(1 - \pi_u)}] \right\}, \tag{12.6.10}$$

where $[y]$ is the nearest integer to y, is an approximate $(p, 1-\alpha)$ upper tolerance limit. Similarly, an approximate $(p, 1 - \alpha)$ lower tolerance limit can be obtained as

$$k_p(\pi_l) \simeq \max\left\{ 0, [n\pi_l - z_p\sqrt{n\pi_l(1 - \pi_l)}] \right\}. \tag{12.6.11}$$

If (π_l, π_u) is a $1 - \alpha$ confidence interval for π, then

$$\left(\max\left\{ 0, \left[n\pi_l - z_{\frac{1+p}{2}}\sqrt{n\pi_l(1 - \pi_l)} \right] \right\}, \min\left\{ n, \left[n\pi_u + z_{\frac{1+p}{2}}\sqrt{n\pi_u(1 - \pi_u)} \right] \right\} \right) \tag{12.6.12}$$

is an approximate $(p, 1 - \alpha)$ equal-tailed tolerance interval.

Example 12.1 (Assessment of number of defective chips in a wafer)

To illustrate the methods for constructing binomial tolerance intervals, we shall use the data set given in the NIST webpage[1]. This example concerns defective chips in a sample of 30 wafers. A chip in a wafer is considered defective

---
[1]http://www.itl.nist.gov/div898/handbook/pmc/section3/pmc332.htm

whenever a misregistration, in terms of horizontal and/or vertical distances from the center, is recorded. On each wafer, locations of 50 chips were measured and the proportions of defective chips are given in Table 12.1. The overall proportion of defective chips is $\widehat{\pi} = \frac{\sum_{i=1}^{30} n_i \widehat{\pi}_i}{\sum_{i=1}^{30} n_i} = \frac{347}{1500} = 0.2313$, where $n_i = 50, i = 1, ..., 30$. We shall compute (0.90, 0.95) one-sided tolerance limits, and (0.90, 0.95) equal-tailed tolerance intervals for the number of defective chips in a wafer using different methods described in the preceding sections.

Table 12.1: Fractions of defective chips in 30 wafers

Sample No.	$\widehat{\pi}_i$	Sample No.	$\widehat{\pi}_i$	Sample No.	$\widehat{\pi}_i$
1	.24	11	.10	21	.40
2	.30	12	.12	22	.36
3	.16	13	.34	23	.48
4	.20	14	.24	24	.30
5	.08	15	.44	25	.18
6	.14	16	.16	26	.24
7	.32	17	.20	27	.14
8	.18	18	.10	28	.26
9	.28	19	.26	29	.18
10	.20	20	.22	30	.12

Tolerance Intervals Based on the Exact Confidence Interval: The 95% exact confidence interval (12.6.6) is given by $\pi_l = B_{347,1154,.025} = 0.2102$ and $\pi_u = B_{348,1153,.975} = 0.2535$. Thus, $(\pi_l, \pi_u) = (0.2102, 0.2535)$ is a 95% confidence interval for π. Furthermore, for these values of π_l and π_u, we have

$$\sum_{i=6}^{50} \binom{50}{i} \pi_l^i (1 - \pi_l)^{50-i} = 0.9665 \quad \text{and} \quad \sum_{i=0}^{18} \binom{50}{i} \pi_u^i (1 - \pi_u)^{50-i} = 0.9671.$$

Note that both probabilities are greater than $\frac{1+.90}{2} = 0.95$. It can be verified that 6 is the largest number for which the first sum is at least $\frac{1+p}{2} = 0.95$, and 18 is the smallest number for which the second sum is at least 0.95. Thus, the (0.90, 0.95) tolerance interval is (6, 18), and we can conclude that at least 90% of the wafers contain 6 to 18 defective chips with 95% confidence. To get the one-sided tolerance limits, we found 95% one-sided lower limit for π as $\pi_l = 0.2135$ and 95% upper limit as $\pi_u = 0.2500$. Also, $P(X \geq 7|50, \pi_l = 0.2135) = 0.9315$ and $P(X \geq 8|50, \pi_l = 0.2135) = 0.8654$. So, 7 is the (0.90, 0.95) one-sided lower tolerance limit. To find the upper tolerance limit, we computed $P(X \leq 16|50, \pi_u = 0.2500) = .9017$ and $P(X \leq 15|50, \pi_u = 0.2500) = 0.8369$, so 16 is the (0.90, 0.95) one-sided upper tolerance limit.

Tolerance Intervals Based on the Score Interval: The 95% score confidence interval (12.6.7) for π is $(\pi_l, \pi_u) = (0.2107, 0.2533)$. Along the lines in the preceding paragraph, it can be seen that $(6, 18)$ is the $(0.90, 0.95)$ tolerance interval. The one-sided 95% score confidence limits are $\pi_l = 0.2139$ and $\pi_u = 0.2497$. The one-sided lower tolerance limit based on $\pi_l = 0.2139$ is 7, and the upper limit based on $\pi_u = 0.2497$ is 16.

Tolerance Intervals Based on the Approximate Quantiles: We shall now compute (12.6.10), (12.6.11) and (12.6.12) using the score confidence intervals given in the preceding paragraph. Using $\pi_l = 0.2139$ and $z_{.90} = 1.282$ in (12.6.11), we get $6.98 \simeq 7$. Using $\pi_u = 0.2497$ in (12.6.10), we get $16.41 \simeq 16$. Using the 95% confidence interval $(\pi_l, \pi_u) = (0.2107, 0.2533)$ and $z_{.95} = 1.645$ in (12.6.12), we obtain $(5.79, 17.72) \simeq (6, 18)$ as an approximate $(0.90, 0.95)$ equal-tailed tolerance interval.

Because of the large sample size, we obtained the same tolerance limits based on all three methods.

12.6.2 Poisson Distribution

One-sided tolerance limits and equal-tailed tolerance intervals for a Poisson distribution can be obtained using (12.6.3), (12.6.4) and (12.6.5). Let $X_1, ..., X_n$ be a sample from a Poisson(λ) distribution. Let $S = \sum_{i=1}^{n} X_i$ so that $S \sim$ Poisson$(n\lambda)$. The maximum likelihood estimator of λ is given by $\widehat{\lambda} = \frac{S}{n}$.

The Garwood (1936) exact confidence interval for λ is given by

$$(\lambda_l, \lambda_u) = \left(\frac{1}{2n}\chi^2_{2s; \frac{\alpha}{2}}, \frac{1}{2n}\chi^2_{2s+2; 1-\frac{\alpha}{2}} \right), \tag{12.6.13}$$

where s is an observed value of S and $\chi^2_{m;\alpha}$ is the α quantile of a chi-square distribution with m df. The above interval should be used with the convention that $\chi^2_{0;\alpha} = 0$. One-sided lower (or upper) confidence limit can be obtained by replacing the $\frac{\alpha}{2}$ by α in the lower (or upper) limit in (12.6.13).

The score confidence interval is on the basis of asymptotic normality of the test statistic $T(\widehat{\lambda}, \lambda) = \frac{\widehat{\lambda} - \lambda}{\sqrt{\lambda/n}}$. Specifically, the roots of the quadratic equation $T^2(\widehat{\lambda}, \lambda) = c^2$, where $c = z_{1-\frac{\alpha}{2}}$, form the score confidence interval, and they are given by

$$(\lambda_l, \lambda_u) = \widehat{\lambda} + \frac{c^2}{2n} \pm \frac{c}{\sqrt{n}}\sqrt{\widehat{\lambda} + \frac{c^2}{4n}}. \tag{12.6.14}$$

One-sided lower (or upper) confidence limit can be obtained by replacing $c = z_{1-\frac{\alpha}{2}}$ by $z_{1-\alpha}$ in the lower (or upper) limit in (12.6.14).

As in the binomial case, the $(p, 1 - \alpha)$ upper tolerance limit is the smallest number $k_p(\lambda_u)$ so that $P(X \leq k_p(\lambda_u)|\lambda_u) \geq p$, where λ_u is a $1 - \alpha$ upper confidence limit for λ based on an observed value s of S, and X is a Poisson(λ_u) random variable. Similarly, a $(p, 1-\alpha)$ lower tolerance limit is the largest number $k_{1-p}(\lambda_l)$ so that $P(X \geq k_{1-p}(\lambda_l)|\lambda_l) \geq p$, where λ_l is the $1 - \alpha$ lower confidence limit for λ. If (λ_l, λ_u) is a $1 - \alpha$ confidence interval λ, then $\left(k_{\frac{1-p}{2}}(\lambda_l), k_{\frac{1+p}{2}}(\lambda_u) \right)$ is a $(p, 1 - \alpha)$ equal-tailed tolerance interval.

The equal-tailed tolerance intervals based on the normal approximation to a Poisson quantile are as follows: Let λ_l (λ_u) be a $1 - \alpha$ one-sided lower (upper) confidence limit for λ. Then $\left[\lambda_u + z_p \sqrt{\lambda_u} \right]$ is a $(p, 1 - \alpha)$ upper tolerance limit, and $\max \left\{ 0, [\lambda_l - z_p \sqrt{\lambda_l}] \right\}$ is a $(p, 1-\alpha)$ lower tolerance limit. If (λ_l, λ_u) is a $1-\alpha$ confidence interval for λ, then $\left(\max \left\{ 0, \left[\lambda_l - z_{\frac{1+p}{2}} \sqrt{\lambda_l} \right] \right\}, \left[\lambda_u + z_{\frac{1+p}{2}} \sqrt{\lambda_u} \right] \right)$ is a $(p, 1 - \alpha)$ equal-tailed tolerance interval.

Example 12.2 (Estimating the number of shutdowns of a system)

We shall use the example given in Hahn and Chandra (1981) for illustrating the construction of tolerance intervals for a Poisson distribution. Suppose the number of unscheduled shutdowns per year in a large population of systems follows a Poisson distribution with mean λ. Suppose there were 24 shutdowns over a period of 5 years, and we like to find (0.90, 0.95) tolerance intervals for the population of system shutdowns per year. Here $n = 5$ and $s = 24$, which is the total number of shutdowns. The MLE of λ is given by $\widehat{\lambda} = \frac{24}{5} = 4.8$.

Tolerance Intervals Based on the Exact Confidence Intervals: The exact 95% confidence interval (12.6.13) is given by $(\lambda_l, \lambda_u) = \left(\frac{1}{10}\chi^2_{2s;.025}, \frac{1}{10}\chi^2_{2s+2;.975} \right) = (3.075, 7.142)$. Further, for these values of λ_l and λ_u, we have

$$\sum_{i=1}^{\infty} \frac{e^{-\lambda_l} \lambda_l^i}{i!} = 0.9538 \quad \text{and} \quad \sum_{i=0}^{12} \frac{e^{-\lambda_u} \lambda_u^i}{i!} = 0.9691.$$

Also, it can be verified that 1 is the largest integer so that the first sum is at least 0.95, and 12 is the smallest integer so that the second sum is at least 0.95. So, (1, 12) is the desired (0.90, 0.95) equal-tailed tolerance interval. Thus, at least 90% of the population of systems shutdowns per year range from 1 to 12 with confidence 95%.

To find the (0.90, 0.95) one-sided tolerance limits, we obtained the exact 95% lower limit for λ as $\lambda_l = 3.310$, and the exact 95% upper limit as $\lambda_u = 6.750$.

To find the lower tolerance limit, we note that $P(X \geq 1|\lambda_l = 3.310) = 0.9635$ and $P(X \geq 2|\lambda_l = 3.310) = 0.8426$. Also, $P(X \leq 10|\lambda_u = 6.750) = 0.9183$ and $P(X \leq 9|\lambda_u = 6.750) = 0.8549$. Thus, 1 is the (0.90, 0.95) one-sided lower tolerance limit, and 10 is the one-sided upper tolerance limit. Furthermore, the upper tolerance limit of 10 means that at least 90% of population of systems shutdowns are at most 10 with confidence 95%.

Tolerance Intervals Based on the Score Intervals: Recall that MLE $\widehat{\lambda} = \frac{24}{5} = 4.8$. The 95% score confidence interval (12.6.14) for λ is (3.226, 7.142). Proceeding as above, it can be readily checked that the (0.90, 0.95) equal-tailed tolerance interval based on (3.226, 7.142) is (1, 12). Thus, the equal-tailed tolerance intervals based on the exact and score confidence intervals are the same. The 95% one-sided lower confidence limit for λ is $\lambda_l = 3.437$ which yields the (0.90, 0.95) lower tolerance limit as 1; the 95% upper confidence limit is $\lambda_u = 6.705$ which yields the (0.90, 0.95) upper tolerance limit as 10.

Tolerance Intervals Based on the Approximate Quantiles: We shall use the score confidence limits for λ to obtain confidence limits for the approximate quantiles. Using the one-sided confidence limit $\lambda_l = 3.437$ and $z_{.90} = 1.282$, we get $\lambda_l - 1.282\sqrt{\lambda_l} = 1.06 \simeq 1$; using $\lambda_u = 6.705$, we get $\lambda_u + 1.282\sqrt{\lambda_u} = 10.02 \simeq 10$. Thus, the one-sided tolerance limits are 1 and 10. Using the two-sided confidence interval $(\lambda_l, \lambda_u) = (3.226, 7.142)$ and $z_{.95} = 1.645$, we get

$$\left(\max\left\{ 0, \left[\lambda_l - 1.645\sqrt{\lambda_l} \right] \right\}, \left[\lambda_u + 1.645\sqrt{\lambda_u} \right] \right) \simeq (0, 12).$$

We once again observe that the results based on all the methods are in agreement, except that the equal-tailed tolerance interval based on the approximate quantile is slightly different from others.

12.6.3   Two-Sided Tolerance Intervals for Binomial and Poisson Distributions

Note that the equal-tailed tolerance interval $(L_e(\boldsymbol{X}), U_e(\boldsymbol{X}))$ defined in (12.6.2) not only includes a proportion p of the distribution, but also controls the proportions of the population fall in both tails. As a result, the equal-tailed tolerance interval is expected to be wider than the two-sided tolerance interval $(L(\boldsymbol{X}), U(\boldsymbol{X}))$. Recall that a $(p, 1-\alpha)$ two-sided tolerance interval $(L(\boldsymbol{X}), U(\boldsymbol{X}))$ includes at least a proportion p of the distribution with confidence $1 - \alpha$. That is,

$$P_{\boldsymbol{X}} \left\{ P_X \left(L(\boldsymbol{X}) \leq X \leq U(\boldsymbol{X}) | \boldsymbol{X} \right) \geq p \right\} \geq 1 - \alpha. \tag{12.6.15}$$

Furthermore, in applications, especially in discrete quality assessment, a two-sided tolerance interval could serve the intended purpose, rather than an equal-tailed tolerance interval. A way to find a two-sided tolerance interval is to determine the value of α^* so that the $(p, 1 - \alpha^*)$ equal-tailed tolerance interval includes at least a proportion p of the distribution with coverage probability at least $1 - \alpha$. For a binomial or Poisson distribution with parameter θ, Wang and Tsung (2009) proposed a numerical approach to find the value of α^* so that

$$\left(k_{\frac{1-p}{2}}(\theta_l), k_{\frac{1+p}{2}}(\theta_u) \right), \tag{12.6.16}$$

where (θ_l, θ_u) is a $1 - \alpha^*$ confidence interval for θ (θ is the binomial parameter π or the Poisson mean λ), includes at least a proportion p of the population with minimum coverage probability close to the nominal level $1 - \alpha$. Krishnamoorthy and Xia (2009) noted that a $(p, 1 - 2\alpha)$ score confidence interval for θ could produce a tolerance interval of the form (12.6.16) with minimum content p and minimum coverage probability close to $1 - \alpha$. In general, for a moderate sample size, the tolerance intervals based on the approximate quantiles along with the score confidence intervals are satisfactory in both binomial and Poisson cases. For more details and accuracy studies, see Krishnamoorthy and Xia (2009).

Example 12.1 (continued)

We shall compute $(p, 1 - \alpha) = (0.90, 0.95)$ two-sided tolerance intervals for the data in Example 12.1. If we use Krishnamoorthy and Xia's (2009) approach, then the required (0.90, 0.95) two-sided tolerance interval is simply a $(p, 1-2\alpha) = (.90, .90)$ equal-tailed tolerance interval. Note that 95% one-sided lower and upper confidence limits form a 90% confidence interval for π. If we choose to use the 90% score confidence interval for π, then the 90% confidence interval for the proportion of defective chips in Example 12.1 is (0.2139, 0.2497). So the left endpoint the two-sided tolerance interval is the largest k that satisfies $P(X \geq k|n = 50, \pi_l = 0.2139) \geq \frac{1+.90}{2} = 0.95$, which is 6. Similarly, using $\pi_u = 0.2497$, we get the right endpoint as 18. Note that this two-sided tolerance interval is the same as the equal-tailed tolerance interval computed earlier. By estimating the approximate quantiles using the score confidence limits, we obtain

$$n\pi_l - 1.645\sqrt{n\pi_l(1 - \pi_l)} = 5.93 \simeq 6,$$

and

$$n\pi_u + 1.645\sqrt{n\pi_u(1 - \pi_u)} = 17.52 \simeq 18.$$

Thus, the (0.90, 0.95) two-sided tolerance interval is $(6, 18)$.

12.7   Tolerance Intervals Based on Censored Samples

Censored samples commonly arise in life-time data analysis, and in environmental and exposure data analysis. There are different types of censoring schemes,
and we consider only the simple cases, namely, type I and type II singly censored
samples. In a type I singly left-censored sample, a censoring value x_0 is specified,
and the measurements that fall below x_0 are not included in the sample while
in a type I right-censored sample the values above x_0 are not recorded. Type I
left-censored samples are often encountered in exposure and pollution data analysis where contaminant levels are observed or measured only if they are above
a threshold value usually referred to as the *detection limit* (of a sampling device
or a laboratory method). In life-time studies, sample items are tested until a
pre-specified period of time x_0, and the survival times of items that fail before
x_0 are recorded, and the test will be stopped at time x_0. This strategy of life
testing leads to type I singly right-censored samples. If n items are subjected to
a stress test until a fixed number of items fail, then the resulting measurements
constitute a type II right-censored sample. A type II left-censored sample is
similarly defined. Note that in a type I censored sample, the observed number
of failures is a random variable, whereas it is a fixed number in a type II censored
sample.

   A popular method for analyzing censored data is the likelihood approach.
Specifically, using the asymptotic normality of the maximum likelihood estimators (MLEs), normal based approaches can be used to develop inferential
procedures for the sampled population. Other finite sample approaches are
based on available pivotal quantities, especially for distributions belonging to a
location-scale family. To describe the pivotal based approach, let $X_1, ..., X_{n-l}$
be a censored sample (l observations are censored) from a location-scale distri-
bution with the location parameter μ and the scale parameter σ. Let $\widehat{\mu}$ and $\widehat{\sigma}$
be the MLEs of μ and σ respectively. Furthermore, let us assume that these
MLEs are equivariant (see Section 1.4.2). If the sample is type II censored, then
it is known that

$$\frac{\widehat{\mu} - \mu}{\widehat{\sigma}} \quad \text{and} \quad \frac{\widehat{\sigma}}{\sigma} \qquad (12.7.1)$$

are pivotal quantities; see Lawless (2003, p. 562). Thus, exact methods for
constructing confidence intervals (or hypothesis tests) for μ and σ can be readily
obtained if a type II censored sample is from a location-scale distribution. If the
sample is type I censored, then the distributions of the aforementioned quantities
may depend on the parameters, and so they may no longer be valid pivotal
quantities. Nevertheless, for some cases they can be used as approximate pivotal
quantities to get satisfactory results, as will be seen in the sequel.

In the following, we use the MLEs and the generalized variable approach for setting tolerance limits for normal, exponential, Weibull and other related distributions.

12.7.1   Normal and Related Distributions

MLEs for Left-Censored Samples

For a type I singly left-censored sample from a normal distribution, Cohen (1959, 1961) derived the MLEs for the mean μ and variance σ^2; these can be computed numerically as solutions of the following equations. Let l denote the number of observations below the censoring value x_0 in a sample of size n, and let X_i ($i = 1, 2, ..., n - l$) denote the observations above x_0. Define

$$\xi = \frac{x_0 - \mu}{\sigma}, \quad Z(\xi) = \frac{\phi(\xi)}{1 - \Phi(\xi)}, \quad \text{and} \quad Y(h, \xi) = \frac{hZ(-\xi)}{1 - h}, \tag{12.7.2}$$

where ϕ and Φ denote respectively the density function and the distribution function of a standard normal random variable, and $h = \frac{l}{n}$. Let

$$\bar{X}_l = \frac{1}{n - l} \sum_{i=1}^{n-l} X_i \quad \text{and} \quad S_l^2 = \frac{1}{n - l} \sum_{i=1}^{n-l} (X_i - \bar{X}_l)^2. \tag{12.7.3}$$

The MLEs of μ, σ^2 and ξ are the solutions of

$$\begin{aligned}
\mu &= \bar{X}_l - \lambda(h, \xi)(\bar{X}_l - x_0) \\
\sigma^2 &= S_l^2 + \lambda(h, \xi)(\bar{X}_l - x_0)^2 \\
\frac{1 - Y(h, \xi)(Y(h, \xi) - \xi)}{(Y(h, \xi) - \xi)^2} &= \frac{S_l^2}{(\bar{X}_l - x_0)^2},
\end{aligned} \tag{12.7.4}$$

where $\lambda(h, \xi) = \frac{Y(h,\xi)}{[Y(h,\xi)-\xi]}$. Let $\widehat{\xi}$ be the solution of the third equation in (12.7.4). Then, the MLEs of μ and σ^2 can be computed by substituting $\lambda(h, \widehat{\xi})$ for $\lambda(h, \xi)$ in the first two equations of (12.7.4). Recall that $h = \frac{l}{n}$.

If the sample is type II left-censored, then the MLEs can be obtained using the equations (12.7.4) with x_0 replaced by $X_{(1)}$, which is the smallest measured value.

A convenient approximation to the MLEs are as follows. Let $g = \frac{S_l^2}{(\bar{X}_l - x_0)^2}$. As $\widehat{\xi}$ is implicitly a function of g, we can write the first two equations of (12.7.4) as

$$\widehat{\mu} = \bar{X}_l - (\bar{X}_l - x_0)\lambda(g, h) \quad \text{and} \quad \widehat{\sigma}^2 = S_l^2 + (\bar{X}_l - x_0)^2\lambda(g, h). \tag{12.7.5}$$

Schmee, Gladstein and Nelson (1985) provided table values of $\lambda(g,h)$ for values of g up to 10, and $h = \frac{l}{n} = 1(.1).9$. Haas and Scheff (1990) developed an approximation to λ that fits the table values of Schmee et. al. (1985) within 6% relative error. Letting $y = \frac{h}{1-h}$ and

$$
\begin{aligned}
D &= \frac{0.182344 - 0.3756}{g+1} + 0.10017g + 0.78079y \\
&\quad - 0.00581g^2 - 0.06642y^2 - 0.0234gy + 0.000174g^3 \\
&\quad + 0.001663g^2 y - 0.00086gy^2 - 0.00653y^3,
\end{aligned} \tag{12.7.6}
$$

the approximation is given by $\lambda(g,h) \simeq \exp(D)$. Krishnamoorthy, Mallick and Mathew (2009) also found that the approximation is quite satisfactory.

Pivotal Quantities and One-Sided Tolerance Limits

As mentioned earlier, $\frac{\widehat{\mu}-\mu}{\widehat{\sigma}}$ and $\frac{\widehat{\sigma}}{\sigma}$ are pivotal quantities, and so their distributions are free of the parameters μ and σ if the sample is type II censored. As the distributions of the pivotal quantities are free of the parameters, without loss of generality, we can assume $\mu = 0$ and $\sigma = 1$ to evaluate their distributions by Monte Carlo simulation. In other words, $\frac{\widehat{\mu}-\mu}{\widehat{\sigma}}$ is distributed as $\frac{\widehat{\mu}^*}{\widehat{\sigma}^*}$ and $\frac{\widehat{\sigma}}{\sigma}$ is distributed as $\widehat{\sigma}^*$, where $\widehat{\mu}^*$ and $\widehat{\sigma}^*$ are the MLEs based on a censored sample from $N(0,1)$ distribution (that is, $\widehat{\mu}^*$ and $\widehat{\sigma}^*$ are the solutions of the equations in (12.7.4) when $X_1, ..., X_{n-l}$ are from a $N(0,1)$ distribution; see Algorithm 12.1 for the computational procedure). Thus, we can estimate the distributions of the pivotal quantities using Monte Carlo simulation (see Algorithm 12.1). As an example, let c_1 and c_2 satisfy $P\left(c_1 \le \frac{\widehat{\mu}^*}{\widehat{\sigma}^*} \le c_2\right) = 1-\alpha$. Then, $(\widehat{\mu}-c_2\widehat{\sigma}, \widehat{\mu}-c_1\widehat{\sigma})$ is a $1-\alpha$ confidence interval for μ. Notice that c_1 and c_2 can be obtained using Monte Carlo simulation. If the sample is type II censored, then this confidence interval is exact. Schmee et al. (1985) used such Monte Carlo approach for constructing confidence intervals for a normal mean and variance.

If the sample is type I censored, then the pivotal quantities in the preceding paragraph can be used as approximate pivotal quantities for making inference, and the results based on them seem to be satisfactory even for small samples; see Schmee et al. (1985) for constructing confidence intervals for a normal mean and variance based on a type I censored sample.

Generalized Variable Approach

As one-sided tolerance limits are confidence limits for appropriate quantiles, we shall develop a generalized pivotal quantity (GPQ) for the p quantile of a

$N(\mu, \sigma^2)$ distribution, based on a type II singly left-censored data. Specifically, we shall describe a procedure for constructing $1 - \alpha$ upper confidence limit for the p quantile $Q_p = \mu + z_p\sigma$, where z_p is the p quantile of the standard normal distribution. Let $(\widehat{\mu}_0, \widehat{\sigma}_0)$ be an observed value of the MLE $(\widehat{\mu}, \widehat{\sigma})$. A GPQ for Q_p (see Section 1.4.2) is given by

$$
\begin{aligned}
G_{Q_p} &= G_\mu + z_p G_\sigma \\
&= \widehat{\mu}_0 - \frac{\widehat{\mu} - \mu}{\widehat{\sigma}}\widehat{\sigma}_0 + z_p\frac{\sigma}{\widehat{\sigma}}\widehat{\sigma}_0 \\
&= \widehat{\mu}_0 + \frac{z_p - \widehat{\mu}^*}{\widehat{\sigma}^*}\widehat{\sigma}_0 \\
&= \widehat{\mu}_0 + Z_p^*\widehat{\sigma}_0,
\end{aligned}
\tag{12.7.7}
$$

where $\widehat{\mu}^* = \frac{\widehat{\mu}-\mu}{\sigma}$, $\widehat{\sigma}^* = \frac{\widehat{\sigma}}{\sigma}$ and $Z_p^* = \frac{z_p - \widehat{\mu}^*}{\widehat{\sigma}^*}$. Let $z_{p;\beta}^*$ denote the β quantile of Z_p^*. Then,

$$
\widehat{\mu}_0 + z_{p;1-\alpha}^*\widehat{\sigma}_0
\tag{12.7.8}
$$

is a $(p, 1 - \alpha)$ upper tolerance limit for the normal distribution. It is easy to check that

$$
\widehat{\mu}_0 + z_{1-p;\alpha}^*\widehat{\sigma}_0
\tag{12.7.9}
$$

is a $(p, 1 - \alpha)$ lower tolerance limit.

The following algorithm describes a Monte Carlo method to estimate the percentiles of Z_p^*. Recall that the sample is type II singly left-censored, and is from a normal distribution. Also, l observations are censored, and $n - l$ observations are uncensored.

Algorithm 12.1

1. Generate a sample $Z_1, ..., Z_n$ from a $N(0, 1)$ distribution, and sort them in ascending order as $Z_{(1)}, ..., Z_{(n)}$ so that $Z_{(1)}$ is the smallest observation.
2. Compute the MLEs $\widehat{\mu}^*$ and $\widehat{\sigma}^*$ based on $Z_{(l+1)}, ..., Z_{(n)}$. These MLEs are the solutions of the equations in (12.7.4) with x_0 replaced by $Z_{(l+1)}$.
3. Compute $Z_p^* = \frac{z_p - \widehat{\mu}^*}{\widehat{\sigma}^*}$.
4. Repeat steps $1 - 3$, for a large number of times, say, 10,000.

The 100α percentile of these 10,000 generated Z_p^*'s is an estimate of $z_{p;\alpha}^*$.

The percentiles of Z_p^* are estimated using Algorithm 12.1, and they are reported in Table 12.2 for $n = 10, 20, 30$ and 50. As an example, if one wants to find a (0.90, 0.95) lower tolerance limit based on a sample of size $n = 30$ with the number of censored values $l = 10$, then the required critical value

$z^*_{1-p;\alpha} = z^*_{10;.05} = -2.096$, and $\hat{\mu}_0 - 2.096\hat{\sigma}_0$ is the desired lower tolerance limit. If it is desired to find $(0.90, 0.95)$ upper tolerance limit when $n = 30$ and $l = 10$, then $z^*_{.90;.95} = 1.867$and $\hat{\mu}_0 + 1.867\hat{\sigma}_0$ is the desired upper tolerance limit.

If a sample is type II singly right-censored, then the procedure for a left-censored sample can be easily modified on the basis of the symmetry of a normal distribution. For example, in order to obtain a $(p, 1 - \alpha)$ upper tolerance limit, we first multiply the data by -1, find a $(p, 1 - \alpha)$ lower tolerance limit using the approach for a left-censored sample, and then multiply the lower tolerance limit so obtained by -1, to get the desired upper tolerance limit. This procedure leads to the $(p, 1 - \alpha)$ upper tolerance limit as

$$\hat{\mu}_0 - z^*_{1-p;\alpha}\hat{\sigma}_0, \qquad (12.7.10)$$

where $\hat{\mu}_0$ and $\hat{\sigma}_0$ are the MLEs based on Algorithm 12.1. In other words, MLEs are computed as if the sample is left-censored, and the formula in (12.7.10) should be used to compute the upper tolerance limit. A $(p, 1-\alpha)$ lower tolerance limit for a type II right-censored sample can be similarly obtained, and is given by

$$\hat{\mu}_0 - z^*_{p;1-\alpha}\hat{\sigma}_0. \qquad (12.7.11)$$

In other words, to compute one-sided tolerance limits for a right-censored sample, we can use the same algorithm that computes the MLEs for a left-censored sample. Furthermore, note that $z^*_{1-p;\alpha}$ $(z^*_{p;1-\alpha})$ is used for computing lower (upper) tolerance limit based on a left-censored sample whereas $-z^*_{1-p;\alpha}$ $(-z^*_{p;1-\alpha})$ is used to find the upper (lower) tolerance limit based on a right-censored sample.

Approximate Pivotal Quantities for Type I Censored Samples

As mentioned earlier, the pivotal quantities given for type II censored case are no longer valid if the sample is type I censored. However, as noted before, these pivotal quantities can serve as good approximations. Thus, the percentiles $z^*_{p;\beta}$ can be used to find approximate tolerance limits when the sample is type I censored. Specifically, if we have a type I censored sample of size n with $n - l$ uncensored observations above the censoring value x_0, then we can find the MLEs by solving the equations in (12.7.4). Using the percentiles of Z^*_p (or of Z^*_{1-p}) corresponding to n and l for the given sample, we can find an approximate one-sided tolerance limit. Recently, Krishnamoorthy, Mallick and Mathew (2009) found that this approximate procedure for type I censored samples is quite satisfactory for constructing lower tolerance limits, whereas it is slightly conservative for constructing upper tolerance limits.

Table 12.2: Percentiles of Z_p^* for constructing one-sided tolerance limits for a normal distribution

n	l	$z_{.05;.05}^*$	$z_{.1;.05}^*$	$z_{.9;.95}^*$	$z_{.95;.95}^*$
10	1	-3.309	-2.674	2.577	3.184
	3	-4.105	-3.327	2.788	3.528
	6	-8.711	-7.266	3.596	4.905
20	1	-2.504	-2.016	1.995	2.506
	3	-2.626	-2.120	2.013	2.529
	6	-2.899	-2.340	2.082	2.619
	10	-3.588	-2.935	2.184	2.786
	15	-7.350	-6.221	2.451	3.413
30	1	-2.285	-1.813	1.809	2.279
	3	-2.327	-1.867	1.829	2.285
	6	-2.422	-1.958	1.856	2.320
	10	-2.644	-2.096	1.867	2.375
	15	-2.991	-2.439	1.928	2.458
	20	-3.877	-3.178	1.991	2.627
50	1	-2.100	-1.667	1.676	2.093
	5	-2.132	-1.706	1.669	2.099
	10	-2.216	-1.751	1.685	2.123
	20	-2.403	-1.909	1.707	2.165
	25	-2.552	-2.073	1.727	2.192
	40	-4.220	-3.560	1.747	2.358

A Likelihood Ratio Test for a Quantile

Suppose we have a type I censored sample of size n with $n - l$ uncensored observations above the censoring value x_0. We shall now explain a likelihood based procedure for inference concerning the normal quantiles, based on such censored data. More specifically, we shall describe a procedure based on the *directed likelihood* or the *signed likelihood ratio*; see Barndorff-Nielsen and Cox (1994, p. 82) or Brazzale, Davison and Reid (2007, p. 139). As is typical with likelihood based methodologies, large sample approximations will be used to implement the procedure described below. We shall use the notations $\eta = \mu + z_p \sigma$ and $\lambda = \sigma^2$. Our problem is that of testing $H_0 : \eta = \eta_0$, for a specified η_0.

The log-likelihood function is given by

$$L(\mu, \lambda) \propto l \ln (\Phi(\xi)) - \frac{n-l}{2} \ln \lambda - \frac{(n-l)(S_l^2 + (\bar{X}_l - \mu)^2)}{2\lambda}, \qquad (12.7.12)$$

where ξ is defined in (12.7.2) and $(\bar{X}_l, S_l^2)$ is defined in (12.7.3). Let $\hat{\mu}$ and $\hat{\lambda} = \hat{\sigma}^2$ denote the MLEs of μ and σ^2, respectively. Then $\hat{\eta} = \hat{\mu} + z_p \sqrt{\hat{\lambda}}$ is the MLE of η. In order to develop the signed likelihood ratio test (SLRT), we need to find the constrained MLE, denoted by $\hat{\lambda}_{\eta_0}$, of λ under $H_0 : \eta = \eta_0$. To find this constrained MLE, we write the log-likelihood function in terms of η and λ as

$$L(\eta, \lambda) \propto l \ln \Phi(\xi_\lambda) - \frac{n-l}{2} \ln \lambda - \frac{(n-l)}{2\lambda} \left(S_l^2 + \left(\bar{X}_l - \eta + z_p \sqrt{\lambda} \right)^2 \right), \quad (12.7.13)$$

where $\xi_\lambda = \frac{x_0 - \eta}{\sqrt{\lambda}} + z_p$. The constrained MLE of λ is the solution of the equation

$$
\begin{aligned}
\left. \frac{\partial L(\eta, \lambda)}{\partial \lambda} \right|_{\eta = \eta_0} &= -l\sqrt{\lambda}(x_0 - \eta_0)w(\xi_\lambda) \\
&\quad + (n-l)[S_l^2 + (\bar{X}_l - \eta_0)^2 + z_p \sqrt{\lambda}(\bar{X}_l - \eta_0) - \lambda] \\
&= 0. \quad (12.7.14)
\end{aligned}
$$

The SLRT is based on the asymptotic result that

$$\psi(\eta_0; \hat{\eta}, \hat{\lambda}_{\eta_0}) = \text{sign}(\hat{\eta} - \eta_0) \left\{ 2(l(\hat{\eta}, \hat{\lambda}) - l(\eta_0, \hat{\lambda}_{\eta_0})) \right\}^{\frac{1}{2}} \sim N(0, 1), \quad (12.7.15)$$

under H_0, where $\text{sign}(x)$ is 1 if $x \geq 0$ and is -1 if $x < 0$. Thus, the SLRT rejects $H_0 : \eta = \eta_0$ if $|\psi(\eta_0; \hat{\eta}, \hat{\lambda}_{\eta_0})| > z_{1-\frac{\alpha}{2}}$.

A confidence limit for η can be obtained by inverting the test. For example, a $1 - \alpha$ upper confidence set for η, which is the acceptance region of the test for $H_0 : \eta \geq \eta_0$ vs. $H_a : \eta < \eta_0$, is given by

$$\left\{ \eta_0 : \psi(\eta_0; \hat{\eta}, \hat{\lambda}_{\eta_0}) \geq -z_\alpha \right\}. \quad (12.7.16)$$

Numerical evidence indicates that the above set is an interval.

Notice that the above SLRT is described for a type I left-censored sample. It can be also used for a type II left-censored sample by replacing x_0 by $X_{(1)}$, the smallest order statistic among the $n - l$ uncensored observations. However, we do not recommend this SLRT for a type II censored sample since an exact method is available, as described earlier.

Estimation of Survival Probabilities

To find a lower confidence limit for the survival probability $S_t = P(X > t) = 1 - \Phi \left(\frac{t - \mu}{\sigma} \right)$, where t is a specified number, it is enough to find a suitable upper confidence limit for $S_t^* = \frac{t-\mu}{\sigma}$. This can be done using the generalized variable approach (see Section 1.4). Specifically, replacing the parameters by their GPQs, we see that

$$G_{S_t^*} = \frac{t - G_\mu}{G_\sigma} = \frac{t - \left(\widehat{\mu}_0 - \frac{\widehat{\mu}^*}{\widehat{\sigma}^*}\widehat{\sigma}_0\right)}{\frac{\widehat{\sigma}_0}{\widehat{\sigma}^*}}, \tag{12.7.17}$$

is a GPQ for $G_{S_t^*}$, where G_μ and G_σ are as defined in (12.7.7). Notice that, for a given $(\widehat{\mu}_0, \widehat{\sigma}_0)$, the distribution of $G_{S_t^*}$ does not depend on any unknown parameters, and the percentiles of $G_{S_t^*}$ can be estimated using Monte Carlo simulation.

For a right-censored sample, we first note that $S_t = P(-X < -t) = \Phi\left(\frac{-t-\mu}{\sigma}\right)$. Therefore, after multiplying the sample by -1, we can apply the generalized variable procedure described above. For instance, to find a lower confidence limit for S_t, it is enough to find a lower confidence limit for $\frac{-t-\mu}{\sigma}$. The GPQ can be constructed using the MLEs based on the "negative transformed" data, and using $\widehat{\mu}^*$ and $\widehat{\sigma}^*$ (the MLEs based on a sample from the standard normal distribution). It is easy to check that the resulting GPQ for $\frac{-t-\mu}{\sigma}$ is

$$\frac{-t + \left(\widehat{\mu}_0 + \frac{\widehat{\mu}^*}{\widehat{\sigma}^*}\widehat{\sigma}_0\right)}{\frac{\widehat{\sigma}_0}{\widehat{\sigma}^*}}, \tag{12.7.18}$$

where $\widehat{\mu}_0$ and $\widehat{\sigma}_0$ are the MLEs based on a left-censored data with censoring time x_0.

Example 12.3 (Survival times of mice)

The following data, given in Schmee and Nelson (1977), represent the number of days to death of 10 mice after inoculation with a culture of human tuberculosis. Out of these 10 mice, seven died at log-times 1.613, 1.643, 1.663, 1.732, 1.740, 1.763, and 1.778, and three survived a log-time of 1.778; so these three observations are censored. In particular, we have a type II right-censored sample with $n = 10$ and $l = 3$. It was noted that a normal distribution gave a good fit to the uncensored lifetimes.

To compute one-sided tolerance limits, the MLEs are obtained as $\widehat{\mu}_0 = 1.742$ and 0.0793. To find a $(0.90, 0.95)$ lower tolerance limit using (12.7.11), the percentile $z_{.90;.95}^*$ has value 2.788 (see Table 12.2). The lower tolerance limit in terms of log-time is $1.742 - 2.788 \times 0.0793 = 1.521$. Thus, at least 90% of mice survived $\exp(1.521) = 4.577$ days with confidence 95%. Similarly, to compute a $(0.90, 0.95)$ upper tolerance limit, the percentile $z_{.1;.05}^*$ is -3.327, and the upper tolerance limit in log-time is $1.742 + 3.327 \times 0.0793 = 2.006$. Since exp(2.006) = 7.434, we conclude that at most 10% of the mice had survival times more that 7.434 days, with 95% confidence.

Example 12.4 (Failure mileages of locomotive controls)

The data in Table 12.3 represent failure mileages (in units of 1000 miles) of different locomotive controls in a life test involving 96 locomotive controls. The test was terminated after 135,000 miles, and by then 37 controls had failed. This example is discussed in Schmee and Nelson (1977), and also in Lawless (2003, Section 5.3). These authors noted that a lognormal distribution gives a good fit to the data. In this type of examples, it is of interest to find a lower tolerance

Table 12.3: Failure mileages (in 1000) of locomotive controls

22.5	37.5	46.0	48.5	51.5	53.0	54.5	57.5	66.5	68.0
69.5	76.5	77.0	78.5	80.0	81.5	82.0	83.0	84.0	91.5
93.5	102.5	107.0	108.5	112.5	113.5	116.0	117.0	118.5	119.0
120.0	122.5	123.0	127.5	131.0	132.5	134.0			

limit to assess the reliability of the controls, and to estimate the lifetime at certain mileages.

We shall first compute a (0.90, 0.95) lower tolerance limit. Notice that the sample is type I right-censored with censoring mileage of 135,000 or $x_0 = 135$ (in 1000 mile unit). Since the lognormal distribution is applicable here, the MLEs based on the log-transformed data were computed as $\hat{\mu}_0 = 5.117$ and $\hat{\sigma}_0 = 0.705$. To compute the tolerance limit using (12.7.11), the percentile $z^*_{.90;.95}$ is evaluated using Algorithm 12.1 (with $n = 96$ and $l = 59$) as 1.579. Thus, using (12.7.11), we have $5.117 - 1.579 \times 0.705 = 4.004$, and $\exp(4.004) = 54.82$ is the desired (0.90, 0.95) lower tolerance limit. Thus with 95% confidence we can assert that at least 90% of the controls survive 54,820 miles.

To compute a (0.90, 0.95) upper tolerance limit using (12.7.10), the percentile $z^*_{1-p;\alpha} = z^*_{.1;.05}$ is evaluated using Algorithm 12.1 as -1.840. Substituting this percentile and the MLEs in (12.7.10), we get $5.117 - (-1.922) \times 0.705 = 6.472$, and $\exp(6.472) = 646.78$. Thus we are 95% confident that at least 90% of locomotive controls fail before 646,780 miles; i.e., at most 10% survive 646,780 miles.

To find the 95% lower confidence limit for the lower 0.1 quantile (that is, to find the (0.90, 0.95) lower tolerance limit) using the SLRT approach, we have to find the 95% upper confidence limit for the upper quantile $\mu + z_{.90}\sigma$ based on negative log-transformed data. That is, we need to find the value of η_0 for which $\psi(-\ln(\eta_0); \widehat{\eta}, \widehat{\lambda}_{\eta_0}) = z_{.05} = -1.645$ (see (12.7.16)). Using the negative log-transformed data, we found that

$$\psi(-\ln(\eta_0); \widehat{\mu}_0 + z_{0.90}\widehat{\sigma}_0, \widehat{\lambda}_{\eta_0}) = \psi(-4.021; -5.117 + 1.282 \times 0.705, 0.678) = -1.645.$$

Notice that $\widehat{\mu}_0 = -5.117$ since this is the MLE based on negative log-transformed data. Thus, $\eta_0 = \exp(\ln(\eta_0)) = \exp(4.021) = 55.77$ is the desired $(0.90, 0.95)$ lower tolerance limit. The value $-\ln(\eta_0)$ that satisfies the above equation was found by trial-error using the generalized limit as a starting value. To compute the above statistic, the interested reader can verify that $l(\widehat{\eta}, \widehat{\lambda}) = -39.458$ and $l(-\ln(\eta_0), \widehat{\lambda}_{\eta_0}) = -40.811$.

We now compute the $(0.90, 0.95)$ upper tolerance limit by inverting the SLRT. That is, we first find the 95% lower limit for the lower quantile $\mu + z_{.10}\sigma$ based on the negative log-transformed samples, and then take the inverse transformation to get the upper tolerance limit. Towards this, we computed

$$\psi(-\ln(\eta_0); \widehat{\mu}_0 + z_{0.10}\widehat{\sigma}_0, \widehat{\lambda}_{\eta_0}) = \psi(-6.413; -5.117 - 1.282 \times 0.705, 0.764) = 1.645.$$

Thus, the $(0.90, 0.95)$ upper tolerance limit is $\exp(-(-6.413)) = 609.72$.

Notice that the lower tolerance limits based both approaches are in good agreement while the upper tolerance limit based on the approximate pivotal quantities is larger than the one based on the likelihood approach.

Suppose it is desired to estimate the survival probability at the eighty thousandth mile, that is, S_{80}. Following (12.7.18), we find the GPQ based on the log-transformed data as

$$\frac{-\ln(80) + \left(5.117 + \frac{\widehat{\mu}^*}{\widehat{\sigma}^*}0.705\right)}{\frac{0.705}{\widehat{\sigma}^*}}. \qquad (12.7.19)$$

Using Monte Carlo simulation with 10,000 runs, the 2.5th and 97.5th quantiles of the above GPQ are estimated as 0.782 and 0.902, respectively. Thus, a 95% confidence interval for S_{80} is given by $(0.782, 0.902)$. That is, the probability that a locomotive control survives 80,000 miles or more is between 0.782 and 0.902, with 95% confidence. Lawless (2003, p. 234) found a 95% confidence interval for S_{80} as $(0.785, 0.902)$ which is practically the same as the above generalized confidence interval. Lawless obtained the confidence interval using the asymptotic normality of the MLEs, and estimated Fisher information matrix.

A $1 - \alpha$ confidence interval for a survival probability S_t can also be obtained by setting the appropriate one-sided tolerance limits to t, and then solving for the content; see Exercise 1.5.2.

Lognormal and Gamma Distributions

The preceding procedures for censored-samples can be readily extended to a distribution that has one-one relation with the normal distribution. As seen in Examples 12.3 and 12.4, if a censored sample is from a lognormal distribution, then we simply apply the normal based methods to log-transformed samples. Similarly, if a censored sample is from a gamma distribution, tolerance limits can be obtained by applying the ones for the normal case to the cube root transformed sample; see Chapter 7. The results from the normal based procedures for a gamma distribution are only approximate, and as seen for the case of uncensored samples in Section 7.3.2, the results are expected to be satisfactory. An illustrative example for the gamma case is given below.

Example 12.5 (Example 7.1 continued)

Let us consider the alkalinity concentration data in Table 7.1, assuming that there is a detection limit $x_0 = 45$. That is, we assume that the values below 45 were undetected (or left-censored), and use the normal based procedures to develop one-sided tolerance limits. We first note that for this example with the above assumption, we have $n = 27$, and $l = 8$ observations are type I left-censored. Here is the remaining data, after cube root transformation.

| 3.66 | 3.71 | 3.71 | 3.73 | 3.78 | 3.78 | 3.80 | 3.87 | 3.89 |
| 3.89 | 3.91 | 3.98 | 4.04 | 4.12 | 4.29 | 4.34 | 4.46 | 4.58 | 4.90 |

The MLEs for the cube root transformed samples are $\widehat{\mu}_0 = 3.798$ and $\widehat{\sigma}_0 = 0.4663$. The required percentile to compute a $(0.90, 0.95)$ lower tolerance limit is $z^*_{.10;.05} = -2.115$ and using (12.7.9) we get $3.798 - 2.115 \times 0.4663 = 2.8118$. Thus, $(2.8118)^3 = 22.23$ is the desired $(0.90,0.95)$ lower tolerance limit for alkalinity concentrations. It should be noted that the same lower tolerance limit based on the complete sample is 28.341 (see Example 7.1). To compute a $(0.90, 0.95)$ upper tolerance limit, the percentile $z^*_{.90;.95}$ is computed as 1.916, and $\widehat{\mu}_0 + z^*_{p;1-\alpha}\widehat{\mu}_0 = 3.798 + 1.916 \times 0.4663 = 4.691$. Thus, the desired upper tolerance limit is $(4.691)^3 = 103.23$. The one based on the complete sample is 97.71 (see Example 7.1).

12.7.2   Two-Parameter Exponential Distribution

Consider failure times that follow a two-parameter exponential distribution with the pdf $\frac{1}{\theta}\exp\left(-\frac{(x-\mu)}{\theta}\right)$, where $x > \mu$ and $\theta > 0$. Suppose we have a type

II right-censored sample of size n, with m uncensored observations. Let the corresponding ordered observations be $X_{(1)}, ..., X_{(m)}$, where $X_{(1)}$ is the smallest. We shall first describe the procedures for constructing one-sided tolerance limits.

The MLEs are given by

$$\widehat{\mu} = X_{(1)} \quad \text{and} \quad \widehat{\theta} = \frac{1}{m} \left[\sum_{i=1}^{m} X_i + (n-m)X_{(m)} - nX_{(1)} \right]. \tag{12.7.20}$$

It is well-known that the MLEs are independent with

$$\frac{(\widehat{\mu} - \mu)}{\theta} \sim \frac{\chi_2^2}{2n} \quad \text{and} \quad \frac{\widehat{\theta}}{\theta} \sim \frac{\chi_{2m-2}^2}{2m}. \tag{12.7.21}$$

Since a one-sided tolerance limit is a confidence limit for a quantile, we first note that the p quantile of the exponential distribution is given by $q_p = \mu - \theta \ln(1-p)$. A $1 - \alpha$ upper confidence limit for q_p is the desired $(p, 1 - \alpha)$ upper tolerance limit. As the distributional results of the MLEs are similar to those for the uncensored sample case considered in Section 7.4, exact method and generalized variable approach can be readily obtained for constructing confidence limits for q_p. We shall describe the generalized variable approach in the following.

One-Sided Tolerance Limits

It follows from Section 7.4.1 that a generalized pivotal quantity for $q_p = \mu - \theta \ln(1 - p)$ is

$$\begin{aligned}
G_{Q_p} &= G_\mu - G_\theta \ln(1-p) \\
&= \widehat{\mu}_0 - \frac{\widehat{\mu} - \mu}{\widehat{\theta}} \widehat{\theta}_0 - \frac{\widehat{\theta}_0}{(\widehat{\theta}/\theta)} \ln(1-p) \\
&= \widehat{\mu}_0 - \frac{m}{n} \frac{\chi_2^2}{\chi_{2m-2}^2} \widehat{\theta}_0 - \frac{2m}{\chi_{2m-2}^2} \ln(1-p) \\
&= \widehat{\mu}_0 - \frac{m}{n} \left(\frac{\chi_2^2 + 2n \ln(1-p)}{\chi_{2m-2}^2} \right) \widehat{\theta}_0 \\
&= \widehat{\mu}_0 - \frac{m}{n} E_p^* \widehat{\theta}_0, \tag{12.7.22}
\end{aligned}$$

where G_μ and G_θ are GPQs for μ and θ respectively, χ_2^2 and χ_{2m-2}^2 represent independent chi-square random variables with df 2 and $2m - 2$, respectively, $(\widehat{\mu}_0, \widehat{\theta}_0)$ is an observed value of $(\widehat{\mu}, \widehat{\theta})$ and $E_p^* = \left(\frac{\chi_2^2 + 2n \ln(1-p)}{\chi_{2m-2}^2} \right)$. Let $E_{p;\alpha}^*$ denote the α quantile of E_p^*. Then

$$\widehat{\mu}_0 - E_{p;\alpha}^* \frac{m}{n} \widehat{\theta}_0 \tag{12.7.23}$$

is a $1-\alpha$ upper confidence limit for Q_p, which is also a $(p, 1-\alpha)$ upper tolerance limit for the exponential(μ, θ) distribution. Similarly, we see that

$$\widehat{\mu}_0 - E^*_{1-p, 1-\alpha} \frac{m}{n} \widehat{\theta}_0 \qquad (12.7.24)$$

is a $(p, 1 - \alpha)$ lower tolerance limit for the exponential(μ, θ) distribution. For $n \le \frac{\ln(\alpha)}{\ln(p)}$, it is not difficult (see Exercise 12.8.11) to show that

$$-E^*_{1-p; 1-\alpha} = 1 - \left(\frac{p^n}{\alpha}\right)^{\frac{1}{m-1}}. \qquad (12.7.25)$$

As noted in Section 7.4, it can be shown that the one-sided tolerance limits obtained based on the GPQ are actually exact.

Estimation of Survival Probability

A $1 - \alpha$ lower confidence limit for the survival probability $S_t = P(X > t)$ can be deduced from the exact lower tolerance limit (see Section 7.4.3). Specifically, setting the exact lower tolerance limit $\widehat{\mu}_0 - E^*_{1-p; 1-\alpha} \frac{m}{n} \widehat{\theta}_0$ to t, and solving the resulting equation for p (see Section 1.1.3), we get

$$\alpha^{\frac{1}{n}} \left[1 - \frac{n}{m}\left(\frac{t - \widehat{\mu}}{\widehat{\theta}}\right)\right]^{\frac{m-1}{n}} \qquad (12.7.26)$$

as a $1 - \alpha$ lower confidence limit for S_t when $n \le \frac{\ln(\alpha)}{\ln(p)}$.

To apply the generalized variable method, we first note that $S_t = \exp(-(t - \mu)/\theta)$. Replacing the parameters by their GPQs, and after some simplification, we get

$$G_{S_t} = \exp\left\{-\frac{1}{2}\left[\frac{1}{m}\left(\frac{t - \widehat{\mu}_0}{\widehat{\theta}_0}\right)\chi^2_{2n-2} + \frac{1}{n}\chi^2_2\right]\right\} = \exp\left\{-\frac{1}{2}A^*\right\}, \qquad (12.7.27)$$

where $A^* = \frac{1}{m}\left(\frac{t - \widehat{\mu}_0}{\widehat{\theta}_0}\right)\chi^2_{2m-2} + \frac{1}{n}\chi^2_2$. If A^*_α is the α quantile of A^*, then $\exp\left(-\frac{1}{2}A^*_\alpha\right)$ is a $1 - \alpha$ upper confidence limit for S_t, and $\exp\left(-\frac{1}{2}A^*_{1-\alpha}\right)$ is a $1 - \alpha$ lower confidence limit for S_t. The lower confidence limits so obtained are also exact, as noted in Section 7.4.

It is possible to develop an approximation for the percentiles of A^*, similar to the approximation given in Chapter 7; see (7.4.26). The approximation is due to Roy and Mathew (2005). In order to describe the approximation, let

$$a_1 = \frac{1}{2m\widehat{\theta}_0} \times \frac{(t-\widehat{\mu}_0)^2(k-1)+(\frac{m}{n}\widehat{\theta}_0)^2}{(t-\widehat{\mu}_0)(m-1)+\frac{m}{n}\widehat{\theta}_0}$$

$$a_2 = \frac{2\left[(t-\widehat{\mu}_0)(m-1)+\frac{m}{n}\widehat{\theta}_0\right]^2}{(t-\widehat{\mu}_0)^2(m-1)+(\frac{m}{n}\widehat{\theta}_0)^2}.$$

Then an approximation to the $(1-\alpha$ percentile of A^*, say $\tilde{A}^*_{1-\alpha}$, is given by

$$\tilde{A}^*_{1-\alpha} = \begin{cases} -\frac{1}{n}\left[\ln\alpha + (m-1)\ln\left\{1+\frac{n}{m}\frac{(\widehat{\mu}_0-t)}{\widehat{\theta}_0}\right\}\right], & \text{if } t \le \widehat{\mu}_0 \\ a_1\chi^2_{a_2;1-\alpha}, & \text{if } t > \widehat{\mu}_0, \end{cases}$$

where a_1 and a_2 are as defined above. For further details and a derivation of the above approximation, we refer to Roy and Mathew (2005).

Example 12.6 (Failure mileages of military careers)

Let us compute one-sided tolerance limits and a confidence limit for a survival probability using failure mileage data given in Example 7.3. For the sake of illustration, we shall assume that the data above the mileage 1182 are censored, and only the following mileages are observed.

$$\begin{array}{ccccccccccc} 162 & 200 & 271 & 302 & 393 & 508 & 539 & 629 & 706 & 777 \\ 884 & 1008 & 1101 & 1182 \end{array}$$

Thus, we have $n = 19$ and $m = 14$. The MLEs are $\widehat{\mu}_0 = 162$ and $\widehat{\theta}_0 = 822.29$. To find an exact $(0.90, 0.95)$ lower tolerance limit, we first check that $n = 19 < \frac{\ln(0.05)}{\ln(0.10)} = 28.43$, and so the formula (12.7.25) for the tolerance factor is valid, and the tolerance factor simplifies to -0.0795. Substituting the relevant quantities in (12.7.24), we get $162 - 0.0795 \times 14 \times 822.29/19 = 114.10$. It is interesting note here that the $(0.90, 0.95)$ lower tolerance limit based on all $n = 19$ observations is 114.39 (see Example 7.3). To compute a $(0.90, 0.95)$ upper tolerance limit, the required percentile is $E^*_{p;\alpha} = E^*_{0.90;0.05} = -5.5717$ (computed using simulation), and (12.7.23) simplifies to $162 + 5.5717 \times 14 \times 822.29/19 = 3537.88$. We once again observe that the same upper tolerance limit based on all 19 observations is 3235.83 (see Example 7.3).

A 95% lower confidence limit for the survival probability $S_{200} = P(X > 200)$ based on the exact formula (12.7.26) is

$$0.05^{\frac{1}{19}}\left[1 - \frac{19}{14}\left(\frac{200-162}{822.29}\right)\right]^{\frac{13}{19}} = 0.817.$$

In order to compute the generalized lower confidence limit using (12.7.27), the percentile $A^*_{0.05} = 0.4039$, and $\exp(-0.4039/2) = 0.817$. Thus, both exact and the generalized variable method produced the same 95% lower confidence limit for the survival probability S_{200}. We note here that the exact 95% lower confidence limit based on all 19 observations is also 0.817 (see Example 7.3).

12.7.3   Weibull and Extreme Value Distributions

Let $X_1, \ldots, X_n$ be a sample from a Weibull(b, c) distribution with pdf

$$f(x|b, c) = \frac{c}{b}\left(\frac{x}{b}\right)^{c-1} \exp\left\{-\left[\frac{x}{b}\right]^c\right\}, \quad x > 0, \; b > 0, \; c > 0. \tag{12.7.28}$$

Suppose that $n-m$ observations are type II right-censored. Cohen (1965) derived the likelihood equations; the MLE of c can be obtained as the solution of

$$\frac{1}{\widehat{c}} - \frac{\sum\limits_{i=1}^{n} X_{i*}^{\widehat{c}} \ln(X_{i*})}{\sum\limits_{i=1}^{n} X_{i*}^{\widehat{c}}} + \frac{1}{m}\sum_{i=1}^{m} \ln(X_{i*}) = 0, \tag{12.7.29}$$

where $X_{i*} = X_{(i)}$ for $i = 1, \ldots, m$, $X_{i*} = X_{(m)}$ for $i = m+1, \ldots, n$, and $X_{(1)}, \ldots, X_{(m)}$ are ordered uncensored observations (see Exercise 12.8.15). The MLE of b is given by $\widehat{b} = \left(\sum_{i=1}^{n} X_{i*}^{\widehat{c}}/m\right)^{1/\widehat{c}}$. The same formulas are used to find the MLEs when the samples are type I censored except that we take $X_{i*} = x_0$ for $i = m+1, \ldots, n$, where x_0 is the censoring point.

Generalized Variable Approach for Estimating Quantiles

It follows from Lemma 7.1 that the pivotal quantities based on the MLEs are $\frac{\widehat{c}}{c}$ and $\widehat{c}\ln\left(\frac{\widehat{b}}{b}\right)$. These pivotal quantities are valid if the sample is type II censored, and they can be used as approximate pivots if the sample is type I censored. As shown in Section 7.5.2, empirical distributions of these pivotal quantities can be obtained using Monte Carlo simulation. Specifically, from (7.5.5) and (7.5.6), we have the GPQs for b and c as

$$G_b = \left(\frac{1}{\widehat{b}^*}\right)^{\frac{\widehat{c}^*}{\widehat{c}_0}} \widehat{b}_0 \quad \text{and} \quad G_c = \frac{\widehat{c}_0}{\widehat{c}^*}, \tag{12.7.30}$$

where $\widehat{c}^*$ is the MLE satisfying (12.7.29) when X_i's are from a Weibull$(1, 1)$ distribution, and $\widehat{b}^* = \left(\sum_{i=1}^{n} X_{i*}^{\widehat{c}^*}/m\right)^{1/\widehat{c}^*}$. Furthermore, $(\widehat{b}_0, \widehat{c}_0)$ is an observed value of $(\widehat{b}, \widehat{c})$.

Confidence limits for a quantile can be readily obtained using the approach in Section 7.5.4 for the case of an uncensored sample. In particular, we note that the p quantile is $q_p = b\left(-\ln(1-p)\right)^{\frac{1}{c}} = b\theta_p^{\frac{1}{c}}$, where $\theta_p = -\ln(1-p)$. By substituting the GPQs for the parameters, a GPQ for q_p can be obtained as

$$G_{q_p} = \left(\frac{1}{\widehat{b}^*}\right)^{\frac{\widehat{c}^*}{\widehat{c}_0}} \widehat{b}_0\, \theta_p^{\frac{\widehat{c}^*}{\widehat{c}_0}}, \tag{12.7.31}$$

where $\widehat{b}^*$ and $\widehat{c}^*$ are as defined above. The $1-\alpha$ quantile of G_{q_p} is a $(p, 1-\alpha)$ upper tolerance limit. Furthermore, the α quantile of $G_{q_{1-p}}$ is a $(p, 1-\alpha)$ lower tolerance limit.

The one-sided tolerance limits can be expressed in simpler forms. Let $w_{p;\gamma}^*$ be the γ quantile of $w_p^* = \widehat{c}^*(-\ln(\widehat{b}^*)+\ln(\theta_p))$. Then, a $(p, 1-\alpha)$ upper tolerance limit is given by

$$\widehat{b}_0 \exp\left(w_{p;1-\alpha}^*/\widehat{c}_0\right) \tag{12.7.32}$$

which is a $(p, 1-\alpha)$ upper tolerance limit for the Weibull(b, c) distribution. A $(p, 1-\alpha)$ lower tolerance limit, i.e., a $1-\alpha$ lower confidence limit for q_{1-p}, is given by

$$\widehat{b}_0 \exp\left(w_{1-p;\alpha}^*/\widehat{c}_0\right). \tag{12.7.33}$$

The percentiles of w_p^* are given in Table12.4 for a few values of n and m.

A GPQ for a Survival Probability

A GPQ for the survival probability $S_t = P(X > t) = \exp(-(t/b)^c)$, where t is a specified number, can be readily obtained from Section 7.5.5. Let $S^*(t) = \ln\left(-\ln S_t\right)$. Then

$$G_{S^*(t)} = \widehat{c}^{*-1} \ln(-\ln(\widehat{S}_t)) + \ln(\widehat{b}^*) \tag{12.7.34}$$

is a GPQ for $S^*(t)$, where $\widehat{S}_t$ is the MLE of S_t, which can be obtained by replacing the parameters in the expression for S_t by their MLEs. If U is a $1-\alpha$ upper confidence limit for S_t^*, then $\exp(-\exp(U))$ is a $1-\alpha$ lower confidence limit for S_t. Notice that, for a given $\widehat{S}_t$, a confidence limit for S_t^* can be obtained using Monte Carlo simulation.

All the procedures given above are applicable to find tolerance limits, and to find confidence limits for a survival probability, if the censored sample is from an extreme value distribution because of the one-one relation between the Weibull and extreme value distributions. Specifically, tolerance limits for an extreme value distribution can be obtained by applying the Weibull procedures after taking log-transformation of the sample from an extreme value distribution.

Table 12.4: Percentiles of $w_p^* = \widehat{c}^*(-\ln(\widehat{b}^*) + \ln(\theta_p))$ required for constructing one-sided tolerance limits for a Weibull Distribution

n	m	$w_{.05;.05}^*$	$w_{.1;.05}^*$	$w_{.9;.95}^*$	$w_{.95;.95}^*$
10	9	−5.962	−4.572	2.044	2.493
	7	−6.760	−5.088	2.825	3.428
	4	−10.07	−7.038	7.672	8.903
20	19	−4.561	−3.511	1.439	1.781
	17	−4.706	−3.608	1.561	1.935
	14	−4.934	−3.762	1.854	2.260
	10	−5.373	−4.019	2.612	3.098
	5	−7.010	−4.749	6.973	7.831
30	29	−4.151	−3.195	1.270	1.590
	27	−4.213	−3.240	1.326	1.654
	24	−4.312	−3.283	1.423	1.767
	20	−4.444	−3.376	1.622	1.996
	15	−4.674	−3.506	2.052	2.463
	10	−5.089	−3.672	3.109	3.601
50	49	−3.805	−2.920	1.139	1.439
	45	−3.856	−2.956	1.184	1.491
	40	−3.917	−2.986	1.247	1.569
	30	−4.043	−3.069	1.463	1.812
	25	−4.141	−3.112	1.643	2.016
	10	−4.610	−3.262	3.710	4.230

Example 12.7 (Number of million revolutions before failure for ball bearings)

Let us consider the ball-bearing data given in Example 7.4. Assume that the test was terminated after the 16th failure. That is, we have a type II right-censored sample with $n = 23$ and $m = 16$. The observed numbers of millions of revolutions for these 16 ball bearings are given below.

$$17.88 \quad 28.92 \quad 33.00 \quad 41.52 \quad 42.12 \quad 45.60 \quad 48.40 \quad 51.84$$
$$51.96 \quad 54.12 \quad 55.56 \quad 67.80 \quad 68.64 \quad 68.64 \quad 68.88 \quad 84.12$$

The MLEs are $\widehat{c}_0 = 2.469$ and $\widehat{b}_0 = 76.694$.

Tolerance Limits: To compute a $(0.90, 0.95)$ lower tolerance limit using (12.7.33), we obtained $w_{1-p;\alpha}^* = w_{.10;.05}^* = -3.603$; this value is obtained using Monte Carlo simulation consisting of 100,000 runs. Thus, the $(0.90, 0.95)$ lower tolerance limit is given by $\widehat{b}_0 \exp\left(w^*_{.1;.05}/\widehat{c}_0\right) = 76.694\exp(-3.603/2.469) = 17.82$. Hence we conclude with 95% confidence that at least 90% of ball bearings survive 17.82 million revolutions. Notice that the same lower tolerance limit based on all 23 observations is 16.65.

To compute a (0.90, 0.95) upper tolerance limit using (12.7.32), we obtained $w^*_{p;1-\alpha} = w^*_{.90;.95} = 1.746$. The (0.90, 0.95) upper tolerance limit is given by $\widehat{b}_0 \exp\left(w^*_{.90;.95}/\widehat{c}_0\right) = 76.694\exp(1.746/2.469) = 155.55$. Hence we can conclude with 95% confidence that at most 10% of ball bearings survive 155.55 millions of revolutions. The same upper tolerance limit based on all 23 observations is 154.42.

Estimation of Survival Probability: Let us compute a 95% lower confidence limit for the probability that a bearing will last at least 50 million revolutions. Substituting the MLEs $\widehat{b}_0$ and $\widehat{c}_0$, we obtained $\widehat{S}_{50} = 0.706$. Using this value in (12.7.34), and using Monte Carlo simulation with 100,000 runs, we computed a 95% upper limit for S^*_{50} as -0.541. Thus, a 95% lower limit for S_{50} is $\exp(-\exp(-0.541)) = 0.559$. We here note that the 95% lower confidence limit for S_{50} based on all 23 observations is 0.545.

Results Under Type I Censoring: We now assume that the samples were censored at 89 million revolutions, that is $x_0 = 89$. For this censoring value, the uncensored observations are the same 16 observations given above. The MLEs are $\widehat{c} = 2.286$ and $\widehat{b} = 79.463$. The required percentile $w^*_{.10,.05}$ computed for the type II censoring case is -3.603. Thus, the (0.90, 0.95) lower tolerance limit is given by $\widehat{b}_0 \exp\left(w^*_{.10;.05}/\widehat{c}_0\right) = 79.463\exp(-3.603/2.286) = 16.39$. To compute a (0.90, 0.95) upper tolerance limit, we note that $w^*_{.90,.95} = 1.746$, and $79.463 \times \exp(1.746/2.286) = 170.56$.

To find a 95% lower confidence limit for the probability that a bearing will last at least 50 million revolutions, we note that $\widehat{S}_{50} = 0.707$. Using this value in (12.7.34), and using Monte Carlo simulation with 100,000 runs, we computed a 95% upper limit for S^*_{50} as -0.542. Thus, a 95% lower limit for S_{50} is $\exp(-\exp(-0.542)) = 0.559$.

12.8   Exercises

12.8.1. For $N(\mu, \sigma^2)$, is a β-expectation tolerance interval wider or narrower compared to a $100\beta\%$ confidence interval for μ? Clearly explain.

12.8.2. For $X_1 \sim N(\mu_1, \sigma_1^2)$ and $X_2 \sim N(\mu_2, \sigma_2^2)$, where X_1 and X_2 are also independent, explain how you will compute a β-expectation tolerance interval for $X_1 - X_2$. Compute a two-sided β-expectation tolerance interval for $X_1 - X_2$ for $\beta = 0.95$, using the data in Table 2.6 of Chapter 2.

12.8.3. Explain how you will use the Wilson-Hilferty normal approximation (see Chapter 7) to compute approximate one-sided and two-sided β-expectation tolerance intervals for a gamma distribution. Estimate the coverage probabilities of the resulting procedures using simulation, and using the sample size-parameter combinations in Table 7.1, for $\beta = 0.90$, 0.95 and 0.99. Comment on the accuracy of the approximation for the purpose of computing β-expectation tolerance intervals for a gamma distribution.

12.8.4. For the alkalinity concentration data in Table 7.3, compute a β-expectation upper tolerance limit using the Wilson-Hilferty normal approximation, for $\beta = 0.95$.

12.8.5. For Example 4.2 dealing with the study of breaking strengths of cement briquettes, compute β-expectation lower tolerance limits for the observable random variable and the unobservable true values using Mee's approach and using the Lin-Liao approach, for $\beta = 0.90$.

12.8.6. Show that under the non-informative prior distribution (11.2.1), a Bayesian β-expectation tolerance intervals (one-sided as well as two-sided) for $N(\mu, \sigma^2)$ coincide with the frequentist solutions given in Section 12.1.1.

12.8.7. Derive Bayesian β-expectation tolerance intervals (one-sided as well as two-sided) for $N(\mu, \sigma^2)$ under the conjugate prior distribution (11.2.2). Compute such an upper tolerance limit based on the data in Example 2.1, for $\beta = 0.95$, and for the values of the prior parameters given in Section 11.2.2.

12.8.8. Let $X_1, ..., X_{n-l}$ be the uncensored measurements in a type II right-censored sample of size n from a normal distribution. Show that $\widehat{\mu}_0 - z^*_{1-p}\widehat{\sigma}_0$ in (12.7.10) is a $(p, 1 - \alpha)$ upper tolerance limit for the normal distribution.

12.8.9. Consider a sample of size n from a normal distribution with mean μ and variance σ^2. Assume that the sample is type II left-censored with l censored observations. That is, $X_{(1)}, ..., X_{(l)}$ are not observed.

   (a) Find a $1 - \alpha$ generalized confidence interval for μ.
   (b) Find a $1 - \alpha$ generalized confidence interval for σ^2.
                                                               [Schmee et al. 1985]

12.8.10. Let $X_{i1}, ..., X_{in_i}$ be a sample from $N(\mu_i, \sigma^2)$ distribution, $i = 1, 2$, where the two populations are independent. Furthermore, let l_i be the number of type II left-censored observations for the sample i, $i = 1, 2$. Find a generalized confidence interval for $\mu_1 - \mu_2$.

12.8.11. Let $X_{(1)}, ..., X_{(m)}$ be a type II right-censored sample from an exponential distribution with the pdf $\frac{1}{\theta} \exp\left(-\frac{(x-\mu)}{\theta}\right)$, $x > \mu$, $\theta > 0$. Show that the factor for constructing a $(p, 1-\alpha)$ lower tolerance limit is given by (12.7.25), provided $n < \frac{\ln(\alpha)}{\ln(p)}$.

12.8.12. Let $X \sim \text{binomial}(n, \theta)$. Let j be such that $P(X \geq n - j) \geq 1 - p$ and $P(X \geq n - j + 1) < 1 - p$. Show that $n - j$ is the smallest number so that

$$P(X \leq n - j) \geq p.$$

12.8.13. The following is a sample generated from a binomial(n, π) distribution with $n = 50$ and $\pi = 0.40$

14	19	19	16	17	15	22	16	17	15
18	18	16	16	13	17	17	16	9	11
17	11	23	23	11	13	19	13	19	17

Compute the following tolerance intervals using the exact confidence intervals for π, score confidence intervals for π, and using the approximate quantiles.

(a) A $(0.95, 0.95)$ one-sided lower and upper tolerance limits for the binomial$(50, \pi)$ distribution. Interpret their meaning.

(b) A $(0.90, 0.95)$ equal-tailed tolerance interval for the binomial(n, π) distribution, and compare the endpoints of the tolerance interval with the appropriate one-sided limits in part (a).

(c) A $(0.90, 0.95)$ two-sided tolerance interval, and find its actual content.

(d) Find the actual contents of the tolerance intervals in parts (a) and (b).

12.8.14. The following data represent the reported number of automobile accidents per day in a sample 45 days from a city police record.

5	4	3	8	12	3	8	10	10	7	10	8	10	7	14
7	11	7	5	9	6	11	12	5	6	11	15	7	6	13
8	5	7	10	11	5	10	8	3	4	11	8	5	10	3

(a) Show that a Poisson model fits the data.

(b) Find (0.95, 0.95) one-sided lower and upper tolerance limits for the number of accidents per day in the city.

(c) Compute a (0.90, 0.95) equal-tailed tolerance interval and compute a (0.90, 0.95) two-sided tolerance interval. Interpret the meanings of these two tolerance intervals.

12.8.15. Consider a type I right-censored sample of size n (with censoring time x_0) from a Weibull distribution with the pdf

$$\frac{\gamma}{\theta} x^{\gamma-1} e^{-x^\gamma/\theta}, \quad x \geq 0, \ \gamma > 0, \ \theta > 0.$$

Let $X_1, ..., X_m$ be recorded measurements below x_0.

(a) Verify that the likelihood function can be expressed as

$$\frac{n!}{(n-m)!} \left[\prod_{i=1}^{m} \left(\frac{\gamma}{\theta}\right) x_i^{\gamma-1} \exp(-x_i^\gamma/\theta) \right] (1 - F(x_0))^{n-m},$$

where $F(x)$ is the cdf of the Weibull distribution. Let $X_i^* = X_i$, $i = 1, ..., m$ and $X_i^* = x_0$ for $i = m+1, ..., n$.

(b) Show that the MLEs are determined by the equations

$$\left[\frac{\sum_{i=1}^n X_i^{*\widehat{\gamma}} \ln X_i^*}{\sum_{i=1}^n X_i^{*\widehat{\gamma}}} - \frac{1}{\widehat{\gamma}} \right] = \frac{1}{m} \sum_{i=1}^m \ln X_i$$

$$\widehat{\theta} = \frac{1}{m} \sum_{i=1}^n X_i^{*\widehat{\gamma}}.$$

(c) Using part (b), show that the MLEs for a Weibull(b, c) distribution with the pdf (12.7.28) are as given in Section 12.7.4. [Cohen, 1965]

12.8.16. Consider the data in Exercise 7.6.10 of Chapter 7. Assume that the censoring value is 19.50. That is, the values above 19.50 are discarded to create the following type I right-censored sample from a Pareto distribution with $\sigma = 4$ and $\lambda = 1.2$.

6.43	4.27	5.97	4.02	5.58	4.28	12.28	18.33	9.56
10.93	6.51	11.45	4.09	4.15	7.84	4.06	9.30	14.69
12.20	5.86	12.28	4.89	29.99	6.07	7.85		

(a) Find (0.90, 0.95) lower tolerance limit for the Pareto distribution, and compare it with the one in part (a) of Exercise 7.6.10.

(b) Find $(0.95, 0.99)$ upper tolerance limit for the Pareto distribution, and compare it with the one in part (c) of Exercise 7.6.10.

12.8.17. Give procedures for finding a $(p, 1 - \alpha)$ one-sided tolerance limit, and for finding a $1 - \alpha$ lower limit for a survival probability $P(X > t)$ for the power distribution (with the pdf (7.4.3) of Chapter 7) based on a type II right-censored sample.

Appendix A

Data Sets

Table A1: A sample of data from the glucose monitoring meter experiment in Example 6.1

Meter type	Meter number	Blood sample	Lot	Responses	Meter type	Meter number	Blood sample	Lot	Responses
T	1	1	1	44, 46, 47	R	1	1	1	47, 47, 48
T	1	1	2	53, 52, 47	R	1	1	2	51, 53, 49
T	1	1	3	46, 49, 44	R	1	1	3	47, 50, 46
T	1	2	1	47, 50, 49	R	1	2	1	49, 50, 48
T	1	2	2	48, 49, 49	R	1	2	2	47, 53, 52
T	1	2	3	49, 51, 47	R	1	2	3	51, 52, 49
T	1	3	1	48, 47, 49	R	1	3	1	51, 50, 50
T	1	3	2	48, 52, 52	R	1	3	2	53, 52, 55
T	1	3	3	52, 47, 50	R	1	3	3	52, 49, 50
T	2	1	1	45, 47, 47	R	2	1	1	48, 49, 49
T	2	1	2	48, 51, 47	R	2	1	2	50, 52, 48
T	2	1	3	47, 49, 42	R	2	1	3	49, 50, 44
T	2	2	1	48, 48, 47	R	2	2	1	52, 50, 48
T	2	2	2	48, 49, 50	R	2	2	2	50, 48, 49
T	2	2	3	48, 51, 49	R	2	2	3	49, 53, 48
T	2	3	1	51, 49, 49	R	2	3	1	50, 52, 51
T	2	3	2	52, 51, 52	R	2	3	2	53, 53, 53
T	2	3	3	53, 46, 49	R	2	3	3	52, 47, 50
T	3	1	1	47, 48, 48	R	3	1	1	47, 47, 50
T	3	1	2	49, 53, 48	R	3	1	2	50, 51, 48
T	3	1	3	47, 49, 43	R	3	1	3	46, 49, 42
T	3	2	1	49, 51, 48	R	3	2	1	49, 47, 44
T	3	2	2	50, 49, 50	R	3	2	2	49, 52, 50
T	3	2	3	48, 52, 49	R	3	2	3	55, 51, 51
T	3	3	1	49, 52, 47	R	3	3	1	49, 50, 50
T	3	3	2	51, 50, 54	R	3	3	2	53, 52, 53
T	3	3	3	51, 47, 51	R	3	3	3	55, 49, 50

Courtesy: Hari K. Iyer

To download the complete data, visit the Wiley ftp site

ftp://ftp.wiley.com/public/sci_tech_med/statistical_tolerance

Table A2: Data from a bioequivalence study reported in Example 6.5

Sequence	Subject	Period	Treatment	C_{max}
RTTR	1	2	T	13.4481
RTTR	1	1	R	12.0276
RTTR	1	3	T	14.6708
RTTR	1	4	R	21.7649
RTTR	2	2	T	16.2387
RTTR	2	1	R	14.5254
RTTR	2	3	T	16.1629
RTTR	2	4	R	17.0446
TRRT	3	1	T	14.7002
TRRT	3	2	R	16.4133
TRRT	3	4	T	13.0392
TRRT	3	3	R	12.5214
RTTR	4	2	T	16.4272
RTTR	4	1	R	17.0416
RTTR	4	3	T	17.2711
RTTR	4	4	R	14.3423
RTTR	5	2	T	21.2888
RTTR	5	1	R	18.4951
RTTR	5	3	T	23.6546
RTTR	5	4	R	26.6882
TRRT	6	1	T	14.2801
TRRT	6	2	R	15.1318
TRRT	6	4	T	21.8449
TRRT	6	3	R	18.7057
RTTR	7	2	T	18.8196
RTTR	7	1	R	19.1911
RTTR	7	3	T	19.2031
RTTR	7	4	R	17.8359
RTTR	8	1	T	20.0005
TRRT	8	2	R	11.0799
TRRT	8	4	T	22.1917
TRRT	8	3	R	21.7314
TRRT	9	1	T	14.2437
TRRT	9	2	R	13.7229
TRRT	9	4	T	14.8358
TRRT	9	3	R	12.4715
TRRT	10	1	T	10.7469
TRRT	10	2	R	12.1024
TRRT	10	4	T	13.3457
TRRT	10	3	R	12.7732
TRRT	11	1	T	19.9780
TRRT	11	2	R	21.1899

Source: www.fda.gov/cder/bioequivdata/drug14b.txt

Sequence	Subject	Period	Treatment	C_{max}
TRRT	11	4	T	25.3532
TRRT	11	3	R	20.1001
TRRT	12	1	T	33.8299
TRRT	12	2	R	35.7511
TRRT	12	4	T	32.2236
TRRT	12	3	R	29.8041
TRRT	13	1	T	24.0893
TRRT	13	2	R	25.3548
TRRT	13	4	T	27.3516
TRRT	13	3	R	24.1224
RTTR	16	2	T	17.3160
RTTR	16	1	R	18.1459
RTTR	16	3	T	14.7359
RTTR	16	4	R	14.6257
TRRT	17	1	T	9.1800
TRRT	17	2	R	11.5268
TRRT	17	4	T	12.0653
TRRT	17	3	R	12.7756
TRRT	18	1	T	14.7935
TRRT	18	2	R	14.4807
TRRT	18	4	T	13.8565
TRRT	18	3	R	12.3518
TRRT	19	1	T	13.8787
TRRT	19	2	R	18.8532
TRRT	19	4	T	27.1953
TRRT	19	3	R	19.4058
RTTR	20	2	T	26.5864
RTTR	20	1	R	22.7744
RTTR	20	3	T	21.6330
RTTR	20	4	R	18.3490
TRRT	21	1	T	18.7630
TRRT	21	2	R	14.6189
TRRT	21	4	T	13.9584
TRRT	21	3	R	19.0528
RTTR	22	2	T	10.9703
RTTR	22	1	R	10.2247
RTTR	22	3	T	12.6435
RTTR	22	4	R	9.5695
RTTR	23	2	T	15.2883
RTTR	23	1	R	24.7679
RTTR	23	3	T	23.2319
RTTR	23	4	R	18.5578

Sequence	Subject	Period	Treatment	C_{max}
\multicolumn{5}{c}{Table A2 continued}

Sequence	Subject	Period	Treatment	C_{max}
TRRT	24	1	T	14.3020
TRRT	24	2	R	19.0388
TRRT	24	4	T	17.9381
TRRT	24	3	R	12.9558
RTTR	25	2	T	14.1964
RTTR	25	1	R	13.9130
RTTR	25	3	T	12.2957
RTTR	25	4	R	14.2661
TRRT	26	1	T	19.8866
TRRT	26	2	R	28.3235
TRRT	26	4	T	27.9804
TRRT	26	3	R	22.6038
RTTR	27	2	T	20.8374
RTTR	27	1	R	18.2589
RTTR	27	3	T	18.8469
RTTR	27	4	R	17.3252
RTTR	28	2	T	14.8162
RTTR	28	1	R	10.7835
RTTR	28	3	T	7.5955
RTTR	28	4	R	14.9183
TRRT	29	1	T	20.8129
TRRT	29	2	R	28.7001
TRRT	29	4	T	17.3599
TRRT	29	3	R	22.3080
RTTR	30	2	T	16.0436
RTTR	30	1	R	15.7004
RTTR	30	3	T	21.6636
RTTR	30	4	R	20.0429
RTTR	31	2	T	14.9095
RTTR	31	1	R	14.6172
RTTR	31	3	T	14.4230
RTTR	31	4	R	13.4043
TRRT	32	1	T	14.4853
TRRT	32	2	R	26.4230
TRRT	32	4	T	23.4636
TRRT	32	3	R	13.4968
RTTR	33	2	T	13.8006
RTTR	33	1	R	10.5561
RTTR	33	3	T	8.1311

Table A2 continued

Sequence	Subject	Period	Treatment	C_{max}
RTTR	33	4	R	9.1687
TRRT	34	1	T	24.4197
TRRT	34	2	R	27.3871
TRRT	34	4	T	25.5714
TRRT	34	3	R	21.3814
RTTR	35	2	T	27.5151
RTTR	35	1	R	30.0800
RTTR	35	3	T	35.1856
RTTR	35	4	R	38.2588
RTTR	36	2	T	19.9215
RTTR	36	1	R	19.4226
RTTR	36	3	T	14.3518
RTTR	36	4	R	21.9982
TRRT	37	1	T	10.0174
TRRT	37	2	R	13.0770
TRRT	37	4	T	11.5961
TRRT	37	3	R	9.3507
RTTR	38	2	T	20.6108
RTTR	38	1	R	24.7480
RTTR	38	3	T	25.3391
RTTR	38	4	R	21.2389
RTTR	39	2	T	17.0882
RTTR	39	1	R	19.6846
RTTR	39	3	T	13.6475
RTTR	39	4	R	14.3690
RTTR	40	2	T	12.7704
RTTR	40	1	R	13.9277
RTTR	40	3	T	12.2176
RTTR	40	4	R	14.2982

Appendix B

Tables

Table B1: The $(p, 1 - \alpha)$ one-sided tolerance factors given in (2.2.3) for a normal distribution

				$1 - \alpha = 0.90$			
				p			
n	0.50	0.75	0.80	0.90	0.95	0.99	0.999
2	2.176	5.842	6.987	10.25	13.09	18.50	24.58
3	1.089	2.603	3.039	4.258	5.311	7.340	9.651
4	0.819	1.972	2.295	3.188	3.957	5.438	7.129
5	0.686	1.698	1.976	2.742	3.400	4.666	6.111
6	0.603	1.540	1.795	2.494	3.092	4.243	5.556
7	0.544	1.435	1.676	2.333	2.894	3.972	5.202
8	0.500	1.360	1.590	2.219	2.754	3.783	4.955
9	0.466	1.302	1.525	2.133	2.650	3.641	4.771
10	0.437	1.257	1.474	2.066	2.568	3.532	4.629
11	0.414	1.219	1.433	2.011	2.503	3.443	4.514
12	0.394	1.188	1.398	1.966	2.448	3.371	4.420
13	0.376	1.162	1.368	1.928	2.402	3.309	4.341
14	0.361	1.139	1.343	1.895	2.363	3.257	4.273
15	0.347	1.119	1.321	1.867	2.329	3.212	4.215
16	0.335	1.101	1.301	1.842	2.299	3.172	4.164
17	0.324	1.085	1.284	1.819	2.272	3.137	4.119
18	0.314	1.071	1.268	1.800	2.249	3.105	4.078
19	0.305	1.058	1.254	1.782	2.227	3.077	4.042
20	0.297	1.046	1.241	1.765	2.208	3.052	4.009
21	0.289	1.035	1.229	1.750	2.190	3.028	3.979
22	0.282	1.025	1.218	1.737	2.174	3.007	3.952
23	0.275	1.016	1.208	1.724	2.159	2.987	3.927
24	0.269	1.007	1.199	1.712	2.145	2.969	3.903
25	0.264	1.000	1.190	1.702	2.132	2.952	3.882
26	0.258	0.992	1.182	1.691	2.120	2.937	3.862
27	0.253	0.985	1.174	1.682	2.109	2.922	3.843
28	0.248	0.979	1.167	1.673	2.099	2.909	3.826
29	0.244	0.973	1.160	1.665	2.089	2.896	3.810
30	0.239	0.967	1.154	1.657	2.080	2.884	3.794
31	0.235	0.961	1.148	1.650	2.071	2.872	3.780
32	0.231	0.956	1.143	1.643	2.063	2.862	3.766
33	0.228	0.951	1.137	1.636	2.055	2.852	3.753
34	0.224	0.947	1.132	1.630	2.048	2.842	3.741
35	0.221	0.942	1.127	1.624	2.041	2.833	3.729
36	0.218	0.938	1.123	1.618	2.034	2.824	3.718
37	0.215	0.934	1.118	1.613	2.028	2.816	3.708
38	0.212	0.930	1.114	1.608	2.022	2.808	3.698
39	0.209	0.926	1.110	1.603	2.016	2.800	3.688
40	0.206	0.923	1.106	1.598	2.010	2.793	3.679

Table B1 continued

| | | | | $1-\alpha=0.90$ | | | |
n	0.50	0.75	0.80	0.90	0.95	0.99	0.999
41	0.204	0.919	1.103	1.593	2.005	2.786	3.670
42	0.201	0.916	1.099	1.589	2.000	2.780	3.662
43	0.199	0.913	1.096	1.585	1.995	2.773	3.654
44	0.196	0.910	1.092	1.581	1.990	2.767	3.646
45	0.194	0.907	1.089	1.577	1.986	2.761	3.638
46	0.192	0.904	1.086	1.573	1.981	2.756	3.631
47	0.190	0.901	1.083	1.570	1.977	2.750	3.624
48	0.188	0.899	1.080	1.566	1.973	2.745	3.617
49	0.186	0.896	1.078	1.563	1.969	2.740	3.611
50	0.184	0.894	1.075	1.559	1.965	2.735	3.605
51	0.182	0.891	1.072	1.556	1.962	2.730	3.599
52	0.180	0.889	1.070	1.553	1.958	2.726	3.593
53	0.178	0.887	1.067	1.550	1.955	2.721	3.587
54	0.177	0.884	1.065	1.547	1.951	2.717	3.582
55	0.175	0.882	1.063	1.545	1.948	2.713	3.577
56	0.173	0.880	1.061	1.542	1.945	2.709	3.571
57	0.172	0.878	1.059	1.539	1.942	2.705	3.566
58	0.170	0.876	1.056	1.537	1.939	2.701	3.562
59	0.169	0.875	1.054	1.534	1.936	2.697	3.557
60	0.167	0.873	1.052	1.532	1.933	2.694	3.552
61	0.166	0.871	1.051	1.530	1.931	2.690	3.548
62	0.165	0.869	1.049	1.527	1.928	2.687	3.544
63	0.163	0.868	1.047	1.525	1.925	2.683	3.539
64	0.162	0.866	1.045	1.523	1.923	2.680	3.535
65	0.161	0.864	1.043	1.521	1.920	2.677	3.531
66	0.159	0.863	1.042	1.519	1.918	2.674	3.527
67	0.158	0.861	1.040	1.517	1.916	2.671	3.524
68	0.157	0.860	1.038	1.515	1.913	2.668	3.520
69	0.156	0.858	1.037	1.513	1.911	2.665	3.516
70	0.155	0.857	1.035	1.511	1.909	2.662	3.513
71	0.154	0.855	1.034	1.509	1.907	2.660	3.509
72	0.152	0.854	1.032	1.508	1.905	2.657	3.506
73	0.151	0.853	1.031	1.506	1.903	2.654	3.503
74	0.150	0.851	1.029	1.504	1.901	2.652	3.499
75	0.149	0.850	1.028	1.503	1.899	2.649	3.496
76	0.148	0.849	1.027	1.501	1.897	2.647	3.493
77	0.147	0.848	1.025	1.499	1.895	2.644	3.490
78	0.146	0.846	1.024	1.498	1.893	2.642	3.487
79	0.145	0.845	1.023	1.496	1.892	2.640	3.484
80	0.144	0.844	1.022	1.495	1.890	2.638	3.482

Table B1 continued

				p			
				$1-\alpha=0.90$			
n	0.50	0.75	0.80	0.90	0.95	0.99	0.999
85	0.140	0.839	1.016	1.488	1.882	2.627	3.468
90	0.136	0.834	1.011	1.481	1.874	2.618	3.456
95	0.132	0.829	1.006	1.475	1.867	2.609	3.445
100	0.129	0.825	1.001	1.470	1.861	2.601	3.435
125	0.115	0.808	0.983	1.448	1.836	2.569	3.395
150	0.105	0.796	0.970	1.433	1.818	2.546	3.366
175	0.097	0.786	0.960	1.421	1.804	2.528	3.343
200	0.091	0.779	0.952	1.411	1.793	2.514	3.326
225	0.086	0.773	0.945	1.403	1.784	2.503	3.311
250	0.081	0.767	0.940	1.397	1.777	2.493	3.299
275	0.077	0.763	0.935	1.391	1.770	2.485	3.289
300	0.074	0.759	0.931	1.386	1.765	2.477	3.280
350	0.069	0.753	0.924	1.378	1.755	2.466	3.265
400	0.064	0.747	0.919	1.372	1.748	2.456	3.253
450	0.061	0.743	0.914	1.366	1.742	2.448	3.243
500	0.057	0.740	0.910	1.362	1.736	2.442	3.235
600	0.052	0.734	0.904	1.355	1.728	2.431	3.222
700	0.048	0.729	0.899	1.349	1.722	2.423	3.211
1000	0.041	0.720	0.890	1.338	1.709	2.407	3.191

Table B1 continued

				p			
n	0.50	0.75	0.80	0.90	0.95	0.99	0.999
2	4.464	11.763	14.051	20.581	26.26	37.094	49.276
3	1.686	3.806	4.424	6.155	7.656	10.553	13.857
4	1.177	2.618	3.026	4.162	5.144	7.042	9.214
5	0.953	2.150	2.483	3.407	4.203	5.741	7.502
6	0.823	1.895	2.191	3.006	3.708	5.062	6.612
7	0.734	1.732	2.005	2.755	3.399	4.642	6.063
8	0.670	1.618	1.875	2.582	3.187	4.354	5.688
9	0.620	1.532	1.779	2.454	3.031	4.143	5.413
10	0.580	1.465	1.703	2.355	2.911	3.981	5.203
11	0.546	1.411	1.643	2.275	2.815	3.852	5.036
12	0.518	1.366	1.593	2.210	2.736	3.747	4.900
13	0.494	1.328	1.551	2.155	2.671	3.659	4.787
14	0.473	1.296	1.514	2.109	2.614	3.585	4.690
15	0.455	1.268	1.483	2.068	2.566	3.520	4.607
16	0.438	1.243	1.455	2.033	2.524	3.464	4.535
17	0.423	1.220	1.431	2.002	2.486	3.414	4.471
18	0.410	1.201	1.409	1.974	2.453	3.370	4.415
19	0.398	1.183	1.389	1.949	2.423	3.331	4.364
20	0.387	1.166	1.371	1.926	2.396	3.295	4.318
21	0.376	1.152	1.355	1.905	2.371	3.263	4.277
22	0.367	1.138	1.340	1.886	2.349	3.233	4.239
23	0.358	1.125	1.326	1.869	2.328	3.206	4.204
24	0.350	1.114	1.313	1.853	2.309	3.181	4.172
25	0.342	1.103	1.302	1.838	2.292	3.158	4.142
26	0.335	1.093	1.291	1.824	2.275	3.136	4.115
27	0.328	1.083	1.280	1.811	2.260	3.116	4.089
28	0.322	1.075	1.271	1.799	2.246	3.098	4.066
29	0.316	1.066	1.262	1.788	2.232	3.080	4.043
30	0.310	1.058	1.253	1.777	2.220	3.064	4.022
31	0.305	1.051	1.245	1.767	2.208	3.048	4.002
32	0.300	1.044	1.237	1.758	2.197	3.034	3.984
33	0.295	1.037	1.230	1.749	2.186	3.020	3.966
34	0.290	1.031	1.223	1.740	2.176	3.007	3.950
35	0.286	1.025	1.217	1.732	2.167	2.995	3.934
36	0.282	1.019	1.211	1.725	2.158	2.983	3.919
37	0.278	1.014	1.205	1.717	2.149	2.972	3.904
38	0.274	1.009	1.199	1.710	2.141	2.961	3.891
39	0.270	1.004	1.194	1.704	2.133	2.951	3.878
40	0.266	0.999	1.188	1.697	2.125	2.941	3.865

$1 - \alpha = 0.95$

Table B1 continued

				$1-\alpha = 0.95$			
				p			
n	0.50	0.75	0.80	0.90	0.95	0.99	0.999
41	0.263	0.994	1.183	1.691	2.118	2.932	3.854
42	0.260	0.990	1.179	1.685	2.111	2.923	3.842
43	0.256	0.986	1.174	1.680	2.105	2.914	3.831
44	0.253	0.982	1.170	1.674	2.098	2.906	3.821
45	0.250	0.978	1.165	1.669	2.092	2.898	3.811
46	0.248	0.974	1.161	1.664	2.086	2.890	3.801
47	0.245	0.971	1.157	1.659	2.081	2.883	3.792
48	0.242	0.967	1.154	1.654	2.075	2.876	3.783
49	0.240	0.964	1.150	1.650	2.070	2.869	3.774
50	0.237	0.960	1.146	1.646	2.065	2.862	3.766
51	0.235	0.957	1.143	1.641	2.060	2.856	3.758
52	0.232	0.954	1.140	1.637	2.055	2.850	3.750
53	0.230	0.951	1.136	1.633	2.051	2.844	3.742
54	0.228	0.948	1.133	1.630	2.046	2.838	3.735
55	0.226	0.945	1.130	1.626	2.042	2.833	3.728
56	0.224	0.943	1.127	1.622	2.038	2.827	3.721
57	0.222	0.940	1.125	1.619	2.034	2.822	3.714
58	0.220	0.938	1.122	1.615	2.030	2.817	3.708
59	0.218	0.935	1.119	1.612	2.026	2.812	3.701
60	0.216	0.933	1.116	1.609	2.022	2.807	3.695
61	0.214	0.930	1.114	1.606	2.019	2.802	3.689
62	0.212	0.928	1.111	1.603	2.015	2.798	3.684
63	0.210	0.926	1.109	1.600	2.012	2.793	3.678
64	0.209	0.924	1.107	1.597	2.008	2.789	3.673
65	0.207	0.921	1.104	1.594	2.005	2.785	3.667
66	0.205	0.919	1.102	1.591	2.002	2.781	3.662
67	0.204	0.917	1.100	1.589	1.999	2.777	3.657
68	0.202	0.915	1.098	1.586	1.996	2.773	3.652
69	0.201	0.913	1.096	1.584	1.993	2.769	3.647
70	0.199	0.911	1.094	1.581	1.990	2.765	3.643
71	0.198	0.910	1.092	1.579	1.987	2.762	3.638
72	0.196	0.908	1.090	1.576	1.984	2.758	3.633
73	0.195	0.906	1.088	1.574	1.982	2.755	3.629
74	0.194	0.904	1.086	1.572	1.979	2.751	3.625
75	0.192	0.903	1.084	1.570	1.976	2.748	3.621
76	0.191	0.901	1.082	1.568	1.974	2.745	3.617
77	0.190	0.899	1.081	1.565	1.971	2.742	3.613
78	0.189	0.898	1.079	1.563	1.969	2.739	3.609
79	0.187	0.896	1.077	1.561	1.967	2.736	3.605
80	0.186	0.895	1.076	1.559	1.964	2.733	3.601

Table B1 continued

				p			
n	0.50	0.75	0.80	0.90	0.95	0.99	0.999

$1 - \alpha = 0.95$

n	0.50	0.75	0.80	0.90	0.95	0.99	0.999
85	0.180	0.888	1.068	1.550	1.954	2.719	3.583
90	0.175	0.881	1.061	1.542	1.944	2.706	3.567
95	0.170	0.875	1.055	1.534	1.935	2.695	3.553
100	0.166	0.870	1.049	1.527	1.927	2.684	3.539
125	0.148	0.848	1.025	1.498	1.894	2.642	3.486
150	0.135	0.832	1.008	1.478	1.870	2.611	3.448
175	0.125	0.819	0.995	1.462	1.852	2.588	3.419
200	0.117	0.809	0.984	1.450	1.837	2.570	3.395
225	0.110	0.801	0.976	1.439	1.825	2.555	3.376
250	0.104	0.795	0.969	1.431	1.815	2.542	3.361
275	0.100	0.789	0.962	1.423	1.807	2.531	3.347
300	0.095	0.784	0.957	1.417	1.800	2.522	3.335
350	0.088	0.775	0.948	1.406	1.787	2.506	3.316
400	0.082	0.769	0.941	1.398	1.778	2.494	3.300
450	0.078	0.763	0.935	1.391	1.770	2.484	3.288
500	0.074	0.758	0.930	1.385	1.763	2.475	3.277
600	0.067	0.751	0.922	1.376	1.752	2.462	3.260
700	0.062	0.745	0.916	1.368	1.744	2.451	3.247
1000	0.052	0.733	0.904	1.354	1.727	2.430	3.220

Table B1 continued

				$1 - \alpha = 0.99$			
				p			
n	0.50	0.75	0.80	0.90	0.95	0.99	0.999
2	22.50	58.94	70.38	103.0	131.4	185.6	246.6
3	4.021	8.728	10.11	14.00	17.37	23.90	31.35
4	2.270	4.715	5.417	7.380	9.083	12.39	16.18
5	1.676	3.454	3.958	5.362	6.578	8.939	11.65
6	1.374	2.848	3.262	4.411	5.406	7.335	9.550
7	1.188	2.491	2.854	3.859	4.728	6.412	8.346
8	1.060	2.253	2.584	3.497	4.285	5.812	7.564
9	0.965	2.083	2.391	3.240	3.972	5.389	7.014
10	0.892	1.954	2.246	3.048	3.738	5.074	6.605
11	0.833	1.853	2.131	2.898	3.556	4.829	6.288
12	0.785	1.771	2.039	2.777	3.410	4.633	6.035
13	0.744	1.703	1.963	2.677	3.290	4.472	5.827
14	0.708	1.645	1.898	2.593	3.189	4.337	5.652
15	0.678	1.595	1.843	2.521	3.102	4.222	5.504
16	0.651	1.552	1.795	2.459	3.028	4.123	5.377
17	0.627	1.514	1.753	2.405	2.963	4.037	5.265
18	0.605	1.481	1.716	2.357	2.905	3.960	5.167
19	0.586	1.450	1.682	2.314	2.854	3.892	5.079
20	0.568	1.423	1.652	2.276	2.808	3.832	5.001
21	0.552	1.399	1.625	2.241	2.766	3.777	4.931
22	0.537	1.376	1.600	2.209	2.729	3.727	4.867
23	0.523	1.355	1.577	2.180	2.694	3.681	4.808
24	0.510	1.336	1.556	2.154	2.662	3.640	4.755
25	0.498	1.319	1.537	2.129	2.633	3.601	4.706
26	0.487	1.303	1.519	2.106	2.606	3.566	4.660
27	0.477	1.287	1.502	2.085	2.581	3.533	4.618
28	0.467	1.273	1.486	2.065	2.558	3.502	4.579
29	0.458	1.260	1.472	2.047	2.536	3.473	4.542
30	0.450	1.247	1.458	2.030	2.515	3.447	4.508
31	0.441	1.236	1.445	2.014	2.496	3.421	4.476
32	0.434	1.225	1.433	1.998	2.478	3.398	4.445
33	0.426	1.214	1.422	1.984	2.461	3.375	4.417
34	0.419	1.204	1.411	1.970	2.445	3.354	4.390
35	0.413	1.195	1.400	1.957	2.430	3.334	4.364
36	0.406	1.186	1.391	1.945	2.415	3.315	4.340
37	0.400	1.177	1.381	1.934	2.402	3.297	4.317
38	0.394	1.169	1.372	1.922	2.389	3.280	4.296
39	0.389	1.161	1.364	1.912	2.376	3.264	4.275
40	0.384	1.154	1.356	1.902	2.364	3.249	4.255

Table B1 continued

				$1 - \alpha = 0.99$				
				p			
n	0.50	0.75	0.80	0.90	0.95	0.99	0.999
41	0.378	1.147	1.348	1.892	2.353	3.234	4.236
42	0.374	1.140	1.341	1.883	2.342	3.220	4.218
43	0.369	1.133	1.333	1.874	2.331	3.206	4.201
44	0.364	1.127	1.327	1.865	2.321	3.193	4.184
45	0.360	1.121	1.320	1.857	2.312	3.180	4.168
46	0.356	1.115	1.314	1.849	2.303	3.168	4.153
47	0.352	1.110	1.308	1.842	2.294	3.157	4.138
48	0.348	1.104	1.302	1.835	2.285	3.146	4.124
49	0.344	1.099	1.296	1.828	2.277	3.135	4.110
50	0.340	1.094	1.291	1.821	2.269	3.125	4.097
51	0.337	1.089	1.285	1.814	2.261	3.115	4.084
52	0.333	1.084	1.280	1.808	2.254	3.105	4.072
53	0.330	1.080	1.275	1.802	2.247	3.096	4.060
54	0.326	1.075	1.270	1.796	2.240	3.087	4.049
55	0.323	1.071	1.266	1.790	2.233	3.078	4.038
56	0.320	1.067	1.261	1.785	2.226	3.070	4.027
57	0.317	1.063	1.257	1.779	2.220	3.061	4.017
58	0.314	1.059	1.253	1.774	2.214	3.053	4.007
59	0.311	1.055	1.249	1.769	2.208	3.046	3.997
60	0.309	1.052	1.245	1.764	2.202	3.038	3.987
61	0.306	1.048	1.241	1.759	2.197	3.031	3.978
62	0.303	1.045	1.237	1.755	2.191	3.024	3.969
63	0.301	1.041	1.233	1.750	2.186	3.017	3.960
64	0.298	1.038	1.230	1.746	2.181	3.010	3.952
65	0.296	1.035	1.226	1.741	2.176	3.004	3.944
66	0.294	1.032	1.223	1.737	2.171	2.998	3.936
67	0.291	1.028	1.219	1.733	2.166	2.991	3.928
68	0.289	1.025	1.216	1.729	2.162	2.985	3.920
69	0.287	1.023	1.213	1.725	2.157	2.980	3.913
70	0.285	1.020	1.210	1.722	2.153	2.974	3.906
71	0.283	1.017	1.207	1.718	2.148	2.968	3.899
72	0.280	1.014	1.204	1.714	2.144	2.963	3.892
73	0.278	1.012	1.201	1.711	2.140	2.958	3.885
74	0.276	1.009	1.198	1.707	2.136	2.952	3.878
75	0.275	1.006	1.196	1.704	2.132	2.947	3.872
76	0.273	1.004	1.193	1.701	2.128	2.942	3.866
77	0.271	1.002	1.190	1.698	2.125	2.938	3.860
78	0.269	0.999	1.188	1.694	2.121	2.933	3.854
79	0.267	0.997	1.185	1.691	2.117	2.928	3.848
80	0.265	0.995	1.183	1.688	2.114	2.924	3.842

Table B1 continued

| n | \multicolumn{7}{c}{$1-\alpha = 0.99$} |
| | \multicolumn{7}{c}{p} |
	0.50	0.75	0.80	0.90	0.95	0.99	0.999
85	0.257	0.984	1.171	1.674	2.097	2.902	3.815
90	0.250	0.974	1.161	1.661	2.082	2.883	3.791
95	0.243	0.965	1.151	1.650	2.069	2.866	3.769
100	0.236	0.957	1.142	1.639	2.056	2.850	3.748
125	0.211	0.924	1.107	1.596	2.007	2.786	3.668
150	0.192	0.901	1.082	1.566	1.971	2.740	3.610
175	0.177	0.883	1.062	1.542	1.944	2.706	3.567
200	0.166	0.868	1.047	1.524	1.923	2.679	3.532
225	0.156	0.856	1.034	1.509	1.905	2.656	3.504
250	0.148	0.847	1.024	1.496	1.891	2.638	3.481
275	0.141	0.838	1.015	1.485	1.878	2.622	3.461
300	0.135	0.831	1.007	1.476	1.868	2.608	3.443
350	0.125	0.819	0.994	1.461	1.850	2.585	3.415
400	0.117	0.809	0.984	1.448	1.836	2.567	3.392
450	0.110	0.801	0.975	1.438	1.824	2.553	3.374
500	0.104	0.794	0.968	1.430	1.814	2.540	3.358
600	0.095	0.783	0.957	1.416	1.799	2.520	3.333
700	0.088	0.775	0.948	1.406	1.787	2.505	3.314
1000	0.074	0.758	0.930	1.385	1.762	2.475	3.276

Table B2: The exact $(p, 1 - \alpha)$ two-sided tolerance factors for a normal distribution

| | | | | $1 - \alpha = 0.90$ | | | |
| | | | | p | | | |
n	0.50	0.75	0.80	0.90	0.95	0.99	0.999
2	6.808	11.17	12.33	15.51	18.22	23.42	29.36
3	2.492	4.134	4.577	5.788	6.823	8.819	11.10
4	1.766	2.954	3.276	4.157	4.913	6.372	8.046
5	1.473	2.477	2.750	3.499	4.142	5.387	6.816
6	1.314	2.217	2.464	3.141	3.723	4.850	6.146
7	1.213	2.053	2.282	2.913	3.456	4.508	5.720
8	1.143	1.938	2.156	2.754	3.270	4.271	5.423
9	1.092	1.854	2.062	2.637	3.132	4.094	5.203
10	1.053	1.788	1.990	2.546	3.026	3.958	5.033
11	1.021	1.736	1.932	2.473	2.941	3.849	4.897
12	0.996	1.694	1.885	2.414	2.871	3.759	4.785
13	0.974	1.658	1.846	2.364	2.812	3.684	4.691
14	0.956	1.628	1.812	2.322	2.762	3.620	4.611
15	0.941	1.602	1.783	2.285	2.720	3.565	4.541
16	0.927	1.579	1.758	2.254	2.682	3.517	4.481
17	0.915	1.559	1.736	2.226	2.649	3.474	4.428
18	0.905	1.541	1.717	2.201	2.620	3.436	4.380
19	0.895	1.526	1.699	2.178	2.593	3.402	4.338
20	0.887	1.511	1.683	2.158	2.570	3.372	4.299
21	0.879	1.498	1.669	2.140	2.548	3.344	4.264
22	0.872	1.487	1.656	2.123	2.528	3.318	4.232
23	0.866	1.476	1.644	2.108	2.510	3.295	4.203
24	0.860	1.466	1.633	2.094	2.494	3.274	4.176
25	0.855	1.457	1.623	2.081	2.479	3.254	4.151
26	0.850	1.448	1.613	2.069	2.464	3.235	4.128
27	0.845	1.440	1.604	2.058	2.451	3.218	4.106
28	0.841	1.433	1.596	2.048	2.439	3.202	4.086
29	0.837	1.426	1.589	2.038	2.427	3.187	4.068
30	0.833	1.420	1.581	2.029	2.417	3.173	4.050
31	0.829	1.414	1.575	2.020	2.406	3.160	4.033
32	0.826	1.408	1.568	2.012	2.397	3.148	4.018
33	0.823	1.403	1.562	2.005	2.388	3.136	4.003
34	0.820	1.397	1.557	1.997	2.379	3.125	3.989
35	0.817	1.393	1.551	1.991	2.371	3.114	3.975
36	0.814	1.388	1.546	1.984	2.363	3.104	3.963
37	0.812	1.384	1.541	1.978	2.356	3.095	3.951
38	0.809	1.380	1.537	1.972	2.349	3.086	3.939

Table B2 continued

				$1 - \alpha = 0.90$			
				p			
n	0.50	0.75	0.80	0.90	0.95	0.99	0.999
39	0.807	1.376	1.532	1.966	2.343	3.077	3.928
40	0.805	1.372	1.528	1.961	2.336	3.069	3.918
41	0.802	1.368	1.524	1.956	2.330	3.061	3.908
42	0.800	1.365	1.52	1.951	2.324	3.053	3.898
43	0.798	1.362	1.517	1.946	2.319	3.046	3.889
44	0.797	1.358	1.513	1.942	2.313	3.039	3.880
45	0.795	1.355	1.510	1.938	2.308	3.032	3.872
46	0.793	1.352	1.507	1.933	2.303	3.026	3.864
47	0.791	1.350	1.504	1.929	2.299	3.020	3.856
48	0.790	1.347	1.501	1.926	2.294	3.014	3.848
49	0.788	1.344	1.498	1.922	2.290	3.008	3.841
50	0.787	1.342	1.495	1.918	2.285	3.003	3.834
51	0.785	1.339	1.492	1.915	2.281	2.997	3.827
52	0.784	1.337	1.490	1.912	2.277	2.992	3.821
53	0.783	1.335	1.487	1.908	2.274	2.987	3.815
54	0.781	1.333	1.485	1.905	2.270	2.982	3.808
55	0.780	1.331	1.482	1.902	2.266	2.978	3.803
56	0.779	1.328	1.480	1.899	2.263	2.973	3.797
57	0.778	1.326	1.478	1.896	2.259	2.969	3.791
58	0.777	1.325	1.476	1.894	2.256	2.964	3.786
59	0.776	1.323	1.474	1.891	2.253	2.960	3.781
60	0.775	1.321	1.471	1.888	2.250	2.956	3.775
61	0.774	1.319	1.470	1.886	2.247	2.952	3.771
62	0.773	1.317	1.468	1.884	2.244	2.949	3.766
63	0.772	1.316	1.466	1.881	2.241	2.945	3.761
64	0.771	1.314	1.464	1.879	2.239	2.941	3.756
65	0.770	1.313	1.462	1.877	2.236	2.938	3.752
66	0.769	1.311	1.461	1.874	2.233	2.934	3.748
67	0.768	1.309	1.459	1.872	2.231	2.931	3.743
68	0.767	1.308	1.457	1.870	2.228	2.928	3.739
69	0.766	1.307	1.456	1.868	2.226	2.925	3.735
70	0.765	1.305	1.454	1.866	2.223	2.922	3.731
71	0.765	1.304	1.453	1.864	2.221	2.919	3.728
72	0.764	1.303	1.451	1.862	2.219	2.916	3.724
73	0.763	1.301	1.450	1.860	2.217	2.913	3.720
74	0.762	1.300	1.448	1.859	2.215	2.910	3.717
75	0.762	1.299	1.447	1.857	2.213	2.907	3.713
76	0.761	1.298	1.446	1.855	2.210	2.905	3.710
77	0.760	1.296	1.444	1.854	2.209	2.902	3.707
78	0.759	1.295	1.443	1.852	2.207	2.900	3.703
79	0.759	1.294	1.442	1.850	2.205	2.897	3.700
80	0.758	1.293	1.440	1.849	2.203	2.895	3.697

Table B2 continued

				p			
n	0.50	0.75	0.80	0.90	0.95	0.99	0.999
85	0.755	1.288	1.435	1.841	2.194	2.883	3.683
90	0.752	1.283	1.430	1.835	2.186	2.873	3.669
95	0.750	1.279	1.425	1.829	2.179	2.863	3.657
100	0.748	1.275	1.421	1.823	2.172	2.855	3.646
125	0.739	1.260	1.403	1.801	2.146	2.820	3.603
150	0.732	1.249	1.391	1.785	2.127	2.796	3.571
175	0.727	1.240	1.382	1.774	2.113	2.777	3.548
200	0.723	1.234	1.375	1.764	2.102	2.763	3.529
225	0.720	1.228	1.369	1.757	2.093	2.751	3.514
250	0.718	1.224	1.364	1.750	2.085	2.741	3.501
275	0.715	1.220	1.359	1.745	2.079	2.732	3.490
300	0.714	1.217	1.356	1.740	2.073	2.725	3.481
350	0.710	1.211	1.350	1.732	2.064	2.713	3.465
400	0.708	1.207	1.345	1.726	2.057	2.703	3.453
450	0.706	1.204	1.341	1.721	2.051	2.695	3.443
500	0.704	1.201	1.338	1.717	2.046	2.689	3.435
600	0.701	1.196	1.332	1.710	2.038	2.678	3.421
700	0.699	1.192	1.328	1.705	2.032	2.670	3.411
1000	0.695	1.185	1.320	1.695	2.019	2.654	3.390

$1 - \alpha = 0.90$

Table B2 continued

				p			
				$1 - \alpha = 0.95$			
n	0.50	0.75	0.80	0.90	0.95	0.99	0.999
2	13.65	22.38	24.72	31.09	36.52	46.94	58.84
3	3.585	5.937	6.572	8.306	9.789	12.65	15.92
4	2.288	3.818	4.233	5.368	6.341	8.221	10.38
5	1.812	3.041	3.375	4.291	5.077	6.598	8.345
6	1.566	2.638	2.930	3.733	4.422	5.758	7.294
7	1.415	2.391	2.657	3.390	4.020	5.241	6.647
8	1.313	2.223	2.472	3.156	3.746	4.889	6.206
9	1.239	2.101	2.337	2.986	3.546	4.633	5.885
10	1.183	2.008	2.234	2.856	3.393	4.437	5.640
11	1.139	1.934	2.152	2.754	3.273	4.282	5.446
12	1.103	1.874	2.086	2.670	3.175	4.156	5.287
13	1.073	1.825	2.031	2.601	3.093	4.051	5.156
14	1.048	1.783	1.985	2.542	3.024	3.962	5.044
15	1.027	1.747	1.945	2.492	2.965	3.885	4.949
16	1.008	1.716	1.911	2.449	2.913	3.819	4.865
17	0.992	1.689	1.881	2.410	2.868	3.761	4.792
18	0.978	1.665	1.854	2.376	2.828	3.709	4.727
19	0.965	1.643	1.830	2.346	2.793	3.663	4.669
20	0.953	1.624	1.809	2.319	2.760	3.621	4.616
21	0.943	1.607	1.789	2.294	2.731	3.583	4.569
22	0.934	1.591	1.772	2.272	2.705	3.549	4.526
23	0.925	1.576	1.755	2.251	2.681	3.518	4.486
24	0.917	1.563	1.741	2.232	2.658	3.489	4.450
25	0.910	1.551	1.727	2.215	2.638	3.462	4.416
26	0.903	1.539	1.714	2.199	2.619	3.437	4.385
27	0.897	1.529	1.703	2.184	2.601	3.415	4.356
28	0.891	1.519	1.692	2.170	2.585	3.393	4.330
29	0.886	1.510	1.682	2.157	2.569	3.373	4.304
30	0.881	1.501	1.672	2.145	2.555	3.355	4.281
31	0.876	1.493	1.663	2.134	2.541	3.337	4.259
32	0.871	1.486	1.655	2.123	2.529	3.320	4.238
33	0.867	1.478	1.647	2.113	2.517	3.305	4.218
34	0.863	1.472	1.639	2.103	2.505	3.290	4.199
35	0.859	1.465	1.632	2.094	2.495	3.276	4.182
36	0.856	1.459	1.626	2.086	2.484	3.263	4.165
37	0.852	1.454	1.619	2.077	2.475	3.250	4.149
38	0.849	1.448	1.613	2.070	2.466	3.238	4.134
39	0.846	1.443	1.607	2.062	2.457	3.227	4.119
40	0.843	1.438	1.602	2.055	2.448	3.216	4.105

Table B2 continued

				$1 - \alpha = 0.95$			
				p			
n	0.50	0.75	0.80	0.90	0.95	0.99	0.999
41	0.840	1.433	1.596	2.049	2.44	3.205	4.092
42	0.838	1.429	1.591	2.042	2.433	3.196	4.080
43	0.835	1.424	1.587	2.036	2.425	3.186	4.068
44	0.833	1.420	1.582	2.030	2.418	3.177	4.056
45	0.830	1.416	1.578	2.024	2.412	3.168	4.045
46	0.828	1.412	1.573	2.019	2.405	3.160	4.034
47	0.826	1.409	1.569	2.014	2.399	3.151	4.024
48	0.824	1.405	1.565	2.009	2.393	3.144	4.014
49	0.822	1.402	1.561	2.004	2.387	3.136	4.004
50	0.820	1.398	1.558	1.999	2.382	3.129	3.995
51	0.818	1.395	1.554	1.994	2.376	3.122	3.986
52	0.816	1.392	1.551	1.990	2.371	3.115	3.978
53	0.815	1.389	1.547	1.986	2.366	3.108	3.969
54	0.813	1.386	1.544	1.982	2.361	3.102	3.961
55	0.811	1.383	1.541	1.978	2.356	3.096	3.953
56	0.810	1.381	1.538	1.974	2.352	3.090	3.946
57	0.808	1.378	1.535	1.970	2.347	3.084	3.939
58	0.807	1.376	1.533	1.967	2.343	3.079	3.932
59	0.805	1.373	1.530	1.963	2.339	3.073	3.925
60	0.804	1.371	1.527	1.960	2.335	3.068	3.918
61	0.803	1.369	1.525	1.957	2.331	3.063	3.912
62	0.801	1.366	1.522	1.953	2.327	3.058	3.905
63	0.800	1.364	1.520	1.950	2.324	3.053	3.899
64	0.799	1.362	1.517	1.947	2.320	3.048	3.893
65	0.797	1.360	1.515	1.944	2.317	3.044	3.887
66	0.796	1.358	1.513	1.941	2.313	3.039	3.882
67	0.795	1.356	1.511	1.939	2.310	3.035	3.876
68	0.794	1.354	1.509	1.936	2.307	3.031	3.871
69	0.793	1.352	1.506	1.933	2.304	3.027	3.866
70	0.792	1.350	1.504	1.931	2.300	3.023	3.861
71	0.791	1.349	1.502	1.928	2.297	3.019	3.856
72	0.790	1.347	1.501	1.926	2.295	3.015	3.851
73	0.789	1.345	1.499	1.923	2.292	3.011	3.846
74	0.788	1.344	1.497	1.921	2.289	3.008	3.841
75	0.787	1.342	1.495	1.919	2.286	3.004	3.837
76	0.786	1.341	1.493	1.917	2.284	3.001	3.832
77	0.785	1.339	1.492	1.914	2.281	2.997	3.828
78	0.784	1.337	1.490	1.912	2.278	2.994	3.824
79	0.783	1.336	1.488	1.910	2.276	2.991	3.820
80	0.783	1.335	1.487	1.908	2.274	2.988	3.816
85	0.779	1.328	1.479	1.899	2.262	2.973	3.797
90	0.775	1.322	1.473	1.890	2.252	2.959	3.780

Table B2 continued

	$1-\alpha = 0.95$						
				p			
n	0.50	0.75	0.80	0.90	0.95	0.99	0.999
95	0.772	1.316	1.466	1.882	2.242	2.947	3.764
100	0.769	1.311	1.461	1.875	2.234	2.936	3.750
125	0.757	1.291	1.439	1.846	2.200	2.891	3.693
150	0.749	1.277	1.423	1.826	2.176	2.859	3.652
175	0.742	1.266	1.411	1.811	2.157	2.835	3.622
200	0.737	1.258	1.401	1.798	2.143	2.816	3.598
225	0.733	1.251	1.394	1.789	2.131	2.801	3.578
250	0.730	1.245	1.387	1.780	2.121	2.788	3.561
275	0.727	1.240	1.382	1.773	2.113	2.777	3.547
300	0.725	1.236	1.377	1.767	2.106	2.767	3.535
350	0.721	1.229	1.369	1.757	2.094	2.752	3.515
400	0.717	1.223	1.363	1.749	2.084	2.739	3.499
450	0.715	1.219	1.358	1.743	2.077	2.729	3.486
500	0.712	1.215	1.354	1.737	2.070	2.721	3.476
600	0.709	1.209	1.347	1.729	2.060	2.707	3.458
700	0.706	1.204	1.342	1.722	2.052	2.697	3.445
1000	0.701	1.195	1.331	1.709	2.036	2.676	3.418

Table B2 continued

				p			
				$1-\alpha=0.99$			
n	0.50	0.75	0.80	0.90	0.95	0.99	0.999
2	68.32	112.0	123.7	155.6	182.7	234.9	294.4
3	8.122	13.44	14.87	18.78	22.13	28.59	35.98
4	4.028	6.706	7.431	9.416	11.12	14.41	18.18
5	2.824	4.724	5.240	6.655	7.870	10.22	12.92
6	2.270	3.812	4.231	5.383	6.373	8.292	10.50
7	1.954	3.291	3.656	4.658	5.520	7.191	9.114
8	1.750	2.955	3.284	4.189	4.968	6.479	8.220
9	1.608	2.720	3.024	3.860	4.581	5.980	7.593
10	1.503	2.546	2.831	3.617	4.294	5.610	7.127
11	1.422	2.411	2.682	3.429	4.073	5.324	6.768
12	1.357	2.304	2.563	3.279	3.896	5.096	6.481
13	1.305	2.216	2.466	3.156	3.751	4.909	6.246
14	1.261	2.144	2.386	3.054	3.631	4.753	6.050
15	1.224	2.082	2.317	2.967	3.529	4.621	5.883
16	1.193	2.029	2.259	2.893	3.441	4.507	5.740
17	1.165	1.983	2.207	2.828	3.364	4.408	5.615
18	1.141	1.942	2.163	2.771	3.297	4.321	5.505
19	1.120	1.907	2.123	2.720	3.237	4.244	5.408
20	1.101	1.875	2.087	2.675	3.184	4.175	5.321
21	1.084	1.846	2.055	2.635	3.136	4.113	5.242
22	1.068	1.820	2.026	2.598	3.092	4.056	5.171
23	1.054	1.796	2.000	2.564	3.053	4.005	5.106
24	1.042	1.775	1.976	2.534	3.017	3.958	5.047
25	1.030	1.755	1.954	2.506	2.984	3.915	4.993
26	1.019	1.736	1.934	2.480	2.953	3.875	4.942
27	1.009	1.720	1.915	2.456	2.925	3.838	4.896
28	1.000	1.704	1.898	2.434	2.898	3.804	4.853
29	0.991	1.689	1.882	2.413	2.874	3.772	4.812
30	0.983	1.676	1.866	2.394	2.851	3.742	4.775
31	0.976	1.663	1.852	2.376	2.829	3.715	4.739
32	0.969	1.651	1.839	2.359	2.809	3.688	4.706
33	0.962	1.640	1.826	2.343	2.790	3.664	4.675
34	0.956	1.629	1.815	2.328	2.773	3.640	4.646
35	0.950	1.619	1.803	2.314	2.756	3.618	4.618
36	0.944	1.610	1.793	2.300	2.740	3.598	4.592
37	0.939	1.601	1.783	2.287	2.725	3.578	4.567
38	0.934	1.592	1.773	2.275	2.710	3.559	4.543
39	0.929	1.584	1.764	2.264	2.697	3.541	4.520
40	0.925	1.576	1.756	2.253	2.684	3.524	4.499

Table B2 continued

				p			
				$1-\alpha = 0.99$			
n	0.50	0.75	0.80	0.90	0.95	0.99	0.999
41	0.920	1.569	1.748	2.242	2.671	3.508	4.478
42	0.916	1.562	1.740	2.232	2.659	3.493	4.459
43	0.912	1.555	1.732	2.223	2.648	3.478	4.440
44	0.908	1.549	1.725	2.214	2.637	3.464	4.422
45	0.905	1.543	1.718	2.205	2.627	3.450	4.405
46	0.901	1.537	1.712	2.196	2.617	3.437	4.388
47	0.898	1.531	1.705	2.188	2.607	3.425	4.372
48	0.895	1.526	1.699	2.181	2.598	3.412	4.357
49	0.892	1.520	1.694	2.173	2.589	3.401	4.342
50	0.889	1.515	1.688	2.166	2.580	3.390	4.328
51	0.886	1.510	1.683	2.159	2.572	3.379	4.314
52	0.883	1.506	1.677	2.152	2.564	3.369	4.301
53	0.880	1.501	1.672	2.146	2.557	3.359	4.288
54	0.878	1.497	1.667	2.140	2.549	3.349	4.276
55	0.875	1.493	1.663	2.134	2.542	3.339	4.264
56	0.873	1.488	1.658	2.128	2.535	3.330	4.253
57	0.871	1.484	1.654	2.122	2.528	3.322	4.241
58	0.868	1.481	1.649	2.117	2.522	3.313	4.231
59	0.866	1.477	1.645	2.111	2.516	3.305	4.220
60	0.864	1.473	1.641	2.106	2.509	3.297	4.210
61	0.862	1.470	1.637	2.101	2.503	3.289	4.200
62	0.860	1.466	1.634	2.096	2.498	3.282	4.191
63	0.858	1.463	1.630	2.092	2.492	3.274	4.181
64	0.856	1.460	1.626	2.087	2.487	3.267	4.172
65	0.854	1.457	1.623	2.083	2.481	3.260	4.164
66	0.852	1.454	1.620	2.078	2.476	3.254	4.155
67	0.851	1.451	1.617	2.075	2.472	3.248	4.148
68	0.849	1.448	1.613	2.071	2.467	3.241	4.139
69	0.848	1.445	1.610	2.067	2.462	3.235	4.132
70	0.846	1.443	1.607	2.063	2.458	3.229	4.124
71	0.844	1.440	1.604	2.059	2.453	3.223	4.116
72	0.843	1.437	1.601	2.055	2.449	3.217	4.109
73	0.841	1.435	1.599	2.052	2.444	3.212	4.102
74	0.840	1.432	1.596	2.048	2.440	3.206	4.095
75	0.839	1.430	1.593	2.045	2.436	3.201	4.088
76	0.837	1.428	1.590	2.041	2.432	3.196	4.081
77	0.836	1.425	1.588	2.038	2.428	3.190	4.075
78	0.834	1.423	1.585	2.035	2.424	3.185	4.068
79	0.833	1.421	1.583	2.032	2.421	3.181	4.062
80	0.832	1.419	1.581	2.028	2.417	3.176	4.056

Table B2 continued

				p			
				$1 - \alpha = 0.99$			
n	0.50	0.75	0.80	0.90	0.95	0.99	0.999
85	0.826	1.409	1.569	2.014	2.400	3.153	4.027
90	0.821	1.400	1.559	2.001	2.384	3.133	4.002
95	0.816	1.391	1.550	1.989	2.370	3.115	3.978
100	0.811	1.384	1.542	1.979	2.357	3.098	3.957
125	0.794	1.354	1.509	1.936	2.307	3.032	3.873
150	0.782	1.333	1.485	1.906	2.271	2.985	3.813
175	0.772	1.317	1.468	1.884	2.244	2.949	3.768
200	0.765	1.305	1.454	1.866	2.223	2.922	3.732
225	0.759	1.295	1.442	1.851	2.206	2.899	3.703
250	0.754	1.286	1.433	1.839	2.192	2.880	3.679
275	0.750	1.279	1.425	1.829	2.179	2.864	3.659
300	0.746	1.273	1.418	1.820	2.169	2.850	3.641
350	0.740	1.263	1.407	1.806	2.152	2.828	3.612
400	0.736	1.255	1.398	1.794	2.138	2.810	3.589
450	0.732	1.248	1.391	1.785	2.127	2.795	3.571
500	0.729	1.243	1.385	1.777	2.117	2.783	3.555
600	0.724	1.234	1.375	1.765	2.103	2.763	3.530
700	0.720	1.227	1.367	1.755	2.091	2.748	3.511
1000	0.712	1.214	1.352	1.736	2.068	2.718	3.473

Table B3: The $(p, 1 - \alpha)$ two-sided tolerance factors controlling both tails of a normal distribution

| | | | | $1 - \alpha = 0.90$ | | | |
| | | | | p | | | |
n	0.50	0.75	0.80	0.90	0.95	0.99	0.999
2	9.847	13.64	14.68	17.57	20.08	24.99	30.67
3	3.617	5.050	5.448	6.554	7.516	9.402	11.60
4	2.557	3.600	3.890	4.700	5.405	6.788	8.398
5	2.121	3.009	3.257	3.948	4.550	5.733	7.110
6	1.879	2.683	2.908	3.535	4.082	5.156	6.407
7	1.722	2.474	2.685	3.271	3.783	4.789	5.960
8	1.611	2.327	2.527	3.086	3.574	4.531	5.647
9	1.527	2.217	2.410	2.948	3.418	4.340	5.415
10	1.461	2.131	2.318	2.840	3.296	4.192	5.235
11	1.408	2.061	2.244	2.754	3.199	4.072	5.090
12	1.364	2.004	2.183	2.682	3.118	3.974	4.971
13	1.326	1.955	2.132	2.622	3.050	3.891	4.871
14	1.294	1.914	2.087	2.571	2.993	3.821	4.786
15	1.266	1.878	2.049	2.526	2.942	3.760	4.712
16	1.241	1.846	2.015	2.487	2.898	3.706	4.647
17	1.219	1.818	1.986	2.452	2.859	3.659	4.590
18	1.200	1.793	1.959	2.421	2.825	3.616	4.539
19	1.182	1.770	1.935	2.393	2.793	3.578	4.493
20	1.166	1.750	1.913	2.368	2.765	3.544	4.451
21	1.151	1.731	1.893	2.345	2.739	3.513	4.413
22	1.137	1.714	1.875	2.324	2.715	3.484	4.379
23	1.125	1.698	1.858	2.304	2.694	3.458	4.347
24	1.113	1.683	1.843	2.287	2.674	3.433	4.318
25	1.102	1.670	1.828	2.270	2.655	3.411	4.291
26	1.092	1.657	1.815	2.254	2.638	3.390	4.265
27	1.083	1.645	1.802	2.240	2.621	3.370	4.242
28	1.074	1.634	1.791	2.226	2.606	3.352	4.220
29	1.066	1.624	1.780	2.214	2.592	3.335	4.199
30	1.058	1.614	1.770	2.202	2.579	3.319	4.180
31	1.051	1.605	1.760	2.191	2.566	3.303	4.161
32	1.044	1.596	1.751	2.180	2.554	3.289	4.144
33	1.037	1.588	1.742	2.170	2.543	3.275	4.128
34	1.031	1.580	1.734	2.160	2.533	3.263	4.112
35	1.025	1.573	1.726	2.151	2.522	3.250	4.098
36	1.019	1.566	1.718	2.143	2.513	3.239	4.084
37	1.014	1.559	1.711	2.135	2.504	3.228	4.070
38	1.009	1.553	1.704	2.127	2.495	3.217	4.058
39	1.004	1.546	1.698	2.119	2.486	3.207	4.045
40	0.999	1.540	1.692	2.112	2.478	3.197	4.034

Table B3 continued

				$1 - \alpha = 0.90$			
				p			
n	0.50	0.75	0.80	0.90	0.95	0.99	0.999
41	0.994	1.535	1.686	2.105	2.471	3.188	4.023
42	0.990	1.529	1.680	2.099	2.463	3.179	4.012
43	0.986	1.524	1.674	2.092	2.456	3.171	4.002
44	0.982	1.519	1.669	2.086	2.450	3.162	3.992
45	0.978	1.514	1.664	2.080	2.443	3.155	3.983
46	0.974	1.510	1.659	2.075	2.437	3.147	3.973
47	0.970	1.505	1.654	2.069	2.431	3.140	3.965
48	0.967	1.501	1.650	2.064	2.425	3.133	3.956
49	0.963	1.497	1.645	2.059	2.419	3.126	3.948
50	0.960	1.493	1.641	2.054	2.414	3.119	3.940
51	0.957	1.489	1.637	2.049	2.408	3.113	3.933
52	0.954	1.485	1.633	2.045	2.403	3.107	3.925
53	0.951	1.481	1.629	2.040	2.398	3.101	3.918
54	0.948	1.478	1.625	2.036	2.394	3.095	3.911
55	0.945	1.474	1.622	2.032	2.389	3.089	3.904
56	0.943	1.471	1.618	2.028	2.385	3.084	3.898
57	0.940	1.468	1.615	2.023	2.380	3.079	3.892
58	0.937	1.465	1.612	2.020	2.376	3.074	3.886
59	0.935	1.462	1.608	2.016	2.372	3.069	3.880
60	0.932	1.459	1.605	2.013	2.368	3.064	3.874
61	0.930	1.456	1.602	2.009	2.364	3.059	3.868
62	0.928	1.453	1.599	2.006	2.360	3.055	3.863
63	0.926	1.450	1.596	2.003	2.356	3.050	3.857
64	0.923	1.448	1.593	1.999	2.353	3.046	3.852
65	0.921	1.445	1.591	1.996	2.349	3.042	3.847
66	0.919	1.442	1.588	1.993	2.346	3.038	3.842
67	0.917	1.440	1.585	1.990	2.343	3.034	3.837
68	0.915	1.438	1.583	1.987	2.339	3.030	3.833
69	0.913	1.435	1.580	1.984	2.336	3.026	3.828
70	0.911	1.433	1.578	1.982	2.333	3.022	3.824
71	0.910	1.431	1.576	1.979	2.330	3.019	3.819
72	0.908	1.428	1.573	1.976	2.327	3.015	3.815
73	0.906	1.426	1.571	1.974	2.324	3.012	3.811
74	0.904	1.424	1.569	1.971	2.322	3.008	3.807
75	0.903	1.422	1.567	1.969	2.319	3.005	3.803
76	0.901	1.420	1.565	1.966	2.316	3.002	3.799
77	0.899	1.418	1.562	1.964	2.313	2.998	3.795
78	0.898	1.416	1.560	1.962	2.311	2.995	3.792
79	0.896	1.414	1.558	1.959	2.308	2.992	3.788
80	0.895	1.412	1.557	1.957	2.306	2.989	3.784

Table B3 continued

				p			
				$1 - \alpha = 0.90$			
n	0.50	0.75	0.80	0.90	0.95	0.99	0.999
85	0.888	1.404	1.547	1.947	2.294	2.975	3.768
90	0.881	1.396	1.539	1.937	2.284	2.963	3.752
95	0.875	1.389	1.531	1.928	2.274	2.951	3.738
100	0.870	1.382	1.524	1.920	2.265	2.940	3.726
125	0.848	1.355	1.496	1.888	2.230	2.898	3.675
150	0.832	1.336	1.476	1.865	2.204	2.867	3.638
175	0.819	1.321	1.461	1.848	2.184	2.844	3.610
200	0.809	1.309	1.448	1.834	2.169	2.825	3.587
225	0.801	1.300	1.438	1.822	2.156	2.809	3.569
250	0.795	1.292	1.430	1.812	2.145	2.797	3.554
275	0.789	1.285	1.422	1.804	2.136	2.786	3.541
300	0.784	1.279	1.416	1.797	2.128	2.776	3.529
350	0.775	1.269	1.405	1.785	2.115	2.760	3.511
400	0.769	1.261	1.397	1.775	2.104	2.748	3.495
450	0.763	1.254	1.390	1.768	2.096	2.737	3.483
500	0.758	1.249	1.384	1.761	2.088	2.729	3.473
600	0.751	1.240	1.375	1.750	2.077	2.715	3.456
700	0.745	1.233	1.368	1.742	2.068	2.704	3.443
800	0.740	1.227	1.362	1.736	2.060	2.696	3.433
1000	0.733	1.219	1.353	1.726	2.049	2.682	3.418

Table B3 continued

				p			
n	0.50	0.75	0.80	0.90	0.95	0.99	0.999
2	19.75	27.34	29.43	35.23	40.25	50.07	61.47
3	5.214	7.258	7.826	9.408	10.79	13.48	16.63
4	3.330	4.663	5.035	6.074	6.980	8.759	10.83
5	2.631	3.705	4.007	4.847	5.582	7.025	8.707
6	2.264	3.206	3.471	4.209	4.855	6.125	7.606
7	2.035	2.896	3.139	3.815	4.407	5.571	6.928
8	1.877	2.684	2.911	3.546	4.100	5.192	6.465
9	1.760	2.528	2.744	3.348	3.876	4.915	6.128
10	1.670	2.408	2.616	3.197	3.704	4.704	5.869
11	1.598	2.313	2.514	3.076	3.568	4.535	5.664
12	1.539	2.234	2.430	2.978	3.456	4.398	5.497
13	1.489	2.169	2.360	2.895	3.363	4.284	5.357
14	1.446	2.113	2.301	2.825	3.284	4.187	5.239
15	1.409	2.065	2.250	2.765	3.216	4.103	5.137
16	1.377	2.023	2.205	2.713	3.157	4.030	5.049
17	1.349	1.986	2.165	2.666	3.104	3.966	4.971
18	1.323	1.953	2.130	2.625	3.058	3.909	4.901
19	1.300	1.923	2.099	2.588	3.016	3.858	4.839
20	1.279	1.897	2.070	2.555	2.978	3.812	4.783
21	1.260	1.872	2.044	2.524	2.944	3.770	4.732
22	1.243	1.850	2.020	2.497	2.913	3.732	4.686
23	1.227	1.829	1.999	2.471	2.884	3.697	4.644
24	1.212	1.811	1.979	2.448	2.858	3.664	4.604
25	1.199	1.793	1.960	2.426	2.833	3.634	4.568
26	1.186	1.777	1.943	2.406	2.811	3.607	4.535
27	1.174	1.762	1.927	2.387	2.790	3.581	4.504
28	1.163	1.748	1.912	2.370	2.770	3.557	4.474
29	1.153	1.734	1.898	2.353	2.752	3.535	4.447
30	1.143	1.722	1.884	2.338	2.734	3.513	4.422
31	1.133	1.710	1.872	2.323	2.718	3.494	4.397
32	1.125	1.699	1.860	2.310	2.703	3.475	4.375
33	1.116	1.689	1.849	2.297	2.688	3.457	4.353
34	1.109	1.679	1.838	2.285	2.674	3.440	4.333
35	1.101	1.669	1.828	2.273	2.661	3.424	4.314
36	1.094	1.660	1.819	2.262	2.649	3.409	4.295
37	1.087	1.652	1.810	2.251	2.637	3.395	4.278
38	1.081	1.644	1.801	2.241	2.626	3.381	4.261
39	1.074	1.636	1.793	2.232	2.615	3.368	4.246
40	1.069	1.628	1.785	2.223	2.605	3.356	4.230

$1-\alpha = 0.95$

Table B3 continued

				$1 - \alpha = 0.95$			
				p			
n	0.50	0.75	0.80	0.90	0.95	0.99	0.999
41	1.063	1.621	1.778	2.214	2.595	3.344	4.216
42	1.057	1.614	1.770	2.206	2.586	3.332	4.202
43	1.052	1.608	1.763	2.198	2.577	3.321	4.189
44	1.047	1.602	1.757	2.190	2.568	3.311	4.176
45	1.042	1.595	1.750	2.182	2.560	3.300	4.164
46	1.038	1.590	1.744	2.175	2.552	3.291	4.152
47	1.033	1.584	1.738	2.168	2.544	3.281	4.141
48	1.029	1.579	1.732	2.162	2.536	3.272	4.130
49	1.025	1.573	1.727	2.155	2.529	3.264	4.119
50	1.021	1.568	1.722	2.149	2.522	3.255	4.109
51	1.017	1.563	1.716	2.143	2.516	3.247	4.099
52	1.013	1.559	1.711	2.137	2.509	3.239	4.089
53	1.009	1.554	1.707	2.132	2.503	3.231	4.080
54	1.006	1.550	1.702	2.126	2.497	3.224	4.071
55	1.002	1.545	1.697	2.121	2.491	3.217	4.063
56	0.999	1.541	1.693	2.116	2.485	3.210	4.054
57	0.996	1.537	1.689	2.111	2.480	3.203	4.046
58	0.993	1.533	1.685	2.106	2.474	3.197	4.038
59	0.989	1.530	1.681	2.102	2.469	3.190	4.031
60	0.986	1.526	1.677	2.097	2.464	3.184	4.023
61	0.984	1.522	1.673	2.093	2.459	3.178	4.016
62	0.981	1.519	1.669	2.088	2.454	3.173	4.009
63	0.978	1.515	1.666	2.084	2.450	3.167	4.002
64	0.975	1.512	1.662	2.080	2.445	3.161	3.996
65	0.973	1.509	1.659	2.076	2.441	3.156	3.989
66	0.970	1.506	1.655	2.072	2.436	3.151	3.983
67	0.968	1.503	1.652	2.069	2.432	3.146	3.977
68	0.965	1.500	1.649	2.065	2.428	3.141	3.971
69	0.963	1.497	1.646	2.061	2.424	3.136	3.965
70	0.961	1.494	1.643	2.058	2.420	3.131	3.959
71	0.958	1.491	1.640	2.055	2.416	3.127	3.953
72	0.956	1.488	1.637	2.051	2.413	3.122	3.948
73	0.954	1.486	1.634	2.048	2.409	3.118	3.943
74	0.952	1.483	1.631	2.045	2.405	3.113	3.938
75	0.950	1.480	1.629	2.042	2.402	3.109	3.932
76	0.948	1.478	1.626	2.039	2.399	3.105	3.928
77	0.946	1.476	1.624	2.036	2.395	3.101	3.923
78	0.944	1.473	1.621	2.033	2.392	3.097	3.918
79	0.942	1.471	1.619	2.030	2.389	3.093	3.913
80	0.940	1.469	1.616	2.027	2.386	3.089	3.909

Table B3 continued

				$1 - \alpha = 0.95$			
				p			
n	0.50	0.75	0.80	0.90	0.95	0.99	0.999
85	0.931	1.458	1.605	2.014	2.371	3.072	3.887
90	0.923	1.448	1.594	2.002	2.358	3.056	3.868
95	0.916	1.439	1.585	1.991	2.346	3.041	3.850
100	0.909	1.431	1.576	1.982	2.335	3.028	3.834
125	0.883	1.398	1.542	1.942	2.290	2.974	3.769
150	0.863	1.375	1.517	1.913	2.258	2.935	3.723
175	0.848	1.357	1.498	1.891	2.234	2.906	3.687
200	0.836	1.342	1.483	1.874	2.215	2.883	3.659
225	0.827	1.330	1.471	1.860	2.199	2.863	3.636
250	0.818	1.321	1.460	1.848	2.186	2.847	3.617
275	0.811	1.312	1.451	1.838	2.174	2.834	3.601
300	0.805	1.305	1.444	1.829	2.165	2.822	3.586
350	0.795	1.293	1.431	1.815	2.148	2.802	3.563
400	0.787	1.283	1.421	1.803	2.136	2.787	3.544
450	0.780	1.275	1.412	1.793	2.125	2.774	3.529
500	0.775	1.268	1.405	1.785	2.116	2.763	3.516
600	0.766	1.258	1.394	1.773	2.102	2.746	3.495
700	0.759	1.249	1.385	1.763	2.091	2.733	3.479
800	0.753	1.243	1.378	1.755	2.082	2.722	3.466
1000	0.745	1.233	1.368	1.743	2.069	2.706	3.447

Table B3 continued

				$1-\alpha=0.99$			
				p			
n	0.50	0.75	0.80	0.90	0.95	0.99	0.999
2	98.83	136.7	147.2	176.2	201.4	250.5	307.6
3	11.83	16.43	17.71	21.28	24.39	30.48	37.58
4	5.891	8.203	8.852	10.66	12.24	15.35	18.98
5	4.136	5.776	6.238	7.531	8.660	10.89	13.48
6	3.323	4.655	5.031	6.084	7.008	8.827	10.95
7	2.856	4.013	4.340	5.258	6.063	7.650	9.503
8	2.551	3.596	3.892	4.723	5.451	6.888	8.567
9	2.336	3.303	3.577	4.346	5.021	6.353	7.909
10	2.174	3.084	3.342	4.066	4.702	5.955	7.421
11	2.049	2.914	3.160	3.849	4.454	5.649	7.045
12	1.948	2.778	3.014	3.675	4.256	5.403	6.743
13	1.865	2.666	2.894	3.533	4.094	5.201	6.496
14	1.795	2.573	2.794	3.414	3.958	5.033	6.290
15	1.735	2.493	2.708	3.312	3.843	4.890	6.114
16	1.683	2.424	2.635	3.225	3.743	4.767	5.963
17	1.638	2.364	2.570	3.149	3.657	4.659	5.831
18	1.598	2.311	2.513	3.081	3.580	4.564	5.715
19	1.562	2.264	2.463	3.022	3.512	4.480	5.612
20	1.529	2.221	2.417	2.968	3.451	4.405	5.520
21	1.500	2.183	2.377	2.920	3.396	4.337	5.436
22	1.474	2.148	2.339	2.876	3.346	4.275	5.361
23	1.449	2.116	2.305	2.836	3.301	4.219	5.292
24	1.427	2.087	2.274	2.799	3.259	4.168	5.229
25	1.406	2.060	2.246	2.765	3.221	4.120	5.171
26	1.387	2.036	2.219	2.734	3.186	4.077	5.118
27	1.370	2.013	2.195	2.705	3.153	4.036	5.068
28	1.353	1.991	2.172	2.678	3.122	3.998	5.022
29	1.338	1.971	2.150	2.653	3.094	3.963	4.979
30	1.323	1.952	2.130	2.629	3.067	3.930	4.939
31	1.309	1.935	2.111	2.607	3.042	3.899	4.901
32	1.297	1.918	2.094	2.586	3.018	3.870	4.865
33	1.284	1.902	2.077	2.567	2.996	3.843	4.832
34	1.273	1.888	2.061	2.548	2.975	3.817	4.801
35	1.262	1.874	2.046	2.531	2.955	3.793	4.771
36	1.252	1.860	2.032	2.514	2.936	3.769	4.742
37	1.242	1.848	2.019	2.498	2.919	3.747	4.716
38	1.232	1.836	2.006	2.483	2.902	3.727	4.690
39	1.223	1.824	1.994	2.469	2.885	3.707	4.666
40	1.215	1.813	1.982	2.455	2.870	3.688	4.642

Table B3 continued

				$1 - \alpha = 0.99$			
				p			
n	0.50	0.75	0.80	0.90	0.95	0.99	0.999
41	1.207	1.803	1.971	2.442	2.855	3.669	4.620
42	1.199	1.793	1.960	2.430	2.841	3.652	4.599
43	1.191	1.783	1.950	2.418	2.828	3.636	4.579
44	1.184	1.774	1.940	2.406	2.815	3.620	4.559
45	1.177	1.765	1.931	2.395	2.802	3.604	4.541
46	1.170	1.757	1.922	2.385	2.790	3.590	4.523
47	1.164	1.749	1.913	2.375	2.779	3.576	4.506
48	1.158	1.741	1.905	2.365	2.768	3.562	4.489
49	1.152	1.733	1.897	2.355	2.757	3.549	4.473
50	1.146	1.726	1.889	2.346	2.747	3.536	4.458
51	1.141	1.719	1.882	2.338	2.737	3.524	4.443
52	1.135	1.712	1.874	2.329	2.728	3.512	4.429
53	1.130	1.705	1.867	2.321	2.718	3.501	4.415
54	1.125	1.699	1.861	2.313	2.709	3.490	4.401
55	1.120	1.693	1.854	2.305	2.701	3.480	4.389
56	1.115	1.687	1.848	2.298	2.692	3.469	4.376
57	1.111	1.681	1.842	2.291	2.684	3.459	4.364
58	1.106	1.675	1.836	2.284	2.676	3.450	4.352
59	1.102	1.670	1.830	2.277	2.669	3.440	4.341
60	1.098	1.665	1.824	2.270	2.661	3.431	4.330
61	1.094	1.660	1.819	2.264	2.654	3.422	4.319
62	1.090	1.654	1.813	2.258	2.647	3.414	4.308
63	1.086	1.650	1.808	2.252	2.640	3.406	4.298
64	1.082	1.645	1.803	2.246	2.634	3.397	4.288
65	1.078	1.640	1.798	2.240	2.627	3.390	4.279
66	1.075	1.636	1.793	2.235	2.621	3.382	4.270
67	1.071	1.631	1.789	2.229	2.615	3.374	4.260
68	1.068	1.627	1.784	2.224	2.609	3.367	4.252
69	1.065	1.623	1.780	2.219	2.603	3.360	4.243
70	1.061	1.619	1.776	2.214	2.598	3.353	4.234
71	1.058	1.615	1.771	2.209	2.592	3.346	4.226
72	1.055	1.611	1.767	2.204	2.587	3.340	4.218
73	1.052	1.607	1.763	2.200	2.582	3.333	4.210
74	1.049	1.604	1.759	2.195	2.576	3.327	4.203
75	1.046	1.600	1.756	2.191	2.571	3.321	4.195
76	1.043	1.597	1.752	2.186	2.567	3.315	4.188
77	1.041	1.593	1.748	2.182	2.562	3.309	4.181
78	1.038	1.590	1.745	2.178	2.557	3.303	4.174
79	1.035	1.586	1.741	2.174	2.552	3.298	4.167
80	1.033	1.583	1.738	2.170	2.548	3.292	4.160

Table B3 continued

				$1 - \alpha = 0.99$				
				p				
n	0.50	0.75	0.80	0.90	0.95	0.99	0.999
85	1.021	1.568	1.722	2.151	2.527	3.266	4.129
90	1.010	1.554	1.707	2.134	2.508	3.243	4.101
95	1.000	1.542	1.694	2.119	2.491	3.222	4.075
100	0.990	1.530	1.682	2.105	2.475	3.203	4.051
125	0.953	1.485	1.634	2.049	2.412	3.126	3.958
150	0.927	1.452	1.599	2.009	2.367	3.071	3.891
175	0.906	1.427	1.573	1.979	2.333	3.029	3.840
200	0.890	1.408	1.552	1.955	2.306	2.996	3.800
225	0.877	1.392	1.535	1.935	2.284	2.970	3.768
250	0.866	1.378	1.521	1.919	2.266	2.947	3.741
275	0.857	1.367	1.509	1.905	2.250	2.928	3.717
300	0.848	1.357	1.498	1.893	2.237	2.912	3.697
350	0.835	1.340	1.481	1.873	2.214	2.885	3.665
400	0.824	1.327	1.467	1.857	2.197	2.863	3.638
450	0.815	1.316	1.456	1.844	2.182	2.845	3.617
500	0.807	1.307	1.446	1.833	2.170	2.830	3.599
600	0.795	1.293	1.431	1.816	2.150	2.807	3.570
700	0.786	1.282	1.420	1.802	2.135	2.789	3.548
800	0.779	1.273	1.410	1.792	2.124	2.774	3.531
1000	0.767	1.259	1.396	1.775	2.105	2.752	3.504

Table B4: Critical values k_h satisfying (2.3.13) for testing hypotheses about the normal quantiles $\mu - z_{\frac{1+p}{2}}\sigma$ and $\mu + z_{\frac{1+p}{2}}\sigma$, at level α

			$\alpha = 0.10$			
			p			
n	0.70	0.80	0.90	0.95	0.98	0.99
2	4.101	5.853	8.629	11.11	14.02	16.00
3	2.026	2.716	3.779	4.722	5.830	6.589
4	1.639	2.153	2.937	3.629	4.442	4.999
5	1.481	1.924	2.597	3.191	3.887	4.364
6	1.394	1.799	2.413	2.953	3.587	4.020
7	1.339	1.721	2.297	2.804	3.397	3.803
8	1.301	1.666	2.217	2.700	3.266	3.653
9	1.273	1.626	2.157	2.623	3.169	3.542
10	1.252	1.595	2.112	2.564	3.094	3.456
11	1.235	1.570	2.075	2.517	3.034	3.388
12	1.221	1.550	2.045	2.479	2.985	3.332
13	1.209	1.533	2.020	2.446	2.944	3.285
14	1.199	1.519	1.999	2.419	2.910	3.245
15	1.190	1.506	1.981	2.395	2.880	3.211
16	1.183	1.496	1.965	2.374	2.853	3.181
17	1.176	1.486	1.950	2.356	2.830	3.154
18	1.170	1.478	1.938	2.340	2.810	3.130
19	1.165	1.470	1.927	2.325	2.791	3.109
20	1.160	1.463	1.916	2.312	2.775	3.090
21	1.156	1.457	1.907	2.300	2.760	3.073
22	1.152	1.451	1.899	2.290	2.746	3.057
23	1.149	1.446	1.891	2.280	2.733	3.043
24	1.145	1.441	1.884	2.270	2.721	3.029
25	1.142	1.437	1.877	2.262	2.711	3.017
26	1.139	1.433	1.871	2.254	2.701	3.005
27	1.137	1.429	1.866	2.247	2.691	2.995
28	1.134	1.425	1.860	2.240	2.683	2.985
29	1.132	1.422	1.855	2.233	2.674	2.975
30	1.130	1.419	1.851	2.227	2.667	2.967
31	1.128	1.416	1.846	2.222	2.659	2.958
32	1.126	1.413	1.842	2.216	2.653	2.950
33	1.124	1.411	1.838	2.211	2.646	2.943
34	1.122	1.408	1.835	2.206	2.640	2.936
35	1.121	1.406	1.831	2.202	2.634	2.930
36	1.119	1.403	1.828	2.198	2.629	2.923
37	1.118	1.401	1.825	2.193	2.624	2.917
38	1.116	1.399	1.822	2.190	2.619	2.912
39	1.115	1.397	1.819	2.186	2.614	2.906
40	1.114	1.396	1.816	2.182	2.609	2.901

Table B4 continued

			$\alpha = 0.10$			
			p			
n	0.70	0.80	0.90	0.95	0.98	0.99
41	1.113	1.394	1.813	2.179	2.605	2.896
42	1.111	1.392	1.811	2.176	2.601	2.891
43	1.110	1.390	1.808	2.172	2.597	2.887
44	1.109	1.389	1.806	2.169	2.593	2.882
45	1.108	1.387	1.804	2.166	2.589	2.878
46	1.107	1.386	1.802	2.164	2.586	2.874
47	1.106	1.385	1.800	2.161	2.582	2.870
48	1.105	1.383	1.798	2.158	2.579	2.866
49	1.105	1.382	1.796	2.156	2.576	2.863
50	1.104	1.381	1.794	2.154	2.573	2.859
51	1.103	1.380	1.792	2.151	2.570	2.856
52	1.102	1.378	1.790	2.149	2.567	2.852
53	1.101	1.377	1.789	2.147	2.564	2.849
54	1.101	1.376	1.787	2.145	2.562	2.846
55	1.100	1.375	1.785	2.143	2.559	2.843
56	1.099	1.374	1.784	2.141	2.557	2.840
57	1.098	1.373	1.782	2.139	2.554	2.837
58	1.098	1.372	1.781	2.137	2.552	2.835
59	1.097	1.371	1.780	2.135	2.549	2.832
60	1.097	1.370	1.778	2.133	2.547	2.829
61	1.096	1.369	1.777	2.132	2.545	2.827
62	1.095	1.369	1.776	2.130	2.543	2.825
63	1.095	1.368	1.774	2.128	2.541	2.822
64	1.094	1.367	1.773	2.127	2.539	2.820
65	1.094	1.366	1.772	2.125	2.537	2.818
66	1.093	1.365	1.771	2.124	2.535	2.816
67	1.093	1.365	1.770	2.122	2.533	2.813
68	1.092	1.364	1.769	2.121	2.531	2.811
69	1.092	1.363	1.768	2.119	2.530	2.809
70	1.091	1.362	1.766	2.118	2.528	2.807
71	1.091	1.362	1.765	2.117	2.526	2.805
72	1.090	1.361	1.764	2.116	2.525	2.804
73	1.090	1.360	1.763	2.114	2.523	2.802
74	1.089	1.360	1.763	2.113	2.521	2.800
75	1.089	1.359	1.762	2.112	2.520	2.798
76	1.089	1.359	1.761	2.111	2.518	2.796
77	1.088	1.358	1.760	2.110	2.517	2.795
78	1.088	1.357	1.759	2.108	2.516	2.793
79	1.087	1.357	1.758	2.107	2.514	2.792
80	1.087	1.356	1.757	2.106	2.513	2.790

Table B4 continued

			$\alpha = 0.10$			
			p			
n	0.70	0.80	0.90	0.95	0.98	0.99
85	1.085	1.354	1.753	2.101	2.506	2.783
90	1.084	1.351	1.750	2.097	2.500	2.776
95	1.082	1.349	1.747	2.092	2.495	2.770
100	1.081	1.347	1.744	2.089	2.490	2.764
125	1.075	1.339	1.732	2.073	2.471	2.742
150	1.072	1.334	1.723	2.062	2.457	2.726
175	1.069	1.329	1.717	2.054	2.446	2.713
200	1.066	1.326	1.712	2.047	2.438	2.704
225	1.065	1.323	1.708	2.042	2.431	2.696
250	1.063	1.321	1.704	2.037	2.425	2.689
275	1.062	1.319	1.701	2.033	2.420	2.683
300	1.060	1.317	1.699	2.030	2.416	2.678
350	1.059	1.314	1.694	2.024	2.409	2.670
400	1.057	1.312	1.691	2.020	2.403	2.664
450	1.056	1.310	1.688	2.016	2.398	2.658
500	1.055	1.309	1.686	2.013	2.394	2.654
600	1.053	1.306	1.682	2.008	2.388	2.647
700	1.052	1.304	1.679	2.005	2.383	2.641
800	1.051	1.303	1.677	2.002	2.379	2.637
1000	1.049	1.300	1.673	1.997	2.374	2.630

Table B4 continued

			$\alpha = 0.05$			
			p			
n	0.70	0.80	0.90	0.95	0.98	0.99
2	8.239	11.746	17.308	22.274	28.101	32.078
3	2.927	3.913	5.433	6.783	8.368	9.454
4	2.133	2.793	3.801	4.691	5.738	6.454
5	1.830	2.371	3.192	3.916	4.767	5.349
6	1.670	2.149	2.875	3.514	4.263	4.776
7	1.571	2.012	2.680	3.266	3.954	4.424
8	1.503	1.919	2.547	3.098	3.743	4.185
9	1.453	1.850	2.449	2.975	3.590	4.011
10	1.415	1.798	2.375	2.881	3.473	3.877
11	1.385	1.757	2.316	2.806	3.380	3.772
12	1.360	1.723	2.268	2.746	3.304	3.686
13	1.339	1.695	2.228	2.695	3.242	3.615
14	1.322	1.671	2.194	2.653	3.188	3.554
15	1.307	1.650	2.165	2.616	3.143	3.502
16	1.294	1.632	2.140	2.584	3.103	3.457
17	1.282	1.617	2.118	2.556	3.068	3.417
18	1.272	1.603	2.098	2.531	3.037	3.382
19	1.263	1.590	2.080	2.508	3.009	3.350
20	1.255	1.579	2.064	2.488	2.984	3.322
21	1.247	1.569	2.050	2.470	2.961	3.296
22	1.240	1.559	2.037	2.454	2.940	3.273
23	1.234	1.551	2.025	2.438	2.921	3.251
24	1.228	1.543	2.013	2.424	2.904	3.231
25	1.223	1.536	2.003	2.411	2.888	3.213
26	1.218	1.529	1.994	2.399	2.873	3.196
27	1.213	1.523	1.985	2.388	2.859	3.180
28	1.209	1.517	1.977	2.378	2.846	3.166
29	1.205	1.511	1.969	2.368	2.834	3.152
30	1.201	1.506	1.962	2.359	2.823	3.139
31	1.198	1.501	1.955	2.350	2.812	3.127
32	1.194	1.497	1.948	2.342	2.802	3.115
33	1.191	1.492	1.942	2.335	2.792	3.105
34	1.188	1.488	1.937	2.327	2.783	3.094
35	1.185	1.485	1.931	2.320	2.775	3.085
36	1.183	1.481	1.926	2.314	2.767	3.075
37	1.180	1.477	1.921	2.308	2.759	3.067
38	1.178	1.474	1.916	2.302	2.752	3.058
39	1.175	1.471	1.912	2.296	2.745	3.050
40	1.173	1.468	1.908	2.291	2.738	3.043

Table B4 continued

			$\alpha = 0.05$			
			p			
n	0.70	0.80	0.90	0.95	0.98	0.99
41	1.171	1.465	1.903	2.286	2.731	3.036
42	1.169	1.462	1.900	2.281	2.725	3.029
43	1.167	1.460	1.896	2.276	2.719	3.022
44	1.165	1.457	1.892	2.271	2.714	3.015
45	1.163	1.455	1.889	2.267	2.708	3.009
46	1.162	1.452	1.885	2.263	2.703	3.003
47	1.160	1.450	1.882	2.259	2.698	2.998
48	1.158	1.448	1.879	2.255	2.693	2.992
49	1.157	1.446	1.876	2.251	2.689	2.987
50	1.155	1.444	1.873	2.248	2.684	2.982
51	1.154	1.442	1.871	2.244	2.680	2.977
52	1.153	1.440	1.868	2.241	2.676	2.972
53	1.151	1.438	1.865	2.238	2.672	2.968
54	1.150	1.436	1.863	2.234	2.668	2.963
55	1.149	1.434	1.860	2.231	2.664	2.959
56	1.147	1.433	1.858	2.228	2.660	2.955
57	1.146	1.431	1.856	2.225	2.656	2.951
58	1.145	1.430	1.853	2.223	2.653	2.947
59	1.144	1.428	1.851	2.220	2.650	2.943
60	1.143	1.427	1.849	2.217	2.646	2.939
61	1.142	1.425	1.847	2.215	2.643	2.935
62	1.141	1.424	1.845	2.212	2.640	2.932
63	1.140	1.422	1.843	2.210	2.637	2.928
64	1.139	1.421	1.841	2.208	2.634	2.925
65	1.138	1.420	1.840	2.205	2.631	2.922
66	1.137	1.419	1.838	2.203	2.629	2.919
67	1.136	1.417	1.836	2.201	2.626	2.916
68	1.135	1.416	1.834	2.199	2.623	2.913
69	1.134	1.415	1.833	2.197	2.621	2.910
70	1.134	1.414	1.831	2.195	2.618	2.907
71	1.133	1.413	1.830	2.193	2.616	2.904
72	1.132	1.412	1.828	2.191	2.613	2.901
73	1.131	1.411	1.827	2.189	2.611	2.899
74	1.130	1.410	1.825	2.187	2.609	2.896
75	1.130	1.408	1.824	2.185	2.606	2.894
76	1.129	1.408	1.822	2.183	2.604	2.891
77	1.128	1.407	1.821	2.182	2.602	2.889
78	1.128	1.406	1.820	2.180	2.600	2.886
79	1.127	1.405	1.818	2.178	2.598	2.884
80	1.126	1.404	1.817	2.177	2.596	2.882

Table B4 continued

			$\alpha = 0.05$			
			p			
n	0.70	0.80	0.90	0.95	0.98	0.99
85	1.123	1.400	1.811	2.169	2.587	2.871
90	1.120	1.396	1.806	2.162	2.578	2.861
95	1.118	1.392	1.801	2.156	2.570	2.853
100	1.115	1.389	1.796	2.150	2.563	2.845
125	1.106	1.376	1.778	2.128	2.535	2.812
150	1.099	1.367	1.765	2.111	2.514	2.789
175	1.094	1.360	1.755	2.099	2.499	2.772
200	1.090	1.354	1.747	2.089	2.487	2.758
225	1.087	1.350	1.741	2.081	2.477	2.746
250	1.084	1.346	1.736	2.074	2.468	2.737
275	1.081	1.343	1.731	2.068	2.461	2.729
300	1.079	1.340	1.727	2.063	2.455	2.722
350	1.076	1.335	1.720	2.055	2.445	2.710
400	1.073	1.332	1.715	2.049	2.436	2.701
450	1.071	1.328	1.711	2.043	2.430	2.693
500	1.069	1.326	1.707	2.039	2.424	2.687
600	1.066	1.322	1.702	2.031	2.415	2.677
700	1.064	1.319	1.697	2.026	2.408	2.669
800	1.062	1.316	1.694	2.021	2.403	2.662
1000	1.059	1.312	1.688	2.015	2.394	2.653

Table B4 continued

				$\alpha = 0.01$		
				p		
n	0.70	0.80	0.90	0.95	0.98	0.99
2	41.27	58.80	86.62	111.46	140.61	160.61
3	6.650	8.875	12.31	15.35	18.93	21.38
4	3.770	4.922	6.681	8.237	10.07	11.32
5	2.863	3.696	4.963	6.081	7.393	8.291
6	2.430	3.117	4.157	5.074	6.148	6.884
7	2.178	2.781	3.692	4.493	5.433	6.075
8	2.013	2.561	3.389	4.116	4.968	5.551
9	1.895	2.406	3.175	3.850	4.641	5.182
10	1.808	2.290	3.016	3.652	4.398	4.907
11	1.739	2.200	2.892	3.499	4.209	4.695
12	1.685	2.127	2.793	3.376	4.059	4.525
13	1.639	2.068	2.712	3.276	3.935	4.386
14	1.602	2.018	2.644	3.191	3.832	4.269
15	1.569	1.976	2.586	3.120	3.744	4.171
16	1.542	1.939	2.536	3.058	3.668	4.085
17	1.517	1.907	2.493	3.004	3.602	4.011
18	1.496	1.879	2.454	2.957	3.544	3.945
19	1.477	1.854	2.420	2.914	3.492	3.887
20	1.459	1.832	2.389	2.877	3.446	3.835
21	1.444	1.812	2.362	2.843	3.404	3.788
22	1.430	1.793	2.337	2.812	3.366	3.745
23	1.417	1.776	2.314	2.783	3.332	3.706
24	1.405	1.761	2.293	2.758	3.300	3.671
25	1.394	1.747	2.274	2.734	3.271	3.638
26	1.384	1.734	2.256	2.712	3.244	3.608
27	1.375	1.722	2.240	2.692	3.219	3.580
28	1.367	1.710	2.224	2.673	3.196	3.554
29	1.358	1.700	2.210	2.655	3.175	3.530
30	1.351	1.690	2.197	2.639	3.155	3.507
31	1.344	1.681	2.184	2.623	3.136	3.485
32	1.337	1.672	2.172	2.609	3.118	3.465
33	1.331	1.664	2.161	2.595	3.101	3.446
34	1.325	1.656	2.151	2.582	3.085	3.429
35	1.319	1.649	2.141	2.570	3.070	3.412
36	1.314	1.642	2.131	2.558	3.056	3.396
37	1.309	1.635	2.122	2.547	3.042	3.381
38	1.304	1.629	2.114	2.536	3.030	3.366
39	1.299	1.623	2.106	2.527	3.017	3.352
40	1.295	1.617	2.098	2.517	3.006	3.339

Table B4 continued

| | | | $\alpha = 0.01$ | | | |
| | | | p | | | |
n	0.70	0.80	0.90	0.95	0.98	0.99
41	1.291	1.612	2.091	2.508	2.995	3.327
42	1.287	1.607	2.083	2.499	2.984	3.315
43	1.283	1.602	2.077	2.491	2.974	3.303
44	1.279	1.597	2.070	2.483	2.964	3.292
45	1.276	1.592	2.064	2.475	2.955	3.282
46	1.273	1.588	2.058	2.468	2.946	3.272
47	1.269	1.584	2.052	2.461	2.937	3.262
48	1.266	1.579	2.047	2.454	2.929	3.253
49	1.263	1.576	2.042	2.447	2.921	3.244
50	1.260	1.572	2.036	2.441	2.913	3.235
51	1.257	1.568	2.031	2.435	2.906	3.227
52	1.255	1.565	2.027	2.429	2.898	3.219
53	1.252	1.561	2.022	2.423	2.891	3.211
54	1.250	1.558	2.018	2.418	2.885	3.203
55	1.247	1.555	2.013	2.413	2.878	3.196
56	1.245	1.552	2.009	2.407	2.872	3.189
57	1.242	1.549	2.005	2.402	2.866	3.182
58	1.240	1.546	2.001	2.398	2.860	3.175
59	1.238	1.543	1.997	2.393	2.854	3.169
60	1.236	1.540	1.994	2.388	2.849	3.163
61	1.234	1.537	1.990	2.384	2.843	3.157
62	1.232	1.535	1.986	2.380	2.838	3.151
63	1.230	1.532	1.983	2.375	2.833	3.145
64	1.228	1.530	1.980	2.371	2.828	3.139
65	1.226	1.528	1.977	2.367	2.823	3.134
66	1.224	1.525	1.973	2.364	2.818	3.129
67	1.223	1.523	1.970	2.360	2.814	3.123
68	1.221	1.521	1.967	2.356	2.809	3.118
69	1.219	1.519	1.965	2.353	2.805	3.114
70	1.218	1.517	1.962	2.349	2.801	3.109
71	1.216	1.515	1.959	2.346	2.797	3.104
72	1.215	1.513	1.956	2.343	2.793	3.100
73	1.213	1.511	1.954	2.339	2.789	3.095
74	1.212	1.509	1.951	2.336	2.785	3.091
75	1.210	1.507	1.949	2.333	2.781	3.087
76	1.209	1.505	1.946	2.330	2.777	3.083
77	1.208	1.503	1.944	2.327	2.774	3.079
78	1.206	1.502	1.941	2.324	2.770	3.075
79	1.205	1.500	1.939	2.321	2.767	3.071
80	1.204	1.498	1.937	2.319	2.764	3.067

Table B4 continued

			$\alpha = 0.01$			
			p			
n	0.70	0.80	0.90	0.95	0.98	0.99
85	1.198	1.491	1.926	2.306	2.748	3.049
90	1.192	1.484	1.917	2.294	2.734	3.033
95	1.187	1.477	1.908	2.283	2.721	3.019
100	1.183	1.471	1.900	2.274	2.709	3.005
125	1.165	1.448	1.869	2.235	2.662	2.952
150	1.152	1.431	1.847	2.208	2.628	2.915
175	1.143	1.419	1.830	2.187	2.603	2.886
200	1.135	1.409	1.816	2.170	2.583	2.864
225	1.129	1.401	1.805	2.157	2.566	2.845
250	1.123	1.394	1.796	2.146	2.553	2.830
275	1.119	1.388	1.789	2.136	2.541	2.817
300	1.115	1.383	1.782	2.128	2.531	2.806
350	1.109	1.375	1.771	2.114	2.514	2.787
400	1.104	1.369	1.762	2.104	2.501	2.772
450	1.100	1.363	1.755	2.095	2.490	2.760
500	1.096	1.359	1.749	2.087	2.481	2.750
600	1.091	1.352	1.739	2.075	2.467	2.734
700	1.086	1.346	1.732	2.066	2.456	2.721
800	1.083	1.342	1.726	2.059	2.447	2.711
1000	1.078	1.335	1.717	2.048	2.434	2.696

Table B5: Factors k_1^* satisfying (2.5.3) to compute $(p, 1-\alpha)$ simultaneous one-sided tolerance limits for l normal populations

$$p = 0.90, \ 1-\alpha = 0.90$$

n	2	3	4	5	6	7	8	9	10
2	5.419	4.485	4.111	3.914	3.795	3.716	3.661	3.621	3.591
3	3.358	3.137	3.045	2.998	2.972	2.956	2.946	2.941	2.938
4	2.800	2.705	2.668	2.652	2.646	2.644	2.644	2.646	2.649
5	2.525	2.475	2.460	2.456	2.457	2.460	2.464	2.469	2.475
6	2.356	2.329	2.323	2.325	2.329	2.335	2.341	2.347	2.353
7	2.239	2.224	2.225	2.230	2.236	2.243	2.250	2.256	2.263
8	2.152	2.145	2.149	2.156	2.163	2.171	2.178	2.185	2.192
9	2.085	2.083	2.089	2.097	2.105	2.113	2.121	2.128	2.134
10	2.030	2.032	2.040	2.049	2.057	2.065	2.073	2.080	2.087
11	1.986	1.990	1.998	2.008	2.016	2.025	2.032	2.039	2.046
12	1.948	1.954	1.963	1.973	1.981	1.990	1.997	2.004	2.011
13	1.915	1.922	1.932	1.942	1.951	1.959	1.967	1.974	1.980
14	1.886	1.895	1.905	1.915	1.924	1.932	1.940	1.947	1.953
15	1.861	1.871	1.881	1.891	1.900	1.908	1.916	1.922	1.929
16	1.839	1.849	1.860	1.870	1.879	1.887	1.894	1.901	1.907
17	1.819	1.830	1.841	1.851	1.859	1.867	1.874	1.881	1.887
18	1.801	1.812	1.823	1.833	1.842	1.850	1.857	1.863	1.869
19	1.785	1.796	1.807	1.817	1.826	1.833	1.840	1.847	1.852
20	1.770	1.782	1.793	1.802	1.811	1.818	1.825	1.832	1.837
21	1.756	1.768	1.779	1.789	1.797	1.805	1.811	1.818	1.823
22	1.744	1.756	1.766	1.776	1.784	1.792	1.799	1.805	1.810
23	1.732	1.744	1.755	1.764	1.773	1.780	1.787	1.793	1.798
24	1.721	1.733	1.744	1.753	1.762	1.769	1.775	1.781	1.787
25	1.711	1.723	1.734	1.743	1.751	1.758	1.765	1.771	1.776
26	1.701	1.714	1.724	1.734	1.742	1.749	1.755	1.761	1.766
27	1.692	1.705	1.715	1.725	1.732	1.740	1.746	1.751	1.757
28	1.684	1.696	1.707	1.716	1.724	1.731	1.737	1.743	1.748
29	1.676	1.688	1.699	1.708	1.716	1.723	1.729	1.734	1.739
30	1.668	1.681	1.691	1.700	1.708	1.715	1.721	1.726	1.731
40	1.611	1.623	1.633	1.641	1.648	1.654	1.660	1.665	1.669
50	1.573	1.585	1.594	1.601	1.608	1.614	1.619	1.623	1.627
60	1.546	1.557	1.565	1.572	1.579	1.584	1.588	1.593	1.596
70	1.525	1.535	1.543	1.550	1.556	1.561	1.565	1.569	1.572
80	1.508	1.518	1.526	1.532	1.538	1.542	1.546	1.550	1.553
90	1.494	1.504	1.511	1.517	1.523	1.527	1.531	1.534	1.537
100	1.483	1.492	1.499	1.505	1.510	1.514	1.518	1.521	1.524

Table B5 continued

$p = 0.95,\ 1 - \alpha = 0.90$

					l				
n	2	3	4	5	6	7	8	9	10
2	6.496	5.275	4.786	4.527	4.369	4.263	4.188	4.132	4.090
3	4.024	3.706	3.568	3.494	3.449	3.420	3.401	3.387	3.378
4	3.365	3.210	3.144	3.110	3.091	3.080	3.074	3.070	3.069
5	3.044	2.951	2.913	2.895	2.886	2.882	2.881	2.882	2.883
6	2.848	2.786	2.762	2.752	2.749	2.748	2.750	2.752	2.755
7	2.714	2.669	2.654	2.649	2.648	2.650	2.653	2.656	2.660
8	2.614	2.581	2.571	2.569	2.571	2.574	2.577	2.581	2.586
9	2.537	2.512	2.506	2.506	2.508	2.512	2.516	2.521	2.525
10	2.476	2.456	2.452	2.454	2.457	2.461	2.466	2.471	2.475
11	2.424	2.409	2.408	2.410	2.414	2.419	2.423	2.428	2.433
12	2.381	2.370	2.369	2.372	2.377	2.382	2.387	2.392	2.396
13	2.345	2.335	2.336	2.340	2.345	2.350	2.355	2.360	2.364
14	2.313	2.305	2.307	2.311	2.316	2.321	2.326	2.331	2.336
15	2.284	2.279	2.281	2.286	2.291	2.296	2.301	2.306	2.311
16	2.259	2.255	2.258	2.263	2.268	2.274	2.279	2.284	2.288
17	2.237	2.234	2.238	2.242	2.248	2.253	2.258	2.263	2.268
18	2.217	2.215	2.219	2.224	2.229	2.235	2.240	2.244	2.249
19	2.198	2.197	2.202	2.207	2.212	2.218	2.223	2.227	2.232
20	2.182	2.181	2.186	2.191	2.197	2.202	2.207	2.212	2.216
21	2.166	2.167	2.171	2.177	2.182	2.187	2.192	2.197	2.202
22	2.152	2.153	2.158	2.163	2.169	2.174	2.179	2.184	2.188
23	2.139	2.141	2.145	2.151	2.156	2.162	2.167	2.171	2.175
24	2.127	2.129	2.134	2.139	2.145	2.150	2.155	2.159	2.164
25	2.116	2.118	2.123	2.128	2.134	2.139	2.144	2.148	2.153
26	2.105	2.108	2.113	2.118	2.124	2.129	2.134	2.138	2.142
27	2.095	2.098	2.103	2.109	2.114	2.119	2.124	2.128	2.133
28	2.086	2.089	2.094	2.100	2.105	2.110	2.115	2.119	2.123
29	2.077	2.080	2.086	2.091	2.097	2.102	2.106	2.111	2.115
30	2.069	2.072	2.078	2.083	2.089	2.094	2.098	2.102	2.106
40	2.005	2.010	2.015	2.021	2.026	2.031	2.035	2.039	2.042
50	1.963	1.968	1.974	1.979	1.984	1.988	1.992	1.996	1.999
60	1.933	1.938	1.944	1.949	1.953	1.957	1.961	1.964	1.967
70	1.910	1.915	1.920	1.925	1.929	1.933	1.937	1.940	1.943
80	1.891	1.897	1.902	1.906	1.910	1.914	1.917	1.920	1.923
90	1.876	1.881	1.886	1.891	1.895	1.898	1.901	1.904	1.907
100	1.864	1.869	1.873	1.878	1.881	1.885	1.888	1.891	1.893

Table B5 continued

$p = 0.99, \ 1 - \alpha = 0.90$

					l				
n	2	3	4	5	6	7	8	9	10
2	8.547	6.776	6.069	5.692	5.459	5.302	5.188	5.103	5.037
3	5.298	4.790	4.563	4.437	4.357	4.302	4.264	4.235	4.213
4	4.447	4.175	4.051	3.981	3.938	3.909	3.889	3.875	3.864
5	4.037	3.857	3.775	3.729	3.701	3.683	3.671	3.663	3.657
6	3.789	3.657	3.597	3.564	3.544	3.532	3.524	3.519	3.516
7	3.620	3.516	3.470	3.445	3.430	3.422	3.416	3.413	3.411
8	3.496	3.411	3.374	3.354	3.343	3.336	3.333	3.331	3.330
9	3.400	3.329	3.298	3.282	3.273	3.268	3.266	3.265	3.265
10	3.323	3.262	3.236	3.222	3.215	3.212	3.210	3.210	3.210
11	3.260	3.207	3.184	3.173	3.167	3.164	3.163	3.163	3.164
12	3.207	3.160	3.140	3.130	3.126	3.124	3.123	3.124	3.125
13	3.162	3.120	3.102	3.093	3.090	3.088	3.088	3.089	3.090
14	3.123	3.084	3.068	3.061	3.058	3.057	3.057	3.058	3.060
15	3.089	3.053	3.039	3.033	3.030	3.029	3.030	3.031	3.033
16	3.058	3.026	3.013	3.007	3.005	3.005	3.005	3.007	3.008
17	3.031	3.001	2.989	2.984	2.982	2.982	2.983	2.985	2.986
18	3.006	2.978	2.968	2.963	2.962	2.962	2.963	2.964	2.966
19	2.984	2.958	2.948	2.944	2.943	2.943	2.945	2.946	2.948
20	2.964	2.939	2.930	2.927	2.926	2.926	2.928	2.929	2.931
21	2.945	2.922	2.914	2.911	2.910	2.911	2.912	2.914	2.915
22	2.928	2.906	2.898	2.896	2.895	2.896	2.897	2.899	2.901
23	2.912	2.892	2.884	2.882	2.882	2.882	2.884	2.886	2.888
24	2.898	2.878	2.871	2.869	2.869	2.870	2.871	2.873	2.875
25	2.884	2.865	2.859	2.857	2.857	2.858	2.859	2.861	2.863
26	2.871	2.853	2.847	2.846	2.846	2.847	2.848	2.850	2.852
27	2.859	2.842	2.837	2.835	2.835	2.836	2.838	2.840	2.842
28	2.848	2.832	2.826	2.825	2.825	2.826	2.828	2.830	2.832
29	2.837	2.822	2.817	2.816	2.816	2.817	2.819	2.821	2.823
30	2.827	2.813	2.808	2.807	2.807	2.808	2.810	2.812	2.814
40	2.751	2.741	2.738	2.738	2.738	2.740	2.742	2.744	2.746
50	2.701	2.693	2.691	2.691	2.693	2.694	2.696	2.698	2.700
60	2.665	2.659	2.657	2.658	2.659	2.661	2.662	2.664	2.666
70	2.638	2.632	2.631	2.632	2.633	2.635	2.637	2.638	2.640
80	2.616	2.611	2.611	2.611	2.613	2.614	2.616	2.618	2.619
90	2.598	2.594	2.593	2.594	2.596	2.597	2.599	2.600	2.602
100	2.583	2.579	2.579	2.580	2.581	2.583	2.584	2.586	2.588

Table B5 continued

$p = 0.90, \ 1 - \alpha = 0.95$

					l				
n	2	3	4	5	6	7	8	9	10
2	7.812	5.828	5.078	4.691	4.457	4.301	4.190	4.108	4.045
3	4.160	3.695	3.494	3.385	3.318	3.273	3.242	3.219	3.203
4	3.302	3.079	2.982	2.929	2.898	2.878	2.865	2.857	2.851
5	2.902	2.768	2.709	2.679	2.662	2.653	2.647	2.644	2.642
6	2.665	2.573	2.535	2.516	2.506	2.501	2.499	2.499	2.500
7	2.504	2.437	2.410	2.398	2.393	2.391	2.391	2.392	2.394
8	2.386	2.336	2.316	2.308	2.306	2.306	2.307	2.309	2.312
9	2.296	2.256	2.242	2.237	2.236	2.237	2.240	2.242	2.246
10	2.224	2.192	2.181	2.178	2.179	2.181	2.184	2.187	2.190
11	2.165	2.138	2.131	2.129	2.131	2.133	2.136	2.140	2.144
12	2.115	2.093	2.087	2.087	2.089	2.092	2.096	2.100	2.103
13	2.072	2.054	2.050	2.051	2.053	2.057	2.061	2.064	2.068
14	2.035	2.020	2.017	2.019	2.022	2.026	2.029	2.033	2.037
15	2.003	1.990	1.989	1.990	1.994	1.998	2.002	2.006	2.010
16	1.974	1.964	1.963	1.965	1.969	1.973	1.977	1.981	1.985
17	1.949	1.940	1.939	1.942	1.946	1.950	1.954	1.958	1.962
18	1.926	1.918	1.918	1.922	1.925	1.930	1.934	1.938	1.942
19	1.905	1.898	1.899	1.903	1.907	1.911	1.915	1.919	1.923
20	1.886	1.880	1.882	1.885	1.889	1.894	1.898	1.902	1.906
21	1.868	1.864	1.865	1.869	1.873	1.878	1.882	1.886	1.890
22	1.852	1.848	1.850	1.854	1.859	1.863	1.867	1.871	1.875
23	1.837	1.834	1.837	1.841	1.845	1.849	1.853	1.858	1.861
24	1.823	1.821	1.824	1.828	1.832	1.836	1.841	1.845	1.849
25	1.811	1.809	1.812	1.816	1.820	1.824	1.829	1.833	1.836
26	1.799	1.797	1.800	1.804	1.809	1.813	1.817	1.821	1.825
27	1.787	1.786	1.790	1.794	1.798	1.803	1.807	1.811	1.815
28	1.777	1.776	1.779	1.784	1.788	1.793	1.797	1.801	1.804
29	1.767	1.766	1.770	1.774	1.779	1.783	1.787	1.791	1.795
30	1.757	1.757	1.761	1.765	1.770	1.774	1.778	1.782	1.786
40	1.686	1.688	1.692	1.696	1.701	1.705	1.709	1.712	1.716
50	1.638	1.641	1.646	1.650	1.654	1.658	1.662	1.665	1.668
60	1.604	1.608	1.612	1.617	1.621	1.624	1.628	1.631	1.635
70	1.578	1.582	1.586	1.591	1.594	1.598	1.601	1.605	1.608
80	1.558	1.561	1.566	1.570	1.574	1.577	1.581	1.584	1.586
90	1.541	1.545	1.549	1.553	1.556	1.561	1.564	1.566	1.569
100	1.526	1.530	1.535	1.538	1.543	1.546	1.549	1.551	1.554

Table B5 continued

$p = 0.95,\ 1 - \alpha = 0.95$

					l				
n	2	3	4	5	6	7	8	9	10
2	9.349	6.839	5.897	5.412	5.117	4.920	4.780	4.675	4.593
3	4.967	4.347	4.078	3.929	3.836	3.772	3.727	3.693	3.667
4	3.951	3.638	3.498	3.420	3.371	3.339	3.316	3.299	3.287
5	3.482	3.283	3.193	3.143	3.113	3.093	3.080	3.071	3.064
6	3.205	3.063	2.998	2.964	2.943	2.930	2.921	2.916	2.913
7	3.019	2.910	2.861	2.835	2.820	2.811	2.806	2.802	2.801
8	2.883	2.796	2.757	2.737	2.726	2.720	2.716	2.714	2.714
9	2.779	2.707	2.676	2.660	2.651	2.647	2.644	2.644	2.644
10	2.697	2.635	2.609	2.596	2.590	2.587	2.585	2.585	2.586
11	2.629	2.576	2.554	2.543	2.538	2.536	2.535	2.536	2.537
12	2.572	2.526	2.507	2.498	2.494	2.493	2.493	2.493	2.495
13	2.524	2.483	2.466	2.459	2.456	2.455	2.455	2.456	2.458
14	2.482	2.445	2.431	2.424	2.422	2.422	2.422	2.424	2.426
15	2.446	2.412	2.399	2.394	2.392	2.392	2.393	2.395	2.397
16	2.413	2.383	2.371	2.367	2.365	2.366	2.367	2.369	2.371
17	2.384	2.356	2.346	2.342	2.341	2.342	2.343	2.345	2.347
18	2.358	2.333	2.323	2.320	2.320	2.320	2.322	2.324	2.326
19	2.335	2.311	2.303	2.300	2.300	2.301	2.302	2.304	2.307
20	2.313	2.291	2.284	2.281	2.281	2.282	2.284	2.286	2.289
21	2.293	2.273	2.266	2.264	2.264	2.266	2.268	2.270	2.272
22	2.275	2.256	2.250	2.248	2.249	2.250	2.252	2.254	2.257
23	2.259	2.241	2.235	2.234	2.234	2.236	2.238	2.240	2.242
24	2.243	2.226	2.221	2.220	2.221	2.222	2.224	2.227	2.229
25	2.229	2.213	2.208	2.207	2.208	2.210	2.212	2.214	2.217
26	2.215	2.200	2.196	2.195	2.196	2.198	2.200	2.202	2.205
27	2.203	2.188	2.184	2.184	2.185	2.187	2.189	2.191	2.194
28	2.191	2.177	2.174	2.173	2.174	2.176	2.179	2.181	2.183
29	2.180	2.167	2.163	2.163	2.164	2.166	2.169	2.171	2.173
30	2.169	2.157	2.154	2.154	2.155	2.157	2.159	2.162	2.164
40	2.089	2.081	2.080	2.081	2.082	2.084	2.087	2.089	2.091
50	2.036	2.031	2.030	2.032	2.034	2.036	2.038	2.040	2.042
60	1.999	1.994	1.995	1.996	1.998	2.000	2.002	2.005	2.007
70	1.970	1.967	1.967	1.969	1.971	1.973	1.975	1.977	1.981
80	1.947	1.944	1.945	1.947	1.949	1.951	1.953	1.956	1.958
90	1.928	1.926	1.927	1.929	1.931	1.933	1.936	1.938	1.940
100	1.913	1.911	1.912	1.914	1.916	1.917	1.921	1.923	1.924

Table B5 continued

$$p = 0.99, \; 1 - \alpha = 0.95$$

					l				
n	2	3	4	5	6	7	8	9	10
2	12.278	8.765	7.459	6.784	6.374	6.099	5.901	5.752	5.636
3	6.515	5.596	5.193	4.967	4.823	4.723	4.651	4.595	4.552
4	5.196	4.709	4.484	4.356	4.273	4.215	4.174	4.142	4.118
5	4.594	4.269	4.116	4.028	3.971	3.932	3.903	3.882	3.866
6	4.241	3.999	3.883	3.817	3.774	3.745	3.724	3.708	3.696
7	4.005	3.812	3.720	3.667	3.633	3.610	3.593	3.581	3.572
8	3.835	3.675	3.598	3.554	3.526	3.507	3.493	3.484	3.476
9	3.704	3.567	3.502	3.464	3.440	3.424	3.413	3.405	3.399
10	3.601	3.481	3.424	3.391	3.371	3.357	3.347	3.341	3.336
11	3.517	3.410	3.359	3.330	3.312	3.300	3.292	3.286	3.282
12	3.446	3.350	3.305	3.278	3.262	3.252	3.245	3.240	3.236
13	3.386	3.299	3.257	3.234	3.219	3.210	3.203	3.199	3.196
14	3.335	3.254	3.216	3.195	3.181	3.173	3.167	3.163	3.161
15	3.290	3.215	3.180	3.160	3.148	3.140	3.135	3.132	3.130
16	3.250	3.180	3.148	3.129	3.118	3.111	3.106	3.103	3.101
17	3.214	3.149	3.118	3.101	3.091	3.085	3.080	3.078	3.076
18	3.182	3.121	3.092	3.076	3.067	3.061	3.057	3.054	3.053
19	3.153	3.096	3.068	3.053	3.044	3.039	3.035	3.033	3.032
20	3.127	3.072	3.047	3.032	3.024	3.019	3.015	3.014	3.012
21	3.103	3.051	3.026	3.013	3.005	3.000	2.997	2.995	2.995
22	3.081	3.031	3.008	2.995	2.988	2.983	2.980	2.979	2.978
23	3.060	3.013	2.991	2.979	2.972	2.967	2.965	2.963	2.962
24	3.041	2.996	2.975	2.963	2.956	2.952	2.950	2.949	2.948
25	3.024	2.980	2.960	2.949	2.942	2.939	2.936	2.935	2.935
26	3.007	2.965	2.946	2.935	2.929	2.926	2.924	2.922	2.922
27	2.992	2.952	2.933	2.923	2.917	2.913	2.911	2.910	2.910
28	2.978	2.939	2.920	2.911	2.905	2.902	2.900	2.899	2.899
29	2.964	2.926	2.909	2.899	2.894	2.891	2.889	2.888	2.888
30	2.951	2.915	2.898	2.889	2.884	2.881	2.879	2.878	2.878
40	2.854	2.826	2.814	2.807	2.803	2.801	2.800	2.800	2.800
50	2.791	2.768	2.758	2.752	2.750	2.748	2.748	2.748	2.748
60	2.746	2.726	2.717	2.713	2.711	2.710	2.709	2.709	2.710
70	2.711	2.694	2.686	2.682	2.681	2.680	2.680	2.680	2.680
80	2.684	2.668	2.661	2.658	2.657	2.656	2.656	2.656	2.657
90	2.661	2.647	2.641	2.638	2.637	2.636	2.636	2.637	2.637
100	2.643	2.630	2.624	2.621	2.620	2.620	2.620	2.620	2.621

Table B5 continued

$p = 0.90, \ 1 - \alpha = 0.99$

					l				
n	2	3	4	5	6	7	8	9	10
2	17.729	10.289	7.943	6.837	6.199	5.786	5.499	5.287	5.125
3	6.529	5.156	4.601	4.302	4.117	3.992	3.902	3.834	3.782
4	4.616	3.981	3.705	3.551	3.455	3.389	3.341	3.306	3.278
5	3.830	3.440	3.265	3.167	3.105	3.062	3.032	3.010	2.993
6	3.394	3.120	2.995	2.925	2.880	2.851	2.830	2.815	2.803
7	3.113	2.904	2.809	2.755	2.722	2.699	2.684	2.673	2.665
8	2.914	2.747	2.671	2.628	2.601	2.584	2.572	2.564	2.558
9	2.765	2.627	2.563	2.528	2.507	2.493	2.483	2.477	2.473
10	2.648	2.530	2.477	2.447	2.429	2.418	2.410	2.405	2.402
11	2.553	2.452	2.405	2.380	2.365	2.355	2.349	2.345	2.343
12	2.475	2.385	2.345	2.323	2.310	2.302	2.297	2.294	2.293
13	2.409	2.329	2.293	2.273	2.262	2.255	2.251	2.250	2.249
14	2.352	2.280	2.247	2.230	2.221	2.215	2.211	2.211	2.210
15	2.302	2.237	2.208	2.192	2.184	2.179	2.177	2.176	2.175
16	2.259	2.199	2.172	2.158	2.151	2.146	2.145	2.144	2.144
17	2.220	2.165	2.141	2.128	2.121	2.119	2.117	2.116	2.116
18	2.185	2.134	2.112	2.100	2.094	2.092	2.091	2.090	2.090
19	2.154	2.107	2.086	2.075	2.071	2.068	2.067	2.066	2.067
20	2.126	2.081	2.062	2.052	2.049	2.046	2.045	2.045	2.045
21	2.100	2.058	2.040	2.031	2.028	2.026	2.025	2.025	2.025
22	2.076	2.037	2.020	2.012	2.009	2.007	2.006	2.006	2.007
23	2.054	2.017	2.001	1.995	1.991	1.989	1.989	1.989	1.990
24	2.034	1.999	1.984	1.978	1.975	1.973	1.973	1.973	1.974
25	2.015	1.982	1.968	1.962	1.959	1.958	1.958	1.958	1.959
26	1.998	1.966	1.953	1.948	1.945	1.944	1.943	1.944	1.945
27	1.981	1.951	1.938	1.934	1.931	1.930	1.930	1.931	1.932
28	1.966	1.937	1.926	1.921	1.918	1.917	1.918	1.918	1.919
29	1.951	1.924	1.914	1.909	1.906	1.906	1.906	1.906	1.907
30	1.938	1.911	1.902	1.897	1.895	1.894	1.894	1.895	1.896
40	1.835	1.817	1.810	1.807	1.807	1.807	1.808	1.809	1.810
50	1.769	1.755	1.750	1.748	1.748	1.748	1.749	1.750	1.752
60	1.721	1.710	1.706	1.705	1.705	1.706	1.707	1.708	1.709
70	1.685	1.675	1.672	1.671	1.672	1.673	1.674	1.675	1.677
80	1.656	1.648	1.645	1.645	1.645	1.646	1.648	1.649	1.650
90	1.633	1.625	1.623	1.623	1.624	1.625	1.626	1.627	1.629
100	1.613	1.607	1.605	1.605	1.606	1.607	1.608	1.609	1.610

Table B5 continued

$$p = 0.95, \ 1 - \alpha = 0.99$$

					l				
n	2	3	4	5	6	7	8	9	10
2	21.194	12.045	9.201	7.862	7.091	6.594	6.246	5.990	5.793
3	7.763	6.036	5.339	4.964	4.731	4.571	4.456	4.369	4.301
4	5.491	4.674	4.316	4.116	3.989	3.901	3.837	3.789	3.751
5	4.564	4.051	3.818	3.686	3.601	3.542	3.500	3.468	3.443
6	4.052	3.685	3.514	3.417	3.354	3.311	3.280	3.257	3.239
7	3.724	3.439	3.306	3.229	3.180	3.147	3.122	3.104	3.091
8	3.492	3.261	3.152	3.089	3.049	3.022	3.002	2.988	2.977
9	3.319	3.125	3.033	2.980	2.946	2.923	2.907	2.895	2.886
10	3.184	3.017	2.937	2.891	2.862	2.843	2.829	2.819	2.811
11	3.075	2.928	2.858	2.818	2.792	2.775	2.763	2.755	2.749
12	2.985	2.854	2.791	2.755	2.733	2.718	2.708	2.700	2.695
13	2.909	2.790	2.734	2.702	2.682	2.668	2.659	2.653	2.648
14	2.844	2.736	2.684	2.655	2.637	2.625	2.617	2.611	2.609
15	2.787	2.688	2.641	2.614	2.597	2.586	2.579	2.574	2.572
16	2.737	2.646	2.602	2.577	2.562	2.552	2.545	2.543	2.540
17	2.693	2.608	2.567	2.544	2.530	2.521	2.517	2.513	2.510
18	2.654	2.574	2.536	2.515	2.502	2.493	2.490	2.486	2.483
19	2.618	2.543	2.507	2.488	2.475	2.468	2.465	2.461	2.459
20	2.586	2.515	2.482	2.463	2.451	2.446	2.441	2.438	2.436
21	2.556	2.489	2.458	2.440	2.429	2.425	2.420	2.417	2.415
22	2.529	2.466	2.436	2.419	2.409	2.405	2.401	2.398	2.396
23	2.505	2.444	2.415	2.400	2.392	2.386	2.382	2.380	2.378
24	2.482	2.424	2.397	2.382	2.375	2.369	2.365	2.363	2.361
25	2.460	2.405	2.379	2.365	2.358	2.353	2.349	2.347	2.346
26	2.441	2.387	2.363	2.349	2.343	2.338	2.334	2.332	2.331
27	2.422	2.371	2.347	2.334	2.328	2.323	2.320	2.318	2.317
28	2.405	2.356	2.333	2.322	2.315	2.310	2.307	2.305	2.304
29	2.388	2.341	2.319	2.309	2.302	2.297	2.295	2.293	2.292
30	2.373	2.327	2.306	2.297	2.290	2.285	2.283	2.281	2.280
40	2.257	2.223	2.209	2.201	2.196	2.193	2.191	2.191	2.190
50	2.182	2.157	2.144	2.137	2.133	2.131	2.130	2.130	2.130
60	2.131	2.107	2.097	2.091	2.088	2.087	2.086	2.086	2.086
70	2.090	2.070	2.061	2.056	2.053	2.052	2.052	2.052	2.052
80	2.058	2.040	2.032	2.028	2.026	2.025	2.024	2.024	2.025
90	2.032	2.015	2.008	2.005	2.003	2.002	2.002	2.002	2.002
100	2.010	1.995	1.988	1.985	1.984	1.983	1.983	1.983	1.983

Table B5 continued

$p = 0.99, \; 1 - \alpha = 0.99$

					l				
n	2	3	4	5	6	7	8	9	10
2	27.789	15.401	11.598	9.818	8.798	8.135	7.674	7.333	7.070
3	10.140	7.728	6.759	6.235	5.908	5.683	5.519	5.395	5.297
4	7.178	6.007	5.492	5.201	5.015	4.884	4.788	4.715	4.657
5	5.979	5.227	4.882	4.683	4.553	4.462	4.395	4.343	4.302
6	5.321	4.772	4.512	4.361	4.262	4.193	4.141	4.101	4.070
7	4.901	4.468	4.261	4.139	4.059	4.002	3.960	3.928	3.903
8	4.606	4.249	4.076	3.973	3.906	3.858	3.823	3.796	3.775
9	4.387	4.082	3.933	3.845	3.787	3.745	3.715	3.692	3.674
10	4.216	3.950	3.819	3.741	3.690	3.654	3.627	3.606	3.591
11	4.078	3.842	3.725	3.655	3.609	3.577	3.553	3.535	3.521
12	3.965	3.752	3.646	3.583	3.541	3.512	3.490	3.474	3.461
13	3.870	3.675	3.578	3.521	3.483	3.456	3.436	3.421	3.409
14	3.788	3.609	3.520	3.467	3.431	3.407	3.389	3.375	3.364
15	3.717	3.552	3.468	3.419	3.386	3.363	3.346	3.334	3.324
16	3.655	3.501	3.423	3.377	3.346	3.325	3.309	3.297	3.288
17	3.600	3.455	3.382	3.339	3.310	3.290	3.275	3.264	3.256
18	3.551	3.414	3.346	3.304	3.277	3.258	3.245	3.234	3.231
19	3.507	3.377	3.312	3.273	3.248	3.230	3.217	3.207	3.204
20	3.467	3.344	3.282	3.245	3.221	3.204	3.191	3.186	3.179
21	3.430	3.313	3.254	3.219	3.196	3.179	3.168	3.163	3.156
22	3.397	3.285	3.229	3.195	3.173	3.157	3.146	3.142	3.135
23	3.366	3.259	3.205	3.173	3.151	3.137	3.130	3.122	3.115
24	3.338	3.235	3.183	3.152	3.131	3.117	3.112	3.103	3.097
25	3.312	3.213	3.162	3.133	3.113	3.099	3.094	3.086	3.080
26	3.287	3.192	3.143	3.115	3.096	3.083	3.077	3.070	3.064
27	3.265	3.172	3.125	3.098	3.079	3.071	3.062	3.055	3.049
28	3.243	3.154	3.109	3.082	3.064	3.056	3.047	3.040	3.035
29	3.223	3.137	3.093	3.067	3.050	3.042	3.033	3.027	3.021
30	3.204	3.120	3.078	3.053	3.036	3.029	3.020	3.014	3.009
40	3.062	2.997	2.964	2.949	2.936	2.926	2.920	2.915	2.911
50	2.971	2.916	2.894	2.877	2.866	2.858	2.852	2.848	2.845
60	2.906	2.859	2.839	2.825	2.815	2.808	2.804	2.800	2.798
70	2.857	2.820	2.798	2.785	2.776	2.770	2.766	2.763	2.761
80	2.823	2.785	2.765	2.753	2.745	2.740	2.736	2.734	2.732
90	2.791	2.756	2.738	2.727	2.720	2.715	2.712	2.709	2.708
100	2.764	2.732	2.715	2.705	2.699	2.694	2.691	2.689	2.687

Table B6: Factors k' satisfying (2.5.8) to compute $(p, 1 - \alpha)$ simultaneous two-sided tolerance intervals for l normal populations

$$p = 0.90, \ 1 - \alpha = 0.90$$

					l				
n	2	3	4	5	6	7	8	9	10
2	6.626	5.187	4.637	4.353	4.182	4.069	3.988	3.929	3.884
3	3.947	3.536	3.367	3.278	3.225	3.192	3.169	3.153	3.142
4	3.238	3.015	2.923	2.876	2.850	2.835	2.825	2.820	2.816
5	2.895	2.743	2.681	2.651	2.636	2.627	2.623	2.621	2.621
6	2.689	2.572	2.526	2.504	2.493	2.488	2.486	2.486	2.487
7	2.549	2.453	2.415	2.398	2.390	2.386	2.385	2.386	2.388
8	2.447	2.365	2.332	2.318	2.311	2.309	2.308	2.309	2.311
9	2.369	2.296	2.268	2.255	2.249	2.247	2.247	2.248	2.250
10	2.307	2.242	2.215	2.204	2.198	2.197	2.197	2.198	2.200
11	2.257	2.196	2.172	2.161	2.156	2.155	2.155	2.156	2.158
12	2.215	2.159	2.136	2.126	2.121	2.119	2.120	2.121	2.122
13	2.180	2.127	2.105	2.095	2.091	2.089	2.089	2.090	2.092
14	2.149	2.099	2.078	2.069	2.064	2.063	2.063	2.063	2.065
15	2.123	2.075	2.055	2.046	2.041	2.040	2.039	2.040	2.041
16	2.099	2.054	2.034	2.025	2.021	2.019	2.019	2.019	2.021
17	2.079	2.035	2.016	2.007	2.003	2.001	2.001	2.001	2.002
18	2.060	2.018	2.000	1.991	1.987	1.985	1.984	1.985	1.985
19	2.044	2.003	1.985	1.976	1.972	1.970	1.969	1.970	1.970
20	2.029	1.989	1.972	1.963	1.959	1.957	1.956	1.956	1.957
21	2.015	1.977	1.959	1.951	1.947	1.944	1.944	1.944	1.944
22	2.003	1.965	1.948	1.940	1.936	1.933	1.932	1.932	1.933
23	1.991	1.955	1.938	1.930	1.925	1.923	1.922	1.922	1.922
24	1.981	1.945	1.929	1.920	1.916	1.914	1.913	1.912	1.913
25	1.971	1.936	1.920	1.912	1.907	1.905	1.904	1.904	1.904
26	1.962	1.928	1.912	1.904	1.899	1.897	1.896	1.895	1.895
27	1.954	1.920	1.904	1.896	1.892	1.889	1.888	1.888	1.888
28	1.946	1.913	1.897	1.889	1.885	1.882	1.881	1.880	1.880
29	1.938	1.906	1.891	1.883	1.878	1.876	1.874	1.874	1.874
30	1.932	1.899	1.885	1.877	1.872	1.870	1.868	1.868	1.867
40	1.880	1.852	1.839	1.831	1.827	1.824	1.822	1.821	1.821
50	1.848	1.823	1.810	1.803	1.799	1.796	1.794	1.792	1.792
60	1.825	1.802	1.790	1.784	1.779	1.776	1.774	1.773	1.772
70	1.808	1.787	1.776	1.769	1.765	1.762	1.760	1.758	1.757
90	1.784	1.765	1.755	1.749	1.745	1.742	1.740	1.739	1.738
100	1.776	1.758	1.748	1.742	1.738	1.735	1.733	1.732	1.731
300	1.714	1.703	1.697	1.693	1.690	1.688	1.687	1.685	1.684
500	1.697	1.688	1.683	1.680	1.678	1.676	1.675	1.674	1.673
1000	1.680	1.674	1.671	1.669	1.667	1.666	1.665	1.664	1.663

Table B6 continued

$p = 0.95, \ 1 - \alpha = 0.90$

					l				
n	2	3	4	5	6	7	8	9	10
2	7.699	5.982	5.320	4.974	4.764	4.623	4.522	4.447	4.390
3	4.613	4.109	3.894	3.779	3.708	3.661	3.628	3.604	3.586
4	3.800	3.521	3.401	3.337	3.299	3.274	3.258	3.247	3.239
5	3.409	3.216	3.134	3.091	3.066	3.050	3.041	3.035	3.031
6	3.172	3.025	2.961	2.929	2.911	2.900	2.894	2.890	2.889
7	3.012	2.891	2.839	2.813	2.799	2.791	2.787	2.784	2.784
8	2.896	2.792	2.747	2.725	2.714	2.707	2.704	2.702	2.702
9	2.806	2.714	2.675	2.656	2.646	2.640	2.638	2.637	2.637
10	2.735	2.652	2.617	2.600	2.591	2.586	2.584	2.583	2.583
11	2.678	2.601	2.569	2.553	2.545	2.540	2.538	2.538	2.538
12	2.629	2.559	2.528	2.513	2.506	2.502	2.500	2.499	2.500
13	2.588	2.522	2.494	2.480	2.472	2.468	2.467	2.466	2.467
14	2.553	2.491	2.464	2.450	2.443	2.439	2.438	2.437	2.438
15	2.523	2.463	2.437	2.424	2.417	2.414	2.412	2.412	2.412
16	2.496	2.439	2.414	2.401	2.395	2.391	2.390	2.389	2.389
17	2.472	2.417	2.393	2.381	2.374	2.371	2.369	2.369	2.369
18	2.450	2.398	2.375	2.363	2.356	2.353	2.351	2.351	2.351
19	2.431	2.380	2.358	2.346	2.340	2.336	2.335	2.334	2.334
20	2.414	2.365	2.343	2.331	2.325	2.321	2.320	2.319	2.319
21	2.398	2.350	2.329	2.317	2.311	2.308	2.306	2.305	2.305
22	2.383	2.337	2.316	2.305	2.299	2.295	2.293	2.293	2.293
23	2.370	2.325	2.304	2.293	2.287	2.284	2.282	2.281	2.281
24	2.358	2.314	2.293	2.283	2.277	2.273	2.271	2.270	2.270
25	2.346	2.303	2.283	2.273	2.267	2.263	2.261	2.260	2.260
26	2.336	2.294	2.274	2.264	2.258	2.254	2.252	2.251	2.251
27	2.326	2.285	2.265	2.255	2.249	2.246	2.244	2.242	2.242
28	2.317	2.276	2.257	2.247	2.241	2.238	2.236	2.234	2.234
29	2.308	2.268	2.250	2.240	2.234	2.230	2.228	2.227	2.226
30	2.300	2.261	2.242	2.232	2.227	2.223	2.221	2.220	2.219
40	2.239	2.206	2.189	2.180	2.174	2.171	2.168	2.167	2.166
50	2.201	2.171	2.156	2.147	2.142	2.138	2.135	2.134	2.133
60	2.174	2.147	2.133	2.124	2.119	2.115	2.113	2.111	2.110
70	2.154	2.129	2.115	2.107	2.102	2.099	2.096	2.094	2.093
80	2.139	2.115	2.102	2.094	2.089	2.086	2.083	2.082	2.080
100	2.116	2.094	2.083	2.075	2.071	2.067	2.065	2.063	2.062
300	2.042	2.029	2.022	2.017	2.014	2.011	2.010	2.008	2.007
500	2.022	2.012	2.006	2.002	2.000	1.998	1.996	1.995	1.994
1000	2.002	1.995	1.991	1.988	1.986	1.985	1.984	1.983	1.982

Table B6 continued

$p = 0.99, \ 1 - \alpha = 0.90$

					l				
n	2	3	4	5	6	7	8	9	10
2	9.744	7.493	6.614	6.150	5.865	5.671	5.532	5.427	5.345
3	5.888	5.199	4.898	4.730	4.624	4.551	4.498	4.459	4.428
4	4.879	4.488	4.312	4.213	4.151	4.109	4.078	4.056	4.039
5	4.395	4.122	3.997	3.928	3.884	3.855	3.835	3.820	3.809
6	4.104	3.892	3.795	3.741	3.708	3.686	3.670	3.659	3.652
7	3.906	3.731	3.651	3.607	3.580	3.562	3.550	3.542	3.536
8	3.762	3.612	3.543	3.506	3.483	3.468	3.458	3.451	3.446
9	3.652	3.519	3.459	3.425	3.405	3.392	3.384	3.378	3.374
10	3.564	3.445	3.390	3.360	3.342	3.331	3.323	3.318	3.315
11	3.492	3.383	3.333	3.306	3.289	3.279	3.272	3.268	3.265
12	3.432	3.331	3.285	3.259	3.244	3.235	3.229	3.225	3.222
13	3.381	3.287	3.243	3.220	3.205	3.197	3.191	3.187	3.185
14	3.337	3.249	3.208	3.185	3.172	3.163	3.158	3.154	3.152
15	3.299	3.215	3.176	3.155	3.142	3.134	3.129	3.126	3.124
16	3.265	3.185	3.148	3.128	3.115	3.108	3.103	3.100	3.098
17	3.235	3.159	3.123	3.103	3.092	3.084	3.080	3.077	3.075
18	3.208	3.135	3.100	3.082	3.070	3.063	3.059	3.056	3.054
19	3.184	3.113	3.080	3.062	3.051	3.044	3.039	3.037	3.035
20	3.162	3.094	3.062	3.044	3.033	3.026	3.022	3.019	3.018
21	3.142	3.076	3.045	3.027	3.017	3.010	3.006	3.003	3.002
22	3.124	3.060	3.029	3.012	3.002	2.996	2.991	2.989	2.987
23	3.107	3.044	3.015	2.998	2.988	2.982	2.978	2.975	2.973
24	3.091	3.030	3.002	2.985	2.975	2.969	2.965	2.963	2.961
25	3.077	3.018	2.989	2.973	2.964	2.957	2.953	2.951	2.949
26	3.063	3.005	2.978	2.962	2.953	2.947	2.943	2.940	2.938
27	3.051	2.994	2.967	2.952	2.942	2.936	2.932	2.930	2.928
28	3.039	2.984	2.957	2.942	2.933	2.927	2.923	2.920	2.918
29	3.028	2.974	2.948	2.933	2.924	2.918	2.914	2.911	2.909
30	3.018	2.965	2.939	2.924	2.915	2.909	2.905	2.903	2.901
40	2.940	2.895	2.872	2.859	2.851	2.845	2.842	2.839	2.837
50	2.891	2.850	2.830	2.818	2.810	2.805	2.801	2.799	2.797
60	2.856	2.819	2.800	2.789	2.782	2.777	2.773	2.770	2.768
70	2.830	2.796	2.778	2.768	2.761	2.756	2.752	2.749	2.747
80	2.810	2.778	2.761	2.751	2.744	2.739	2.736	2.733	2.731
100	2.780	2.751	2.736	2.727	2.720	2.716	2.712	2.710	2.708
300	2.684	2.667	2.657	2.651	2.647	2.643	2.641	2.639	2.637
500	2.657	2.644	2.636	2.631	2.628	2.625	2.623	2.621	2.620
1000	2.632	2.622	2.617	2.613	2.611	2.609	2.607	2.606	2.605

Table B6 continued

$p = 0.90, \ 1 - \alpha = 0.95$

					l				
n	2	3	4	5	6	7	8	9	10
2	9.515	6.711	5.703	5.195	4.890	4.689	4.546	4.439	4.357
3	4.851	4.135	3.838	3.679	3.581	3.515	3.468	3.434	3.408
4	3.780	3.403	3.242	3.155	3.103	3.068	3.044	3.027	3.015
5	3.291	3.039	2.930	2.872	2.837	2.815	2.801	2.791	2.784
6	3.005	2.814	2.732	2.689	2.663	2.648	2.638	2.631	2.627
7	2.815	2.660	2.593	2.559	2.539	2.527	2.519	2.515	2.512
8	2.679	2.547	2.490	2.461	2.444	2.434	2.429	2.425	2.423
9	2.575	2.460	2.410	2.384	2.370	2.361	2.356	2.354	2.353
10	2.494	2.390	2.345	2.322	2.309	2.302	2.298	2.295	2.294
11	2.428	2.333	2.292	2.271	2.259	2.252	2.249	2.247	2.246
12	2.373	2.286	2.247	2.228	2.217	2.210	2.207	2.205	2.205
13	2.327	2.245	2.209	2.191	2.180	2.175	2.171	2.170	2.169
14	2.287	2.211	2.177	2.159	2.149	2.143	2.140	2.139	2.138
15	2.253	2.180	2.148	2.131	2.121	2.116	2.113	2.111	2.111
16	2.223	2.154	2.123	2.106	2.097	2.092	2.089	2.087	2.086
17	2.196	2.130	2.100	2.084	2.075	2.070	2.067	2.066	2.065
18	2.173	2.109	2.080	2.065	2.056	2.051	2.048	2.046	2.045
19	2.152	2.090	2.062	2.047	2.038	2.033	2.030	2.029	2.028
20	2.132	2.073	2.046	2.031	2.023	2.018	2.015	2.013	2.012
21	2.115	2.057	2.031	2.016	2.008	2.003	2.000	1.998	1.997
22	2.099	2.043	2.017	2.003	1.995	1.990	1.987	1.985	1.984
23	2.084	2.030	2.005	1.991	1.983	1.978	1.975	1.973	1.972
24	2.071	2.018	1.993	1.980	1.971	1.967	1.963	1.962	1.961
25	2.058	2.007	1.983	1.969	1.961	1.956	1.953	1.951	1.950
26	2.047	1.997	1.973	1.959	1.951	1.947	1.943	1.942	1.940
27	2.036	1.987	1.963	1.950	1.943	1.938	1.934	1.933	1.931
28	2.026	1.978	1.955	1.942	1.934	1.929	1.926	1.924	1.923
29	2.017	1.970	1.947	1.934	1.926	1.921	1.918	1.916	1.915
30	2.008	1.962	1.939	1.927	1.919	1.914	1.911	1.909	1.908
40	1.943	1.903	1.883	1.872	1.865	1.860	1.857	1.854	1.853
50	1.902	1.866	1.848	1.838	1.831	1.826	1.823	1.821	1.819
60	1.873	1.841	1.824	1.814	1.808	1.803	1.800	1.797	1.796
70	1.852	1.822	1.806	1.797	1.791	1.786	1.783	1.781	1.779
80	1.835	1.807	1.793	1.784	1.777	1.773	1.770	1.768	1.766
100	1.811	1.786	1.773	1.764	1.759	1.754	1.751	1.749	1.747
300	1.732	1.718	1.710	1.705	1.701	1.698	1.696	1.694	1.693
500	1.711	1.700	1.693	1.689	1.686	1.684	1.682	1.681	1.679
1000	1.690	1.682	1.678	1.675	1.673	1.671	1.670	1.669	1.668

Table B6 continued

$p = 0.95, \ 1 - \alpha = 0.95$

					l				
n	2	3	4	5	6	7	8	9	10
2	11.050	7.732	6.535	5.927	5.561	5.318	5.144	5.015	4.914
3	5.663	4.795	4.430	4.231	4.106	4.021	3.960	3.914	3.879
4	4.431	3.967	3.764	3.651	3.581	3.534	3.500	3.475	3.456
5	3.870	3.555	3.416	3.339	3.291	3.259	3.237	3.221	3.209
6	3.542	3.303	3.196	3.137	3.101	3.078	3.061	3.050	3.042
7	3.324	3.129	3.042	2.995	2.966	2.947	2.934	2.925	2.919
8	3.167	3.002	2.927	2.887	2.863	2.847	2.837	2.830	2.825
9	3.048	2.903	2.838	2.803	2.781	2.768	2.759	2.753	2.749
10	2.954	2.825	2.766	2.734	2.715	2.703	2.696	2.690	2.687
11	2.878	2.760	2.707	2.678	2.660	2.650	2.643	2.638	2.635
12	2.815	2.706	2.657	2.630	2.614	2.604	2.597	2.593	2.591
13	2.761	2.660	2.614	2.589	2.574	2.565	2.559	2.555	2.552
14	2.716	2.621	2.577	2.553	2.539	2.530	2.525	2.521	2.519
15	2.676	2.586	2.545	2.522	2.509	2.500	2.495	2.491	2.489
16	2.641	2.556	2.516	2.495	2.482	2.473	2.468	2.465	2.463
17	2.610	2.529	2.491	2.470	2.457	2.450	2.445	2.441	2.439
18	2.583	2.505	2.468	2.448	2.436	2.428	2.423	2.420	2.418
19	2.558	2.483	2.448	2.428	2.416	2.409	2.404	2.401	2.399
20	2.536	2.463	2.429	2.410	2.398	2.391	2.386	2.383	2.381
21	2.516	2.445	2.412	2.393	2.382	2.375	2.370	2.367	2.365
22	2.497	2.429	2.396	2.378	2.367	2.360	2.356	2.353	2.351
23	2.480	2.414	2.382	2.364	2.354	2.347	2.342	2.339	2.337
24	2.464	2.400	2.369	2.351	2.341	2.334	2.330	2.327	2.325
25	2.450	2.387	2.357	2.340	2.329	2.322	2.318	2.315	2.313
26	2.436	2.375	2.345	2.328	2.318	2.311	2.307	2.304	2.302
27	2.424	2.364	2.335	2.318	2.308	2.301	2.297	2.294	2.292
28	2.412	2.353	2.325	2.309	2.298	2.292	2.288	2.285	2.283
29	2.401	2.343	2.315	2.299	2.290	2.283	2.279	2.276	2.274
30	2.390	2.334	2.307	2.291	2.281	2.275	2.270	2.267	2.265
40	2.314	2.266	2.242	2.228	2.219	2.213	2.208	2.205	2.203
50	2.265	2.223	2.201	2.188	2.180	2.174	2.170	2.167	2.164
60	2.231	2.193	2.173	2.161	2.153	2.147	2.143	2.140	2.138
70	2.206	2.170	2.152	2.140	2.133	2.127	2.123	2.120	2.118
80	2.186	2.153	2.136	2.125	2.117	2.112	2.108	2.105	2.103
100	2.157	2.128	2.112	2.102	2.095	2.090	2.086	2.083	2.081
300	2.064	2.047	2.038	2.031	2.027	2.023	2.021	2.019	2.017
500	2.039	2.025	2.018	2.013	2.009	2.006	2.004	2.003	2.001
1000	2.014	2.005	1.999	1.996	1.993	1.991	1.989	1.988	1.987

Table B6 continued

$p = 0.99, \ 1 - \alpha = 0.95$

				l					
n	2	3	4	5	6	7	8	9	10
2	13.978	9.675	8.113	7.316	6.834	6.511	6.279	6.105	5.969
3	7.220	6.057	5.558	5.281	5.105	4.983	4.894	4.826	4.773
4	5.681	5.046	4.759	4.596	4.492	4.419	4.366	4.325	4.293
5	4.982	4.546	4.345	4.230	4.156	4.104	4.067	4.038	4.016
6	4.574	4.240	4.084	3.995	3.937	3.897	3.868	3.847	3.830
7	4.303	4.030	3.902	3.828	3.781	3.748	3.725	3.707	3.694
8	4.108	3.876	3.766	3.703	3.663	3.635	3.615	3.600	3.589
9	3.961	3.757	3.660	3.605	3.569	3.545	3.528	3.515	3.505
10	3.844	3.662	3.575	3.525	3.493	3.472	3.456	3.445	3.436
11	3.749	3.583	3.504	3.459	3.430	3.410	3.396	3.386	3.378
12	3.670	3.518	3.445	3.403	3.376	3.358	3.345	3.336	3.329
13	3.604	3.462	3.394	3.355	3.330	3.313	3.301	3.292	3.286
14	3.547	3.414	3.350	3.313	3.289	3.274	3.262	3.254	3.248
15	3.497	3.372	3.311	3.276	3.254	3.239	3.228	3.221	3.215
16	3.453	3.334	3.277	3.243	3.222	3.208	3.198	3.191	3.185
17	3.415	3.301	3.246	3.214	3.194	3.180	3.171	3.164	3.159
18	3.380	3.271	3.219	3.188	3.169	3.155	3.146	3.140	3.135
19	3.349	3.245	3.194	3.164	3.145	3.133	3.124	3.117	3.113
20	3.320	3.220	3.171	3.143	3.124	3.112	3.103	3.097	3.093
21	3.295	3.198	3.150	3.123	3.105	3.093	3.085	3.079	3.074
22	3.271	3.178	3.131	3.105	3.087	3.076	3.068	3.062	3.057
23	3.250	3.159	3.114	3.088	3.071	3.060	3.052	3.046	3.042
24	3.230	3.141	3.098	3.072	3.056	3.045	3.037	3.031	3.027
25	3.211	3.125	3.083	3.058	3.042	3.031	3.023	3.018	3.014
26	3.194	3.110	3.069	3.044	3.029	3.018	3.011	3.005	3.001
27	3.178	3.096	3.056	3.032	3.016	3.006	2.999	2.993	2.989
28	3.163	3.083	3.043	3.020	3.005	2.995	2.987	2.982	2.978
29	3.149	3.071	3.032	3.009	2.994	2.984	2.977	2.972	2.968
30	3.136	3.060	3.021	2.999	2.984	2.974	2.967	2.962	2.958
40	3.037	2.973	2.940	2.921	2.908	2.899	2.892	2.888	2.884
50	2.975	2.918	2.889	2.871	2.859	2.851	2.845	2.840	2.837
60	2.931	2.880	2.853	2.836	2.825	2.817	2.812	2.807	2.804
70	2.898	2.851	2.826	2.810	2.800	2.793	2.787	2.783	2.780
80	2.872	2.829	2.805	2.790	2.780	2.773	2.768	2.764	2.761
100	2.835	2.796	2.774	2.761	2.752	2.745	2.740	2.736	2.733
300	2.713	2.690	2.678	2.669	2.663	2.659	2.655	2.653	2.650
500	2.679	2.662	2.652	2.645	2.640	2.637	2.634	2.632	2.630
1000	2.647	2.635	2.628	2.623	2.619	2.617	2.615	2.613	2.611

Table B6 continued

$p = 0.90, \; 1 - \alpha = 0.99$

				l					
n	2	3	4	5	6	7	8	9	10
2	21.528	11.793	8.877	7.530	6.764	6.272	5.931	5.680	5.488
3	7.544	5.718	5.009	4.636	4.408	4.253	4.142	4.059	3.994
4	5.219	4.350	3.986	3.788	3.665	3.581	3.520	3.475	3.440
5	4.280	3.728	3.490	3.359	3.276	3.220	3.180	3.151	3.128
6	3.766	3.365	3.188	3.091	3.030	2.989	2.959	2.938	2.921
7	3.439	3.123	2.983	2.905	2.856	2.824	2.801	2.784	2.772
8	3.211	2.949	2.832	2.767	2.726	2.700	2.681	2.667	2.658
9	3.042	2.817	2.715	2.659	2.625	2.602	2.586	2.575	2.566
10	2.911	2.712	2.623	2.573	2.543	2.522	2.509	2.499	2.492
11	2.806	2.628	2.547	2.502	2.475	2.457	2.445	2.436	2.430
12	2.720	2.558	2.483	2.442	2.418	2.401	2.390	2.382	2.377
13	2.648	2.498	2.430	2.392	2.369	2.353	2.343	2.336	2.331
14	2.587	2.448	2.383	2.348	2.326	2.312	2.302	2.296	2.291
15	2.534	2.403	2.343	2.310	2.289	2.275	2.266	2.260	2.256
16	2.488	2.365	2.307	2.276	2.256	2.243	2.235	2.229	2.224
17	2.448	2.331	2.276	2.245	2.227	2.214	2.206	2.201	2.197
18	2.412	2.300	2.248	2.218	2.200	2.189	2.181	2.175	2.171
19	2.380	2.273	2.223	2.194	2.177	2.165	2.158	2.152	2.149
20	2.351	2.248	2.200	2.172	2.155	2.144	2.137	2.132	2.128
21	2.325	2.226	2.179	2.152	2.136	2.125	2.118	2.113	2.109
22	2.301	2.205	2.160	2.134	2.118	2.107	2.100	2.095	2.092
23	2.279	2.187	2.142	2.117	2.101	2.091	2.084	2.079	2.076
24	2.259	2.169	2.126	2.101	2.086	2.076	2.069	2.064	2.061
25	2.240	2.153	2.111	2.087	2.072	2.062	2.055	2.051	2.047
26	2.223	2.139	2.097	2.074	2.059	2.049	2.042	2.038	2.034
27	2.207	2.125	2.084	2.061	2.047	2.037	2.031	2.026	2.023
28	2.192	2.112	2.072	2.050	2.035	2.026	2.019	2.015	2.011
29	2.179	2.100	2.061	2.039	2.025	2.015	2.009	2.004	2.001
30	2.166	2.089	2.051	2.029	2.015	2.006	1.999	1.995	1.991
40	2.070	2.006	1.973	1.954	1.941	1.933	1.927	1.922	1.919
50	2.010	1.954	1.925	1.907	1.896	1.888	1.882	1.878	1.874
60	1.969	1.918	1.891	1.875	1.864	1.857	1.851	1.847	1.844
70	1.938	1.892	1.867	1.852	1.841	1.834	1.829	1.825	1.821
80	1.914	1.871	1.848	1.834	1.824	1.817	1.812	1.808	1.804
100	1.880	1.841	1.820	1.807	1.798	1.792	1.787	1.783	1.780
300	1.768	1.747	1.735	1.727	1.721	1.717	1.714	1.711	1.709
500	1.738	1.722	1.712	1.706	1.702	1.698	1.695	1.693	1.691
1000	1.709	1.698	1.691	1.687	1.683	1.681	1.679	1.677	1.676

Table B6 continued

$p = 0.95, \ 1 - \alpha = 0.99$

n	2	3	4	5	6	7	8	9	10
2	24.993	13.576	10.156	8.575	7.675	7.096	6.693	6.396	6.169
3	8.796	6.616	5.764	5.313	5.034	4.844	4.707	4.604	4.523
4	6.106	5.055	4.609	4.364	4.209	4.102	4.024	3.966	3.920
5	5.021	4.348	4.052	3.886	3.780	3.707	3.654	3.614	3.583
6	4.429	3.936	3.714	3.589	3.509	3.454	3.414	3.384	3.361
7	4.052	3.662	3.484	3.383	3.319	3.275	3.243	3.219	3.201
8	3.788	3.464	3.315	3.231	3.177	3.140	3.113	3.094	3.079
9	3.593	3.314	3.185	3.112	3.065	3.034	3.011	2.994	2.981
10	3.442	3.196	3.082	3.017	2.976	2.948	2.928	2.913	2.902
11	3.321	3.100	2.997	2.938	2.901	2.876	2.858	2.845	2.835
12	3.222	3.021	2.926	2.873	2.839	2.816	2.799	2.787	2.778
13	3.138	2.953	2.866	2.816	2.785	2.764	2.748	2.737	2.729
14	3.068	2.896	2.814	2.767	2.738	2.718	2.704	2.694	2.686
15	3.007	2.845	2.769	2.725	2.697	2.678	2.665	2.656	2.648
16	2.953	2.801	2.729	2.687	2.661	2.643	2.631	2.622	2.615
17	2.906	2.762	2.693	2.654	2.629	2.612	2.600	2.591	2.585
18	2.865	2.727	2.662	2.624	2.600	2.583	2.572	2.564	2.557
19	2.827	2.696	2.633	2.596	2.573	2.558	2.547	2.539	2.533
20	2.794	2.668	2.607	2.572	2.549	2.534	2.524	2.516	2.510
21	2.763	2.642	2.583	2.549	2.528	2.513	2.503	2.495	2.490
22	2.735	2.618	2.562	2.529	2.508	2.493	2.483	2.476	2.471
23	2.710	2.597	2.542	2.510	2.489	2.475	2.465	2.458	2.453
24	2.686	2.577	2.523	2.492	2.472	2.459	2.449	2.442	2.437
25	2.665	2.559	2.506	2.476	2.456	2.443	2.434	2.427	2.422
26	2.645	2.542	2.491	2.461	2.442	2.429	2.420	2.413	2.408
27	2.626	2.526	2.476	2.447	2.428	2.415	2.406	2.400	2.395
28	2.609	2.511	2.462	2.434	2.415	2.403	2.394	2.387	2.383
29	2.592	2.497	2.449	2.421	2.403	2.391	2.382	2.376	2.371
30	2.577	2.484	2.437	2.410	2.392	2.380	2.371	2.365	2.360
40	2.465	2.387	2.348	2.324	2.308	2.297	2.290	2.284	2.280
50	2.394	2.326	2.291	2.270	2.256	2.246	2.238	2.233	2.229
60	2.345	2.284	2.252	2.233	2.219	2.210	2.203	2.198	2.194
70	2.309	2.253	2.224	2.205	2.193	2.184	2.177	2.172	2.168
80	2.281	2.229	2.201	2.184	2.172	2.164	2.157	2.152	2.148
100	2.239	2.194	2.169	2.153	2.142	2.134	2.128	2.124	2.120
300	2.107	2.082	2.067	2.058	2.051	2.046	2.042	2.039	2.036
500	2.071	2.052	2.040	2.033	2.028	2.023	2.020	2.018	2.015
1000	2.036	2.023	2.015	2.010	2.006	2.003	2.000	1.999	1.997

Table B6 continued

$$p = 0.99, \ 1 - \alpha = 0.99$$

					l				
n	2	3	4	5	6	7	8	9	10
2	31.602	16.968	12.587	10.561	9.405	8.661	8.142	7.759	7.465
3	11.197	8.334	7.206	6.604	6.229	5.973	5.786	5.644	5.533
4	7.810	6.407	5.802	5.465	5.249	5.099	4.988	4.903	4.836
5	6.448	5.538	5.128	4.895	4.744	4.638	4.560	4.499	4.452
6	5.705	5.032	4.722	4.543	4.427	4.345	4.284	4.237	4.200
7	5.232	4.697	4.446	4.300	4.205	4.138	4.088	4.050	4.020
8	4.903	4.456	4.244	4.121	4.039	3.982	3.940	3.908	3.883
9	4.658	4.273	4.089	3.981	3.910	3.860	3.824	3.796	3.773
10	4.469	4.129	3.965	3.869	3.806	3.762	3.729	3.704	3.684
11	4.317	4.012	3.864	3.777	3.720	3.680	3.650	3.627	3.609
12	4.192	3.914	3.779	3.699	3.647	3.610	3.583	3.562	3.546
13	4.088	3.832	3.707	3.633	3.585	3.550	3.525	3.506	3.491
14	3.999	3.761	3.644	3.575	3.530	3.498	3.475	3.457	3.443
15	3.922	3.700	3.590	3.526	3.483	3.453	3.430	3.414	3.400
16	3.855	3.645	3.542	3.481	3.440	3.412	3.391	3.375	3.363
17	3.796	3.597	3.499	3.441	3.402	3.375	3.355	3.340	3.329
18	3.743	3.554	3.461	3.405	3.368	3.343	3.324	3.309	3.298
19	3.696	3.516	3.427	3.373	3.338	3.313	3.295	3.281	3.270
20	3.654	3.481	3.395	3.343	3.309	3.286	3.268	3.255	3.244
21	3.615	3.449	3.366	3.316	3.284	3.261	3.244	3.231	3.221
22	3.580	3.420	3.340	3.292	3.260	3.238	3.221	3.209	3.199
23	3.547	3.394	3.316	3.269	3.238	3.217	3.201	3.189	3.180
24	3.517	3.369	3.293	3.248	3.218	3.197	3.182	3.170	3.161
25	3.490	3.346	3.272	3.228	3.199	3.179	3.164	3.153	3.144
26	3.464	3.324	3.253	3.210	3.182	3.162	3.147	3.136	3.128
27	3.441	3.305	3.235	3.193	3.166	3.146	3.132	3.121	3.113
28	3.419	3.286	3.218	3.177	3.150	3.131	3.117	3.107	3.098
29	3.398	3.269	3.202	3.162	3.136	3.117	3.104	3.093	3.085
30	3.378	3.252	3.187	3.148	3.122	3.104	3.091	3.081	3.073
40	3.235	3.130	3.076	3.043	3.021	3.005	2.994	2.985	2.978
50	3.144	3.053	3.005	2.976	2.956	2.942	2.932	2.923	2.917
60	3.080	2.999	2.956	2.929	2.911	2.898	2.888	2.881	2.875
70	3.033	2.959	2.919	2.894	2.877	2.865	2.856	2.849	2.843
80	2.996	2.928	2.891	2.867	2.851	2.840	2.831	2.824	2.819
100	2.942	2.882	2.849	2.828	2.813	2.803	2.795	2.788	2.783
300	2.769	2.736	2.717	2.704	2.695	2.689	2.683	2.679	2.676
500	2.722	2.696	2.681	2.672	2.665	2.659	2.655	2.652	2.649
1000	2.676	2.659	2.648	2.641	2.636	2.632	2.629	2.627	2.624

Table B7: Values of λ for the simultaneous tolerance factor $k_{1s}(d)$ in (3.2.18) with $m - m_1 = 1$ and $n^{-\frac{1}{2}} \le d \le \left(\frac{1+\tau^2}{n}\right)^{\frac{1}{2}}$

$1 - \alpha = 0.90$

	$p = 0.90$			$p = 0.95$			$p = 0.99$		
n	$\tau = 2$	$\tau = 3$	$\tau = 4$	$\tau = 2$	$\tau = 3$	$\tau = 4$	$\tau = 2$	$\tau = 3$	$\tau = 4$
4	2.1352	2.1553	2.1658	2.2370	2.2530	2.2615	2.3832	2.3948	2.4012
5	1.6448	1.6589	1.6664	1.7108	1.7222	1.7284	1.8062	1.8146	1.8194
6	1.4524	1.4641	1.4704	1.5041	I.5136	1.5189	1.5788	1.5859	1.5901
7	1.3505	1.3608	1.3665	1.3943	1.4027	1.4075	1.4576	1.4639	I.4677
8	1.2874	1.2968	1.3021	1.3261	1.3339	1.3383	1.3820	1.3878	1.3913
9	1.2444	1.2533	1.2582	1.2796	1.2868	1.2910	1.3302	1.3357	1.3390
10	1.2133	1.2216	1.2263	1.2458	1.2526	1.2566	1.2924	1.2976	1.3008
11	1.1897	1.1976	1.2021	1.2200	1.2266	1.2304	1.2635	1.2685	1.2715
12	1.1711	1.1787	1.1830	1.1997	1.2060	1.2097	1.2407	1.2454	1.2484
13	1.1561	1.1634	1.1676	1.1833	1.1893	1.1929	1.2221	1.2267	1.2296
14	1.1437	1.1508	1.1549	1.1697	1.1756	1.1790	1.2067	1,2i11	1.2139
15	1.1334	1.1402	1.1442	1.1583	1.1639	1.1673	1.1937	1.1980	1.2007
16	1.1245	1.1312	1.1350	1.1485	1.1540	1.1573	1.1826	1.1867	1.1894
17	1.1169	1.1234	1.1272	1.1400	1.1454	1.1487	1.1729	1.1769	1.1796
18	1.1102	1.1166	1.1203	1.1326	1.1379	1.1411	1.1644	1.1684	1.1709
19	1.1044	1.1106	1.1142	1.1261	1.1313	1.1344	1.1569	1.1608	1.1633
20	1.0992	1.1053	1.1088	1.1203	1.1254	1.1284	1.1502	1.1540	1.1565
21	1.0945	1.1005	1.1040	1.1151	1.1201	1.1231	1.1442	1.1479	1.1504
22	1.0903	1.0962	1.0996	1.1105	1.1153	1.1183	1.1388	1.1424	1.1448
23	1.0866	1.0923	1.0957	1.1062	1.1110	1.1139	1.1338	1.1374	1.1398
25	1.0800	1.0855	1.0888	1.0988	1.1034	1.1062	1.1252	1.1287	1.1309
30	1.0676	1.0728	1.0759	1.0848	1.0891	1.0917	1.1087	1.1119	1.1141
35	1.0590	1.0638	1.0667	1.0749	1.0789	1.0814	1.0969	1.0999	1.1020
40	1.0525	1.0571	1,0599	1.0674	1.0712	1.0736	1.0880	1.0909	1.0928
50	1.0435	1.0477	1.0503	1.0569	1.0604	1.0626	1.0753	1.0779	1.0797
55	1.0403	1.0443	1.0468	1.0530	1.0564	1.0585	1.0706	1.0731	1.0748
65	1.0352	1.0390	1.0413	1,0470	1.0501	1.0521	1.0631	1.0655	1.0671
75	1.0314	1.0350	1.0372	1.0425	1.0454	1.0473	1.0575	1.0597	1.0613
90	1.0273	1.0306	1.0327	1.0375	1.0402	1.0420	1.0512	1.0532	1.0547
120	1.0220	1.0250	1.0268	1.0309	1.0333	1.0349	1.0428	1.0446	1.0459
150	1.0187	1.0214	1.0231	1.0267	1.0289	1.0304	1.0374	1.0390	1.0402
200	1.0153	1.0177	1.0191	1.0223	1.0242	1.0255	1.0316	1.0330	1.0340
300	1.0116	1.0136	1.0148	1.0174	1.0190	1.0201	1.0250	1.0262	1.0271
400	1.0096	1.0113	1.0124	1.0146	1.0161	1.0170	1.0213	1.0223	1.0231
500	1.0083	1,0099	1.0109	1.0128	1 0141	1.0150	1.0188	1.0197	1.0204
1000	1.0054	1.0066	1.0073	1.0087	1.0096	1.0102	1.0129	1.0136	1.0141

Table B7 continued

$$1 - \alpha = 0.95$$

	$p = 0.90$			$p = 0.95$			$p = 0.99$		
n	$\tau = 2$	$\tau = 3$	$\tau = 4$	$\tau = 2$	$\tau = 3$	$\tau = 4$	$\tau = 2$	$\tau = 3$	$\tau = 4$
4	3.0774	3.1068	3.1222	3.2190	3.2424	3.2549	3.4237	3.4407	3.4502
5	2.1404	2.1594	2.1697	2.2202	2.2357	2.2443	2.3374	2.3488	2.3554
6	1.7990	1.8144	1.8229	1.8565	1.8692	1,8763	1.9417	1.9511	1.9567
7	1.6242	1.6377	1.6453	1.6702	1.6813	1.6878	1.7387	1.7470	1.7521
8	1.5179	1.5303	1.5374	1.5567	1.5670	1 5730	1.6150	1.6226	1.6274
9	1.4463	1.4579	1.4646	1.4802	1.4898	1.4956	1.5314	1.5386	1.5432
10	1.3946	1.4056	1.4121	1.4250	1.4341	1.4397	1.4710	1.4778	1.4822
11	1.3554	1.3660	1.3722	1.3831	1.3918	1.3972	1.4251	1.4316	1.4358
12	1.3246	1.3348	1.3409	1.3501	1.3585	1.3638	1.3889	1.3952	1.3993
13	1.2997	1.3096	1.3155	1.3234	1.3316	1.3367	1.3596	1.3657	1.3697
14	1.2791	1.2888	1.2946	1.3013	1.3093	1.3143	1.3353	1.3412	1.3452
15	1.2618	1.2712	1.2769	1.2827	1.2905	1.2954	1.3149	1.3206	1.3245
16	1.2470	1.2562	1.2618	1.2668	1.2744	1.2793	1.2973	1.3029	1.3068
17	1.2341	1.2431	1.2487	1.2530	1.2605	1.2653	1.2821	1.2876	1 2914
18	1.2229	1.2317	1.2372	1.2409	1.2482	1.2530	1.2688	1.2742	1.2779
19	1.2130	1.2217	1.2270	1.2303	1.2374	1.2421	1.2570	1.2623	1.2660
20	1.2041	1.2127	1.2180	1.2207	1.2278	1.2324	1.2465	1.2517	1.2553
21	1.1962	1.2046	1.2099	1.2122	1.2191	1.2237	1.2370	1.2421	1 2457
22	1.1890	1.1973	1.2025	1.2044	1.2113	1.2158	1.2285	1.2335	1.2370
23	1.1825	1.1907	1.1958	1.1974	1.2042	1.2086	1.2207	1.2256	1.2291
25	1.1711	1.1790	1.1841	1.1851	1.1917	1.1960	1.2071	1.2119	1.2153
30	1.1493	1.1568	1.1617	1.1616	1.1678	1.1720	1.1810	1.1855	1.1888
35	1.1337	1.1409	1.1456	1.1448	1.1506	1.1547	1.1623	1.1665	1.1697
40	1.1219	1.1288	1.1333	1.1320	1.1376	1.1415	1.1481	1.1521	1.1551
50	1.1050	1.1114	1.1157	1.1137	1.1189	1.1226	1.1276	1.1314	1.1342
55	1.0987	1.1049	1.1091	1.1069	1.1119	1.1155	1.1200	1.1236	1.1264
65	1.0887	1.0947	1.0986	1-0961	1.1009	1.1043	1.1080	1.1114	1.1140
75	1.0812	1.0868	1.0907	1.0879	1.0924	1.0957	1.0988	1.1020	1.1045
90	1.0726	1.0780	1.0816	1.0786	1.0829	1.0860	1.0884	1.0914	1.0938
120	1.0612	1.0661	1.0694	1.0662	1.0701	1.0729	1 0745	1 0772	1.0794
150	1.0538	1.0583	1.0614	1.0582	1.0617	1.0644	1.0655	1.0680	1.0699
200	1.0457	1.0497	1.0526	1.0494	1.0526	1.0550	1.0556	1.0578	1.0596
300	1.0366	1.0400	1.0424	1.0394	1.0421	1.0442	1.0444	1.0463	1.0478
400	1.0313	1.0343	1.0365	1.0337	1.0361	1.0379	1.0380	1.0396	1.0409
500	1.0277	1.0305	1.0326	1.0299	1.0321	1.0338	1.0337	1.0351	1.0364
1000	1.0192	1.0213	1.0229	1.0207	1.0223	1.0236	1.0233	1.0244	1.0753

Table B7 continued

$1 - \alpha = 0.99$

	$p = 0.90$			$p = 0.95$			$p = 0.99$		
n	$\tau = 2$	$\tau = 3$	$\tau = 4$	$\tau = 2$	$\tau = 3$	$\tau = 4$	$\tau = 2$	$\tau = 3$	$\tau = 4$
4	6.9820	7.0493	7.0847	7.2943	7.3482	7.3771	7.7484	7.7876	7.8096
5	3.7835	3.8184	3.8376	3.9139	3.9425	3.9586	4.1086	4.1298	4.1423
6	2.8229	2.8489	2.8637	2.9015	2.9230	2.9356	3.0218	3.0378	3.0477
7	2.3771	2.3993	2.4124	2.4323	2.4507	2.4619	2.5188	2.5325	2.5413
8	2.1219	2.1421	2.1542	2.1638	2.1806	2.1911	2.2312	2.2436	2.2518
9	1.9567	1.9756	1.9872	1.9900	2.0058	2.0159	2.0452	2.0568	2.0647
10	1.8407	1.8588	1.8702	1.8682	1.8832	1.8931	1.9148	1.9258	1.9335
11	1.7547	1.7722	1.7834	1.7777	1.7923	1.8020	1.8180	1.8286	1.8361
12	1.6881	1.7052	1.7163	1.7078	1.7220	1.7316	1.7432	1.7535	1.7609
13	1.6349	1.6517	1.6627	1.6520	1.6659	1.6754	1.6835	I.6935	1.7009
14	1.5913	1.6079	1.6189	1.6062	1.6199	1.6294	1.6346	1.6445	1.6517
15	1.5549	1.5713	1.5822	1.5680	1.5815	1.5910	1.5938	1.6035	1.6107
16	1.5240	1.5402	1.5511	1.5356	1.5489	1.5584	1.5592	1.5686	1.5759
17	1.4974	1.5134	1,5243	1.5076	1.5208	1.5303	1.5293	1.5387	1.5458
18	1.4741	1.4900	1.5010	1.4833	1.4963	1.5058	1.5033	1.5125	1.5197
19	1.4536	1.4694	1.4804	1.4618	1.4747	1.4842	1.4805	1.4895	1.4967
20	1.4354	1.4511	1.4621	1.4428	1.4556	1.4650	1.4601	1.4691	1.4762
21	1.4191	1.4347	1.4457	1.4257	1.4384	1.4479	1.4420	1.4508	1.4579
22	1.4044	1.4199	1.4309	1.4103	1.4229	1.4324	1.4256	1.4344	1.4414
23	1.3911	1.4065	1.4175	1.3964	1.4089	1.4184	1.4107	1.4194	1.4265
25	1.3678	1.3830	1.3941	1.3720	1.3844	1.3938	1.3848	1.3934	1.4003
30	1.3234	1.3383	1.3494	1.3257	1.3377	1.3472	1.3356	1.3438	1.3507
35	1.2916	1.3063	1.3174	1.2926	1.3043	1.3137	1.3005	1.3084	1.3152
40	1.2675	1.2819	1.2931	1.2675	1.2790	1.2884	1.2740	1.2817	1.2884
50	1.2328	1.2469	1.2580	1.2316	1.2427	1.2520	1.2361	1.2434	1.2500
55	1.2198	1.2337	1.2448	1.2182	1.2291	1.2383	1.2220	1.2292	1.2356
65	1.1992	1.2127	1.2238	1.1970	1.2076	1.2166	1.1998	1.2066	1.2129
90	1.1655	1.1784	1.1892	1.1625	1.1724	1.1811	1.1638	1.1700	1.1759
100	1.1561	1.1687	1.1795	1.1529	1.1626	1.1712	1.1538	1.1599	1.1657
120	1.1413	1.1535	1.1640	1.1379	1.1471	1.1555	1.1382	1.1440	1.1496
150	1.1253	1.1369	1.1472	1.1218	1.1305	1.1386	1.1216	1.1270	1.1323
200	1.1077	1.1185	1.1284	1.1041	1.1121	1.1197	1.1034	1.1083	1.1132
300	1.0873	1.0970	1.1061	1.0837	1.0909	1.0978	1.0828	1.0870	1.0914
400	1.0753	1.0842	1.0928	1.0720	1.0784	1.0849	1.0709	1.0747	1.0786
500	1.0672	1.0756	1.0837	1.0640	1.0700	1.0761	1.0629	1.0664	1.0701
1000	1.0474	1.0540	1.0607	1.0448	1.0494	1.0543	1.0437	1.0463	1.0492

Table B8: Values of λ for the simultaneous tolerance factor $k_{2s}(d)$ in (3.3.18) with $m = 2$ and $n^{-\frac{1}{2}} \le d \le \left(\frac{1+\tau^2}{n}\right)^{\frac{1}{2}}$

					$1 - \alpha = 0.90$				
	$p = 0.90$			$p = 0.95$			$p = 0.99$		
n	$\tau = 2$	$\tau = 3$	$\tau = 4$	$\tau = 2$	$\tau = 3$	$\tau = 4$	$\tau = 2$	$\tau = 3$	$\tau = 4$
4	2.3683	2.3915	2.4079	2.4688	2.4875	2.5010	2.6074	2.6217	2.6324
5	1.7710	1.7907	1.8049	1.8423	1.8580	1.8697	1.9400	1.9517	1.9608
6	1.5341	1.5524	1.5659	1.5936	1.6082	1.6192	1.6745	1.6853	1.6937
7	1.4069	1.4245	1.4376	1.4600	1.4739	1.4846	1.5316	1.5418	1.5499
8	1.3272	1.3442	1.3571	1.3762	1.3896	1.4001	1.4418	1.4515	1.4595
9	1.2723	1.2889	1.3016	1.3184	1.3315	1.3418	1.3798	1.3892	1.3970
10	1.2322	1.2483	1.2609	1.2760	1.2889	1.2990	1.3342	1.3434	1.3510
11	1.2014	1.2172	1.2296	1.2436	1.2561	1.2661	1.2991	1.3081	1.3156
12	1.1771	1.1925	1.2047	1.2178	1.2301	1.2400	1.2713	1.2801	1.2874
13	1.1573	1.1723	1.1844	1.1969	1.2088	1.2186	1.2486	1.2571	1.2644
14	1.1410	1.1556	1.1675	1.1794	1.1912	1.2009	1.2296	1.2380	1.2452
15	1.1272	1.1415	1.1532	1.1647	1.1762	1.1858	1.2136	1.2218	1.2289
16	1.1154	1.1293	1.1409	1.1521	1.1634	1.1728	1.1998	1.2079	1.2149
17	1.1053	1.1188	1.1303	1.1412	1.1522	1.1616	1.1878	1.1958	1.2027
18	1.0964	1.1096	1.1209	1.1317	1.1425	1.1517	1.1773	1.1851	1.1919
19	1.0886	1.1015	1.1126	1.1232	1.1338	1.1429	1.1680	1.1757	1.1824
20	1.0817	1.0943	1.1052	1.1157	1.1261	1.1350	1.1597	1.1672	1.1739
21	1.0755	1.0878	1.0985	1.1090	1.1191	1.1280	1.1522	1.1596	1.1662
22	1.0700	1.0819	1.0925	1.1030	1.1129	1.1216	1.1454	1.1527	1.1592
23	1.0650	1.0766	1.0871	1.0975	1.1072	1.1158	1.1393	1.1465	1.1529
24	1.0604	1.0718	1.0821	1.0925	1.1020	1.1105	1.1337	1.1407	1.1471
25	1.0563	1.0674	1.0775	1.0879	1.0973	1.1056	1.1285	1.1355	1.1417
27	1.0491	1.0597	1.0695	1.0799	1.0889	1.0970	1.1194	1.1261	1.1322
29	1.0431	1.0531	1.0625	1.0731	1.0817	1.0896	1.1115	1.1181	1.1241
30	1.0404	1.0502	1.0594	1.0700	1.0784	1.0862	1.1080	1.1144	1.1204
35	1.0295	1.0382	1.0467	1.0575	1.0651	1.0724	1.0934	1.0994	1.1050
40	1.0217	1.0294	1.0373	1.0483	1.0552	1.0621	1.0825	1.0880	1.0934
45	1.0159	1.0228	1.0300	1.0413	1.0476	1.0540	1.0740	1.0791	1.0842
60	1.0053	1.0102	1.0159	1.0278	1.0325	1.0378	1.0568	1.0610	1.0654
80	.9984	1.0017	1.0060	1.0182	1.0216	1.0258	1.0438	1.0471	1.0507
100	.9950	.9972	1.0005	1.0128	1.0153	1.0187	1.0359	1.0385	1.0416
150	.9915	.9925	.9944	1.0063	1.0075	1.0096	1.0253	1.0269	1.0290
200	.9905	.9911	.9921	1.0034	1.0041	1.0054	1.0199	1.0209	1.0224
300	.9903	.9905	.9909	1.0009	1.0012	1.0018	1.0144	1.0149	1.0158
400	.9906	.9907	.9909	.9998	1.0000	1.0003	1.0116	1.0119	1.0124
500	.9910	.9911	.9912	.9993	.9994	.9996	1.0099	1.0100	1.0103
1000	.9926	.9926	.9926	.9986	.9986	.9986	1.0061	1.0061	1.0062

Table B8 continued

$$1 - \alpha = 0.95$$

	$p = 0.90$			$p = 0.95$			$p = 0.99$		
n	$\tau = 2$	$\tau = 3$	$\tau = 4$	$\tau = 2$	$\tau = 3$	$\tau = 4$	$\tau = 2$	$\tau = 3$	$\tau = 4$
4	3.4009	3.4356	3.4602	3.5435	3.5714	3.5917	3.7403	3.7616	3.7777
5	2.2908	2.3184	2.3385	2.3808	2.4028	2.4192	2.5044	2.5207	2.5334
6	1.8860	1.9112	1.9300	1.9567	1.9767	1.9920	2.0532	2.0677	2.0794
7	1.6778	1.7018	1.7201	1.7385	1.7575	1.7723	1.8209	1.8345	1.8456
8	1.5506	1.5739	1.5919	1.6052	1.6236	1.6381	1.6788	1.6919	1.7027
9	1.4644	1.4872	1.5051	1.5149	1.5328	1.5472	1.5825	1.5951	1.6058
10	1.4020	1.4243	1.4420	1.4494	1.4670	1.4813	1.5126	1.5249	1.5354
11	1.3545	1.3764	1.3940	1.3996	1.4168	1.4310	1.4594	1.4714	1.4818
12	1.3171	1.3385	1.3560	1.3604	1.3773	1.3913	1.4174	1.4292	1.4395
13	1.2868	1.3078	1.3252	1.3285	1.3451	1.3591	1.3833	1.3949	1.4050
14	1.2617	1.2822	1.2995	1.3021	1.3184	1.3323	1.3350	1.3664	1.3765
15	1.2405	1.2607	1.2778	1.2799	1.2959	1.3096	1.3311	1.3423	1.3523
16	1.2224	1.2422	1.2591	1.2608	1.2766	1.2902	1.3107	1.3217	1.3315
17	1.2068	1.2261	1.2429	1.2443	1.2598	1.2733	1.2929	1.3038	1.3135
18	1.1931	1.2120	1.2286	1.2299	1.2450	1.2584	1.2773	1.2880	1.2977
19	1.1810	1.1995	1.2160	1.2171	1.2320	1.2452	1.2636	1.2741	1.2836
20	1.1703	1.1884	1.2046	1.2057	1.2203	1.2334	1.2513	1.2616	1.2711
21	1.1607	1.1784	1.1945	1.1955	1.2098	1.2228	1.2402	1.2504	1.2598
22	1.1521	1.1694	1.1852	1.1863	1.2004	1.2132	1.2302	1.2402	1.2495
23	1.1442	1.1612	1.1768	1.1780	1.1918	1.2045	1.2212	1.2310	1.2402
24	1.1371	1.1537	1.1691	1.1703	1.1839	1.1964	1.2129	1.2226	1.2317
25	1.1306	1.1468	1.1620	1.1634	1.1767	1.1891	1.2053	1.2148	1.2238
27	1.1192	1.1346	1.1495	1.1511	1.1638	1.1760	1.1918	1.2010	1.2099
29	1.1094	1.1241	1.1386	1.1405	1,1528	1.1647	1.1802	1.1892	1.1978
30	1.1051	1.1195	1.1337	1.1358	1.1479	1.1596	1.1750	1.1838	1.1924
35	1.0873	1.1001	1.1133	1.1163	1.1273	1.1383	1.1533	1.1616	1.1697
40	1.0742	1.0856	1.0980	1.1018	1.1118	1.1222	1.1370	1.1447	1.1525
45	1.0643	1.0745	1.0859	1.0906	1.0997	1.1095	1.1243	1.1314	1.1388
60	1.0451	1.0524	1.0616	1.0684	1.0753	1.0835	1.0983	1.1042	1.1106
80	1.0315	1.0364	1.0432	1.0520	1.0569	1.0633	1.0783	1.0829	1.0883
100	1.0238	1.0272	1.0324	1.0422	1.0458	1.0510	1.0660	1.0696	1.0742
150	1.0143	1.0158	1,0186	1.0294	1.0312	1.0343	1.0489	1.0510	1.0542
200	1.0099	1.0107	1.0123	1.0230	1.0240	1.0260	1.0399	1.0413	1.0435
300	1.0059	1.0061	1.0068	1.0166	1.0170	1.0179	1.0304	1.0310	1.0322
400	1.0040	1.0041	1.0044	1.0133	1.0315	1.0140	1.0252	1.0256	1.0263
500	1.0029	1.0029	1.0031	1.0113	1.0114	1.0116	1.0220	1.0221	1.0226
1000	1.0009	1.0009	1.0009	1.0069	1.0069	1.0070	1.0145	1.0145	1.0146

Table B9: Coefficients a_1, a_2, a_3 and a_4, for the $(p, 1-\alpha)$ tolerance factor (4.3.13) for a one-way random model involving a levels of a factor with n observations per level

$$p = 0.90, \ 1 - \alpha = 0.95$$

a	n	a_1	a_2	a_3	a_4	a	n	a_1	a_2	a_3	a_4
3	2	1.783	8.360	−10.672	6.773	7	2	1.845	−1.974	4.808	−1.924
3	3	1.355	2.839	2.725	−.763	7	3	1.711	−.911	2.926	−.970
3	4	1.369	1.499	5.960	−2.672	7	4	1.637	−.682	2.682	−.881
3	5	1.403	1.051	6.880	−3.179	7	5	1.598	−.609	2.671	−.905
3	6	1.444	.843	7.118	−3.250	7	6	1.569	−.553	2.660	−.920
3	7	1.450	.925	6.843	−3.063	7	7	1.547	−.504	2.637	−.924
3	8	1.442	.995	6.714	−2.995	7	8	1.530	−.465	2.622	−.931
3	9	1.443	.981	6.748	−3.016	7	9	1.515	−.426	2.589	−.923
3	10	1.426	1.195	6.275	−2.741	7	10	1.498	−.358	2.529	−.914
3	∞	1.255	1.960	5.233	−2.293	7	∞	1.287	.558	1.222	−.311
4	2	1.820	−1.036	5.548	−2.170	8	2	1.749	−1.136	2.979	−1.010
4	3	1.604	−.389	4.887	−1.940	8	3	1.660	−.668	2.260	−.670
4	4	1.559	−.286	4.848	−1.960	8	4	1.607	−.554	2.197	−.668
4	5	1.550	−.307	4.946	−2.028	8	5	1.578	−.530	2.254	−.721
4	6	1.542	−.305	4.986	−2.061	8	6	1.555	−.501	2.281	−.753
4	7	1.531	−.275	4.964	−2.059	8	7	1.536	−.466	2.279	−.767
4	8	1.520	−.241	4.934	−2.051	8	8	1.520	−.431	2.263	−.771
4	9	1.508	−.190	4.868	−2.024	8	9	1.506	−.394	2.236	−.766
4	10	1.484	−.077	4.702	−1.947	8	10	1.485	−.315	2.138	−.727
4	∞	1.281	.940	3.148	−1.208	8	∞	1.286	.520	.996	−.220
5	2	1.860	−1.878	5.814	−2.389	9	2	1.740	−1.068	2.640	−.859
5	3	1.710	−1.042	4.462	−1.723	9	3	1.651	−.611	1.943	−.529
5	4	1.635	−.743	4.074	−1.559	9	4	1.599	−.511	1.905	−.539
5	5	1.598	−.638	3.984	−1.537	9	5	1.569	−.490	1.970	−.596
5	6	1.574	−.575	3.939	−1.531	9	6	1.546	−.463	2.001	−.631
5	7	1.555	−.516	3.884	−1.516	9	7	1.527	−.429	2.003	−.647
5	8	1.539	−.464	3.833	−1.501	9	8	1.510	−.395	1.990	−.652
5	9	1.525	−.419	3.786	−1.485	9	9	1.494	−.347	1.949	−.642
5	10	1.502	−.317	3.645	−1.423	9	10	1.480	−.308	1.912	−.631
5	∞	1.286	.707	2.125	−.712	9	∞	1.286	.490	.837	−.159
6	2	1.861	−2.064	5.431	−2.222	10	2	1.730	−.992	2.343	−.727
6	3	1.721	−1.024	3.607	−1.298	10	3	1.640	−.556	1.689	−.418
6	4	1.644	−.747	3.271	−1.162	10	4	1.590	−.471	1.676	−.440
6	5	1.604	−.654	3.219	−1.163	10	5	1.560	−.452	1.748	−.501
6	6	1.577	−.591	3.186	−1.165	10	6	1.536	−.426	1.782	−.537
6	7	1.555	−.533	3.145	−1.160	10	7	1.517	−.395	1.789	−.556
6	8	1.537	−.486	3.107	−1.152	10	8	1.501	−.363	1.779	−.563
6	9	1.523	−−.447	3.077	−1.147	10	9	1.486	−.322	1.749	−.557
6	10	1.507	−.381	3.005	−1.119	10	10	1.475	−.301	1.740	−.560
6	∞	1.287	.613	1.564	−.457	10	∞	1.285	.466	.720	−.117

Table B9 continued

$p = 0.99, \ 1 - \alpha = 0.95$

a	n	a_1	a_2	a_3	a_4	a	n	a_1	a_2	a_3	a_4
3	2	3.105	4.815	2.357	.276	7	2	2.833	−.830	3.638	−1.000
3	3	2.554	2.311	9.725	−4.038	7	3	2.568	.265	2.138	−.330
3	4	2.543	2.021	10.472	−4.484	7	4	2.588	−.042	2.743	−.647
3	5	2.552	1.857	10.843	−4.699	7	5	2.616	−.336	3.300	−.939
3	6	2.558	1.743	11.104	−4.852	7	6	2.635	−.542	3.701	−1.152
3	7	2.555	1.719	11.184	−4.904	7	7	2.647	−.686	3.989	−1.307
3	8	2.550	1.717	11.214	−4.928	7	8	2.575	−.041	2.770	−.663
3	9	2.501	1.948	10.883	−4.778	7	9	2.575	−.421	3.664	−1.177
3	10	2.488	2.480	9.619	−4.014	7	10	2.562	−.265	3.317	−.972
3	∞	2.269	3.024	9.363	−4.104	7	∞	2.330	.461	2.720	−.824
4	2	2.933	−.544	7.263	−2.610	8	2	2.728	−.005	1.697	−.067
4	3	2.608	.125	6.989	−2.680	8	3	2.557	.465	1.219	.113
4	4	2.613	−.082	7.464	−2.952	8	4	2.576	.103	1.941	−.267
4	5	2.648	−.362	7.984	−3.228	8	5	2.595	−.174	2.493	−.561
4	6	2.671	−.543	8.324	−3.410	8	6	2.608	−.365	2.881	−.770
4	7	2.681	−.646	8.529	−3.523	8	7	2.617	−.503	3.165	−.925
4	8	2.622	−.391	8.201	−3.390	8	8	2.585	−.380	3.006	−.857
4	9	2.622	−.074	7.463	−2.970	8	9	2.624	−.671	3.528	−1.127
4	10	2.554	−.062	7.712	−3.162	8	10	2.506	.535	.822	.492
4	∞	2.310	1.071	6.064	−2.403	8	∞	2.330	.378	2.284	−.638
5	2	2.919	−1.441	6.702	−2.439	9	2	2.642	.758	−.035	.778
5	3	2.608	−.129	4.949	−1.687	9	3	2.551	.620	.527	.446
5	4	2.615	−.310	5.339	−1.902	9	4	2.566	.224	1.338	.014
5	5	2.655	−.617	5.890	−2.187	9	5	2.578	−.038	1.883	−.280
5	6	2.682	−.830	6.283	−2.393	9	6	2.585	−.213	2.256	−.484
5	7	2.696	−.961	6.535	−2.529	9	7	2.573	−.264	2.416	−.583
5	8	2.673	−.895	6.477	−2.514	9	8	2.592	−.435	2.737	−.751
5	9	2.606	−.314	5.540	−2.090	9	9	2.558	−.218	2.390	−.588
5	10	2.542	.192	4.248	−1.241	9	10	2.546	−.323	2.664	−.744
5	∞	2.323	.634	4.351	−1.567	9	∞	2.330	.354	1.968	−.509
6	2	2.905	−1.421	5.433	−1.856	10	2	2.593	1.357	−1.453	1.485
6	3	2.587	.036	3.356	−0.917	10	3	2.549	.735	.001	.697
6	4	2.601	−.198	3.827	−1.168	10	4	2.558	.320	.878	.225
6	5	2.637	−.505	4.387	−1.457	10	5	2.562	.075	1.412	−.068
6	6	2.662	−.723	4.794	−1.671	10	6	2.563	−.085	1.769	−.266
6	7	2.676	−.868	5.075	−1.822	10	7	2.549	−.073	1.792	−.287
6	8	2.556	−.075	3.805	−1.224	10	8	2.545	−.168	2.031	−.427
6	9	2.592	−.350	4.271	−1.451	10	9	2.544	−.171	2.090	−.482
6	10	2.616	−.695	4.918	−1.777	10	10	2.516	−.158	2.123	−.500
6	∞	2.328	.484	3.356	−1.106	10	∞	2.330	.336	1.730	−.415

Table B10: $(p, 1 - \alpha)$ lower tolerance factors for a two-parameter exponential distribution

| | | | $1 - \alpha = 0.90$ | | |
| | | | p | | |
n	0.8	0.9	0.95	0.99	0.999
2	−2.7000	−3.5500	−4.0125	−4.4005	−4.4900
3	−.4291	−.5667	−.6427	−.7050	−.7192
4	−.1500	−.2180	−.2530	−.2814	−.2879
5	−.0691	−.1118	−.1336	−.1512	−.1552
6	−.0354	−.0661	−.0817	−.0943	−.0972
7	−.0188	−.0426	−.0546	−.0644	−.0666
8	−.0096	−.0290	−.0388	−.0467	−.0485
9	−.0042	−.0205	−.0287	−.0354	−.0369
10	−.0000	−.0149	−.0220	−.0277	−.0290
11	.0010	−.0110	−.0173	−.0223	−.0234
12	.0028	−.0082	−.0138	−.0183	−.0193
13	.0037	−.0062	−.0112	−.0153	−.0162
14	.0044	−.0047	−.0093	−.0129	−.0137
15	.0048	−.0035	−.0077	−.0111	−.0118
16	.0051	−.0026	−.0065	−.0096	−.0103
17	.0052	−.0019	−.0055	−.0084	−.0090
18	.0053	−.0013	−.0047	−.0074	−.0080
19	.0054	−.0009	−.0040	−.0066	−.0071
20	.0054	−.0005	−.0035	−.0058	−.0064
21	.0054	−.0002	−.0030	−.0052	−.0058
22	.0053	.0000	−.0026	−.0047	−.0052
23	.0052	.0002	−.0023	−.0043	−.0047
24	.0052	.0004	−.0020	−.0039	−.0043
25	.0051	.0005	−.0017	−.0036	−.0040
26	.0050	.0007	−.0015	−.0033	−.0037
27	.0049	.0008	−.0013	−.0030	−.0034
28	.0049	.0008	−.0012	−.0028	−.0031
29	.0048	.0009	−.0010	−.0026	−.0029
30	.0047	.0010	−.0009	−.0024	−.0027
40	.0039	.0012	−.0002	−.0012	−.0015
50	.0033	.0012	.0001	−.0007	−.0009

			Table B10 continued		
			$1 - \alpha = 0.95$		
			p		
n	0.8	0.9	0.95	0.99	0.999
2	−5.9000	−7.6000	−8.5250	−9.3010	−9.4800
3	−.7333	−.9395	−1.0470	−1.1351	−1.1551
4	−.2540	−.3397	−.3837	−.4196	−.4277
5	−.1200	−.1708	−.1967	−.2177	−.2224
6	−.0655	−.1007	−.1186	−.1331	−.1364
7	−.0386	−.0653	−.0788	−.0898	−.0922
8	−.0236	−.0450	−.0558	−.0646	−.0665
9	−.0146	−.0324	−.0414	−.0487	−.0503
10	−.0089	−.0241	−.0318	−.0379	−.0393
11	−.0051	−.0183	−.0250	−.0304	−.0316
12	−.0024	−.0142	−.0201	−.0249	−.0260
13	.0001	−.0112	−.0165	−.0207	−.0217
14	.0007	−.0089	−.0137	−.0175	−.0184
15	.0017	−.0071	−.0115	−.0150	−.0158
16	.0023	−.0057	−.0098	−.0130	−.0137
17	.0029	−.0046	−.0083	−.0114	−.0120.
18	.0032	−.0037	−.0072	−.0100	−.0106
19	.0035	−.0030	−.0063	−.0089	−.0095
20	.0037	−.0024	−.0055	−.0079	−.0085
21	.0039	−.0019	−.0048	−.0071	−.0076
22	.0040	−.0015	−.0042	−.0064	−.0069
23	.0040	−.0011	−.0037	−.0058	−.0063
24	.0041	−.0009	−.0033	−.0053	−.0057
25	.0041	−.0006	−.0030	−.0048	−.0053
26	.0041	−.0004	−.0026	−.0044	−.0049
27	.0041	−.0002	−.0024	−.0041	−.0045
28	.0041	−.0000	−.0021	−.0038	−.0041
29	.0040	.0001	−.0019	−.0035	−.0039
30	.0040	.0002	−.0017	−.0032	−.0036
40	.0035	.0008	−.0006	−.0017	−.0020
50	.0031	.0009	−.0002	−.0010	−.0012

Table B11: $(p, 1 - \alpha)$ upper tolerance factors for a two-parameter exponential distribution

			$1 - \alpha = 0.90$		
			p		
n	0.8	0.9	0.95	0.99	0.999
2	10.636	17.092	23.650	38.933	60.798
3	2.4500	3.7333	5.0276	8.0452	12.370
4	1.2544	1.8760	2.5014	3.9580	6.0451
5	.8185	1.2123	1.6077	2.5282	3.8467
6	.5995	.8824	1.1662	1.8267	2.7725
7	.4694	.6880	.9073	1.4171	2.1472
8	.3840	.5611	.7386	1.1513	1.7421
9	.3239	.4721	.6206	.9659	1.4602
10	.2794	.4065	.5339	.8298	1.2535
11	.2453	.3563	.4676	.7260	1.0960
12	.2183	.3168	.4153	.6444	.9723
13	.1965	.2848	.3732	.5786	.8726
14	.1785	.2585	.3386	.5246	.7908
15	.1634	.2365	.3096	.4795	.7225
16	.1506	.2178	.2850	.4412	.6647
17	.1396	.2018	.2640	.4084	.6152
18	.1301	.1879	.2457	.3800	.5722
19	.1217	.1757	.2297	.3552	.5347
20	.1143	.1649	.2156	.3333	.5017
21	.1077	.1554	.2031	.3138	.4723
22	.1019	.1469	.1919	.2965	.4462
23	.0966	.1392	.1818	.2809	.4226
24	.0918	.1323	.1728	.2668	.4014
25	.0875	.1260	.1645	.2540	.3821
26	.0835	.1202	.1570	.2424	.3646
27	.0799	.1150	.1501	.2317	.3485
28	.0765	.1102	.1438	.2219	.3338
29	.0735	.1057	.1380	.2129	.3202
30	.0706	.1016	.1326	.2046	.3076
40	.0508	.0729	.0951	.1466	.2203
50	.0395	.0567	.0739	.1139	.1710

Table B11 continued

| | | | $1 - \alpha = 0.95$ | | |
| | | | p | | |
n	0.8	0.9	0.95	0.99	0.999
2	21.9360	35.1770	48.6310	80.0040	124.9040
3	3.6815	5.5984	7.5331	12.0459	18.5165
4	1.6961	2.5327	3.3747	5.3367	8.1490
5	1.0480	1.5502	2.0547	3.2296	4.9129
6	.7418	1.0907	1.4410	2.2561	3.4237
7	.5673	.8307	1.0950	1.7098	2.5903
8	.4559	.6657	.8760	1.3651	2.0655
9	.3793	.5526	.7262	1.1300	1.7080
10	.3237	.4706	.6179	.9603	1.4504
11	.2816	.4088	.5363	.8327	1.2569
12	.2487	.3607	.4728	.7335	1.1066
13	.2224	.3222	.4221	.6544	.9868
14	.2009	.2908	.3808	.5899	.8892
15	.1830	.2647	.3465	.5365	.8084
16	.1679	.2427	.3176	.4915	.7405
17	.1550	.2239	.2929	.4532	.6826
18	.1439	.2078	.2717	.4202	.6327
19	.1342	.1937	.2532	.3914	.5893
20	.1257	.1813	.2369	.3662	.5513
21	.1181	.1703	.2226	.3440	.5176
22	.1114	.1606	.2098	.3241	.4877
23	.1054	.1518	.1983	.3064	.4609
24	.0999	.1440	.1880	.2904	.4368
25	.0950	.1369	.1787	.2759	.4150
26	.0905	.1304	.1702	.2628	.3952
27	.0865	.1245	.1625	.2508	.3772
28	.0827	.1190	.1554	.2398	.3606
29	.0793	.1141	.1489	.2297	.3454
30	.0761	.1095	.1429	.2204	.3314
40	.0541	.077-7	.1013	.1562	.2347
50	.0418	.0600	.0782	.1204	.1809

Table B12: Values of $w_{p;1-\alpha}$ for constructing $(p, 1-\alpha)$ upper tolerance limit (7.5.10) for a Weibull distribution

| | $1-\alpha = 0.90$ | | | $1-\alpha = 0.95$ | | | $1-\alpha = 0.99$ | | |
| | p | | | p | | | p | | |
n	0.90	0.95	0.99	0.90	0.95	0.99	0.90	0.95	0.99
5	2.328	2.837	3.732	2.978	3.606	4.695	4.904	5.856	7.606
6	2.029	2.494	3.298	2.500	3.055	4.015	3.827	4.629	5.985
7	1.841	2.280	3.010	2.225	2.728	3.567	3.225	3.885	5.132
8	1.721	2.130	2.802	2.045	2.506	3.274	2.890	3.489	4.503
9	1.624	2.023	2.669	1.903	2.352	3.094	2.585	3.157	4.141
10	1.554	1.933	2.564	1.800	2.224	2.917	2.411	2.931	3.810
11	1.497	1.858	2.469	1.722	2.117	2.795	2.259	2.753	3.587
12	1.454	1.802	2.408	1.664	2.041	2.699	2.164	2.609	3.434
13	1.413	1.761	2.344	1.607	1.977	2.617	2.058	2.505	3.282
14	1.385	1.716	2.290	1.566	1.917	2.545	1.984	2.401	3.135
15	1.349	1.687	2.257	1.512	1.881	2.495	1.891	2.334	3.049
16	1.325	1.658	2.217	1.479	1.838	2.437	1.838	2.263	2.958
17	1.299	1.632	2.183	1.449	1.803	2.397	1.789	2.185	2.884
18	1.286	1.610	2.157	1.429	1.769	2.358	1.750	2.136	2.813
19	1.270	1.591	2.121	1.409	1.745	2.312	1.718	2.089	2.746
20	1.254	1.571	2.107	1.384	1.717	2.294	1.676	2.054	2.714
21	1.241	1.555	2.087	1.365	1.699	2.261	1.647	2.026	2.674
22	1.224	1.536	2.065	1.346	1.673	2.234	1.614	1.980	2.609
23	1.216	1.525	2.052	1.335	1.659	2.219	1.582	1.957	2.581
24	1.203	1.514	2.034	1.316	1.640	2.191	1.562	1.932	2.534
25	1.191	1.500	2.015	1.301	1.627	2.169	1.540	1.902	2.504
27	1.175	1.477	1.992	1.278	1.596	2.138	1.501	1.845	2.443
30	1.151	1.454	1.962	1.248	1.563	2.092	1.453	1.798	2.372
35	1.130	1.420	1.921	1.210	1.516	2.039	1.392	1.709	2.297
40	1.094	1.385	1.876	1.175	1.472	1.982	1.332	1.657	2.216
45	1.076	1.367	1.860	1.149	1.456	1.961	1.293	1.625	2.167
50	1.064	1.352	1.836	1.135	1.427	1.919	1.280	1.567	2.091
60	1.041	1.319	1.799	1.102	1.391	1.882	1.218	1.524	2.047
80	1.009	1.293	1.763	1.061	1.345	1.831	1.165	1.447	1.971
100	0.983	1.262	1.734	1.031	1.314	1.793	1.127	1.413	1.907
120	0.973	1.247	1.710	1.013	1.286	1.764	1.090	1.371	1.861
140	0.957	1.237	1.690	0.990	1.280	1.738	1.066	1.361	1.839
150	0.955	1.230	1.684	0.989	1.270	1.727	1.056	1.346	1.821
175	0.948	1.219	1.669	0.979	1.253	1.711	1.036	1.323	1.792
200	0.935	1.208	1.661	0.964	1.238	1.699	1.018	1.303	1.785
225	0.930	1.203	1.653	0.956	1.235	1.688	1.010	1.294	1.765
250	0.926	1.198	1.645	0.953	1.225	1.678	1.012	1.276	1.752

Table B13: Values of $w_{1-p;\alpha}$ for constructing the $(p, 1 - \alpha)$ lower tolerance limit (7.5.11) for a Weibull distribution

| | $1 - \alpha = 0.90$ | | | $1 - \alpha = 0.95$ | | | $1 - \alpha = 0.99$ | | |
| | p | | | p | | | p | | |
n	0.90	0.95	0.99	0.90	0.95	0.99	0.90	0.95	0.99
5	−5.411	−7.037	−10.719	−6.678	−8.695	−13.226	−10.437	−13.518	−20.667
6	−4.840	−6.255	−9.520	−5.850	−7.448	−11.425	−8.625	−10.943	−16.729
7	−4.440	−5.756	−8.767	−5.235	−6.782	−10.289	−7.411	−9.415	−14.257
8	−4.176	−5.405	−8.235	−4.871	−6.282	−9.544	−6.548	−8.508	−12.846
9	−3.975	−5.166	−7.869	−4.581	−5.920	−8.979	−6.064	−7.775	−11.735
10	−3.832	−4.962	−7.558	−4.370	−5.634	−8.560	−5.657	−7.266	−10.936
11	−3.714	−4.823	−7.333	−4.209	−5.445	−8.250	−5.373	−6.902	−10.442
12	−3.612	−4.682	−7.140	−4.064	−5.239	−7.967	−5.090	−6.513	−9.895
13	−3.525	−4.578	−6.964	−3.935	−5.102	−7.714	−4.876	−6.289	−9.546
14	−3.453	−4.490	−6.838	−3.827	−4.988	−7.545	−4.715	−6.073	−9.253
15	−3.396	−4.405	−6.721	−3.752	−4.879	−7.407	−4.577	−5.927	−8.902
16	−3.345	−4.343	−6.603	−3.685	−4.787	−7.235	−4.481	−5.757	−8.703
17	−3.291	−4.288	−6.521	−3.618	−4.705	−7.128	−4.346	−5.655	−8.501
18	−3.254	−4.221	−6.434	−3.575	−4.612	−7.015	−4.260	−5.480	−8.306
19	−3.217	−4.179	−6.374	−3.524	−4.554	−6.920	−4.173	−5.371	−8.137
20	−3.180	−4.131	−6.298	−3.460	−4.499	−6.833	−4.091	−5.255	−7.991
21	−3.145	−4.096	−6.247	−3.418	−4.446	−6.750	−4.013	−5.187	−7.849
22	−3.119	−4.057	−6.206	−3.387	−4.394	−6.699	−3.971	−5.136	−7.762
23	−3.092	−4.021	−6.157	−3.350	−4.349	−6.640	−3.913	−5.028	−7.644
24	−3.069	−3.992	−6.105	−3.320	−4.306	−6.562	−3.881	−4.988	−7.567
25	−3.049	−3.956	−6.051	−3.288	−4.257	−6.491	−3.809	−4.911	−7.462
26	−3.024	−3.933	−6.014	−3.255	−4.229	−6.443	−3.758	−4.865	−7.377
27	−3.006	−3.900	−5.970	−3.230	−4.186	−6.388	−3.719	−4.786	−7.289
28	−2.990	−3.890	−5.950	−3.213	−4.164	−6.355	−3.708	−4.736	−7.237
29	−2.972	−3.874	−5.913	−3.191	−4.146	−6.317	−3.646	−4.731	−7.141
30	−2.955	−3.854	−5.876	−3.169	−4.118	−6.262	−3.620	−4.678	−7.061
35	−2.898	−3.758	−5.761	−3.098	−4.002	−6.081	−3.483	−4.447	−6.775
40	−2.835	−3.693	−5.659	−3.013	−3.901	−5.965	−3.352	−4.406	−6.722
50	−2.749	−3.606	−5.536	−2.901	−3.787	−5.812	−3.219	−4.157	−6.331
60	−2.708	−3.537	−5.414	−2.845	−3.704	−5.646	−3.105	−4.022	−6.106
70	−2.663	−3.502	−5.372	−2.786	−3.653	−5.596	−3.033	−3.925	−6.025
90	−2.616	−3.414	−5.246	−2.724	−3.541	−5.438	−2.946	−3.824	−5.798
120	−2.553	−3.350	−5.144	−2.649	−3.451	−5.304	−2.809	−3.674	−5.598
150	−2.525	−3.297	−5.076	−2.598	−3.391	−5.215	−2.777	−3.580	−5.511
200	−2.480	−3.255	−5.007	−2.551	−3.342	−5.133	−2.696	−3.515	−5.373
250	−2.462	−3.216	−4.961	−2.518	−3.286	−5.070	−2.637	−3.421	−5.266

Table B14: Confidence limits for Weibull stress-strength reliability R; $\widehat{R} = $ MLE of R

95% lower confidence limits

$\widehat{R}$					n					
	8	10	12	15	18	20	25	30	40	50
.50	.300	.320	.335	.351	.365	.371	.384	.395	.408	.417
.52	.315	.334	.352	.370	.383	.391	.404	.412	.428	.437
.53	.322	.344	.360	.379	.391	.399	.412	.422	.438	.447
.55	.338	.362	.377	.396	.410	.418	.432	.442	.457	.467
.57	.354	.379	.394	.415	.428	.436	.451	.461	.477	.486
.59	.370	.394	.413	.433	.446	.454	.469	.481	.495	.507
.60	.379	.402	.423	.441	.457	.464	.479	.491	.506	.517
.61	.387	.412	.432	.452	.467	.476	.489	.500	.514	.526
.63	.406	.432	.449	.471	.486	.494	.509	.521	.536	.546
.65	.424	.450	.470	.490	.506	.513	.529	.541	.557	.568
.67	.441	.469	.488	.512	.526	.535	.549	.562	.576	.588
.69	.460	.489	.508	.532	.545	.556	.570	.582	.599	.609
.70	.471	.497	.521	.541	.557	.566	.582	.594	.610	.620
.71	.482	.510	.529	.553	.568	.577	.592	.604	.619	.630
.73	.498	.528	.549	.573	.589	.598	.613	.625	.641	.652
.75	.522	.552	.572	.595	.611	.620	.635	.648	.663	.674
.76	.530	.561	.584	.606	.622	.631	.647	.659	.674	.684
.77	.542	.573	.595	.616	.635	.642	.658	.670	.686	.696
.78	.553	.584	.607	.630	.644	.655	.671	.682	.698	.708
.79	.565	.597	.617	.641	.656	.667	.681	.693	.708	.718
.80	.575	.607	.629	.654	.669	.677	.694	.704	.720	.730
.81	.587	.621	.642	.664	.682	.690	.705	.716	.732	.741
.82	.601	.632	.654	.677	.693	.701	.718	.728	.743	.752
.84	.628	.658	.681	.704	.718	.727	.742	.752	.766	.777
.85	.640	.670	.693	.716	.733	.740	.754	.764	.779	.788
.86	.654	.684	.706	.730	.745	.752	.768	.777	.791	.800
.87	.666	.699	.721	.744	.759	.766	.780	.791	.804	.812
.89	.695	.728	.748	.771	.785	.794	.807	.816	.829	.838
.90	.716	.744	.765	.787	.801	.808	.821	.831	.843	.850
.91	.729	.761	.781	.801	.816	.823	.835	.845	.856	.863
.92	.748	.779	.798	.818	.831	.838	.850	.859	.869	.876
.93	.768	.796	.816	.835	.847	.853	.865	.873	.883	.889
.94	.790	.815	.834	.851	.863	.869	.881	.888	.897	.903
.95	.810	.835	.853	.870	.881	.886	.897	.904	.912	.917
.96	.834	.858	.873	.889	.899	.905	.913	.920	.927	.932
.97	.857	.881	.896	.910	.919	.924	.931	.936	.943	.947
.98	.891	.909	.922	.933	.940	.944	.950	.954	.959	.962

Table B14 continued

99% lower confidence limits

$\widehat{R}$	8	10	12	15	18	20	25	30	40	50
.50	.225	.254	.272	.295	.312	.323	.338	.352	.371	.384
.52	.239	.269	.285	.309	.328	.336	.356	.368	.391	.404
.53	.249	.273	.294	.318	.337	.346	.366	.379	.399	.412
.55	.265	.290	.311	.336	.353	.366	.383	.398	.419	.432
.57	.274	.308	.327	.351	.370	.379	.402	.415	.438	.452
.59	.289	.322	.343	.369	.391	.400	.420	.436	.455	.470
.60	.297	.331	.353	.380	.396	.410	.432	.443	.467	.482
.61	.306	.335	.361	.387	.408	.420	.442	.455	.476	.491
.63	.321	.354	.378	.406	.429	.439	.459	.478	.498	.514
.65	.334	.371	.396	.422	.445	.456	.481	.497	.516	.532
.67	.349	.387	.414	.444	.462	.478	.499	.516	.537	.553
.68	.361	.396	.423	.451	.477	.486	.509	.527	.548	.564
.69	.367	.406	.433	.460	.488	.496	.520	.537	.559	.574
.70	.375	.413	.439	.476	.497	.505	.529	.548	.571	.585
.71	.386	.423	.453	.485	.508	.517	.540	.558	.581	.595
.73	.404	.445	.474	.507	.530	.540	.560	.578	.603	.619
.75	.425	.462	.495	.525	.550	.565	.585	.603	.626	.639
.77	.446	.489	.517	.548	.571	.585	.609	.625	.648	.662
.78	.454	.500	.528	.561	.586	.596	.619	.635	.659	.676
.79	.468	.513	.537	.575	.594	.610	.635	.648	.672	.686
.80	.477	.520	.551	.585	.609	.620	.643	.661	.685	.697
.81	.488	.533	.566	.596	.620	.634	.655	.673	.696	.711
.82	.502	.542	.579	.609	.635	.646	.669	.685	.709	.721
.84	.527	.573	.602	.638	.661	.671	.697	.713	.734	.747
.85	.539	.586	.615	.651	.674	.687	.709	.724	.747	.759
.86	.557	.599	.630	.667	.688	.702	.723	.738	.759	.772
.87	.567	.619	.646	.678	.704	.716	.736	.753	.771	.785
.89	.596	.647	.678	.709	.732	.746	.767	.781	.800	.813
.90	.616	.662	.694	.726	.750	.760	.779	.794	.815	.826
.91	.633	.684	.713	.747	.766	.779	.797	.809	.829	.841
.92	.651	.698	.731	.763	.784	.793	.813	.825	.845	.854
.93	.674	.725	.752	.781	.804	.811	.832	.842	.860	.869
.94	.698	.739	.773	.804	.822	.829	.850	.860	.875	.884
.95	.727	.764	.795	.823	.841	.851	.868	.878	.892	.901
.96	.742	.791	.817	.846	.866	.872	.886	.897	.909	.917
.97	.779	.822	.848	.871	.886	.895	.907	.917	.928	.935
.98	.823	.858	.880	.902	.915	.922	.932	.939	.948	.953

Table B15: Values of n so that (a), (b), and (c) are $(p, 1-\alpha)$ nonparametric tolerance intervals

(a) $\left(X_{\left(\frac{m}{2}\right)}, X_{\left(n-\frac{m}{2}\right)}\right)$ if m is even; (b) $\left(X_{\left(\frac{m+1}{2}\right)}, X_{\left(n-\frac{m+1}{2}+1\right)}\right)$ if m is odd; (c) $(X_{(m)}, X_{(n)})$ for any m

	$p = 0.50$			$p = 0.75$			$p = 0.80$		
	$1 - \alpha$			$1 - \alpha$			$1 - \alpha$		
m	.90	.95	.99	.90	.95	.99	.90	.95	.99
1	7	8	11	15	18	24	18	22	31
2	9	11	14	20	23	31	25	30	39
3	12	13	17	25	29	37	32	37	47
4	14	16	19	30	34	43	38	44	55
5	17	18	22	35	40	49	45	50	62
6	19	21	25	40	45	54	51	57	69
7	21	23	27	45	50	60	57	63	76
8	24	26	30	50	55	65	63	69	83
9	26	28	33	55	60	70	69	76	89
10	28	30	35	59	65	76	75	82	96
11	31	33	38	64	70	81	81	88	102
12	33	35	40	69	74	86	86	94	109
13	35	37	42	73	79	91	92	100	115
14	37	40	45	78	84	96	98	106	122
15	39	42	47	82	89	101	104	112	128
16	42	44	50	87	93	106	109	118	134
17	44	47	52	91	98	111	115	124	141
18	46	49	54	96	103	116	121	129	147
19	48	51	57	100	107	121	126	135	153
20	51	53	59	105	112	126	132	141	159
21	53	56	62	109	117	131	138	147	165
22	55	58	64	114	121	136	143	153	171
23	57	60	66	118	126	141	149	158	177
24	59	62	69	123	130	145	154	164	184
25	62	65	71	127	135	150	160	170	190
26	64	67	73	132	139	155	166	176	196
27	66	69	76	136	144	160	171	181	202
28	68	71	78	141	149	164	177	187	208
29	70	74	80	145	153	169	182	193	213
30	72	76	82	149	158	174	188	198	219
31	75	78	85	154	162	179	193	204	225
32	77	80	87	158	167	183	199	210	231
33	79	82	89	163	171	188	204	215	237
34	81	85	92	167	176	193	210	221	243
35	83	87	94	171	180	197	215	227	249
36	85	89	96	176	185	202	221	232	255
37	87	91	98	180	189	207	226	238	261
38	90	93	101	184	193	211	232	243	266
39	92	96	103	189	198	216	237	249	272
40	94	98	105	193	202	221	243	255	278
41	96	100	107	197	207	225	248	260	284
42	98	102	110	202	211	230	253	266	290
43	100	104	112	206	216	235	259	271	296
44	102	106	114	210	220	239	264	277	301
45	105	109	116	215	225	244	270	282	307
46	107	111	119	219	229	248	275	288	313
47	109	113	121	223	233	253	281	293	319
48	111	115	123	228	238	257	286	299	324
49	113	117	125	232	242	262	291	304	330
50	115	119	128	236	247	267	297	310	336

Table B15 continued

m	p = 0.90 1 − α .90	.95	.99	p = 0.95 1 − α .90	.95	.99	p = 0.99 1 − α .90	.95	.99
1	38	46	64	77	93	130	388	473	662
2	52	61	81	105	124	165	531	628	838
3	65	76	97	132	153	198	667	773	1001
4	78	89	113	158	181	229	798	913	1157
5	91	103	127	184	208	259	926	1049	1307
6	104	116	142	209	234	288	1051	1182	1453
7	116	129	156	234	260	316	1175	1312	1596
8	128	142	170	258	286	344	1297	1441	1736
9	140	154	183	282	311	371	1418	1568	1874
10	152	167	197	306	336	398	1538	1693	2010
11	164	179	210	330	361	425	1658	1818	2144
12	175	191	223	353	386	451	1776	1941	2277
13	187	203	236	377	410	478	1893	2064	2409
14	199	215	249	400	434	504	2010	2185	2539
15	210	227	262	423	458	529	2127	2306	2669
16	222	239	275	446	482	555	2242	2426	2798
17	233	251	287	469	506	580	2358	2546	2925
18	245	263	300	492	530	606	2473	2665	3052
19	256	275	312	515	554	631	2587	2784	3179
20	267	286	325	538	577	656	2701	2902	3304
21	279	298	337	561	601	681	2815	3020	3429
22	290	310	350	583	624	706	2929	3137	3554
23	301	321	362	606	647	730	3042	3254	3678
24	312	333	374	628	671	755	3155	3371	3801
25	324	345	386	651	694	779	3268	3487	3924
26	335	356	398	673	717	804	3380	3603	4047
27	346	368	411	696	740	828	3492	3719	4169
28	357	379	423	718	763	852	3604	3834	4291
29	368	391	435	740	786	877	3716	3949	4412
30	379	402	447	763	809	901	3828	4064	4533
31	390	413	459	785	832	925	3939	4179	4654
32	402	425	471	807	855	949	4050	4293	4774
33	413	436	482	829	877	973	4162	4407	4894
34	424	447	494	851	900	997	4272	4521	5014
35	435	459	506	873	923	1020	4383	4635	5133
36	446	470	518	896	945	1044	4494	4749	5252
37	457	481	530	918	968	1068	4604	4862	5371
38	468	493	542	940	991	1091	4715	4975	5490
39	479	504	553	962	1013	1115	4825	5088	5608
40	490	515	565	984	1036	1139	4935	5201	5727
41	501	526	577	1006	1058	1162	5045	5314	5845
42	511	537	589	1027	1081	1186	5155	5427	5962
43	522	549	600	1049	1103	1209	5264	5539	6080
44	533	560	612	1071	1126	1233	15374	5651	6197
45	544	571	623	1093	1148	1256	15483	5764	6314
46	555	582	635	1115	1170	1279	15593	5876	6431
47	566	593	647	1137	1193	1303	15702	5988	6548
48	577	604	658	1159	1215	1326	15811	6099	6665
49	588	615	670	1180	1237	1349	15920	6211	6781
50	599	627	681	1202	1260	1372	16029	6323	6898

Table B16: Tolerance factors for q-variate normal distributions

$$q = 2$$

	$1 - \alpha = 0.90$			$1 - \alpha = 0.95$			$1 - \alpha = 0.99$		
		p			p			p	
n	0.90	0.95	0.99	0.90	0.95	0.99	0.90	0.95	0.99
5	41.61	57.10	95.62	66.81	93.49	155.4	199.6	289.4	476.9
6	26.00	36.18	60.57	38.33	53.36	89.96	90.53	124.8	209.2
7	19.88	27.23	45.17	27.27	37.59	62.77	54.45	76.82	125.1
8	16.31	22.36	36.89	21.35	29.46	48.77	37.93	52.76	87.86
9	14.16	19.36	31.56	17.94	24.93	40.53	30.06	41.70	69.66
10	12.69	17.20	28.30	15.78	21.40	34.93	24.51	34.26	57.48
11	11.63	15.67	25.73	14.12	19.15	31.65	21.13	29.49	48.88
12	10.82	14.61	23.82	12.88	17.52	28.85	18.77	25.87	43.43
13	10.19	13.75	22.28	12.01	16.31	26.54	17.09	23.44	38.56
14	9.70	13.02	21.09	11.35	15.27	24.83	15.53	21.55	35.25
15	9.30	12.45	20.18	10.77	14.53	23.63	14.52	20.02	32.92
16	8.91	12.03	19.24	10.23	13.86	22.35	13.55	18.52	30.51
17	8.63	11.56	18.57	9.89	13.27	21.44	12.90	17.56	28.50
18	8.41	11.22	17.95	9.53	12.75	20.61	12.31	16.79	27.25
19	8.17	10.92	17.48	9.24	12.37	19.90	11.72	16.04	26.30
20	7.99	10.67	17.02	8.98	12.08	19.33	11.33	15.35	25.18
21	7.81	10.43	16.62	8.71	11.72	18.76	11.03	14.82	24.13
22	7.66	10.21	16.33	8.55	11.38	18.30	10.71	14.34	23.22
23	7.54	10.01	15.94	8.34	11.17	17.82	10.31	13.93	22.50
24	7.40	9.86	15.65	8.20	10.91	17.45	10.04	13.55	21.89
25	7.29	9.72	15.39	8.02	10.70	17.13	9.74	13.17	21.03
26	7.20	9.56	15.11	7.92	10.56	16.78	9.61	12.81	20.72
27	7.11	9.43	14.94	7.81	10.38	16.50	9.35	12.60	20.13
28	7.03	9.31	14.75	7.67	10.22	16.25	9.18	12.38	19.72
30	6.88	9.09	14.37	7.49	9.96	15.77	8.89	11.84	19.07
32	6.74	8.92	14.10	7.32	9.70	15.36	8.59	11.43	18.34
35	6.58	8.71	13.70	7.09	9.40	14.84	8.29	10.99	17.61
37	6.48	8.57	13.50	6.99	9.25	14.58	8.11	10.66	17.14
40	6.36	8.40	13.18	6.84	9.00	14.26	7.84	10.38	16.55
45	6.20	8.16	12.79	6.63	8.72	13.68	7.50	9.96	15.71
50	6.07	7.99	12.47	6.45	8.49	13.33	7.27	9.59	15.11
60	5.87	7.72	12.03	6.21	8.14	12.74	6.89	9.08	14.27
70	5.74	7.52	11.69	6.02	7.90	12.31	6.64	8.72	13.66
80	5.62	7.37	11.47	5.89	7.72	11.99	6.42	8.44	13.14
90	5.55	7.26	11.25	5.79	7.58	11.77	6.29	8.23	12.83
100	5.48	7.16	11.10	5.70	7.46	11.60	6.15	8.07	12.56
150	5.27	6.87	10.62	5.43	7.10	10.98	5.77	7.55	11.68
200	5.15	6.72	10.37	5.30	6.91	10.67	5.57	7.29	11.25
300	5.03	6.56	10.10	5.14	6.71	10.33	5.36	6.99	10.77
500	4.92	6.41	9.86	5.00	6.51	10.03	5.17	6.72	10.35
10^3	4.81	6.27	9.64	4.87	6.34	9.75	4.98	6.49	9.98
∞	4.61	5.99	9.21	4.61	5.99	9.21	4.61	5.99	9.21

Table B16 continued

$q = 3$

| | $1 - \alpha = 0.90$ | | | $1 - \alpha = 0.95$ | | | $1 - \alpha = 0.99$ | | |
| | | p | | | p | | | p | |
n	0.90	0.95	0.99	0.90	0.95	0.99	0.90	0.95	0.99
7	44.05	60.13	99.03	62.21	87.59	146.6	143.4	200.5	345.3
8	31.36	42.92	70.71	42.31	58.47	97.34	84.29	114.9	196.1
9	25.27	34.44	55.51	32.75	44.62	73.80	58.13	80.07	134.7
10	21.53	28.98	46.75	26.76	36.77	59.42	43.66	60.54	99.65
11	18.94	25.33	40.49	23.11	31.06	50.45	35.71	49.24	81.87
12	17.07	22.72	36.37	20.49	27.50	44.04	30.16	41.42	68.30
13	15.75	20.87	33.13	18.64	24.78	39.68	26.43	35.62	58.05
14	14.71	19.38	30.49	17.03	22.79	36.29	23.60	31.87	52.30
15	13.90	18.29	28.70	15.88	21.25	33.67	21.50	29.06	47.21
16	13.19	17.34	27.05	15.12	19.88	31.40	20.06	26.78	42.99
17	12.70	16.59	25.74	14.29	18.92	29.79	18.68	24.67	39.85
18	12.19	15.91	24.75	13.77	17.98	28.29	17.50	23.17	37.69
19	11.81	15.35	23.76	13.16	17.29	27.03	16.72	22.04	34.92
20	11.44	14.89	22.93	12.76	16.61	25.92	15.80	21.02	33.10
21	11.15	14.53	22.22	12.35	16.16	24.93	15.16	20.13	31.72
22	10.90	14.12	21.68	12.01	15.62	24.19	14.61	19.31	30.58
23	10.65	13.75	21.14	11.68	15.23	23.42	14.12	18.57	29.34
24	10.46	13.44	20.60	11.46	14.87	22.82	13.70	18.00	28.10
25	10.28	13.20	20.16	11.20	14.49	22.24	13.36	17.44	27.23
26	10.07	13.01	19.75	10.95	14.17	21.68	12.95	17.16	26.56
27	9.91	12.77	19.42	10.79	13.91	21.32	12.75	16.58	25.70
28	9.78	12.62	19.08	10.60	13.65	20.90	12.43	16.18	25.10
29	9.65	12.39	18.74	10.43	13.43	20.48	12.17	15.83	24.62
30	9.54	12.24	18.56	10.29	13.26	20.08	11.92	15.51	23.97
32	9.33	11.93	18.08	10.03	12.85	19.60	11.49	14.88	23.07
34	9.13	11.73	17.59	9.79	12.55	19.05	11.20	14.47	22.24
35	9.07	11.58	17.40	9.68	12.43	18.78	11.06	14.28	21.84
37	8.93	11.38	17.09	9.48	12.18	18.32	10.72	13.88	21.12
40	8.73	11.08	16.63	9.23	11.84	17.83	10.42	13.42	20.43
45	8.47	10.77	16.04	8.94	11.39	17.10	9.99	12.79	19.35
50	8.26	10.49	15.61	8.70	11.07	16.52	9.63	12.25	18.48
60	7.95	10.07	14.93	8.32	10.59	15.70	9.10	11.59	17.27
70	7.75	9.81	14.49	8.08	10.22	15.13	8.75	11.11	16.46
80	7.60	9.58	14.12	7.90	9.98	14.71	8.48	10.77	15.92
90	7.48	9.42	13.87	7.75	9.77	14.39	8.30	10.46	15.48
100	7.39	9.29	13.65	7.64	9.62	14.13	8.14	10.27	15.13
150	7.09	8.90	13.02	7.28	9.14	13.37	7.66	9.62	14.09
200	6.93	8.70	12.70	7.08	8.89	12.99	7.40	9.28	13.57
300	6.77	8.48	12.36	6.89	8.64	12.59	7.14	8.94	13.02
500	6.63	8.29	12.07	6.72	8.40	12.24	6.90	8.64	12.55
10^3	6.50	8.13	11.81	6.56	8.21	11.92	6.68	8.36	12.15
∞	6.25	7.81	11.34	6.25	7.81	11.34	6.25	7.81	11.34

Table B16 continued

$$q = 4$$

n	$1-\alpha=0.90$			$1-\alpha=0.95$			$1-\alpha=0.99$		
	p			p			p		
	0.90	0.95	0.99	0.90	0.95	0.99	0.90	0.95	0.99
9	45.66	61.37	100.4	60.55	82.66	137.3	116.9	162.0	277.9
10	35.42	47.98	76.80	45.78	61.74	100.8	79.56	110.6	182.6
11	29.91	39.58	63.01	36.95	49.83	80.79	59.47	81.04	135.4
12	25.78	34.14	53.89	31.34	41.86	67.02	47.50	63.83	106.3
13	23.14	30.48	47.95	27.49	36.41	57.85	39.95	54.29	87.76
14	21.16	27.61	43.16	24.79	32.63	51.58	34.42	46.79	76.57
15	19.63	25.52	39.46	22.52	29.59	46.44	30.48	41.23	65.74
16	18.36	23.89	36.67	20.93	27.40	42.71	27.76	37.54	59.81
17	17.38	22.45	34.46	19.69	25.53	39.87	25.44	34.01	53.95
18	16.59	21.35	32.70	18.58	24.10	37.31	23.75	31.28	49.67
19	15.91	20.47	31.06	17.71	23.01	35.27	22.33	29.06	46.70
20	15.33	19.67	29.70	17.00	21.95	33.41	21.09	27.55	43.45
21	14.88	18.92	28.60	16.33	21.15	31.99	20.09	26.32	41.14
22	14.42	18.40	27.53	15.85	20.33	31.01	19.14	24.93	38.69
23	14.04	17.86	26.68	15.30	19.64	29.65	18.39	24.11	37.17
24	13.70	17.37	26.02	14.92	19.06	28.74	17.64	22.92	35.44
25	13.39	16.98	25.30	14.54	18.55	27.83	17.21	22.21	34.01
26	13.13	16.65	24.70	14.22	18.10	27.05	16.79	21.46	33.01
27	12.89	16.34	24.13	13.91	17.69	26.43	16.27	20.91	31.85
28	12.67	16.04	23.66	13.66	17.33	25.86	15.86	20.36	30.97
29	12.48	15.74	23.25	13.41	16.92	25.21	15.54	19.73	30.15
30	12.30	15.47	22.81	13.19	16.65	24.71	15.19	19.44	29.29
31	12.14	15.26	22.42	12.99	16.44	24.28	14.86	18.91	28.69
32	11.96	15.07	22.13	12.77	16.17	23.85	14.65	18.59	28.00
34	11.70	14.66	21.47	12.48	15.69	23.12	14.07	17.93	26.78
35	11.58	14.52	21.28	12.30	15.44	22.81	13.92	17.51	26.24
37	11.35	14.21	20.72	12.06	15.13	22.19	13.47	17.07	25.36
39	11.17	13.95	20.33	11.80	14.78	21.59	13.24	16.71	24.71
40	11.08	13.81	20.07	11.71	14.65	21.38	13.02	16.36	24.28
45	10.71	13.32	19.27	11.26	14.07	20.42	12.40	15.63	22.87
50	10.41	12.94	18.66	10.91	13.58	19.62	11.96	14.98	21.83
60	9.99	12.37	17.75	10.41	12.91	18.56	11.26	14.02	20.27
70	9.71	11.99	17.14	10.07	12.45	17.81	10.82	13.43	19.28
80	9.49	11.71	16.68	9.81	12.11	17.28	10.47	12.94	18.60
90	9.33	11.49	16.32	9.61	11.86	16.87	10.20	12.61	18.01
100	9.19	11.31	16.07	9.47	11.65	16.55	10.02	12.35	17.59
150	8.81	10.80	15.26	9.01	11.05	15.62	9.41	11.54	16.36
200	8.61	10.54	14.84	8.78	10.75	15.15	9.11	11.16	15.76
300	8.40	10.27	14.43	8.53	10.43	14.67	8.78	10.73	15.10
500	8.22	10.04	14.08	8.32	10.15	14.25	8.50	10.39	14.58
10^3	8.06	9.84	13.79	8.13	9.92	13.90	8.26	10.07	14.12
∞	7.78	9.49	13.28	7.78	9.49	13.28	7.78	9.49	13.28

Table B16 continued

$$q = 5$$

| | $1 - \alpha = 0.90$ | | | $1 - \alpha = 0.95$ | | | $1 - \alpha = 0.99$ | | |
| | p | | | p | | | p | | |
n	0.90	0.95	0.99	0.90	0.95	0.99	0.90	0.95	0.99
11	47.24	62.95	101.9	59.98	81.75	133.2	104.1	143.8	239.0
12	38.87	51.63	81.56	48.00	64.25	104.7	76.16	107.5	172.3
13	33.46	44.15	69.42	40.30	53.53	85.88	59.84	82.14	134.6
14	29.75	38.87	60.59	35.22	45.86	73.40	50.51	67.97	109.4
15	27.00	35.10	53.95	31.37	41.01	64.45	43.32	57.63	93.16
16	24.90	32.08	49.30	28.49	37.16	57.62	38.61	50.67	81.16
17	23.19	29.85	45.33	26.22	34.22	52.84	34.67	45.64	72.26
18	21.89	28.01	42.20	24.50	31.74	48.56	31.50	41.58	65.68
19	20.77	26.55	39.94	23.24	29.87	45.23	29.40	38.17	59.65
20	19.88	25.18	37.82	21.95	28.27	42.65	27.43	35.77	55.86
21	19.12	24.18	36.01	20.97	26.81	40.43	25.74	33.51	51.46
22	18.43	23.27	34.57	20.20	25.65	38.38	24.56	31.71	48.99
23	17.82	22.44	33.19	19.47	24.67	37.01	23.28	30.03	46.07
24	17.34	21.78	32.05	18.87	23.76	35.48	22.43	28.71	43.80
25	16.91	21.15	31.07	18.28	23.02	34.22	21.68	27.49	41.76
26	16.48	20.60	30.15	17.76	22.42	33.07	20.84	26.71	40.13
27	16.12	20.17	29.41	17.35	21.78	32.07	20.26	25.67	38.60
28	15.80	19.72	28.64	16.99	21.22	31.22	19.61	24.85	37.39
29	15.51	19.33	28.01	16.57	20.81	30.42	19.13	24.14	36.33
30	15.26	18.99	27.46	16.26	20.32	29.62	18.63	23.59	35.00
31	15.00	18.67	26.93	16.00	19.94	29.11	18.29	22.96	34.05
32	14.81	18.32	26.49	15.77	19.64	28.50	17.84	22.53	33.24
34	14.38	17.80	25.60	15.26	18.97	27.49	17.16	21.62	31.66
35	14.23	17.57	25.20	15.07	18.68	26.99	16.87	21.20	31.14
37	13.91	17.15	24.55	14.66	18.19	26.14	16.41	20.49	29.94
39	13.61	16.81	23.93	14.37	17.74	25.45	15.97	19.88	28.88
40	13.49	16.65	23.64	14.23	17.55	25.08	15.73	19.59	28.38
41	13.40	16.48	23.43	14.10	17.36	24.86	15.57	19.35	27.96
43	13.16	16.19	22.97	13.84	17.03	24.27	15.25	18.83	27.15
45	12.99	15.94	22.59	13.62	16.76	23.80	14.93	18.43	26.56
47	12.82	15.72	22.20	13.42	16.47	23.38	14.67	18.14	25.99
50	12.61	15.41	21.71	13.13	16.12	22.84	14.34	17.64	25.17
60	12.03	14.67	20.54	12.49	15.24	21.44	13.41	16.47	23.32
70	11.65	14.16	19.74	12.02	14.65	20.49	12.83	15.69	22.09
80	11.35	13.77	19.17	11.70	14.23	19.83	12.44	15.11	21.16
90	11.14	13.50	18.73	11.46	13.89	19.32	12.12	14.71	20.52
100	10.98	13.29	18.38	11.27	13.64	18.90	11.87	14.39	19.96
150	10.47	12.61	17.38	10.69	12.89	17.75	11.13	13.42	18.50
200	10.22	12.30	16.89	10.40	12.52	17.20	10.75	12.94	17.80
300	9.96	11.97	16.40	10.10	12.14	16.63	10.37	12.48	17.10
500	9.74	11.69	15.98	9.84	11.82	16.15	10.04	12.07	16.49
10^3	9.56	11.46	15.64	9.63	11.55	15.76	9.76	11.71	15.98
∞	9.24	11.07	15.09	9.24	11.07	15.09	9.24	11.07	15.09

Table B16 continued

$$q = 6$$

n	1 − α = 0.90 p			1 − α = 0.95 p			1 − α = 0.99 p		
	0.90	0.95	0.99	0.90	0.95	0.99	0.90	0.95	0.99
13	49.14	65.00	102.9	60.62	80.55	129.9	93.69	132.5	211.6
14	42.00	55.04	85.79	50.67	66.82	105.9	74.33	103.3	167.2
15	37.14	48.01	74.38	43.65	57.03	89.89	61.79	82.15	135.4
16	33.37	42.97	65.80	38.75	50.31	78.30	53.33	70.23	112.3
17	30.59	39.27	59.46	34.89	45.17	69.41	46.81	61.10	98.40
18	28.41	36.28	54.78	31.96	41.34	63.03	41.63	55.03	86.47
19	26.59	33.97	50.80	29.83	38.19	58.20	37.68	49.80	78.08
20	25.21	31.84	47.62	28.04	35.65	53.96	35.11	45.75	70.81
21	24.09	30.35	44.96	26.41	33.63	50.57	32.89	42.35	65.01
22	23.03	28.91	42.66	25.28	31.98	47.42	30.83	39.54	60.51
23	22.13	27.70	40.62	24.24	30.45	45.19	29.08	37.32	57.11
24	21.45	26.75	39.08	23.33	29.33	43.20	27.81	35.69	53.53
25	20.80	25.88	37.56	22.47	28.24	41.42	26.62	33.77	50.78
26	20.20	25.10	36.26	21.79	27.25	39.88	25.62	32.28	48.51
27	19.69	24.38	35.15	21.19	26.33	38.38	24.49	31.03	46.16
28	19.24	23.77	34.18	20.61	25.62	37.23	23.73	30.01	44.50
29	18.81	23.22	33.33	20.08	25.01	36.16	23.01	29.08	43.17
30	18.45	22.75	32.57	19.69	24.34	35.22	22.52	28.13	41.47
31	18.11	22.26	31.75	19.25	23.89	34.30	21.96	27.36	40.40
32	17.81	21.87	31.19	18.90	23.37	33.56	21.32	26.71	39.09
33	17.51	21.49	30.54	18.53	22.89	32.69	20.95	26.00	37.89
34	17.24	21.15	29.94	18.31	22.48	32.18	20.46	25.52	37.11
35	17.03	20.85	29.45	17.99	22.08	31.50	20.10	24.96	36.14
36	16.78	20.52	29.00	17.71	21.74	31.00	19.79	24.44	35.38
37	16.58	20.24	28.57	17.43	21.40	30.37	19.48	24.09	34.75
39	16.22	19.75	27.81	17.04	20.89	29.43	18.85	23.22	33.32
40	16.04	19.53	27.43	16.86	20.60	29.07	18.59	22.86	32.76
41	15.90	19.33	27.09	16.66	20.36	28.59	18.36	22.64	32.31
43	15.60	18.99	26.54	16.35	19.93	28.00	17.91	21.96	31.35
45	15.36	18.63	26.00	16.05	19.56	27.36	17.51	21.41	30.39
47	15.13	18.32	25.50	15.78	19.17	26.84	17.12	20.93	29.67
50	14.83	17.94	24.86	15.46	18.75	26.07	16.74	20.35	28.63
60	14.08	16.98	23.38	14.58	17.61	24.33	15.62	18.90	26.34
70	13.57	16.32	22.35	14.00	16.86	23.15	14.86	17.95	24.83
80	13.22	15.85	21.65	13.61	16.32	22.34	14.36	17.26	23.79
90	12.95	15.51	21.12	13.30	15.92	21.73	13.98	16.77	22.95
100	12.73	15.23	20.68	13.04	15.61	21.23	13.69	16.39	22.35
150	12.10	14.42	19.46	12.34	14.70	19.84	12.79	15.25	20.61
200	11.79	14.02	18.87	11.98	14.25	19.19	12.35	14.70	19.82
300	11.49	13.62	18.29	11.63	13.80	18.53	11.91	14.14	18.99
500	11.22	13.29	17.81	11.33	13.42	17.98	11.54	13.68	18.31
10^3	11.00	13.03	17.41	11.08	13.12	17.54	11.22	13.28	17.77
∞	10.64	13.03	16.81	10.64	13.03	16.81	10.64	13.03	16.81

Table B16 continued

$$q = 7$$

| | $1 - \alpha = 0.90$ | | | $1 - \alpha = 0.95$ | | | $1 - \alpha = 0.99$ | | |
| | p | | | p | | | p | | |
n	0.90	0.95	0.99	0.90	0.95	0.99	0.90	0.95	0.99
5	51.23	66.65	103.8	61.40	80.43	128.3	90.29	122.9	203.0
16	44.97	58.13	89.60	52.54	68.46	106.8	74.08	100.3	161.6
17	40.20	51.78	78.75	46.46	59.84	93.27	63.19	84.16	135.3
18	36.84	46.93	70.77	41.70	53.98	82.53	55.89	73.53	116.3
19	34.06	43.07	64.51	38.16	49.04	74.68	49.41	65.17	101.4
20	31.77	40.10	59.81	35.42	45.23	68.24	44.43	58.53	91.20
21	29.99	37.66	55.80	33.07	42.22	62.69	41.20	53.23	82.34
22	28.47	35.69	52.25	31.40	39.58	58.71	38.49	49.32	76.21
23	27.23	34.00	49.74	29.73	37.42	55.32	35.98	46.22	70.11
24	26.10	32.57	47.29	28.42	35.63	52.45	33.84	43.32	65.76
25	25.20	31.26	45.03	27.25	34.12	49.90	32.28	40.96	61.93
26	24.37	30.19	43.41	26.37	32.73	47.68	30.76	39.15	58.21
27	23.67	29.26	41.77	25.41	31.58	45.88	29.52	37.24	55.39
28	23.07	28.35	40.47	24.64	30.51	44.03	28.63	35.69	52.57
29	22.42	27.60	39.28	24.03	29.67	42.58	27.56	34.46	50.58
30	21.94	26.86	38.18	23.39	28.78	41.31	26.77	33.40	48.85
31	21.47	26.29	37.16	22.82	28.11	40.15	25.93	32.33	47.06
32	21.06	25.72	36.31	22.36	27.43	39.03	25.22	31.44	45.56
33	20.68	25.23	35.49	21.91	26.87	38.09	24.66	30.56	44.13
34	20.34	24.74	34.75	21.50	26.30	37.13	24.16	29.87	42.84
35	20.04	24.31	34.02	21.15	25.81	36.44	23.62	29.08	41.70
36	19.69	23.92	33.44	20.79	25.32	35.75	23.07	28.50	40.75
37	19.46	23.55	32.90	20.46	24.90	35.02	22.73	27.96	39.81
38	19.20	23.22	32.37	20.18	24.57	34.39	22.28	27.31	39.19
39	18.95	22.88	31.88	19.93	24.18	33.83	21.96	26.87	38.19
40	18.74	22.66	31.40	19.65	23.83	33.28	21.57	26.34	37.35
41	18.52	22.36	30.98	19.40	23.50	32.76	21.32	26.03	36.79
42	18.32	22.11	30.58	19.19	23.24	32.32	21.07	25.63	36.16
43	18.15	21.86	30.22	18.97	22.97	31.86	20.69	25.30	35.54
45	17.83	21.47	29.55	18.58	22.46	31.09	20.26	24.61	34.54
47	17.53	21.05	28.96	18.24	21.97	30.40	19.82	24.00	33.52
50	17.15	20.56	28.16	17.84	21.43	29.49	19.25	23.26	32.20
60	16.19	19.33	26.25	16.73	20.03	27.32	17.91	21.44	29.52
70	15.55	18.51	25.02	16.04	19.08	25.90	16.96	20.30	27.70
80	15.10	17.93	24.13	15.49	18.45	24.86	16.35	19.47	26.35
90	14.75	17.49	23.49	15.15	17.95	24.11	15.85	18.84	25.43
100	14.50	17.16	22.96	14.84	17.57	23.55	15.50	18.40	24.67
150	13.73	16.18	21.50	13.98	16.47	21.91	14.44	17.05	22.73
200	13.35	15.71	20.80	13.55	15.95	21.14	13.94	16.41	21.76
300	12.98	15.25	20.13	13.13	15.42	20.37	13.43	15.78	20.86
500	12.66	14.85	19.58	12.78	14.99	19.76	13.00	15.25	20.11
10^3	12.41	14.55	19.13	12.49	14.64	19.26	12.64	14.80	19.48
∞	12.02	14.07	18.48	12.02	14.07	18.48	12.02	14.07	18.48

Table B16 continued

$$q = 8$$

n	\(1-\alpha=0.90\) p			\(1-\alpha=0.95\) p			\(1-\alpha=0.99\) p		
	0.90	0.95	0.99	0.90	0.95	0.99	0.90	0.95	0.99
17	53.56	68.78	105.4	62.26	81.00	126.8	88.26	117.3	188.4
18	47.72	60.86	92.02	54.69	70.45	109.1	74.26	100.1	157.9
19	43.32	54.88	82.67	49.40	62.91	96.63	64.52	84.73	133.5
20	39.98	50.30	75.07	44.87	57.24	86.43	57.52	75.15	119.2
21	37.30	46.86	69.02	41.28	52.55	78.55	52.09	67.36	105.0
22	35.09	43.79	64.29	38.69	48.85	72.38	47.63	61.25	95.05
23	33.23	41.39	60.05	36.43	45.63	67.41	44.53	56.56	86.84
24	31.67	39.22	56.96	34.43	43.09	63.19	41.50	52.77	80.32
25	30.31	37.50	54.05	32.90	41.03	59.83	39.24	49.52	74.41
26	29.18	35.93	51.52	31.57	39.18	56.91	36.99	46.91	69.88
27	28.20	34.62	49.43	30.31	37.54	54.13	35.32	44.45	65.71
28	27.28	33.43	47.58	29.31	36.18	51.97	33.83	42.63	62.72
30	25.83	31.46	44.41	27.54	33.81	48.19	31.29	39.22	57.23
31	25.24	30.69	43.14	26.83	32.80	46.64	30.38	37.62	54.81
32	24.64	29.96	42.00	26.13	31.97	45.29	29.70	36.59	52.92
33	24.15	29.26	40.86	25.61	31.12	43.95	28.86	35.37	51.16
34	23.70	28.64	39.92	25.05	30.47	42.76	28.05	34.53	49.58
35	23.29	28.08	39.14	24.55	29.77	41.82	27.39	33.72	47.86
36	22.85	27.65	38.30	24.10	29.18	40.73	26.79	32.72	46.77
37	22.50	27.13	37.55	23.69	28.65	39.95	26.22	32.11	45.77
38	22.17	26.72	36.86	23.32	28.20	39.06	25.70	31.37	44.42
39	21.85	26.33	36.22	22.95	27.65	38.54	25.19	30.73	43.21
40	21.59	25.93	35.69	22.63	27.28	37.77	24.78	30.06	42.52
43	20.85	24.98	34.09	21.76	26.13	36.00	23.72	28.83	40.09
45	20.42	24.40	33.30	21.25	25.56	35.03	23.13	27.92	38.71
47	20.03	23.95	32.57	20.87	24.99	34.11	22.56	27.18	37.70
50	19.54	23.29	31.57	20.29	24.24	33.06	21.92	26.27	36.14
55	18.89	22.43	30.28	19.56	23.29	31.56	20.91	25.04	34.22
60	18.35	21.76	29.22	18.94	22.50	30.33	20.19	24.04	32.76
65	17.93	21.20	28.41	18.48	21.89	29.42	19.61	23.28	31.47
70	17.57	20.75	27.71	18.08	21.37	28.63	19.11	22.65	30.54
75	17.26	20.37	27.15	17.74	20.96	27.98	18.69	22.15	29.71
80	17.00	20.04	26.64	17.45	20.57	27.42	18.36	21.65	28.98
90	16.59	19.51	25.86	17.00	19.99	26.56	17.77	20.95	27.91
100	16.26	19.09	25.25	16.62	19.53	25.86	17.36	20.42	27.07
150	15.34	17.94	23.54	15.60	18.24	23.96	16.10	18.86	24.79
200	14.89	17.38	22.73	15.10	17.63	23.05	15.51	18.12	23.72
300	14.45	16.83	21.94	14.61	17.02	22.19	14.93	17.40	22.68
500	14.09	16.38	21.30	14.21	16.53	21.48	14.44	16.79	21.84
\(10^3\)	13.80	16.03	20.80	13.88	16.13	20.93	14.03	16.31	21.16
\(\infty\)	13.36	15.51	20.09	13.36	15.51	20.09	13.36	15.51	20.09

Table B16 continued

$q = 9$

| | $1 - \alpha = 0.90$ | | | $1 - \alpha = 0.95$ | | | $1 - \alpha = 0.99$ | | |
| | p | | | p | | | p | | |
n	0.90	0.95	0.99	0.90	0.95	0.99	0.90	0.95	0.99
19	55.67	70.96	106.8	63.69	82.04	125.8	86.61	114.4	180.6
20	50.24	63.78	95.05	56.96	73.48	110.6	75.73	97.05	154.4
21	46.36	58.07	86.46	51.65	65.86	98.97	66.83	86.55	134.7
22	42.98	53.99	78.89	47.77	60.58	89.66	59.96	77.31	120.5
23	40.37	50.38	73.29	44.48	55.82	82.70	54.64	70.10	107.3
24	38.22	47.28	68.60	41.67	52.48	76.86	50.42	64.76	97.92
25	36.27	44.88	64.33	39.55	49.00	71.96	47.30	59.88	90.97
26	34.74	42.74	61.19	37.58	46.73	67.61	44.44	56.33	83.92
27	33.31	40.97	58.11	35.96	44.43	63.92	42.21	53.11	79.11
28	32.24	39.28	55.79	34.47	42.57	61.20	40.00	50.23	73.83
30	30.22	36.75	51.55	32.17	39.49	56.10	36.92	45.73	66.73
31	29.38	35.61	49.89	31.18	38.10	53.98	35.58	43.91	63.68
32	28.59	34.61	48.32	30.38	36.98	52.33	34.62	42.32	61.02
33	27.97	33.78	47.05	29.58	35.92	50.46	33.33	41.05	58.62
34	27.35	32.99	45.67	28.90	35.06	48.99	32.47	39.81	56.91
35	26.77	32.26	44.61	28.28	34.12	47.70	31.53	38.71	54.71
36	26.30	31.60	43.50	27.69	33.42	46.36	30.76	37.67	53.28
37	25.81	31.02	42.60	27.10	32.77	45.37	30.10	36.67	51.75
38	25.40	30.46	41.83	26.63	32.14	44.40	29.48	35.85	50.25
39	25.00	29.95	40.97	26.20	31.56	43.43	28.92	35.07	49.07
40	24.62	29.46	40.29	25.78	30.97	42.61	28.35	34.32	47.93
43	23.68	28.24	38.37	24.71	29.55	40.46	26.92	32.48	44.90
45	23.14	27.57	37.30	24.11	28.80	39.12	26.09	31.41	43.45
47	22.66	26.93	36.37	23.57	28.08	38.12	25.48	30.55	41.99
50	22.08	26.15	35.16	22.91	27.20	36.76	24.62	29.46	40.33
53	21.55	25.50	34.13	22.32	26.48	35.61	23.92	28.42	38.76
55	21.24	25.07	33.60	21.94	26.04	34.93	23.51	27.92	37.79
57	20.97	24.76	33.05	21.65	25.60	34.36	23.06	27.42	37.06
60	20.58	24.27	32.31	21.24	25.06	33.48	22.55	26.76	36.11
65	20.06	23.59	31.27	20.67	24.32	32.39	21.87	25.80	34.63
70	19.62	23.03	30.50	20.17	23.71	31.46	21.22	25.04	33.40
75	19.27	22.57	29.81	19.77	23.18	30.70	20.80	24.45	32.49
80	18.95	22.19	29.21	19.41	22.77	30.01	20.39	23.93	31.72
90	18.44	21.55	28.30	18.86	22.06	28.98	19.72	23.10	30.49
100	18.07	21.05	27.57	18.45	21.53	28.21	19.20	22.43	29.49
150	16.95	19.68	25.55	17.23	20.01	26.00	17.76	20.64	26.89
200	16.43	19.04	24.62	16.65	19.30	24.98	17.10	19.79	25.65
300	15.91	18.40	23.72	16.08	18.61	23.97	16.41	18.98	24.46
500	15.50	17.89	23.00	15.62	18.04	23.18	15.86	18.31	23.54
10^3	15.17	17.49	22.44	15.25	17.59	22.57	15.40	17.78	22.80
∞	14.68	16.92	21.67	14.68	16.92	21.67	14.68	16.92	21.67

Table B16 continued

$$q = 10$$

| | $1 - \alpha = 0.90$ | | | $1 - \alpha = 0.95$ | | | $1 - \alpha = 0.99$ | | |
| | p | | | p | | | p | | |
n	0.90	0.95	0.99	0.90	0.95	0.99	0.90	0.95	0.99
21	57.86	72.95	108.9	65.22	83.56	127.5	84.98	111.5	176.3
22	53.09	66.50	98.19	59.10	74.95	112.5	75.72	98.22	151.6
23	49.24	61.46	89.83	54.51	68.75	102.1	68.13	87.52	135.3
24	46.06	57.17	82.67	50.67	63.55	93.82	62.31	79.21	121.5
25	43.42	53.57	77.21	47.31	59.19	86.36	57.51	72.93	109.7
26	41.24	50.75	72.42	44.72	55.61	80.65	53.74	67.53	101.3
27	39.34	48.26	68.67	42.52	52.61	75.81	50.40	62.77	94.31
28	37.69	46.13	65.21	40.64	49.96	71.61	47.62	59.54	87.26
30	35.07	42.62	59.78	37.52	45.89	64.96	43.03	53.44	77.57
31	33.98	41.18	57.64	36.23	44.13	62.34	41.17	51.44	73.88
32	33.06	39.94	55.43	35.07	42.61	60.21	39.95	48.87	70.83
33	32.13	38.76	53.67	34.07	41.39	57.67	38.58	47.12	67.81
34	31.35	37.81	52.20	33.20	40.18	55.91	37.24	45.75	65.22
35	30.68	36.80	50.72	32.38	39.05	54.25	36.13	44.15	62.70
36	29.99	35.94	49.48	31.60	38.06	52.86	35.16	43.01	60.35
37	29.47	35.20	48.18	30.90	37.20	51.39	34.33	41.78	58.66
38	28.87	34.50	47.10	30.34	36.37	50.24	33.41	40.69	57.05
39	28.35	33.88	46.14	29.75	35.69	48.94	32.72	39.68	55.19
40	27.92	33.26	45.22	29.20	34.89	47.92	32.03	38.69	53.68
43	26.71	31.77	42.92	27.89	33.22	45.17	30.43	36.43	50.30
45	26.05	30.88	41.57	27.13	32.26	43.69	29.38	35.21	48.29
47	25.47	30.13	40.47	26.46	31.38	42.37	28.62	34.00	46.69
50	24.71	29.16	38.97	25.60	30.33	40.69	27.50	32.76	44.46
53	24.05	28.35	37.70	24.91	29.39	39.29	26.59	31.65	42.70
55	23.70	27.86	37.01	24.51	28.82	38.55	26.12	30.98	41.61
57	23.35	27.45	36.34	24.13	28.39	37.77	25.68	30.36	40.77
60	22.91	26.87	35.50	23.63	27.74	36.77	25.07	29.69	39.61
63	22.50	26.36	34.77	23.16	27.19	36.00	24.52	28.83	38.49
65	22.26	26.03	34.30	22.90	26.81	35.46	24.23	28.52	37.89
67	22.06	25.77	33.92	22.66	26.51	35.00	23.92	28.10	37.32
70	21.74	25.39	33.35	22.31	26.09	34.34	23.51	27.58	36.51
75	21.31	24.84	32.51	21.85	25.52	33.44	22.89	26.83	35.48
80	20.94	24.38	31.82	21.44	24.97	32.70	22.45	26.24	34.53
90	20.33	23.61	30.74	20.78	24.16	31.49	21.66	25.27	33.02
100	19.87	23.06	29.91	20.26	23.54	30.54	21.06	24.50	31.91
150	18.56	21.44	27.58	18.85	21.79	28.04	19.42	22.43	28.91
200	17.96	20.69	26.49	18.20	20.96	26.85	18.63	21.49	27.54
300	17.36	19.96	25.47	17.54	20.16	25.74	17.88	20.56	26.26
500	16.89	19.39	24.67	17.02	19.53	24.86	17.26	19.82	25.22
10^3	16.52	18.94	24.05	16.60	19.03	24.18	16.76	19.23	24.42
∞	15.99	18.31	23.21	15.99	18.31	23.21	15.99	18.31	23.21

Table B17: Two-sided $(p, 1-\alpha)$ tolerance intervals for $N(\mu, \sigma^2)$: sample size requirement to cover no more than $p + \delta$ of the distribution with probability $1 - \alpha'$

	$p = 0.75$								
	$1 - \alpha = 0.99$			$1 - \alpha = 0.95$			$1 - \alpha = 0.90$		
	Values of $1 - \alpha'$								
δ	0.99	0.95	0.90	0.99	0.95	0.90	0.99	0.95	0.90
0.240	20	16	15	13	11	10	11	9	8
0.230	25	20	18	16	l4	12	14	11	9
0.220	30	24	21	19	16	14	17	13	11
0.210	36	28	25	22	19	16	20	15	13
0.200	42	33	28	25	22	19	24	17	14
0.190	48	38	33	29	25	21	28	20	16
0.180	56	44	38	34	29	25	32	23	19
0.170	65	50	43	39	34	28	37	27	22
0.160	76	58	50	45	39	32	43	31	25
0.150	88	68	58	53	46	38	51	36	29
0.140	104	79	67	62	53	44	60	42	33
0.130	123	93	79	74	63	52	71	49	39
0.120	147	111	94	88	75	61	85	59	47
0.110	178	134	113	107	90	74	103	71	56
0.100	219	164	138	131	111	90	127	87	68
0.090	274	205	172	164	139	112	160	108	85
0.080	352	262	220	210	177	143	206	139	109
0.070	465	345	289	278	234	188	273	183	143
0.065	542	402	336	324	273	219	319	214	167
0.060	640	474	396	383	322	258	377	252	196
0.055	766	566	472	458	384	308	451	302	235
0.050	932	688	573	557	467	374	550	367	285
0.045	1156	852	709	691	579	463	683	455	353
0.040	1470	1082	900	1061	879	736	587	706	579
0.035	1928	1418	1178	1394	966	769	1144	760	587
0.030	2636	1935	1606	1907	1319	1050	1567	1039	802
0.028	3031	2224	1845	2194	1517	1207	1803	1195	922
0.026	3522	2583	2141	2550	1762	1401	2096	1389	1071
0.024	4140	3034	2515	2999	2071	1646	2466	1633	1258
0.022	4935	3615	2995	3577	2469	1962	2942	1946	1500
0.020	5981	4379	3627	4337	2991	2376	3568	2359	1817
0.018	7395	5412	4481	5365	3699	2937	4415	2918	2246
0.016	9374	6856	5675	6803	4688	3721	5601	3700	2847

Table B17 continued

$$p = 0.90$$

	$1 - \alpha = 0.99$			$1 - \alpha = 0.95$			$1 - \alpha = 0.90$		
				Values of $1 - \alpha'$					
δ	0.99	0.95	0.90	0.99	0.95	0.90	0.99	0.95	0.90
0.095	41	33	28	29	22	19	24	17	14
0.090	57	45	38	41	30	25	33	24	19
0.085	74	57	49	53	39	32	43	30	24
0.080	94	72	61	66	48	40	54	38	30
0.075	117	89	75	83	60	49	67	47	37
0.070	145	109	93	103	74	60	84	58	46
0.065	179	135	114	128	91	74	104	71	56
0.060	223	167	141	159	113	92	130	88	70
0.055	280	210	176	200	142	114	163	111	87
0.050	356	265	222	255	179	145	208	141	110
0.045	460	341	286	330	231	186	269	181	142
0.040	606	449	375	436	304	244	356	239	186
0.035	823	608	507	592	413	330	485	324	252
0.030	1160	856	712	837	582	465	686	457	354
0.028	1351	995	827	975	677	540	799	532	412
0.026	1587	1168	971	1146	795	634	941	626	484
0.024	1887	1388	1153	1364	945	753	1120	744	575
0.022	2275	1671	1387	1645	1139	907	1351	897	693
0.020	2787	2046	1697	2017	1395	1110	1657	1099	848
0.018	3483	2556	2118	2522	1743	1386	2073	1374	1059
0.016	4461	3269	2709	3233	2232	1774	2658	1760	1356
0.014	5896	4317	3576	4275	2949	2342	3517	2326	1791
0.012	8117	5939	4916	5889	4059	3223	4848	3203	2465

Table B17 continued

$$p = 0.95$$

	$1 - \alpha = 0.99$			$1 - \alpha = 0.95$			$1 - \alpha = 0.90$		
	Values of $1 - \alpha'$								
δ	.99	.95	.90	.99	.95	.90	.99	.95	
0.048	56	43	37	39	29	24	32	23	19
0.044	98	75	64	70	51	42	57	40	32
0.040	149	112	95	106	76	62	86	59	47
0.036	215	162	136	154	109	89	125	85	67
0.032	309	231	194	222	156	126	181	122	96
0.028	450	334	280	323	226	182	264	177	139
0.024	671	497	414	482	337	270	395	264	206
0.022	832	615	513	599	418	334	491	328	255
0.020	1047	773	643	755	525	420	619	413	320
0.018	1342	988	822	968	672	537	794	529	409
0.016	1759	1294	1075	1271	881	702	1043	693	536
0.014	2376	1745	1449	1718	1189	947	1411	937	723
0.012	3339	2449	2031	2418	1671	1329	1987	1317	1015
0.010	4958	3632	3009	3593	2480	1971	2956	1956	1507
0.008	7978	5837	4833	5788	3990	3168	4764	3148	2423

Table B18: Two-sided β−expectation tolerance intervals for $N(\mu, \sigma^2)$: minimum sample size requirement to cover no more than $\beta \pm \delta$ of the distribution with probability at least $1 - \alpha'$

| | $\beta = 0.95$ | | | | | $\beta = 0.975$ | | | |
| | $1 - \alpha'$ | | | | | $1 - \alpha'$ | | | |
δ	.99	.975	.950	.90	δ	.99	.975	.950	.90
.0450	109	75	50	28	.0240	160	109	72	37
.0400	133	92	62	38	.0200	216	148	99	59
.0300	217	153	110	73	.0160	315	219	154	100
.0250	302	217	159	108	.0120	526	377	276	187
.0200	457	335	250	172	.0100	738	537	399	274
.0150	795	590	445	310	.0090	901	661	493	341
.0100	1,762	1,323	1,006	705	.0080	1,129	834	626	434
.0090	2,170	1,632	1,242	872	.0070	1,462	1,086	819	570
.0080	2,741	2,064	1.573	1,104	.0060	1,976	1,475	1,117	780
.0070	3,574	2,695	2,055	1,444	.0050	2,829	2,121	1,611	1,128
.0060	4,857	3,666	2,798	1,967	.0040	4,399	3,310	2,520	1,769
.0050	6,985	5,278	4,030	2,835	.0030	7,792	5,879	4,485	3,152

| | $\beta = 0.75$ | | | | | $\beta = 0.90$ | | | |
| | $1 - \alpha'$ | | | | | $1 - \alpha'$ | | | |
δ	.99	.975	.950	.90	δ	.99	.975	.950	.90
.2400	16	12	9	6	.0950	54	37	26	15
.2200	18	14	10	7	.0900	59	41	28	17
.2000	22	16	12	9	.0800	71	50	35	22
.1800	26	19	15	11	.0700	89	63	45	30
.1600	32	24	18	13	.0600	117	84	61	42
.1400	41	31	23	17	.0500	163	119	88	61
.1200	54	41	31	23	.0400	249	184	138	96
.1000	75	58	45	32	.0300	251	324	245	172
.0800	119	90	69	49	.0250	621	465	353	248
.0600	209	158	121	86	.0200	964	725	552	388
.0500	300	227	174	123	.0150	1,707	1,288	982	691
.0400	468	351	271	191	.0100	3,828	2,894	2,211	1,556
.0300	829	628	480	339	.0090	4,724	3,572	2,729	1,922
.0250	1,193	903	691	487	.0080	5,976	4,521	3,454	2,432
.0200	1,863	1,410	1,079	760	.0070	7,803	5,904	4,512	3,177
.0150	3,310	2,506	1,916	1,350					
.0100	7,444	5,636	4,310	3,036					
.0090	9,189	6,958	5,321	3,748					

References

Abt, K. (1982). Scale-independent non-parametric multivariate tolerance regions and their applications in medicine. *Biometrical Journal*, 24, 27-48.

Agresti A and Coull, B. A. (1998). Approximate is better than "exact" for interval estimation of binomial proportion. *The American Statistician*, 52, 119-125.

Aitchison, J. (1964). Bayesian tolerance regions. *Journal of the Royal Statistical Society B*, 26, 161-175.

_____ (1966). Expected-cover and linear utility tolerance intervals. *Journal of the Royal Statistical Society B*, 28, 57-62.

Aitchison, J. and Dunsmore, I. R. (1975). *Statistical Prediction Analysis*. New York: Cambridge University Press.

Aksoy, H. (2000). Use of gamma distribution in hydrological analysis. *Turkey Journal of Engineering and Environmental Sciences*, 24, 419-428.

Anderson, T. W. (1984). *An Introduction to Multivariate Statistical Analysis* (2nd ed.). New York: John Wiley.

Andrew, N. L. and Underwood, A. J. (1993). Density dependent foraging in the sea urchin *Centrostephanus Rodgersii* on shallow reefs in New South Wales, Australia. *Marine Ecology Progress Series*, 99, 89-98.

Andrews, D. F. and Herzberg, A. M., 1985. *Data*. New York: Springer-Verlag.

Aryal, S., Bhaumik, D. K., Mathew, T., and Gibbons, R. (2007). Approximate tolerance limits and prediction limits for the gamma distribution. *Journal of Applied Statistical Science*, 16, 103-111.

Ashkar, F. and Bobée, B. (1988). Confidence intervals for flood events under a Pearson 3 Or Log Pearson 3 distribution. *Water Resources Bulletin*, 24, 639-650.

Ashkar, F. and Ouarda, T. B. M. J. (1998). Approximate confidence intervals for quantiles of gamma and generalized gamma distributions. *Journal of Hydrologic Engineering*, 3, 43-51.

Bagui, S. C., Bhaumik, D. K. and Parnes, M. (1996). One-sided tolerance limits for unbalanced m-way random-effects ANOVA models. *Journal of Applied Statistical Science*, 3, 135-147.

Bain, L. J. and Engelhardt, M. (1981). Simple approximate distributional results for confidence and tolerance limits for the Weibull distribution based on maximum likelihood estimators. *Technometrics*, 23, 15-20.

Bain, L. J., Engelhardt, M. and Shiue, W. (1984). Approximate tolerance limits and confidence limits on reliability for the gamma distribution. *IEEE Transactions on Reliability*, 33, 184-187.

Barndroff-Nielsen, O. E. and Cox, D. R. (1994). *Inference and Asymptotics*. London: Chapman & Hall.

Basu, A. P. (1981). The estimation of $P(X < Y)$ for the distributions useful in life testing. *Naval Research Logistics Quarterly*, 28, 383-392.

Beckman, R. J. and Tietjen, G. L. (1989). Two-sided tolerance limits for balanced random-effects ANOVA models (Correction: 1998, 40, p. 269). *Technometrics*, 31, 185-197.

Bhattacharyya, G. K. and Johnson, R. A. (1974). Estimation of a reliability in a multi-component stress-strength model. *Journal of the American Statistical Association*, 69, 966-970.

Bhaumik, D. K. and Gibbons, R. D. (2006). One-sided approximate prediction intervals for at least p of m observations from a gamma population at each of r locations. *Technometrics*, 48, 112-119.

Bhaumik, D. K. and Kulkarni, P. M. (1991). One-sided tolerance limits for unbalanced one-way ANOVA random effects model. *Communications in Statistics, Part A – Theory and Methods*, 20, 1665-1675.

_____ (1996). A simple and exact method of constructing tolerance intervals for the one-way ANOVA with random effects. *The American Statistician*, 50, 319-323.

Bowden, D. C. and Steinhorst, R. K. (1973). Tolerance bands for growth curves. *Biometrics*, 29, 361-371.

Bowker, A. H. (1946). Computation of factors for tolerance limits on a normal distribution when the sample is large. *Annals of Mathematical Statistics*, 17, 238-240.

Bowker, A. H. and Liberman, G. J. (1972). *Engineering Statistics* (2nd ed.). New York: Prentice Hall.

Brazzale, A. R., Davison, A. C. and Reid, N. (2007). *Applied Asymptotics: Case Studies in Small-Sample Statistics*. Cambridge University Press.

Brown, E. B., Iyer, H. K. and Wang, C. M. (1997). Tolerance intervals for assessing individual bioequivalence. *Statistics in Medicine*, 16, 803-820.

Brown L. D., Cai T. and Das Gupta A. (2001). Interval estimation for a binomial proportion (with discussion). *Statistical Science*, 16, 101-133.

Brown, P. J. (1982). Multivariate calibration (with discussion). *Journal of the Royal Statistical Society, Series B*, 44, 287-321.

Brown, P. J. (1993). *Measurement, Regression and Calibration.* Oxford University Press.

Brown, P. J. and Sundberg, R. (1987). Confidence and conflict in multivariate calibration. *Journal of the Royal Statistical Society, Series B*, 49, 46-57.

Bucchianico, A. D., Einmahl, J. H. and Mushkudiani, N. A. (2001). Smallest nonparametric tolerance regions. *The Annals of Statistics*, 29, 1320-1343.

Burdick, R. K. and Graybill, F. A. (1992). *Confidence Intervals on Variance Components.* New York: Marcel-Dekker.

Cai, T. (2005). One-sided confidence intervals in discrete distributions. *Journal of Statistical Planning and Inference.* 131, 63-88.

Carroll, R. J. and Ruppert, D. (1991). Prediction and tolerance intervals with transformation and/or weighting. *Technometrics*, 33, 197-210.

Chatterjee, S. K. and Patra, N. K. (1976). On optimum Tukey tolerance sets. *Calcutta Statistical Association Bulletin*, 25, 105-110.

Chatterjee, S. K. and Patra, N. K. (1980). Asymptotically minimal multivariate tolerance sets. *Calcutta Statistical Association Bulletin*, 29, 73-93.

Chen, L.-A., Huang, J.-Y. and Chen, H.-C. (2007). Parametric coverage interval. *Metrologia*, 44, L7-L9.

Chen, L.-A. and Hung, H.-N. (2006). Extending the discussion on coverage intervals and statistical coverage intervals. *Metrologia*, 43, L43-L44.

Chen, S.-Y. and Harris, B (2006). On lower tolerance limits with accurate coverage probabilities for the normal random effects model. *Journal of the American Statistical Association*, 101, 1039-1049.

Chew, V. (1966). Confidence, prediction, and tolerance regions for the multivariate normal distribution. *Journal of the American Statistical Association*, 61, 605-617.

Chinchilli, V. M. and Esinhart, J. D. (1996). Design and analysis of intra-subject variability in cross-over experiments. *Statistics in Medicine*, 15, 1619-1634.

Chow, S.-C. and Liu, J.-P. (2000). *Design and Analysis of Bioavailability and Bioequivalence Studies* (2nd ed.). New York: Marcel-Dekker.

Chou, Y.-M. (1984). Sample sizes for β-expectation tolerance limits which control both tails of the normal distribution. *Naval Research Logistics Quarterly*, 31, 601-607.

Chou, Y.-M. and Mee, R. W. (1984). Determination of sample sizes for setting β-content tolerance limits controlling both tails of the normal distribution. *Statistics & Probability Letters*, 2, 311-314.

Clopper, C. J. and Pearson E. S. (1934). The use of confidence or fiducial limits illustrated in the case of the binomial. *Biometrika*, 26, 404-413.

Cohen, A. C. (1959). Simplified estimators for the normal distribution when samples are singly censored or truncated. *Technometrics*, 1, 217-237.

_____ (1961). Tables for maximum likelihood estimates: singly truncated and singly censored samples. *Technometrics*, 3, 535-541.

_____ (1965). Maximum likelihood estimation in the Weibull distribution based on complete and on censored samples. Technometrics, 7, 579–588.

Das, S. C. (1955). Fitting truncated type III curves to rainfall data. *Australian Journal of Physics*, 8, 298-304.

David, H. A. and Nagaraja, H. (2003). *Order Statistics* (3rd ed.). New York: Wiley.

Davis, A. W. and Hayakawa, T. (1987). Some distribution theory relating to confidence regions in multivariate calibration. *Annals of the Institute of Statistical Mathematics*, 39, 141-152.

Eaton, M. L., Muirhead, R. J. and Pickering , E. H. (2006). Assessing a vector of clinical observations. *Journal of Statistical Planning and Inference*, 136, 3383-3414.

Eberhardt, K. R., Mee, R. W. and Reeve, C. P. (1989). Computing factors for exact two-sided tolerance limits for a normal distribution. *Communications in Statistics–Simulation and Computation*, 18, 397-413.

Engelhardt, M. and Bain, L. J. (1977). Simplified statistical procedures for the Weibull or extreme-value distribution. *Technometrics*, 19, 323-331.

Engelhardt, M. and Bain, L. J. (1978). Tolerance limits and confidence limits on reliability for the two-parameter exponential distribution. *Technometrics*, 20, 37-39.

Esinhart, J. D. and Chinchilli, V. M. (1994). Extension to the use of tolerance intervals for the assessment of individual bioequivalence. *Journal of Biopharmaceutical Statistics*, 4, 39-52.

Evans, M. and Fraser, D. A. S. (1980). An optimum tolerance region for multivariate regression. *Journal of Multivariate Analysis*, 10, 268-272.

Faulkenberry, G. D. and Daly, J. C. (1970). Sample size for tolerance limits on a normal distribution. *Technometrics*, 12, 813-821.

Faulkenberry, G. D. and Weeks, D. L. (1968). Sample size determination for tolerance limits. *Technometrics*, 10, 343-348.

Fernholz, L. T. and Gillespie, J. A. (2001). Content-corrected tolerance limits based on the bootstrap. *Technometrics*, 43, 147-155.

Fertig, K. W. and Mann, N. R. (1974). Population percentile estimation for the unbalanced random-effect model: estimation of the strength of a material when sampling from various batches of the material. In *Reliability and Biometry*, F. Proschan and R. J. Serfling (eds.), Philadelphia: SIAM, pp. 205-228.

Fonseca, M., Mathew, T., Mexia, J. T. and Zmyslony, R. (2007). Tolerance intervals in a two-way nested model with mixed or random effects. *Statistics*, 41, 289-300.

Fountain, R. L. and Chou, Y.-M. (1991). Minimum sample sizes for two-sided tolerance intervals for finite populations. *Journal of Quality Technology*, 23, 90-95.

Fraser, D. A. S. and Guttman, I. (1956). Tolerance regions. *Annals of Mathematical Statistics*, 27, 162-179.

Fuchs, C. and Kenett, R. S. (1987). Multivariate tolerance regions and F-tests. *Journal of Quality Technology*, 19, 122-131.

_____ (1988). Appraisal of ceramic substrates by multivariate tolerance regions. *The Statistician*, 37, 401-411.

Fujikoshi, Y. and Nishii, R. (1984). On the distribution of a statistic in multivariate inverse regression analysis. *Hiroshima Mathematical Journal*, 14, 215-225.

Gamage, J., Mathew, T and Weerahandi, S. (2004). Generalized p-values and generalized confidence regions for the multivariate Behrens-Fisher problem. *Journal of Multivariate Analysis*, 88, 177-189.

Garwood, F. (1936). Fiducial limits for the Poisson distribution. *Biometrika*, 28, 437-442.

Gelman, A, Carlin, J. B., Stern, H. S. and Rubin, D. B. (2003). *Bayesian Data Analysis* (2nd ed.). New York: Chapman & Hall.

Gibbons, R. D. (1994). *Statistical Methods for Groundwater Monitoring*. New York: Wiley.

Gibbons, R. D. and Coleman, D. E. (2001). *Statistical Methods for Detection and Quantification of Environmental Contamination*. New York: Wiley.

Graybill, F. A. and Wang, C.M. (1980). Confidence intervals on nonnegative linear combinations of variances. *Journal of the American Statistical Association*, 75, 869-873.

Grubbs, F. E. (1971). Approximate fiducial bounds on reliability for the two parameter negative exponential distribution. *Technometrics*, 13, 873-876.

Guenther, W. C. (1971). On the use of best tests to obtain best β-content tolerance intervals. *Statistica Neerlandica*, 25, 191-202.

_____ (1972). Tolerance intervals for univariate distributions. *Naval Research Logistic Quarterly*, 19, 309-333.

Guenther, W. C., Patil, S. A. and Uppuluri, V. R. R. (1976). One-sided β-content tolerance factors for the two parameter exponential distribution. *Technometrics*, 18, 333-340.

Guo, H. and Krishnamoorthy, K. (2004). New approximate inferential methods for the reliability parameter in a stress-strength model: The normal case. *Communications in Statistics-Theory and Methods*, 33, 1715-1731.

Guttman, I. (1970). *Statistical Tolerance Regions: Classical and Bayesian*. London: Charles Griffin and Company.

_____ (1988). Tolerance regions, statistical. In *Encyclopedia of Statistical Sciences*, Volume 9 (S. Kotz and N. L. Johnson, eds.), New York: Wiley, pp. 272-287.

Guttman, I. and Menzefricke, U. (2003). Posterior distributions for functions of variance components. *Test*, 12, 115-123.

Haas, C. N. and Scheff, P. A. (1970). Estimation of averages in truncated samples. *Environmental Science and Technology*, 24, 912-919.

Hahn, G. J. and Chandra, R. (1981). Tolerance intervals for Poisson and binomial variables. *Journal of Quality Technology*, 13, 100-110.

Hahn, G. and Meeker, W. Q. (1991). *Statistical Intervals: A Guide to Practitioners*. New York: Wiley.

Hahn, G. J. and Ragunathan, T. E. (1988). Combining information from various sources: A prediction problem and other industrial applications. *Technometrics*, 30, 41-52.

Hall, I. J. (1984). Approximate one-sided tolerance limits for the difference or sum of two independent normal variates. *Journal of Quality Technology*, 16, 15-19.

Hall, I. J. and Sampson, C. B. (1973). Tolerance limits for the distribution of the product and quotient of normal variates. *Biometrics*, 29, 109-119.

Hall, I. J. and Sheldon, D. D. (1979). Improved bivariate normal tolerance regions with some applications. *Journal of Quality Technology*, 11, 13-19.

Hamada, M., Johnson, V., Moore, L. and Wendelberger, J. M. (2004). Bayesian prediction intervals and their relationship to tolerance intervals. *Technometrics*, 46, 452-459.

Hannig, J., Iyer, H. K. and Patterson, P. (2006). Fiducial generalized confidence intervals. *Journal of the American Statistical Association*, 101, 254-269.

Harris, E. K. and Boyd, J. C. (1995). *Statistical Bases of Reference Values in Laboratory Medicine*. New York: Marcel-Dekker.

Hauschke, H., Steinijans, V. and Pigeot, I. (2007). *Bioequivalence Studies in Drug Development: Methods and Applications*. New York: Wiley.

Hawkins, D. M. and Wixley, R. A. J. (1986). A note on the transformation of chi-squared variables to normality. *The American Statistician*, 40, 296-298.

Hinkley, D. V. (1969). On the ratio of two correlated normal random variables. *Biometrika*, 56, 635-639.

Howe, W. G. (1969). Two-sided tolerance limits for normal populations – Some improvements. *Journal of the American Statistical Association*, 64, 610-620.

Imhof, J. P. (1961). Computing the distribution of quadratic forms in normal variables. *Biometrika*, 48, 419-426.

Iyer, H. K., Wang, C. M. and Mathew, T. (2004). Models and confidence intervals for true values in interlaboratory trials. *Journal of the American Statistical Association*, 99, 1060-1071.

Jaech, J. L. (1984). Removing the effects of measurement errors in constructing statistical tolerance intervals. *Journal of Quality Technology*, 16, 69-73.

Jilek, M. (1981). A bibliography of statistical tolerance regions. *Mathematische Operationsforschung und Statistik, Series Statistics* [Current Name: Statistics)], 12, 441-456.

Jilek, M. and Ackerman, H. (1989). A bibliography of statistical tolerance regions II. *Statistics*, 20, 165172.

John, S. (1963). A tolerance region for multivariate normal distributions. *Sankhya, Ser. A*, 25, 363-368.

Johnson, D. and Krishnamoorthy, K. (1996). Combining independent studies in a calibration problem. *Journal of the American Statistical Association*, 91, 1707-1715.

Johnson, R. A. and Wichern, D. W. (1992). *Applied Multivariate Statistical Analysis* (3rd ed.). Englewood Cliffs, NJ: Prentice Hall.

Khuri, A. I. (1982). Direct products: a powerful tool for the analysis of balanced data. *Communications in Statistics–Theory and Methods*, 11, 2903-2920.

Khuri, A. I., Mathew, T. and Sinha, B. K. (1998). *Statistical Tests for Mixed Models*. New York: Wiley.

Ko, G., Burge, H. A., Nardell, E. A. and Thompson, K. M. (2001). Estimation of tuberculosis risk and incidence under upper room ultraviolet germicidal irradiation in a waiting room in a hypothetical scenario. *Risk Analysis*, 21, 657-673.

Kotz, S., Lumelskii, Y. and Pensky, M. (2003). *The Stress-Strength Model and Its Generalizations: Theory and Applications*. Singapore: World Scientific.

Krishnamoorthy, K. (2006). *Handbook of Statistical Distributions with Applications*. Boca Raton, FL: Chapman & Hall/CRC Press. [The PC calculator that accompanies the book is available free from http://www.ucs.louisiana.edu/~kxk4695].

Krishnamoorthy, K. and Guo, H. (2005). Assessing occupational exposure via the one-way random effects model with unbalanced data. *Journal of Statistical Planning and Inference*, 128, 219-229.

Krishnamoorthy, K., Kulkarni, P. M. and Mathew, T. (2001). Multiple use one-sided hypotheses testing in univariate linear calibration. *Journal of Statistical Planning and Inference*, 93, 211-223.

Krishnamoorthy, K. and Lin, Y. (2009). Stress-Strength reliability for Weibull distributions. Submitted for publication.

Krishnamoorthy, K., Lin, Y. and Xia, Y. (2009). Confidence limits and prediction limits for a Weibull distribution. To appear in *Journal of Statistical Planning and Inference*.

Krishnamoorthy, K. and Lu, Y. (2003). Inferences on the common mean of several normal populations based on the generalized variable method. *Biometrics*, 59, 237-247.

Krishnamoorthy, K., Mallick, A. and Mathew, T. (2009). Model based imputation approach for data analysis in the presence of non-detectable values: normal and related distributions. *Annals of Occupational Hygiene*, to appear.

Krishnamoorthy, K. and Mathew, T. (1999). Comparison of approximation methods for computing tolerance factors for a multivariate normal population. *Technometrics*, 41, 234-249.

_____ (2002a). Statistical methods for establishing equivalency of a sampling device to the OSHA standard. *American Industrial Hygiene Association Journal*, 63, 567-571.

_____ (2002b). Assessing occupational exposure via the one-way random effects model with balanced data. *Journal of Agricultural, Biological and Environmental Statistics*, 7, 440-451.

_____ (2003). Inferences on the means of lognormal distributions using generalized p-values and generalized confidence intervals. *Journal of Statistical Planning and Inference*, 115, 103-121.

_____ (2004). One-sided tolerance limits in balanced and unbalanced one-way random models based on generalized confidence limits. *Technometrics*, 46, 44-52.

Krishnamoorthy, K., Mathew, T. and Mukherjee, S. (2008). Normal based methods for a gamma distribution: prediction and tolerance intervals and stress-strength reliability. *Technometrics*, 50, 69-78.

Krishnamoorthy, K., Mathew, T. and Ramachandran, G. (2007). Upper limits for the exceedance probabilities in one-way random effects model. *Annals of Occupational Hygiene*, 51, 397-406.

Krishnamoorthy, K. and Mondal, S. (2006). Improved tolerance factors for multivariate normal distributions. *Communications in Statistics – Simulation and Computation*, 35, 461-478.

_____ (2008). Tolerance factors in multiple and multivariate linear regressions. *Communications in Statistics - Simulation and Computation*, 37, 546-559.

Krishnamoorthy, K., Mukherjee, S. and Guo, H. (2007). Inference on reliability in two-parameter exponential stress-strength model. *Metrika*, 65, 261-273.

Krishnamoorthy, K. and Xia, Y. (2009). A simple approximate procedure for constructing binomial and Poisson tolerance intervals. Submitted for publication.

Land, C. E. (1973). Standard confidence limits for linear functions of the normal mean and variance. *Journal of the American Statistical Association*, 68, 960-963.

Lawless, J. F. (1973). On the estimation of safe life when the underlying life distribution is Weibull. *Technometrics*, 15, 857-865.

_____ (1975). Construction of tolerance bounds for the extreme-value and Weibull distributions. *Technometrics*, 17, 255-261.

_____ (1978). Confidence interval estimation for the Weibull and extreme value distributions. *Technometrics*, 20, 355-364.

_____ (2003). *Statistical Models and Methods for Lifetime Data* (2nd ed.). New York: Wiley.

Lee, Y.-T. (1999). *Tolerance Regions in Univariate and Multivariate Linear Models*. Unpublished doctoral dissertation, Department of Mathematics and Statistics, University of Maryland Baltimore County.

Lee, Y.-T. and Mathew, T. (2004). Tolerance regions in multivariate linear regression. *Journal of Statistical Planning and Inference*, 126, 253-271.

Lemon, G. H. (1977). Factors for one-sided tolerance limits for balanced one-way-ANOVA random-effects model *Journal of the American Statistical Association*, 72, 676-680.

Li, J. and Liu, R. (2008). Multivariate spacings based on data depth: I. construction of nonparametric multivariate tolerance regions. *The Annals of Statistics*, 36, 1299-1323.

Liao, C.-T. and Iyer, H. K. (2001). A tolerance interval for assessing the quality of glucose monitoring meters. Department of Statistics, Colorado State University, Technical Report 2001/14.

_____ (2004). A tolerance interval for the normal distribution with several variance components. *Statistica Sinica*, 14, 217-229.

Liao, C.-T., Lin, T. Y. and Iyer, H. K. (2005). One- and two-sided tolerance intervals for general balanced mixed models and unbalanced one-way random models. *Technometrics*, 47, 323-335.

Lieberman, G. J. and Miller, R. J., Jr. (1963). Simultaneous tolerance intervals in regression. *Biometrika*, 50, 155-168.

Lieberman, G. J. Miller, R. G., Jr. and Hamilton, M. A. (1967). Unlimited simultaneous discrimination intervals in regression. *Biometrika* 54, 133-146.

Limam, M. M. T. and Thomas, D. R. (1988a). Simultaneous tolerance intervals for the linear regression model. *Journal of the American Statistical Association*, 83, 801-804.

_____ (1988b). Simultaneous tolerance intervals in the random one-way model with covariates. *Communications in Statistics- Computation and Simulation*, 17, 1007-1019.

Lin, Y. (2009). Generalized Inferences for Weibull Distributions. Ph.D. dissertation, Department of Mathematics, University of Louisiana at Lafayette, Lafayette, LA.

Lin, T.-Y and Liao, C.-T. (2006). A β−expectation tolerance interval for general balanced mixed linear models. *Computational Statistics & Data Analysis*, 50, 911-925.

Lin, T.-Y. and Liao, C.-T. (2008). Prediction intervals for general balanced linear random models. *Journal of Statistical Planning and Inference*, 138, 3164-3175.

Lin, T.-Y, Liao, C.-T. and Iyer, H. K. (2008). Tolerance intervals for unbalanced one-way random effects models with covariates and heterogeneous variances. *Journal of Agricultural, Biological and Environmental Statistics*, 13, 221-241.

Lyles, R. H., Kupper, L. L. and Rappaport, S. M. (1997a). Assessing regulatory compliance of occupational exposures via the balanced one-way random effects ANOVA model. *Journal of Agricultural, Biological, and Environmental Statistics*, 2, 64-86.

_____ (1997b). A lognormal distribution based exposure assessment method for unbalanced data. *Annals of Occupational Hygiene*, 41, 63-76.

Mann, N. R. and Fertig, K. W. (1973). Tables for obtaining Weibull confidence bounds and tolerance bounds based on best linear invariant estimates of parameters of the extreme value distribution. *Technometrics*, 15, 87-101.

_____ (1975). Simplified efficient point estimators for Weibull parameters. *Technometrics*, 17, 361-368.

_____ (1977). Efficient unbiased quantile estimates for moderate size samples from extreme-value and Weibull distributions: Confidence bounds and tolerance and prediction intervals. *Technometrics*, 19, 87-93.

Mathew, T. and Kasala, S. (1994). An exact confidence region in multivariate calibration. *The Annals of Statistics*, 22, 94-105.

Mathew, T. and Sharma, M. K. (2002). Multiple use confidence regions based on combined information in univariate calibration. *Journal of Statistical Planning and Inference*, 103, 151-172.

Mathew, T., Sharma, M. K. and Nordström, K. (1998). Tolerance regions and multiple-use confidence regions in multivariate calibration. *The Annals of Statistics*, 26, 1989-2013.

Mathew, T. and Webb, D. (2005). Generalized p-values and confidence intervals for variance components: applications to Army test and evaluation. *Technometrics*, 47, 312-322.

Mathew, T. and Zha, W. (1996). Conservative confidence regions in multivariate calibration. *The Annals of Statistics*, 24, 707-725.

_____ (1997). Multiple use confidence regions in multivariate calibration. *Journal of the American Statistical Association*, 92, 1141-1150.

Maxim, L.D., Galvin, J. B., Niebo, R., Segrave, A. M., Kampa, O. A. and Utell, M. J. (2006). Occupational exposure to carbon/coke fibers in plants that produce green or calcined petroleum coke and potential health effects: 2. Fiber Concentrations. *Inhalation Toxicology*, 18, 17-32.

Mee, R. W. (1984a). Tolerance limits and bounds for proportions based on data subject to measurement error. *Journal of Quality Technology*, 16, 74-80.

_____ (1984b). β-expectation and β-content tolerance limits for balanced one-way ANOVA random model. *Technometrics*, 26, 251-254.

_____ (1989). Normal distribution tolerance limits for stratified random samples. *Technometrics*, 31, 99-105.

_____ (1990). Simultaneous tolerance intervals for normal populations with common variance. *Technometrics*, 32, 83-92.

Mee, R. W. and Eberhardt, K. R. (1996). A comparison of uncertainty criteria for calibration. *Technometrics*, 38, 221-229.

Mee, R. W., Eberhardt, K. R. and Reeve, C. P. (1991). Calibration and simultaneous tolerance intervals for regression. *Technometrics*, 33, 211-219.

Mee, R. W. and Owen, D. B. (1983). Improved factors for one-sided tolerance limits for balanced one-way ANOVA random model. *Journal of the American Statistical Association*, 78, 901-905.

Menon, M. V. (1963). Estimation of the shape and scale parameters of the Weibull distribution. *Technometrics*, 5, 175-182.

Millard, S. P. and Neerchal, N. K. (2000). *Environmental Statistics With S-Plus*. New York: CRC Press.

Miller, R. G., Jr. (1981). *Simultaneous Statistical Inference* (2nd ed.). New York: Springer-Verlag.

Miller, R. W. (1989). Parametric empirical Bayes tolerance intervals. *Technometrics*, 31, 449-459.

Montgomery, D. C. (2009). *Design and Analysis of Experiments* (7th ed.). New York: Wiley.

Murphy, R. B. (1948). Non-parametric tolerance limits. *Annals of Mathematical Statistics*, 19, 581-589.

Odeh, R. E. (1978). Tables of two-sided tolerance factors for a normal distribution. *Communications in Statistics - Simulation and Computation*, 7, 183-201.

Odeh, R. E., Chou, Y.-M. and Owen, D. B. (1987) The precision for coverages and sample size requirements for normal tolerance intervals. *Communications in Statistics: Simulation and Computation*, 16, 969-985.

Odeh, R. E., Chou, Y.-M. and Owen, D. B. (1989). Sample-size determination for two-sided β-expectation tolerance intervals for a normal distribution. *Technometrics*, 31, 461-468.

Odeh, R. E. and Mee, R. W. (1990). One-sided simultaneous tolerance limits for regression. *Communications in Statistics, Part B - Simulation and Computation*, 19, 663-680.

Odeh, R. E. and Owen, D. B. (1980). *Tables for Normal Tolerance Limits, Sampling Plans, and Screening*. New York: Marcel-Dekker.

Oden, A. (1973). Simultaneous confidence intervals in inverse linear regression. *Biometrika*, 60, 339-343.

Oman, S. D. (1988). Confidence regions in multivariate calibration. *The Annals of Statistics*, 16, 174-187.

Oman, S. D. and Wax, Y. (1984). Estimating fetal age by ultrasound measurements: an example of multivariate calibration. *Biometrics*, 40, 947-960.

Osborne, C. (1991). Statistical calibration: a review. *International Statistical Review*, 59, 309-336.

Ostle, B. and Mensing, R. W. (1975). *Statistics in Research: Basic Concepts and Techniques for Research Workers*, (3rd ed.). Ames, Iowa: Iowa State University Press.

O'Sullivan, J. B. and Mahan, C. M. (1966). Glucose tolerance test: variability in pregnant and nonpregnant women. *American Journal of Clinical Nutrition*, 19, 345-351.

Owen, D. B. (1964). Control of percentages in both tails of the normal distribution (Corr: V8 p570). *Technometrics*, 6, 377-387.

Patel, J. K. (1986). Tolerance limits: a review. *Communications in Statistics, Part A - Theory and Methods*, 15, 2719-2762.

Patel, J. K. (1989). Prediction intervals: a review. *Communications in Statistics, Part A - Theory and Methods*, 18, 2393-2465.

Patterson, S.D. and Jones, B. (2006). *Bioequivalence and Statistics in Clinical Pharmacology*. New York: Chapman & Hall/CRC Press.

Paulson, E. (1943). A note on tolerance limits. *Annals of Mathematical Statistics*, 14, 90-93.

Poulson, O. M., Holst, E. and Christensen, J. M. (1997). Calculation and application of coverage intervals for biological reference values. *Pure and Applied Chemistry*, 69, 1601-1611.

Proschan, F. (1963). Theoretical explanation of observed decreasing failure rate. *Technometrics*, 5, 375-383.

R Development Core Team (2008). R: A language and environment for statistical computing. R Foundation for Statistical Computing, Vienna, Austria. ISBN 3-900051-07-0, URL http://www.R-project.org.

Rao, C. R. (1973). *Linear Statistical Inference and Its Applications* (2nd ed.). New York: Wiley.

Reiser, B. J. and Guttman, I. (1986). Statistical inference for $\Pr(Y < X)$: The normal case, *Technometrics*, 28, 253-257.

Reiser, B. and Rocke, M. D. (1993). Inference for stress-strength problems under the gamma distribution. *IAPQR Transactions*, 18, 1-22.

Rencher, A. C. (1995). *Methods of Multivariate Analysis*. New York: Wiley.

Rockette, H. E. and Wadsworth, H. M. (1985). Equivalency of alternative cotton dust samplers to the vertical elutriator. *American Industrial Hygiene Association Journal*, 46, 716-719.

Rode, R. A. and Chinchilli, V. M. (1988). The use of Box-Cox transformations in the development of multivariate tolerance regions with applications to clinical chemistry. *The American Statistician*, 42, 23-30.

Roy, A. and Mathew, T. (2005). A Generalized confidence limit for the reliability function of a two-parameter exponential distribution. *Journal of Statistical Planning and Inference*, 128, 509-517.

Rukhin, A. L. and Vangel, M. G. (1998). Estimation of a common mean and weighted mean statistics. *Journal of the American Statistical Association*, 93, 303-309.

Schafer, R. and Sheffield, T. (1976). On procedures for comparing two Weibull populations. *Technometrics*, 18, 231-235.

Scheffe, H. (1973). A statistical theory of calibration. *The Annals of Statistics*, 1, 1-37.

Schmee, J., Gladstein, D. and Nelson W. (1985). Confidence limits of a normal distribution from singly censored samples using maximum likelihood. *Technometrics*, 27, 119-128.

Schmee, J. and Nelson, W. B. (1977). Estimates and approximate confidence limits for (log) normal life distributions from singly censored samples by maximum likelihood. General Electric Co. Corp. Research and Development TIS Report 76CRD250.

Searle, S. R. (1971). *Linear Models*. New York: Wiley.

Searle, S. R., Casella, G. and McCulloch, C. E. (1992). *Variance Components*. New York: Wiley.

Siotani, M. (1964). Tolerance regions for a multivariate normal population. *Annals of the Institute of Statistical Mathematics*, 16, 135-153.

Smith, R. L. and Corbett, M. (1987). Measuring marathon courses: an application of statistical calibration theory. *Applied Statistics*, 36, 283-295.

Smith, R. W. (2002). The use of random-model tolerance intervals in environmental monitoring and regulation. *Journal of Agricultural, Biological and Environmental Statistics*, 7, 74-94.

Smith, W.B. and Hocking, R.R. (1972). Wishart variates generator (algorithm AS 53). *Applied Statistics*, 21, 341-345.

Somerville, P. N. (1958). Tables for obtaining non-parametric tolerance Limits. *The Annals of Mathematical Statistics*, 29, 599-601.

Stephenson, D. B., Kumar, K. R., Doblas-Reyes, F. J., Royer, J. F., Chauvin, E. and Pezzulli, S. (1999). Extreme daily rainfall events and their impact on ensemble forecasts of the Indian monsoon. *Monthly Weather Review*, 127, 1954-1966.

Sundberg, R (1994). Most modern calibration is multivariate. *Seventeenth International Biometric Conference*, Hamilton, Canada, 395-405.

Sundberg, R. (1999). Multivariate calibration - direct and indirect regression methodology (with discussion). *Scandinavian Journal of Statistics*, 26, 161-207.

Thoman, D. R. and Bain, L. J. (1969). Two sample tests in the Weibull distribution. Technometrics 11, 805-815.

Thoman, D. R., Bain, L. J. and Antle, C. E. (1969). Inferences on the parameters of the Weibull distribution. *Technometrics* 11, 445-460.

Thoman, D. R., Bain, L. J. and Antle, C. E. (1970). Maximum likelihood estimation, exact confidence intervals for reliability and tolerance limits in the Weibull distribution. *Technometrics*, 12, 363-371.

Thomas, J. D. and Hultquist, R. A. (1978). Interval estimation for the unbalanced case of the one-way random effects model. *The Annals of Statistics*, 6, 582-587.

Trickett, W. H. and Welch, B. L. (1954). On the comparison of two means: further discussion of iterative methods for calculating tables. *Biometrika*, 41, 361-374.

Trost, D. C. (2006). Multivariate probability-based detection of drug-induced hepatic signals. *Toxicological Reviews*, 25, 37-54.

Tsui, K.W. and Weerahandi, S. (1989). Generalized p-values in significance testing of hypotheses in the presence of nuisance parameters. *Journal of the American Statistical Association* 84, 602-607.

Tukey, J.W. (1947). Nonparametric estimation II: Statistical equivalent blocks and tolerance regions-the continuous case. *Annals of Mathematical Statistics*, 18, 529539.

U. S. FDA (2001). *Guidance for Industry: Statistical Approaches to Establishing Bioequivalence*. Available at www.fda.gov/cder/guidance/3616fnl.htm.

Van der Merwe, A. J. and Hugo, J. (2007). Bayesian tolerance intervals for the balanced two-factor nested random effects model. *Test*, 16, 598-612.

Van der Merwe, A. J., Pretorius, A. L. and Meyer, J. H. (2006). Bayesian tolerance intervals for the unbalanced one-way random effects model. *Journal of Quality Technology*, 38, 280-293.

Vangel, M. G. (1992). New methods for one-sided tolerance limits for a one-way balanced random-effects ANOVA model. *Technometrics*, 34, 176-185.

_____ (2008a). Tolerance interval. In *Encyclopedia of Biostatistics*, 2nd ed. (P. Armitage and T. Colton, eds.), New York: Wiley, pp. 5477-5482.

_____ (2008b). Tolerance region. In *Encyclopedia of Biostatistics*, 2nd ed. (P. Armitage and T. Colton, eds.), New York: Wiley, p. 5482.

Vangel, M. G. and Rukhin, A. L. (1999). Maximum likelihood analysis for heteroscedastic one-way random effects ANOVA in interlaboratory studies. *Biometrics*, 55, 129-136.

von Rosen, D. (1988). Moments for the inverted Wishart distribution. *Scandinavian Journal of Statistics*, 15, 97-109.

Wald, A. (1943). An extension of Wilks' method for setting tolerance limits. *Annals of Mathematical Statistics*, 14, 45-55.

Wald, A. and Wolfowitz, J. (1946). Tolerance limits for a normal distribution, *Annals of the Mathematical Statistics*, 17, 208-215.

Wallis, W. A. (1951). Tolerance intervals for linear regression. *Proceedings of the Second Berkeley Symposium on Mathematical Statistics and Probability*, J. Neyman (ed.). Berkeley, CA: University of California Press, pp. 43-51.

Wang, C. M. and Iyer, H. K. (1994). Tolerance intervals for the distribution of true values in the presence of measurement errors. *Technometrics*, 36, 162-170.

Wang, C. M. and Iyer, H. K. (1996). Sampling plans for obtaining tolerance intervals in a balanced one-way random-effects model. *Communications in Statistics - Theory and Methods*, 25, 313-324.

Wang, H. and Tsung, F. (2008). Tolerance intervals with improved coverage probabilities for binomial and Poisson variables. *Technometrics*, to appear.

Weerahandi, S. (1993). Generalized confidence intervals. *Journal of the American Statistical Association* 88, 899-905.

Weerahandi, S., (1995a). *Exact Statistical Methods for Data Analysis*. New York: Springer-Verlag.

Weerahandi, S., (1995b). ANOVA with unequal variances. *Biometrics*, 51, 589-599.

Weerahandi, S. (2004). *Generalized Inference in Repeated Measures: Exact Methods in MANOVA and Mixed Models*. New York: Wiley.

Weerahandi, S. and Berger, V. W. (1999). Exact inference for growth curves with intraclass correlation structure. *Biometrics*, **55**, 921-924.

Weerahandi, S. and Johnson, R. A. (1992). Testing reliability in a stress-strength model when X and Y are normally distributed. *Technometrics*, 34, 83-91.

Weisberg, A. and Beatty, C. H. (1960). Tables of tolerance-limit factors for normal distribution. *Technometrics*, 4, 483-500.

Wilks, S. S. (1941). Determination of sample sizes for setting tolerance limits. *Annals of Mathematical Statistics*, 12, 91-96.

Wilks, S. S. (1942). Statistical prediction with special reference to the problem of tolerance limits. *Annals of Mathematical Statistics*, 13, 400-409.

Willink, R. (2004). Coverage intervals and statistical coverage intervals. *Metrologia*, 41, L5-L6.

_____ (2006). On using the Monte Carlo method to calculate uncertainty intervals. *Metrologia*, 43, L39-L42.

Wilson, A. L. (1967). An approach to simultaneous tolerance intervals in regression. *The Annals of Mathematical Statistics*, 38, 1536-1540.

Wilson, E. B. and Hilferty, M. M. (1931). The distribution of chi-squares. *Proceedings of the National Academy of Sciences*, 17, 684-688.

Wolfinger, R. D. (1998). Tolerance intervals for variance component models using Bayesian simulation. *Journal of Quality Technology*, 30, 18-32.

Yang, H., Zhang, L. and Cho, I. (2006). Evaluating parallelism testing methods in immunoassay. *Proceedings of the American Statistical Association, Biometrics Section*, 191-197.

Zhang, L., Mathew, T., Yang, H., Krishnamoorthy, K. and Cho, I. (2009). Tolerance limits for a ratio of normal random variables. *Journal of Biopharmaceutical Statistics* (to appear).

Zhou, L. and Mathew, T. (1994). Some tests for variance components using generalized p-values. *Technometrics*, 36, 394-402.

Index

WILEY SERIES IN PROBABILITY AND STATISTICS

ESTABLISHED BY WALTER A. SHEWHART AND SAMUEL S. WILKS

Editors: *David J. Balding, Noel A. C. Cressie, Garrett M. Fitzmaurice,*
Iain M. Johnstone, Geert Molenberghs, David W. Scott,
Adrian F. M. Smith, Ruey S. Tsay, Sanford Weisberg
Editors Emeriti: *Vic Barnett, J. Stuart Hunter, Jozef L. Teugels*

The ***Wiley Series in Probability and Statistics*** is well established and authoritative. It covers many topics of current research interest in both pure and applied statistics and probability theory. Written by leading statisticians and institutions, the titles span both state-of-the-art developments in the field and classical methods.

Reflecting the wide range of current research in statistics, the series encompasses applied, methodological and theoretical statistics, ranging from applications and new techniques made possible by advances in computerized practice to rigorous treatment of theoretical approaches.

This series provides essential and invaluable reading for all statisticians, whether in academia, industry, government, or research.

† ABRAHAM and LEDOLTER · Statistical Methods for Forecasting
AGRESTI · Analysis of Ordinal Categorical Data
AGRESTI · An Introduction to Categorical Data Analysis, *Second Edition*
AGRESTI · Categorical Data Analysis, *Second Edition*
ALTMAN, GILL, and McDONALD · Numerical Issues in Statistical Computing for the
   Social Scientist
AMARATUNGA and CABRERA · Exploration and Analysis of DNA Microarray and
   Protein Array Data
ANDĚL · Mathematics of Chance
ANDERSON · An Introduction to Multivariate Statistical Analysis, *Third Edition*
* ANDERSON · The Statistical Analysis of Time Series
ANDERSON, AUQUIER, HAUCK, OAKES, VANDAELE, and WEISBERG ·
   Statistical Methods for Comparative Studies
ANDERSON and LOYNES · The Teaching of Practical Statistics
ARMITAGE and DAVID (editors) · Advances in Biometry
ARNOLD, BALAKRISHNAN, and NAGARAJA · Records
* ARTHANARI and DODGE · Mathematical Programming in Statistics
* BAILEY · The Elements of Stochastic Processes with Applications to the Natural
   Sciences
BALAKRISHNAN and KOUTRAS · Runs and Scans with Applications
BALAKRISHNAN and NG · Precedence-Type Tests and Applications
BARNETT · Comparative Statistical Inference, *Third Edition*
BARNETT · Environmental Statistics
BARNETT and LEWIS · Outliers in Statistical Data, *Third Edition*
BARTOSZYNSKI and NIEWIADOMSKA-BUGAJ · Probability and Statistical Inference
BASILEVSKY · Statistical Factor Analysis and Related Methods: Theory and
   Applications
BASU and RIGDON · Statistical Methods for the Reliability of Repairable Systems
BATES and WATTS · Nonlinear Regression Analysis and Its Applications
BECHHOFER, SANTNER, and GOLDSMAN · Design and Analysis of Experiments for
   Statistical Selection, Screening, and Multiple Comparisons

*Now available in a lower priced paperback edition in the Wiley Classics Library.
†Now available in a lower priced paperback edition in the Wiley–Interscience Paperback Series.

BELSLEY · Conditioning Diagnostics: Collinearity and Weak Data in Regression
† BELSLEY, KUH, and WELSCH · Regression Diagnostics: Identifying Influential
Data and Sources of Collinearity
BENDAT and PIERSOL · Random Data: Analysis and Measurement Procedures,
Third Edition
BERRY, CHALONER, and GEWEKE · Bayesian Analysis in Statistics and
Econometrics: Essays in Honor of Arnold Zellner
BERNARDO and SMITH · Bayesian Theory
BHAT and MILLER · Elements of Applied Stochastic Processes, *Third Edition*
BHATTACHARYA and WAYMIRE · Stochastic Processes with Applications
BILLINGSLEY · Convergence of Probability Measures, *Second Edition*
BILLINGSLEY · Probability and Measure, *Third Edition*
BIRKES and DODGE · Alternative Methods of Regression
BISWAS, DATTA, FINE, and SEGAL · Statistical Advances in the Biomedical Sciences:
Clinical Trials, Epidemiology, Survival Analysis, and Bioinformatics
BLISCHKE AND MURTHY (editors) · Case Studies in Reliability and Maintenance
BLISCHKE AND MURTHY · Reliability: Modeling, Prediction, and Optimization
BLOOMFIELD · Fourier Analysis of Time Series: An Introduction, *Second Edition*
BOLLEN · Structural Equations with Latent Variables
BOLLEN and CURRAN · Latent Curve Models: A Structural Equation Perspective
BOROVKOV · Ergodicity and Stability of Stochastic Processes
BOULEAU · Numerical Methods for Stochastic Processes
BOX · Bayesian Inference in Statistical Analysis
BOX · R. A. Fisher, the Life of a Scientist
BOX and DRAPER · Response Surfaces, Mixtures, and Ridge Analyses, *Second Edition*
* BOX and DRAPER · Evolutionary Operation: A Statistical Method for Process
Improvement
BOX and FRIENDS · Improving Almost Anything, *Revised Edition*
BOX, HUNTER, and HUNTER · Statistics for Experimenters: Design, Innovation,
and Discovery, *Second Editon*
BOX, JENKINS, and REINSEL · Time Series Analysis: Forcasting and Control, *Fourth
Edition*
BOX, LUCEÑO, and PANIAGUA-QUIÑONES · Statistical Control by Monitoring
and Adjustment, *Second Edition*
BRANDIMARTE · Numerical Methods in Finance: A MATLAB-Based Introduction
† BROWN and HOLLANDER · Statistics: A Biomedical Introduction
BRUNNER, DOMHOF, and LANGER · Nonparametric Analysis of Longitudinal Data in
Factorial Experiments
BUCKLEW · Large Deviation Techniques in Decision, Simulation, and Estimation
CAIROLI and DALANG · Sequential Stochastic Optimization
CASTILLO, HADI, BALAKRISHNAN, and SARABIA · Extreme Value and Related
Models with Applications in Engineering and Science
CHAN · Time Series: Applications to Finance
CHARALAMBIDES · Combinatorial Methods in Discrete Distributions
CHATTERJEE and HADI · Regression Analysis by Example, *Fourth Edition*
CHATTERJEE and HADI · Sensitivity Analysis in Linear Regression
CHERNICK · Bootstrap Methods: A Guide for Practitioners and Researchers,
Second Edition
CHERNICK and FRIIS · Introductory Biostatistics for the Health Sciences
CHILÈS and DELFINER · Geostatistics: Modeling Spatial Uncertainty
CHOW and LIU · Design and Analysis of Clinical Trials: Concepts and Methodologies,
Second Edition
CLARKE · Linear Models: The Theory and Application of Analysis of Variance

---

*Now available in a lower priced paperback edition in the Wiley Classics Library.
†Now available in a lower priced paperback edition in the Wiley–Interscience Paperback Series.

*Now available in a lower priced paperback edition in the Wiley Classics Library.
†Now available in a lower priced paperback edition in the Wiley–Interscience Paperback Series.

*Now available in a lower priced paperback edition in the Wiley Classics Library.
†Now available in a lower priced paperback edition in the Wiley–Interscience Paperback Series.

MONTGOMERY, PECK, and VINING · Introduction to Linear Regression Analysis, *Fourth Edition*

MORGENTHALER and TUKEY · Configural Polysampling: A Route to Practical Robustness

MUIRHEAD · Aspects of Multivariate Statistical Theory

MULLER and STOYAN · Comparison Methods for Stochastic Models and Risks

MURRAY · X-STAT 2.0 Statistical Experimentation, Design Data Analysis, and Nonlinear Optimization

MURTHY, XIE, and JIANG · Weibull Models

MYERS, MONTGOMERY, and ANDERSON-COOK · Response Surface Methodology: Process and Product Optimization Using Designed Experiments, *Third Edition*

MYERS, MONTGOMERY, and VINING · Generalized Linear Models. With Applications in Engineering and the Sciences

† NELSON · Accelerated Testing, Statistical Models, Test Plans, and Data Analyses

† NELSON · Applied Life Data Analysis

NEWMAN · Biostatistical Methods in Epidemiology

OCHI · Applied Probability and Stochastic Processes in Engineering and Physical Sciences

OKABE, BOOTS, SUGIHARA, and CHIU · Spatial Tesselations: Concepts and Applications of Voronoi Diagrams, *Second Edition*

OLIVER and SMITH · Influence Diagrams, Belief Nets and Decision Analysis

PALTA · Quantitative Methods in Population Health: Extensions of Ordinary Regressions

PANJER · Operational Risk: Modeling and Analytics

PANKRATZ · Forecasting with Dynamic Regression Models

PANKRATZ · Forecasting with Univariate Box-Jenkins Models: Concepts and Cases

* PARZEN · Modern Probability Theory and Its Applications

PEÑA, TIAO, and TSAY · A Course in Time Series Analysis

PIANTADOSI · Clinical Trials: A Methodologic Perspective

PORT · Theoretical Probability for Applications

POURAHMADI · Foundations of Time Series Analysis and Prediction Theory

POWELL · Approximate Dynamic Programming: Solving the Curses of Dimensionality

PRESS · Bayesian Statistics: Principles, Models, and Applications

PRESS · Subjective and Objective Bayesian Statistics, *Second Edition*

PRESS and TANUR · The Subjectivity of Scientists and the Bayesian Approach

PUKELSHEIM · Optimal Experimental Design

PURI, VILAPLANA, and WERTZ · New Perspectives in Theoretical and Applied Statistics

† PUTERMAN · Markov Decision Processes: Discrete Stochastic Dynamic Programming

QIU · Image Processing and Jump Regression Analysis

* RAO · Linear Statistical Inference and Its Applications, *Second Edition*

RAUSAND and HØYLAND · System Reliability Theory: Models, Statistical Methods, and Applications, *Second Edition*

RENCHER · Linear Models in Statistics

RENCHER · Methods of Multivariate Analysis, *Second Edition*

RENCHER · Multivariate Statistical Inference with Applications

* RIPLEY · Spatial Statistics

* RIPLEY · Stochastic Simulation

ROBINSON · Practical Strategies for Experimenting

ROHATGI and SALEH · An Introduction to Probability and Statistics, *Second Edition*

ROLSKI, SCHMIDLI, SCHMIDT, and TEUGELS · Stochastic Processes for Insurance and Finance

ROSENBERGER and LACHIN · Randomization in Clinical Trials: Theory and Practice

ROSS · Introduction to Probability and Statistics for Engineers and Scientists

ROSSI, ALLENBY, and McCULLOCH · Bayesian Statistics and Marketing

*Now available in a lower priced paperback edition in the Wiley Classics Library.

†Now available in a lower priced paperback edition in the Wiley–Interscience Paperback Series.

† ROUSSEEUW and LEROY · Robust Regression and Outlier Detection
* RUBIN · Multiple Imputation for Nonresponse in Surveys
RUBINSTEIN and KROESE · Simulation and the Monte Carlo Method, *Second Edition*
RUBINSTEIN and MELAMED · Modern Simulation and Modeling
RYAN · Modern Engineering Statistics
RYAN · Modern Experimental Design
RYAN · Modern Regression Methods, *Second Edition*
RYAN · Statistical Methods for Quality Improvement, *Second Edition*
SALEH · Theory of Preliminary Test and Stein-Type Estimation with Applications
* SCHEFFE · The Analysis of Variance
SCHIMEK · Smoothing and Regression: Approaches, Computation, and Application
SCHOTT · Matrix Analysis for Statistics, *Second Edition*
SCHOUTENS · Levy Processes in Finance: Pricing Financial Derivatives
SCHUSS · Theory and Applications of Stochastic Differential Equations
SCOTT · Multivariate Density Estimation: Theory, Practice, and Visualization
† SEARLE · Linear Models for Unbalanced Data
† SEARLE · Matrix Algebra Useful for Statistics
† SEARLE, CASELLA, and McCULLOCH · Variance Components
SEARLE and WILLETT · Matrix Algebra for Applied Economics
SEBER · A Matrix Handbook For Statisticians
† SEBER · Multivariate Observations
SEBER and LEE · Linear Regression Analysis, *Second Edition*
† SEBER and WILD · Nonlinear Regression
SENNOTT · Stochastic Dynamic Programming and the Control of Queueing Systems
* SERFLING · Approximation Theorems of Mathematical Statistics
SHAFER and VOVK · Probability and Finance: It's Only a Game!
SILVAPULLE and SEN · Constrained Statistical Inference: Inequality, Order, and Shape
    Restrictions
SMALL and McLEISH · Hilbert Space Methods in Probability and Statistical Inference
SRIVASTAVA · Methods of Multivariate Statistics
STAPLETON · Linear Statistical Models
STAPLETON · Models for Probability and Statistical Inference: Theory and Applications
STAUDTE and SHEATHER · Robust Estimation and Testing
STOYAN, KENDALL, and MECKE · Stochastic Geometry and Its Applications, *Second
    Edition*
STOYAN and STOYAN · Fractals, Random Shapes and Point Fields: Methods of
    Geometrical Statistics
STREET and BURGESS · The Construction of Optimal Stated Choice Experiments:
    Theory and Methods
STYAN · The Collected Papers of T. W. Anderson: 1943–1985
SUTTON, ABRAMS, JONES, SHELDON, and SONG · Methods for Meta-Analysis in
    Medical Research
TAKEZAWA · Introduction to Nonparametric Regression
TAMHANE · Statistical Analysis of Designed Experiments: Theory and Applications
TANAKA · Time Series Analysis: Nonstationary and Noninvertible Distribution Theory
THOMPSON · Empirical Model Building
THOMPSON · Sampling, *Second Edition*
THOMPSON · Simulation: A Modeler's Approach
THOMPSON and SEBER · Adaptive Sampling
THOMPSON, WILLIAMS, and FINDLAY · Models for Investors in Real World Markets
TIAO, BISGAARD, HILL, PEÑA, and STIGLER (editors) · Box on Quality and
    Discovery: with Design, Control, and Robustness
TIERNEY · LISP-STAT: An Object-Oriented Environment for Statistical Computing
    and Dynamic Graphics

*Now available in a lower priced paperback edition in the Wiley Classics Library.
†Now available in a lower priced paperback edition in the Wiley–Interscience Paperback Series.

TSAY · Analysis of Financial Time Series, *Second Edition*
UPTON and FINGLETON · Spatial Data Analysis by Example, Volume II:
Categorical and Directional Data
† VAN BELLE · Statistical Rules of Thumb, *Second Edition*
VAN BELLE, FISHER, HEAGERTY, and LUMLEY · Biostatistics: A Methodology for
the Health Sciences, *Second Edition*
VESTRUP · The Theory of Measures and Integration
VIDAKOVIC · Statistical Modeling by Wavelets
VINOD and REAGLE · Preparing for the Worst: Incorporating Downside Risk in Stock
Market Investments
WALLER and GOTWAY · Applied Spatial Statistics for Public Health Data
WEERAHANDI · Generalized Inference in Repeated Measures: Exact Methods in
MANOVA and Mixed Models
WEISBERG · Applied Linear Regression, *Third Edition*
WELSH · Aspects of Statistical Inference
WESTFALL and YOUNG · Resampling-Based Multiple Testing: Examples and
Methods for *p*-Value Adjustment
WHITTAKER · Graphical Models in Applied Multivariate Statistics
WINKER · Optimization Heuristics in Economics: Applications of Threshold Accepting
WONNACOTT and WONNACOTT · Econometrics, *Second Edition*
WOODING · Planning Pharmaceutical Clinical Trials: Basic Statistical Principles
WOODWORTH · Biostatistics: A Bayesian Introduction
WOOLSON and CLARKE · Statistical Methods for the Analysis of Biomedical Data,
Second Edition
WU and HAMADA · Experiments: Planning, Analysis, and Parameter Design
Optimization
WU and ZHANG · Nonparametric Regression Methods for Longitudinal Data Analysis
YANG · The Construction Theory of Denumerable Markov Processes
YOUNG, VALERO-MORA, and FRIENDLY · Visual Statistics: Seeing Data with
Dynamic Interactive Graphics
ZACKS · Stage-Wise Adaptive Designs
ZELTERMAN · Discrete Distributions—Applications in the Health Sciences
* ZELLNER · An Introduction to Bayesian Inference in Econometrics
ZHOU, OBUCHOWSKI, and McCLISH · Statistical Methods in Diagnostic Medicine

---

*Now available in a lower priced paperback edition in the Wiley Classics Library.
†Now available in a lower priced paperback edition in the Wiley–Interscience Paperback Series.